CP Violation

Current Physics – Sources and Comments

Coordinating editor

H. RUBINSTEIN

1. String theory in four dimensions, *edited by* M. Dine
2. Finite-size scaling, *edited by* J.L. Cardy
3. Regge theory of low-p_T hadronic interactions, *edited by* L. Caneschi
4. Superstring construction, *edited by* A.N. Schellekens
5. CP Violation, *edited by* L. Wolfenstein

NORTH-HOLLAND
AMSTERDAM · OXFORD · NEW YORK · TOKYO

CP Violation

Editor

L. WOLFENSTEIN

Department of Physics, Carnegie Mellon University
Pittsburgh, Pennsylvania 15213-3890, USA

N·H
P C

1989
NORTH-HOLLAND
AMSTERDAM · OXFORD · NEW YORK · TOKYO

ISBN: 0444 88081 X (hardbound)
0444 88114 X (paperback)

Published by:

North-Holland
Elsevier Science Publishers B.V.
P.O. Box 103
1000 AC Amsterdam
The Netherlands

Sole distributors for the USA and Canada:

Elsevier Science Publishing Company, Inc.

655, Avenue of the Americas
New York, NY 10010
USA

Library of Congress Cataloging-in-Publication Data

CP violation / Lincoln Wolfenstein, editor.
p. cm. -- (Current physics ; 5)
Includes bibliographical references.
ISBN 0-444-88081-X. -- ISBN 0-444-88114-X (pbk.)
1. CP violation (Nuclear physics) I. Wolfenstein, Lincoln.
II. Series.
QC793.3.V5C6 1988b
539.7--dc20 89-23278
CIP

Printed in The Netherlands

Preface

In 1964, it was discovered by an experiment on K^0 decay that CP invariance was violated in nature. From an analysis of that experiment as well as from general theorems requiring CPT invariance it is believed that time-reversal (T) invariance is also violated. The correct theoretical model to explain this result has remained uncertain until this day.

The articles collected here are mainly concerned with the phenomenological description of the observed CP violation in the K^0 system and with theoretical models designed to explain this. Chapter 1 provides the history leading to the 1964 discovery. Chapter 2 contains reprints of some more recent definitive experiments, together with the phenomenological analysis that was developed to interpret CP violation in the K^0 system. Chapter 3 describes the variety of models that were developed to explain this CP violation, all of which go beyond the standard electroweak theory in some respect. Chapter 4 is devoted to the explanation based on the standard electroweak theory with three generations of quarks.

Outside the K^0 system only two types of experiments are covered here:

(a) CP violation in the B^0 system is included in chapter 4, because this is the one place where large effects are predicted in the standard model;

(b) the electric dipole moment of the neutron is covered in chapter 5, because this provides what appears to be the most sensitive direct test for T invariance violation.

A number of important subjects could not be covered. These include

(1) general formalism for T and CP in quantum mechanics;

(2) CP or T violation tests in hyperon decay, nuclear beta decay, very rare K meson decays, and high-energy collisions;

(3) the cosmological origin of the baryon asymmetry.

There exist a large number of reviews and books concerning CP and T violation. Among these are L. Wolfenstein, Present Status of CP Violation, Ann. Rev. Nucl. Sci. 36 (1986) 137; R.G. Sachs, Physics of Time Reversal, University of Chicago Press (1987); C. Jarlskog, CP Violation, World Scientific (1989).

Acknowledgements

The following articles have been reprinted by kind permission of the publisher, The American Institute of Physics:
From Physical Review:
T.D. Lee, R. Oehme and C.N. Yang, Remarks on possible noninvariance under time reversal and charge conjugation, Phys. Rev. 106 (1957) 340–345; P.K. Kabir, What is not invariant under time reversal, Phys. Rev. D 2 (1970) 540–542; G.C. Branco, Spontaneous CP nonconservation and natural flavor conservation: A minimal model, Phys. Rev. D 22 (1980) 2901–2905; H.-Y. Cheng, Weinberg CP-violation model revisited, Phys. Rev. D 34 (1986) 1397–1403; R.N. Mohapatra and J.C. Pati, Left–right gauge symmetry and an "isoconjugate" model of CP violation, Phys. Rev. D 11 (1975) 566–571; J.H. Smith, E.M. Purcell and N.F. Ramsey, Experimental limit to the electric dipole moment of the neutron, Phys. Rev. 108 (1957) 120–122.
From Physical Review Letters:
J.H. Christenson, J.W. Cronin, V.L. Fitch and R. Turlay, Evidence for the 2π decay of the K_2^0 meson, Phys. Rev. Lett. 13 (1964) 138–140; T.T. Wu and C.N. Yang, Phenomenological analysis of violation of CP invariance in decay of K^0 and $\overline{K}^0$, Phys. Rev. Lett. 13 (1964) 380–385; L. Wolfenstein, Violation of CP invariance and the possibility of very weak interactions, Phys. Rev. Lett. 13 (1964) 562–564; S. Weinberg, Gauge Theory of CP nonconservation, Phys. Rev. Lett. 37 (1976) 657–661; R.D. Peccei and H.R. Quinn, CP conservation in the presence of pseudoparticles, Phys. Rev. Lett. 38 (1977) 1440–1443.
From Reviews of Modern Physics:
V.L. Fitch, The discovery of charge-conjugation parity asymmetry, Rev. Mod. Phys. 53 (1981) 367–371; J.W. Cronin, C.P. Symmetry violation – the search for its origin, Rev. Mod. Phys. 53 (1981) 373–383.
From JETP Letters:
I.S. Altarev, Yu.V. Borisov, N.V. Borovikova, A.B. Brandin, A.I. Egorov, S.N. Ivanov, É.A. Kolomenskıı, M.S. Lasakov, V.M. Lobashev, A.N. Pirozhkov, A.P. Serebrov, Yu.V. Sobolev, R.R. Tal'daev, and B.V. Shul'gina, Search for an electric dipole moment of the neutron, JETP Lett. 44 (1987) 460–465.
From Soviet Journal of Nuclear Physics:
L.B. Okun', Note concerning CP parity, Sov. J. Nucl. Phys. 1 (1965) 670.

The following article has been reprinted from Helvetica Physica Acta by kind permission of the publisher, Birkhauser Verlag:
C.P. Enz and R.R. Lewis, On the phenomenological description of CP violation for K-mesons and its consequences, Helv. Phys. Acta 38 (1965) 860–876.

The following article has been reprinted from Progress of Theoretical Physics by kind permission of the publication office of Progress of Theoretical Physics:
M. Kobayashi and T. Maskawa, CP-violation in the renormalizable theory of weak interaction, Prog. Theor. Phys. 49 (1973) 652–657.

The following article has been reprinted from Proceedings of the Oxford International Conference on Elementary Particle Physics by kind permission of the publication office of the Rutherford Appleton Laboratory:
J.S. Bell and J. Steinberger, Weak interaction of kaons, Proc. Oxford International Conference on Elementary Particle Physics (1965) 195–208 and 221.

Contents

1. History

Introduction

Symmetry principles have played an important role in the development of our ideas about fundamental processes. We are interested here in three discrete symmetries which relate a given state or process to one other state or process; these are
P: Parity, or the inversion of all spatial coordinates,
T: Time-reversal, the replacement of t by $-t$,
C: Particle–antiparticle conjugation, the replacement of all particles by their antiparticles.

In first instance, these symmetries were discovered as symmetries of established physical laws. Thus Wigner [1] discovered that P and T are symmetries of the Schrödinger equation applied to atomic and molecular systems. Similarly, C was discovered as a symmetry of quantum electrodynamics [2].

Nuclear physics required the introduction of two new interactions: the weak interaction to describe beta-decay, and the strong interaction for nuclear forces. Because these interactions had no classical analogue, they had to be invented ab initio. Then, it became natural to assume that the same symmetries held for the new interactions. However, the consequences of making this assumption were not analyzed in any systematic fashion for a long time. An important point made by Luders in 1954 [3] was that if P symmetry is assumed, then the consequences of C invariance are identical to those of T invariance. A way of stating this result is that, although it is easy to construct theories that violate C, P or T symmetry, every relativistic local quantum field theory is invariant under the combined symmetry operation CPT.

It was pointed out by Lee and Yang in 1956 [4] that no experiments involving weak interactions tested the parity symmetry P. This led to the discovery in 1957 [5] that parity symmetry was violated as much as possible in nuclear beta-decay as well as in pion and muon decay. The resulting V–A theory that became established after these discoveries involved a maximal violation of both C and P but retained invariance under the CP and the T transformations. This possibility was first emphasized by Landau and others [6] after the Lee–Yang paper but before the experiments.

Even before P violation was discovered, Lee et al. [7] looked at possible ways of testing the C and T symmetries subject to the overall CPT invariance. They found that K^0 decay was of particular interest. Their analysis is in terms of C violation, but the quantities they discuss (α and the charge asymmetry r) are equally measures of CP violation. (To say this in another way, the C-violating observables discussed do not involve any P violation so they are also CP-violating.) Once it was realized that C was violated in the weak interactions, it was the CP symmetry that was important to test with the K^0 system.

If CP invariance is assumed, the K^0 states with definite masses and lifetimes are K_1 and K_2 which are even and odd under CP. It then follows that only K_1 can decay into two pions while K_2 would decay into three pions. This prediction, originally made by Gell-Mann and Pais [8], was verified with the observation of two types of decay, a short-lived K_S going to $\pi^+\pi^-$ or $\pi^0\pi^0$ and a long lived K_L going to three pions. In 1964, Christenson et al. [9] discovered that K_L had a small branching ratio into $\pi^+\pi^-$. This decay was indicated at the same time in an experiment by Abashian et al. [10]. This suggested strongly that CP invariance was not an exact symmetry. The history of this discovery and its consequences are presented in the Nobel lectures of Fitch [11] and Cronin [12]. The possibilities of alternative explanations not involving CP violation are discussed by Cronin. The observation [13] of interference between the K_L and K_S decays into $\pi^+\pi^-$ provides definite evidence for CP violation. Note that the paper of Lee et al. [7] already discusses the CP-violating interference effect.

References

[1] E.P. Wigner, Z. Phys. 43 (1927) 624; Gött. Nachr. (1932) 546.
The latter classic paper on time reversal is reproduced as chapter 26 of E.P. Wigner, Group Theory (Academic Press, 1959).
[2] H.A. Kramers, Proc. Acad. Sci. Amsterdam 40 (1937) 814.
[3] G. Luders, Danske Videns Selskab Mat–Fys Medd. 28, No. 5 (1954); see also Annals of Physics 2 (1957) 1.
[4] T.D. Lee and C.N. Yang, Phys. Rev. 104 (1956) 254.
[5] C.S. Wu, E. Ambler, R.W. Hayward, D.D. Hoppes and R.P. Hudson, Phys. Rev. 105 (1957) 1413.
[6] L. Landau, Nucl. Phys. 3 (1957) 127. Paper 1, this volume.
See also T.D. Lee and C.N. Yang, Phys. Rev. 105 (1957) 1671;
A. Salam, Nuovo Cimento 5 (1957) 249;
R. Gatto, Phys. Rev. 106 (1957) 168.
[7] T.D. Lee, R. Oehme and C.N. Yang, Phys. Rev. 106 (1957) 340. Paper 2, this volume.
[8] M. Gell-Mann and A. Pais, Phys. Rev. 97 (1955) 1387.
[9] J.H. Christenson, J.W. Cronin, V.L. Fitch and R. Turlay, Phys. Rev. Lett. 13 (1964) 138. Paper 3, this volume.
[10] A. Abashian, R.J. Abrams, D.W. Carpenter, G.P. Fischer, B.M.K. Nefkens and J.H. Smith, Phys. Rev. Lett. 13 (1964) 243.
[11] V.L. Fitch, Rev. Mod. Phys. 53 (1981) 367. Paper 4, this volume.
[12] J.W. Cronin, Rev. Mod. Phys. 53 (1981) 373. Paper 5, this volume.
[13] V.L. Fitch, R.F. Roth, J.S. Russ and W. Vernon, Phys. Rev. Lett. 15 (1965) 73.

Nuclear Physics **3** (1957) 127—131; *North-Holland Publishing Co., Amsterdam*

ON THE CONSERVATION LAWS FOR WEAK INTERACTIONS

L. LANDAU

Institute for Physical Problems, USSR Academy of Sciences, Moscow

Received 9 January 1957

Abstract: A variant of the theory is proposed in which non-conservation of parity can be introduced without assuming asymmetry of space with respect to inversion.

Various possible consequences of non-conservation of parity are considered which pertain to the properties of the neutrino and in this connection some processes involving neutrinos are examined on the assumption that the neutrino mass is exactly zero.

1. Combined Parity

As is well known, the unusual properties of K-mesons have created a perplexing situation in modern physics. The correlation between π-mesons in τ-decay ($K^+ \to 2\pi^+ + \pi^-$) leads to the necessity of assigning a 0^- state to K^+-mesons. This kind of system, however, cannot decay into two π-mesons ($K^+ \to \pi^+ + \pi^0$). We are thus faced with the dilemma of either assuming that two different K-mesons exist or that the conservation laws are violated in K-meson decay. In the first case one must then explain the identity of masses (which are equal to within two electron masses)and the near coincidence in lifetime of the θ and τ-decays. One may attempt to explain the equality of K-meson masses by postulating, as Lee and Yang [1]) have done, the existence of some hitherto unknown symmetry property of nuclear forces which transforms the τ-meson into a θ-meson. If, however, decay involving a neutrino ($K^+ \to \mu^+ + \nu$, $K^+ \to \mu^+ + \nu + \pi^0$, $K^+ \to e^+ + \pi^0 + \nu$) is considered to be essentially the same for particles of various parity a difference in lifetime related to the different rate of τ and θ-decay (≈ 8 % and ≈ 25 %) should be anticipated. This discrepancy should be not less than 30—40 %, a result which seems to be inconsistent with experiment[2]).

Thus we come to the conclusion that the hypothesis of the existence of two different K^+-mesons is contrary to the experimental facts and the only alternative is to assume that the generally accepted conservation laws are violated in K-decay. Since there is no reason to think that the law of conservation of angular momentum is untenable, we are apparently dealing here with a direct violation of the law of conservation of parity.

It might seem at first glance that non-conservation of parity implies asymmetry of space with respect to inversion. If however, complete isotropy of space (conservation of angular momentum) is taken into account this type of asymmetry would seem to be extremely strange and in my opinion a simple rejection of parity conservation would create a difficult situation in theoretical physics. I would like to point out a solution of this problem which consists in the following. As is well known, both the law of conservation of parity and charge conjugation invariance undoubtedly hold in strong interactions. Let us now assume that each of these conservation laws does not hold separately in weak interactions. However, invariance with respect to the set of both operations (which we shall call combined inversion) will be assumed to exist. In combined inversion, space inversion and transformation of a particle into an antiparticle occur simultaneously.

It is easy to see that invariance of the interactions with respect to combined inversion leaves space completely symmetrical, and only the electrical charges will be asymmetrical. The effect of this asymmetry on the symmetry of space is no greater than that due to chemical stereo-isomerism.

On the other hand the law of conservation of parity of charged particles will not hold as the operator of combined inversion does not transform charged particles into themselves.

Furthermore, it is easy to see that the constants characterizing the particles and antiparticles (masses, lifetimes) should be identical since, as a result of invariance with respect to combined inversion, all processes involving particles and antiparticles should differ from each other only in regard to space inversion. Graphically speaking, a K^--meson is a mirror reflected K^+-meson.

Truly neutral particles, that is, particles which are identical to their antiparticles, transform into themselves in combined inversion. Consequently, with respect to these particles combined inversion leads to a law of conservation of combined parity. It should be emphasized that conservable parity is the product of ordinary parity and charge parity of the particles. Evidently, in this sense the π^0-meson is an odd particle; the $K_1^0(\theta^0)$-meson which decays into 2 π-mesons is an even particle and the K_2^0-meson predicted by Gell-Mann and Pais [3]) and recently discovered experimentally [4]) is an odd particle. Combined inversion changes the sign of the magnetic field of a photon but does not change that of the electric field. The ordinary parities of electric and magnetic multipoles are reversed for combined inversion.

It is easy to show from the foregoing that despite the absence of ordinary parity the particles cannot possess dipole moments. Indeed, the only vector which can be constructed from ψ-operators for a particle at rest is its spin vector which is even with respect to inversion and odd with respect to charge. It is consequently odd with respect to combined inversion and, in accord with the foregoing regarding the electromagnetic field, it defines only a magnetic but not an electric moment.

Lee and Yang [5]) † have shown that non-conservation of parity leads to correlations in a number of hyperon production and decay processes. It can be shown that a consequence of invariance with respect to combined inversion is that the weak interaction operators in the Lagrangian contain real coefficients. This circumstance, however, does not appreciably modify the qualitative picture which is obtained in the general case of non-conservation of parity. Therefore asymmetry of hyperon decay with respect to the plane of their creation, which has been predicted by Lee and Yang [5]), will also hold in this case.

I would like to express my deep appreciation to L. Okun, B. Ioffe and A. Rudik for discussions from which the idea of this part of the present paper emerged.

2. Properties of the Neutrino

Rejection of the law of conservation of parity entails the possibility of existence of new properties of the neutrino. The Dirac equation for the case of zero mass splits into two independent pairs of equations. It will be recalled that in the usual theory one cannot confine oneself to a single pair of equations since both pairs transform into each other as a result of space inversion. If, however, we restrict our attention to combined inversion we arrive at the possibility of describing the neutrino by a single pair of equations. In the sense of the usual scheme this would signify that the neutrino is always polarized in the direction of its motion (or in the opposite direction). The polarization of the antineutrino is correspondingly reversed. According to this model the neutrino is not a truly neutral particle and this agrees with the fact that double β-decay has not been observed experimentally and especially with the results of experiments on induced β-decay. We shall call this kind of neutrino a longitudinally polarized neutrino or briefly a longitudinal neutrino.

In the usual theory the neutrino mass is zero, so to say, accidentally. Thus, account of neutrino interactions automatically leads to the appearance of a definite, albeit vanishingly small, rest mass. The mass of the longitudinal neutrino, on the other hand, vanishes automatically and this situation cannot be altered by the existence of any type of interaction.

The longitudinal neutrino concept appreciably reduces the possible number of types of weak interaction operators. Consider, for example, the decay of a μ-meson into an electron and two neutrinos. In the usual manner we represent the interaction operator as the product of operators consisting of μ-meson and electron ψ-operators on the one hand and ψ-operators of the two neutrinos on the other. For the longitudinal neutrino only one combination can be made from the two neutrino operators—a scalar (a

† I would like to sincerely thank the authors for sending me a preprint of their paper.

scalar with respect to rotation; the operation of ordinary inversion is not applicable), as it is well known that the tensor combination of two identical operators obeying Fermi statistics is equal to zero. In this case two combinations, scalar and pseudo-scalar (in the usual sense of the word), can be constructed for a μ-meson and electron.

If a neutrino and antineutrino are emitted in μ-meson decay the situation changes. Only a four-dimensional vector can then be constructed from the longitudinal neutrino and antineutrino operators. In this case two combinations — vector and pseudo-vector—can be made from the μ-meson and electron operators. Thus, despite the absence of invariance with respect to inversion in each of the two cases only two interaction operators are possible.

It is easy to calculate the energy spectrum of the μ-meson decay electrons. It is found to be exactly the same as that calculated by Michel [6]). The two-neutrino case thus yields for Michel's constant, ϱ, the value $\varrho = 0$ and for a neutrino and antineutrino $\varrho = 0.75$. The former case is apparently inconsistent with the experiments, whereas the latter agrees with the results obtained in ref. [7, 8]), which yield $\varrho = 0.64 \pm 0.10$ and $\varrho = 0.57 \pm 0.14$. Thus μ-meson disintegration experiments do not contradict the longitudinal neutrino concept and in this case lead to a unique result, namely, that a neutrino and antineutrino are involved in μ-meson decay.

Consider now the reaction $\pi \to \mu + \nu$. Since the π-meson is spinless we are obliged to set up a scalar expression for the μ-meson and neutrino ν-operators in the $\pi \to \mu + \nu$ decay operator. This automatically yields that if the neutrino is longitudinal the μ-mesons produced in $\pi \to \mu + \nu$ decays will be completely polarized in the direction of their motion (or in the opposite direction).

As Lee and Yang [5]) have noted, a possible consequence of non-conservation of parity is the correlation between the directions of the μ-meson and electron involved in the $\pi \to \mu \to$ e decay. Simple calculations based on our scheme give the following energy and angular distribution for the emitted electrons:

$$\frac{\mathrm{d}N}{N} = 2\varepsilon^2[(3-2\varepsilon)+\lambda \cos \vartheta (2\varepsilon-1)]\mathrm{d}\varepsilon. \tag{1}$$

Here ε is the ratio of electron energy to the largest possible energy, ϑ is the angle between the directions of motion of the μ-meson and electron and λ is a constant which depends on the relation between the vector and pseudovector parts in the combination of the μ-meson and electron ψ-operators,

$$\lambda = \frac{2ab}{a^2+b^2}, \tag{2}$$

where a and b are coefficients of the respective terms and, according to the foregoing, are real. Evidently, λ varies between -1 and $+1$. It is possible that λ is in fact equal to zero. The integral electron distribution is obviously proportional to $(1+\frac{1}{3}\lambda \cos \vartheta)$ and this means that the largest possible value of the forward-backward asymmetry is 2. It should be noted that even if λ appreciably differs from zero it may be difficult to observe $\mu \rightarrow e$ correlation because of depolarization of the slowed down mesons and in particular for μ^+-mesons because of formation of mesonium ($\mu^+ + e^-$ system).

Consider now the effect of longitudinality of the neutrino on β-decay. According to experiment the decay operator should be represented as the sum of the scalar and tensor variants. It can be shown that in either case the same electron polarization in the direction of motion will arise, which is equal to v/c (or $-v/c$), the ratio of the electron velocity to that of light. Thus high energy electrons will be totally polarized in the direction of their motion.

References

1) T. Lee and C. Yang, Phys. Rev. **102** (1956) 290
2) Proceedings of the Sixth Annual Rochester Conference, April 1956
2) M. Gell-Mann and A. Pais, Phys. Rev. **97** (1955) 1387
4) K. Larde, E. Booth, J. Impeduglia, L. Lederman and W. Chinovsky, Phys. Rev. **103** (1950) 1901
5) T. Lee and C. Yang, Phys. Rev. **104** (1956) 254
6) L. Michel, Proc. Phys. Soc. A **63** (1950) 514; **63** (1950) 1371
7) C. P. Sargent, M. Rinehart, L. M. Lederman and K. C. Rogers, Phys. Rev. **99** (1955) 885
8) A. Bonetti, R. Levi Setti, M. Panetti, G. Rossi and G. Tomasini, Nuovo Cimento **3** (1956) 33

Remarks on Possible Noninvariance under Time Reversal and Charge Conjugation*

T. D. LEE, *Columbia University, New York, New York*

AND

REINHARD OEHME AND C. N. YANG, *Institute for Advanced Study, Princeton, New Jersey*

(Received January 7, 1957)

[illegible] we in the nonconservation properties of parity, time reversal, and charge conjugation [illegible] The results are stated in two theorems. The experimental implications for the K–$\bar{K}$ complex are discussed in the last section.

IN a recent paper[1] the question has been raised as to whether the weak interactions are invariant under a space inversion. It was also pointed out there that similar to the situation for space inversion there exists at present no experimental proof that weak interactions are invariant under charge conjugation. Consequently the absolute invariance under charge conjugation is also an open question.

The present note is devoted to a study of questions concerning the invariance under charge conjugation C, and under time reversal T (which is defined to be the Wigner time[2] reversal. It *does not* switch a particle into its antiparticle; nor does it change the sign of the spatial coordinates).

1. *CPT* THEOREM

For the discussion of the experimental consequences of possible nonconservation of P, C, and/or T, a theorem[3] which we shall call the CPT theorem proves very important.

To understand the meaning of the theorem one recalls first that the operations P and C in any many-particle system (with possibilities of creation and annihilation) are represented by unitary operators that operate on the state vectors. The operation T, on the other hand, is represented[2] by the *operator* of complex conjugation *multiplied* by a unitary operator. In the Schrödinger representation the transformation of a second quantized spin 0 field described by $\varphi(r)$ and $\pi(r)$ under these operations[4] can be brought into the following form:

$$\begin{aligned} P\phi(r)P^{-1}&=\eta_P\phi(-r), & P\pi(r)P^{-1}&=\eta_P{}^*\pi(-r),\\ C\phi(r)C^{-1}&=\eta_C\phi^\dagger(r), & C\pi(r)C^{-1}&=\eta_C{}^*\pi^\dagger(r),\\ T\phi(r)T^{-1}&=\eta_T\phi(r), & T\pi(r)T^{-1}&=-\eta_T{}^*\pi(r),\end{aligned} \tag{1}$$

where † means Hermitian conjugate and the phases η_P, η_C, and η_T have absolute values equal to 1. For the spin $\frac{1}{2}$ field $\psi(r)$, the transformations are

$$\begin{aligned} P\psi(r)P^{-1}&=\eta_P\gamma_4\psi(-r),\\ C\psi(r)C^{-1}&=\eta_C\psi^\dagger(r),\\ T\psi(r)T^{-1}&=\eta_T\gamma_1\gamma_2\gamma_3\psi(r),\end{aligned} \tag{2}$$

where the γ matrices are so chosen that γ_1, γ_2, and γ_3 are real and γ_4 is pure imaginary. The phases η_P, η_C, and η_T also have absolute values equal to unity. The transformation properties of fields of higher spin are similar.

From the CPT theorem one concludes that for any local Hermitian Hamiltonian H which is invariant under proper Lorentz transformations (i.e., Lorentz transformations that involve neither space nor time inversions), there always exists a choice of the phases η_C, η_P, and η_T for the various fields (usually in more than one way) with the following properties: (a) H commutes with the product of the operators P, C, and T taken in any order; and (b) if this choice of phases does not make H commute with P, then no other choice does, and the theory is not invariant under space inversion. (Of course, if this choice of phases makes H commute with P, then the theory is invariant under space inversion.) The same holds for C and T.

We shall illustrate this theorem by an example where H is invariant under proper Lorentz transformations. Let ψ_p, ψ_n, ψ_e, and ψ_ν be the fields describing the proton, the neutron, the electron, and the neutrino. The neutrino is assumed to be a non-Majorana particle with a nonvanishing mass. Consider

$$\begin{aligned} H=H_{\text{free}}+\int\{&g_1(\psi_p{}^\dagger\gamma_4\psi_n)(\psi_e{}^\dagger\gamma_4\psi_\nu)\\ &+g_2(\psi_p{}^\dagger\gamma_4\psi_n)(\psi_e{}^\dagger\gamma_4\gamma_5\psi_\nu)\\ &+g_3(\psi_p{}^\dagger\gamma_4\gamma_5\psi_n)(\psi_e{}^\dagger\gamma_4\gamma_5\psi_\nu)\\ &+g_4(\psi_p{}^\dagger\gamma_4\gamma_5\psi_n)(\psi_e{}^\dagger\gamma_4\psi_\nu)\\ &+\text{Hermitian conjugate}\}d^3(r),\end{aligned} \tag{3}$$

* *Note added in proof.*—This paper was written in December, 1956, before parity nonconservation was experimentally established.

[1] T. D. Lee and C. N. Yang, Phys. Rev. **104**, 254 (1956).

[2] E. P. Wigner, Gött. Nachr., Math. Naturw. Kl. (1932), p. 546.

[3] See W. Pauli's article in *Niels Bohr and the Development of Physics* (Pergamon Press, London, 1955). G. Lüders, Kgl. Danske Videnskab. Selskab, Mat.-fys. Medd. **28**, No. 5 (1954). J. Schwinger, Phys. Rev. **91**, 720, 723 (1953); **94**, 1366, formula (54) and p. 1576, discussions after formula (208). We are indebted to Professor Pauli for informing us of the work of Schwinger.

[4] We discuss here only the usual "type" of fields. The possibility of the existence of unusual "types" has been pointed out by Wigner. [An account of these unusual types has been given by L. Michel and A. S. Wightman, Princeton University lecture notes (unpublished).] An examination of these unusual "types" would be an important task if space-time conservation laws should indeed be found to break down for the weak interactions.

where $\gamma_5=\gamma_1\gamma_2\gamma_3\gamma_4$. This example is a special case of an example considered by Pauli.[3]

Writing H as

$$H=H(g_1,g_2,g_3,g_4),$$

one easily proves that

$$PHP^{-1}=H(g_1\eta_P,\ -g_2\eta_P,\ g_3\eta_P,\ -g_4\eta_P), \quad (4)$$

$$CHC^{-1}=H(g_1{}^*\eta_C{}^*,\ -g_2{}^*\eta_C{}^*,\ g_3{}^*\eta_C{}^*,\ -g_4{}^*\eta_C{}^*), \quad (5)$$

$$THT^{-1}=H(g_1{}^*\eta_T,\ g_2{}^*\eta_T,\ g_3{}^*\eta_T,\ g_4{}^*\eta_T). \quad (6)$$

In deriving these formulas, use has been made of the fact that $TgT^{-1}=g^*$ and $T\gamma_iT^{-1}=\gamma_i{}^*$. The phases η_P, η_C, and η_T are products of the respective phases of the four interacting fields. They are given by

$$\begin{aligned}\eta_P&=\eta_P{}^*(p)\eta_P(n)\eta_P{}^*(e)\eta_P(\nu),\\ \eta_C&=\eta_C{}^*(p)\eta_C(n)\eta_C{}^*(e)\eta_C(\nu),\\ \eta_T&=\eta_T{}^*(p)\eta_T(n)\eta_T{}^*(e)\eta_T(\nu).\end{aligned} \quad (7)$$

Using (4), (5), and (6), one can calculate the commutation relation between H and the six operators TCP, TPC, $\cdots$, PCT. It is found that with suitable choices of the phases η, the Hamiltonian H commutes with all of the six, as required by the CPT theorem. In fact the conditions on the phases η are simply

$$\eta_P=\eta_C\eta_T=\pm 1. \quad (8)$$

It follows from the CPT theorem that, if one of the three operators P, C, and T is not conserved, at least one other must also be not conserved. It is of course also possible that all three are separately not conserved. In the example above, by assigning suitable values to the coupling constants g, one can construct examples for all the five possibilities of conservation or nonconservation of P, C, and T. These examples are displayed in Table I.

2. LIFETIME OF CHARGE CONJUGATE PARTICLES AGAINST WEAK DECAY

Consider now a Hamiltonian

$$H=H_{\text{strong}}+H_{\text{weak}},$$

where both terms are invariant under a proper Lorentz transformation. In all subsequent discussions we shall assume that H_{strong} is invariant under C, P, and T. The phases η of the fields are defined (up to, possibly, some arbitrary factors) by this invariance; i.e., by the requirements that

$$CH_{\text{strong}}C^{-1}=H_{\text{strong}},\ \text{etc.}$$

On the other hand, the weak interactions may violate the invariance of C, P, and T. One can prove the following theorem.

Theorem 1.—If a particle A decays through the interaction H_{weak}, and if the particle and its antiparticle $\bar{A}$ do not decay into the same final products (as e.g. when A is charged), then to the lowest order of H_{weak} the lifetimes of A and $\bar{A}$ are the same, even if H_{weak} is *not* invariant under charge conjugation.

Proof.—Consider the case that particle A has spin zero. [The proof for the general case follows along the same lines.] Then the final states B and $\bar{B}$ in the decays

$$A\rightarrow B,\quad \bar{A}\rightarrow\bar{B}$$

also have spin zero. Using the identity

$$\langle\psi_1|\psi_2\rangle^*=\langle T\psi_1|T\psi_2\rangle, \quad (9)$$

one obtains

$$\begin{aligned}\langle B|H_{\text{weak}}|A\rangle^*&=\langle TB|TH_{\text{weak}}T^{-1}|TA\rangle\\ &=\langle TB|C^{-1}P^{-1}H_{\text{weak}}PC|TA\rangle,\end{aligned}$$

by the CPT theorem. Consider first the case that H_{weak} commutes (or anticommutes) with P. Then

$$\langle B|H_{\text{weak}}|A\rangle^*=\pm\langle TB|C^{-1}H_{\text{weak}}C|TA\rangle.$$

For a spinless system,

$$|TA\rangle=|A\rangle,\quad |TB\rangle=|B\rangle. \quad (10)$$

Hence

$$\begin{aligned}\langle B|H_{\text{weak}}|A\rangle^*&=\pm\langle B|C^{-1}H_{\text{weak}}C|A\rangle\\ &=\pm\langle CB|H_{\text{weak}}|CA\rangle=\pm\langle\bar{B}|H_{\text{weak}}|\bar{A}\rangle.\end{aligned} \quad (11)$$

This shows that the lifetimes of A and $\bar{A}$ are the same. If H_{weak} does not commute with P, we write

$$H_{\text{weak}}=H_1+H_2, \quad (12)$$

where

$$\begin{aligned}H_1&=\tfrac{1}{2}[H_{\text{weak}}+PH_{\text{weak}}P^{-1}],\\ H_2&=\tfrac{1}{2}[H_{\text{weak}}-PH_{\text{weak}}P^{-1}].\end{aligned} \quad (13)$$

Then

$$PH_1P^{-1}=H_1, \quad (14)$$

$$PH_2P^{-1}=-H_2. \quad (15)$$

The decays of A through H_1 and through H_2 lead to states B_1 and B_2 with opposite parities. They are orthogonal to the order considered, and hence they contribute independently without interference to the decay rate of A. The lifetimes of A and $\bar{A}$ are therefore again the same.

A consequence of this theorem has already been mentioned in a previous paper[1]: The identity of the experimental lifetimes of $\pi^{\pm}$ and of $\mu^{\pm}$ does not con-

TABLE I. Examples of theories with various possible nonconservation properties.

Value of coupling constants	Conserved operators	Nonconserved operators
g_1=real, g_3=real, $g_2=g_4=0$	P, C, T	$\cdots$
g_1=real, g_3=complex, $g_2=g_4=0$	P, CT, TC	C, T
g_1=real, g_2=imaginary, $g_3=g_4=0$	C, PT, TP	P, T
g_1=real, g_2=real, $g_3=g_4=0$	T, CP, PC	C, P
g_1=real, g_2=complex, $g_3=g_4=0$	PCT, *and* permutations	P, C, T

stitute a proof that charge conjugation invariance holds for the weak interactions.

For a discussion of a case where A and $\bar{A}$ may decay into the same final channels, see Sec. 4.

3. DEPENDENCE OF INTERFERENCE EFFECTS ON CONSERVATION OF C AND T

One would like to ask what are the experimentally detectable manifestations of a weak nonconservation of P, C, or T? For the nonconservation of parity, the answer is clearly to be sought in experiments to differentiate the right-handed screw from the left-handed. Some such experiments have been discussed before.[1]

If parity is indeed not strictly conserved, some of these experiments could *also reveal whether C and/or T are or are not conserved.* To illustrate this let us consider the experiment of the angular distribution of β decay from oriented nuclei. The degree of asymmetry was given in the appendix of reference 1 as proportional to

$$\mathrm{Re}\left[C_T C_T'^{*}-C_A C_A'^{*}+i\frac{Ze^2}{\hbar cp}(C_A C_T'^{*}+C_A' C_T^{*})\right]. \quad (16)$$

Applying the arguments of Sec. 1 we recognize that the two terms in (16) are present or absent depending on whether C or T are not conserved. To be more specific: The first term vanishes if C is strictly conserved, the second term vanishes if T is strictly conserved. If this experiment shows any asymmetry, the p dependence and the Z dependence of the asymmetry could therefore reveal whether C and/or T are nonconserved. (The existence of any asymmetry rules out the possibility that both C and T are conserved, a conclusion we already drew on general grounds in Sec. 1.)

We notice that if C is strictly conserved, the asymmetry discussed above vanishes in the absence of the Coulomb distortion of the electron wave function. In fact, when C is conserved the asymmetry is directly dependent on the existence of a difference of the Coulomb phase shifts for opposite parities. It turns out that this is a consequence of a general theorem which we state and prove below:

Theorem 2.—If, in addition to the assumptions stated in Sec. 2 concerning H_{strong} and H_{weak}, we assume that H is strictly invariant under charge conjugation, (i.e., $[H,C]=0$) and if the decay products in the final state B are *free* particles, then to the lowest order of H_{weak} there is no interference between the parity-conserving and the parity-nonconserving parts of H in the decay of A, provided the interference is sought for in experiments measuring a term of the form $\boldsymbol{\sigma}\cdot\mathbf{p}$[1].

Proof.—We again illustrate the proof by considering the case that A is spinless. The general proof follows along the same lines. We perform the decomposition of H_{weak} as in Eqs. (12)–(15). The final state B consists of two states B_1 and B_2 of opposite parities reached from A through H_1 and H_2, respectively. Clearly H_1 commutes with C, and also, by the CPT theorem, commutes with CPT. Hence using identity (9) and Eq. (10) one obtains

$$\begin{aligned}\langle B_1|H_1|A\rangle^* &= \langle TB_1|TH_1T^{-1}|TA\rangle\\ &= \langle B_1|TH_1T^{-1}|A\rangle\\ &= \langle B_1|P^{-1}C^{-1}H_1CP|A\rangle\\ &= \langle B_1|P^{-1}H_1P|A\rangle = \langle B_1|H_1|A\rangle.\end{aligned}$$

Thus $\langle B_1|H_1|A\rangle$ is real. Similarly one easily proves that $\langle B_2|H_2|A\rangle$ is pure imaginary.

In the above the states B_1 and B_2 are taken as stationary states of H_{strong} consisting of standing waves. [Otherwise Eq. (10) does not hold.] Transition amplitudes into them have a relative phase factor which is, according to the above, pure imaginary. The observed final states are equal to these amplitudes multiplied by the *outgoing* part of the stationary states B_1 and B_2. Such outgoing parts always have real relative amplitudes if the stationary states B_1 and B_2 represent free particles. The theorem now follows immediately.

Using this theorem, one concludes that if any left-right asymmetry of the form $\boldsymbol{\sigma}\cdot\mathbf{p}$ is found, the part of this asymmetry that is independent of the distortion of the final-state wave functions can arise only if charge conjugation symmetry breaks down for the weak interactions. In particular, in decays where there is no strong final-state interactions, as, e.g., in $\pi\rightarrow\mu+\nu$ and $\mu\rightarrow e+\nu+\nu$ decays, the detection[1] of parity nonconservation through the observation of $\boldsymbol{\sigma}\cdot\mathbf{p}$ becomes impossible if C is strictly conserved.

4. K^0, $\bar{K}^0$ DECAY MODES

The existence of the particle $\bar{K}^0$ and some properties of its decay were predicted[5] and discussed under the assumption that charge conjugation is strictly conserved. We wish to discuss in this section the decay of K^0 and $\bar{K}^0$ under the assumption that C, P, and T are conserved for the strong interactions, but are not necessarily conserved in the weak decay interactions.

In the first place, the conservation of strangeness with respect to the strong interactions still requires that two particles K^0 and $\bar{K}^0$ with opposite strangeness exist. To understand their decay processes it is interesting to consider the charge conjugation symmetrical and antisymmetrical combinations introduced in reference 5 (compare, however, footnote 11):

$$K_1=\frac{1}{\sqrt{2}}(K^0+\bar{K}^0),\quad K_2=\frac{1}{\sqrt{2}}(K^0-\bar{K}^0). \quad (17)$$

Unlike the situation in reference 5, if C is now not invariant in the decay process, K_1 and K_2 can decay into the same final states:

$$\begin{aligned}&K_1\rightarrow\pi^++\pi^-,\\ &K_2\rightarrow\pi^++\pi^-,\\ &K_1\rightarrow\pi^\pm+e^\mp+\nu,\\ &K_2\rightarrow\pi^\pm+e^\mp+\nu,\\ &K_1\rightarrow\pi^++\pi^-+\pi^0,\\ &K_2\rightarrow\pi^++\pi^-+\pi^0.\end{aligned}$$

[5] M. Gell-Mann and A. Pais, Phys. Rev. **97**, 1387 (1955).

Interference effects would therefore set in in these decay processes. The questions that one would like to ask are then: what the the lifetimes of the particles K^0 and $\bar{K}^0$? What are the branching ratios into various decay modes?

These questions can be answered by using a Weisskopf-Wigner type of treatment[6] of the time-dependent Schrödinger equation. We write the time-dependent amplitudes of the particles[7] K^0 and $\bar{K}^0$ as $a(t)$ and $b(t)$. The various channels of decay are denoted by j. $F_j(\omega)e^{-i\omega t}$ represents the amplitude of the decay product in the channel j with the energy ω.[8] [We choose units such that $\hbar=1$.] The zero of energy is taken to be the rest energy of K. The Schrödinger equations are then

$$i\frac{da}{dt}=\sum_{j,\omega} H_{aj}(\omega)F_j(\omega)e^{-i\omega t}, \tag{18}$$

$$i\frac{db}{dt}=\sum_{j,\omega} H_{bj}(\omega)F_j(\omega)e^{-i\omega t}, \tag{19}$$

$$i\frac{\partial F_j(\omega)}{\partial t}=e^{i\omega t}[H_{ja}(\omega)a+H_{jb}(\omega)b], \tag{20}$$

where $H_{ja}=H_{aj}^*$ are the matrix elements. The Weisskopf-Wigner treatment consists of first assuming an exponential time dependence for a and b, and then in sums over ω neglecting the variation of the matrix elements with ω in the interval $|\omega|\lesssim$uncertainty of energy of the original state. Using this treatment, one obtains

$$\begin{pmatrix} a \\ b \end{pmatrix}=\psi e^{-\frac{1}{2}\lambda t}. \tag{21}$$

The amplitude ψ and the decay constant λ are given by the eigenequation

$$\Gamma\psi=\lambda\psi. \tag{22}$$

Γ is a 2×2 Hermitian matrix with matrix elements given by

$$\begin{aligned} \Gamma_{11}&=\Gamma_{22}=\sum_j \Gamma_{aj}=\sum_j \Gamma_{bj}, \\ \Gamma_{12}&=\sum_j (\Gamma_{aj}\Gamma_{bj})^{\frac{1}{2}}e^{i\delta_j}=\Gamma_{21}^*, \end{aligned} \tag{23}$$

where

$$\begin{aligned} \Gamma_{aj}&=2\pi|H_{aj}|^2 \text{ (density of states per unit } d\omega)_{\omega=0}, \\ \Gamma_{bj}&=2\pi|H_{bj}|^2 \text{ (density of states per unit } d\omega)_{\omega=0}, \end{aligned} \tag{24}$$

and

$$e^{i\delta_j}=(\text{phase of } H_{aj}H_{bj}^{-1})_{\omega=0}. \tag{25}$$

[6] V. F. Weisskopf and E. P. Wigner, Z. Physik **63**, 54 (1930); **65**, 18 (1930).

[7] In this paper we assume that the K^0 particle (strangeness $=+1$) is a single state.

[8] Each channel j represents a possible decay state that is an eigenstate of H_{strong}. Thus it has a definite spin, charge, and parity.

In the foregoing derivation, use has been made of Eq. (11) which leads to

$$H_{aj}^*=\pm H_{bj'},$$

where j' is the charge conjugate channel of j (which may or may not be the same as j). It is important to notice that this equation is a consequence of the CPT theorem.

The two eigenvalues λ_+, λ_- of (22) correspond to the two decay lifetimes. The general solution is a linear superposition of two solutions $\psi_\pm$ of the form (21), each of which is characterized by a pure exponential decay. Since the 2×2 matrix Γ is Hermitian, the two solutions represent linear orthogonal combinations of the states K and $\bar{K}$.

In writing down Eqs. (18) and (19) we did not include a slight difference of mass in the form of a mass operation M for the states of the K particle. This restriction can be easily removed by adding to the right-hand sides of (18) and (19) the terms $\frac{1}{2}(M_{11}a+M_{12}b)$ and $\frac{1}{2}(M_{21}a+M_{22}b)$, respectively. The mathematical treatment is very similar to the above simple case except that we have now

$$-d\psi/dt=(\Gamma+iM)\psi, \tag{26}$$

where Γ is the same Hermitian matrix given by Eq. (23). (iM) is an anti-Hermitian matrix representing the effects of the mass shifts. By using Eq. (11) one can show that, similar to Eq. (23), M is a Hermitian matrix with

$$M_{11}=M_{22}. \tag{27}$$

Equation (26) can now be readily solved. Its eigenstates, defined by

$$(\Gamma+iM)\psi_\pm=\lambda_\pm\psi_\pm,$$

are

$$\psi_\pm=\begin{pmatrix} p \\ \pm q \end{pmatrix}(|p|^2+|q|^2)^{-\frac{1}{2}}, \tag{28}$$

with the corresponding time constants

$$\lambda_\pm=\Gamma_{11}+iM_{11}\pm(pq); \tag{29}$$

where p and q are two complex numbers given by

$$p^2=\Gamma_{12}+iM_{12},\quad q^2=\Gamma_{21}+iM_{21}=\Gamma_{12}^*+iM_{12}^*. \tag{30}$$

If at $t=0$ a K particle is produced, then at a later time the state function ψ can be expressed in terms of these two eigenstates $\psi_\pm$ as

$$\psi(t)=\left(\frac{1}{2p}\right)(|p|^2+|q|^2)^{\frac{1}{2}}[\psi_+e^{-\frac{1}{2}\lambda_+t}+\psi_-e^{-\frac{1}{2}\lambda_-t}]. \tag{31}$$

It is convenient to separate the real and imaginary parts of $\lambda_\pm$. Without loss of generality, we may write

$$\lambda_+=\gamma_+,\quad \lambda_-=\gamma_-+2i\Delta, \tag{32}$$

where γ_+, γ_- are two real numbers representing the reciprocal lifetimes of the short-lived ones and the long-lived ones respectively and Δ is the mass difference between these two eigenstates. One notices that *these two eigenstates* ψ_+ *and* ψ_- *do not in general represent the states* K_1 *and* K_2 introduced in (17). *In fact they even may not be orthogonal to each other* (see footnote 11).

A general discussion of the decay processes is rather involved. We shall make only the following remarks:

(*A*) The fractional number of K mesons that decay at time t after its production is given by

$$N(t)dt=-d[\psi^\dagger\psi]. \quad (33)$$

Using (26), one easily shows that

$$-\frac{d}{dt}[\psi^\dagger\psi]=\psi^\dagger\Gamma\psi.$$

By using (28)–(31), Eq. (33) becomes

$$N(t)=\tfrac{1}{2}(1+\alpha)^{-1}\{\gamma_+e^{-\gamma_+t}+\gamma_-e^{-\gamma_-t}+\alpha e^{-\frac{1}{2}(\gamma_++\gamma_-)t} \times[(\gamma_++\gamma_-)\cos\Delta t-2\Delta\sin\Delta t]\}, \quad (34)$$

where

$$\alpha=\psi_+^\dagger\psi_-=[|p|^2-|q|^2][|p|^2+|q|^2]^{-1} \quad (35)$$

is a real number representing the nonorthogonality of these two eigenstates. The four real numbers γ_+, γ_-, Δ, and α characterize the decay of the K particle. They satisfy the inequalities

$$\gamma_\pm\geqq 0, \quad \alpha^2\leqq\frac{4\gamma_+\gamma_-}{(\gamma_++\gamma_-)^2+4\Delta^2}, \quad (36)$$

which follow from the fact that Γ is a positive Hermitian matrix. These conditions also insure that $N(t)\geqq 0$ for all t.

Experimentally $N(t)$ is measurable. From $N(t)$ one can in principle determine all four constants γ_+, γ_-, Δ, and α. Indications from presently existing experiments[9] show that probably $\gamma_+/\gamma_->100$. Equation (36) then shows that $\alpha^2<4\gamma_-/\gamma_+<0.04$.

(*B*) The above discussion also leads easily to a determination of the branching ratio of the long-lived component (and the short-lived component) into the various decay modes. If charge conjugation is conserved, the long-lived component is an eigenstate of charge conjugation.[5] Consequently its decay into charge conjugate channels such as $\pi^+e^-\nu$ and $\pi^-e^+\bar{\nu}$ must be equally probable, as is well known. If charge conjugation is not strictly conserved, decays into $\pi^+e^-\nu$ and $\pi^-e^+\nu$ may have different probabilities for the long-lived component.

A more complete discussion of the charge asymmetry of the decay of the long-lived K^0 will be given in the appendix. We mention here only that Lederman[10] has kindly informed us that experimental work in this direction is in progress. It is important to notice that if the experiments should yield a large asymmetry, and a small α (as mentioned above), Eq. (A7) would impose very strict conditions on the relative magnitudes of the amplitudes f_1, g_1, f_2 and g_2. (To see this roughly we need only examine the limiting case $\alpha=0$ discussed below.)

(*C*) If $\alpha=0$, the two eigenstates are orthogonal.[11] Also $|p|=|q|$. [See (35).] This is the case if the mass matrix is negligible. In this case $\psi_\pm$ are both 1:1 superpositions of the particle K^0 and $\bar{K}^0$. The fraction of particles decaying in dt, namely $N(t)dt$, becomes the sum of two pure exponentials by (34). Furthermore (A7) shows that the decays of the long-lived component into charge conjugate channels such as $\pi^++e^-+\nu$ and $\pi^-+e^++\bar{\nu}$ are equally probable.

One of us (Reinhard Oehme) would like to express his gratitude to Professor Robert Oppenheimer for his kind hospitality at the Institute for Advanced Study.

APPENDIX

In this appendix we shall show the interrelationship between the parameters p, q and the branching ratio for the decay of, say, the long-lived component of K particle into various charge conjugate states.

Consider first the following decay channel of the K particle

$$K\rightarrow e^-+\pi^++\nu. \quad (A1)$$

The final product may be in states with either parity $=+1$ or parity $=-1$. Let us denote the matrix elements for the decay process into these two types of states by f_1 and f_2. Similarly, we denote the matrix elements for

$$K\rightarrow e^++\pi^-+\bar{\nu}, \quad (A2)$$

with the final state having parity $=+1$ and parity $=-1$, by g_1 and g_2.

By using the CPT theorem and Eq. (11), the corresponding matrix elements for the decay of $\bar{K}$,

$$\bar{K}\rightarrow e^++\pi^-+\bar{\nu}, \quad (A3)$$

are related to that of (A1). These elements are f_1^* and $-f_2^*$. Similarly the matrix elements for

$$\bar{K}\rightarrow e^-+\pi^++\nu \quad (A4)$$

are g_1^* and $-g_2^*$. Let ψ_+ represent the long-lived component K_+ of the K particle. The matrix elements for

[9] K. Lande *et al.*, Phys. Rev. **103**, 1901 (1956).

[10] L. Lederman (private communication).

[11] We recall that since the strangeness S is conserved in the strong interaction, the phase η_c of a K particle ($S=+1$) under charge conjugation is not fixed by the strong interactions. If the weak interaction is not invariant under charge conjugation, the phase η_c is defined only up to a factor $e^{iS\theta}$. If ψ_+ is orthogonal to ψ_-, there exists however a most convenient choice which makes $\psi_\pm$ identical with the K_1, K_2 defined in (17).

the decay of K_+,

$$K_+ \to e^- + \pi^+ + \nu, \quad \text{(A5)}$$

into the two different final parity states are proportional to $pf_1 + qg_1^*$ and $pf_2 - qg_2^*$, respectively, while the corresponding elements for

$$K_+ \to e^+ + \pi^- + \bar{\nu} \quad \text{(A6)}$$

are proportional to $pg_1 + qf_1^*$ and $pg_2 - qf_2^*$. The branching ratio r for the decay of K_+ into $e^- + \pi^+ + \nu$ and $e^+ + \pi^- + \bar{\nu}$ is, therefore,

$$r = \frac{|pf_1 + qg_1^*|^2 + |pf_2 - qg_2^*|^2}{|pg_1 + qf_1^*|^2 + |pg_2 - qf_2^*|^2}. \quad \text{(A7)}$$

EVIDENCE FOR THE 2π DECAY OF THE K_2^0 MESON*†

J. H. Christenson, J. W. Cronin,‡ V. L. Fitch,‡ and R. Turlay§
Princeton University, Princeton, New Jersey
(Received 10 July 1964)

This Letter reports the results of experimental studies designed to search for the 2π decay of the K_2^0 meson. Several previous experiments have served[1,2] to set an upper limit of 1/300 for the fraction of K_2^0's which decay into two charged pions. The present experiment, using spark chamber techniques, proposed to extend this limit.

In this measurement, K_2^0 mesons were produced at the Brookhaven AGS in an internal Be target bombarded by 30-BeV protons. A neutral beam was defined at 30 degrees relative to the circulating protons by a $1\frac{1}{2}$-in.$\times 1\frac{1}{2}$-in.$\times$48-in. collimator at an average distance of 14.5 ft. from the internal target. This collimator was followed by a sweeping magnet of 512 kG-in. at ~20 ft. and a 6-in.$\times$6-in.$\times$48-in. collimator at 55 ft. A $1\frac{1}{2}$-in. thickness of Pb was placed in front of the first collimator to attenuate the gamma rays in the beam.

The experimental layout is shown in relation to the beam in Fig. 1. The detector for the decay products consisted of two spectrometers each composed of two spark chambers for track delineation separated by a magnetic field of 178 kG-in. The axis of each spectrometer was in the horizontal plane and each subtended an average solid angle of 0.7×10^{-2} steradians. The spark chambers were triggered on a coincidence between water Cherenkov and scintillation counters positioned immediately behind the spectrometers. When coherent K_1^0 regeneration in solid materials was being studied, an anticoincidence counter was placed immediately behind the regenerator. To minimize interactions K_2^0 decays were observed from a volume of He gas at nearly STP.

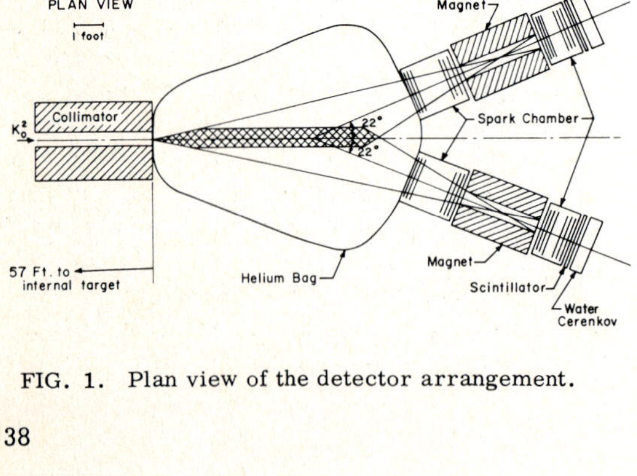

FIG. 1. Plan view of the detector arrangement.

The analysis program computed the vector momentum of each charged particle observed in the decay and the invariant mass, m^*, assuming each charged particle had the mass of the charged pion. In this detector the K_{e3} decay leads to a distribution in m^* ranging from 280 MeV to ~536 MeV; the $K_{\mu 3}$, from 280 to ~516; and the $K_{\pi 3}$, from 280 to 363 MeV. We emphasize that m^* equal to the K^0 mass is not a preferred result when the three-body decays are analyzed in this way. In addition, the vector sum of the two momenta and the angle, θ, between it and the direction of the K_2^0 beam were determined. This angle should be zero for two-body decay and is, in general, different from zero for three-body decays.

An important calibration of the apparatus and data reduction system was afforded by observing the decays of K_1^0 mesons produced by coherent regeneration in 43 gm/cm² of tungsten. Since the K_1^0 mesons produced by coherent regeneration have the same momentum and direction as the K_2^0 beam, the K_1^0 decay simulates the direct decay of the K_2^0 into two pions. The regenerator was successively placed at intervals of 11 in. along the region of the beam sensed by the detector to approximate the spatial distribution of the K_2^0's. The K_1^0 vector momenta peaked about the forward direction with a standard deviation of 3.4 ± 0.3 milliradians. The mass distribution of these events was fitted to a Gaussian with an average mass 498.1 ± 0.4 MeV and standard deviation of 3.6 ± 0.2 MeV. The mean momentum of the K_1^0 decays was found to be 1100 MeV/c. At this momentum the beam region sensed by the detector was 300 K_1^0 decay lengths from the target.

For the K_2^0 decays in He gas, the experimental distribution in m^* is shown in Fig. 2(a). It is compared in the figure with the results of a Monte Carlo calculation which takes into account the nature of the interaction and the form factors involved in the decay, coupled with the detection efficiency of the apparatus. The computed curve shown in Fig. 2(a) is for a vector interaction, form-factor ratio $f^-/f^+=0.5$, and relative abundance 0.47, 0.37, and 0.16 for the K_{e3}, $K_{\mu 3}$, and $K_{\pi 3}$, respectively.[3] The scalar interaction has been computed as well as the vector interaction

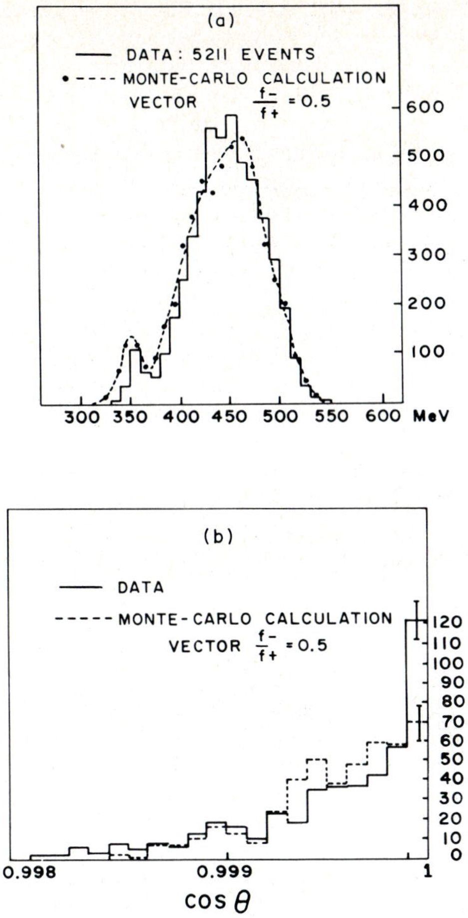

FIG. 2. (a) Experimental distribution in m^* compared with Monte Carlo calculation. The calculated distribution is normalized to the total number of observed events. (b) Angular distribution of those events in the range $490 < m^* < 510$ MeV. The calculated curve is normalized to the number of events in the complete sample.

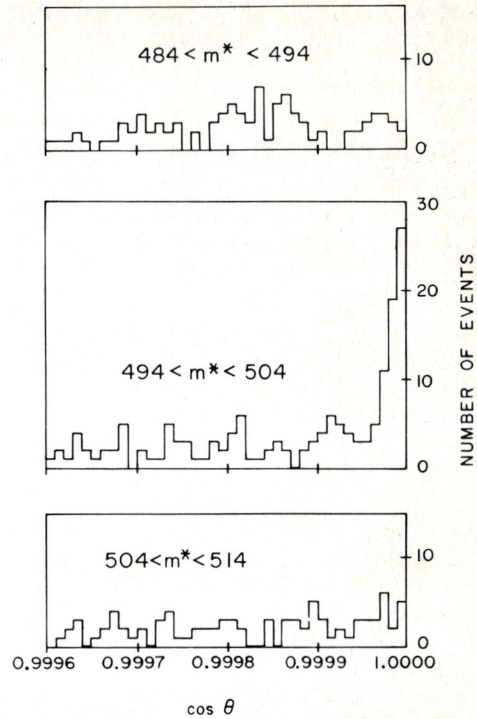

FIG. 3. Angular distribution in three mass ranges for events with $\cos\theta > 0.9995$.

with a form-factor ratio $f^-/f^+ = -6.6$. The data are not sensitive to the choice of form factors but do discriminate against the scalar interaction.

Figure 2(b) shows the distribution in $\cos\theta$ for those events which fall in the mass range from 490 to 510 MeV together with the corresponding result from the Monte Carlo calculation. Those events within a restricted angular range ($\cos\theta > 0.9995$) were remeasured on a somewhat more precise measuring machine and recomputed using an independent computer program. The results of these two analyses are the same within the respective resolutions. Figure 3 shows the results from the more accurate measuring machine. The angular distribution from three mass ranges are shown; one above, one below, and one encompassing the mass of the neutral K meson.

The average of the distribution of masses of those events in Fig. 3 with $\cos\theta > 0.999\,99$ is found to be 499.1 ± 0.8 MeV. A corresponding calculation has been made for the tungsten data resulting in a mean mass of 498.1 ± 0.4. The difference is 1.0 ± 0.9 MeV. Alternately we may take the mass of the K^0 to be known and compute the mass of the secondaries for two-body decay. Again restricting our attention to those events with $\cos\theta > 0.999\,99$ and assuming one of the secondaries to be a pion, the mass of the other particle is determined to be 137.4 ± 1.8. Fitted to a Gaussian shape the forward peak in Fig. 3 has a standard deviation of 4.0 ± 0.7 milliradians to be compared with 3.4 ± 0.3 milliradians for the tungsten. The events from the He gas appear identical with those from the coherent regeneration in tungsten in both mass and angular spread.

The relative efficiency for detection of the three-body K_2^0 decays compared to that for decay to two pions is 0.23. We obtain 45 ± 9 events in the forward peak after subtraction of background out of a total corrected sample of 22 700 K_2^0 decays.

Data taken with a hydrogen target in the beam also show evidence of a forward peak in the $\cos\theta$ distribution. After subtraction of background, 45 ± 10 events are observed in the forward peak at the K^0 mass. We estimate that ~10 events can be expected from coherent regeneration. The number of events remaining (35) is entirely consistent with the decay data when the relative target volumes and integrated beam intensities are taken into account. This number is substantially smaller (by more than a factor of 15) than one would expect on the basis of the data of Adair et al.[4]

We have examined many possibilities which might lead to a pronounced forward peak in the angular distribution at the K^0 mass. These include the following:

(i) K_1^0 coherent regeneration. In the He gas it is computed to be too small by a factor of $\sim 10^6$ to account for the effect observed, assuming reasonable scattering amplitudes. Anomalously large scattering amplitudes would presumably lead to exaggerated effects in liquid H_2 which are not observed. The walls of the He bag are outside the sensitive volume of the detector. The spatial distribution of the forward events is the same as that for the regular K_2^0 decays which eliminates the possibility of regeneration having occurred in the collimator.

(ii) $K_{\mu 3}$ or K_{e3} decay. A spectrum can be constructed to reproduce the observed data. It requires the preferential emission of the neutrino within a narrow band of energy, ±4 MeV, centered at 17 ± 2 MeV ($K_{\mu 3}$) or 39 ± 2 MeV (K_{e3}). This must be coupled with an appropriate angular correlation to produce the forward peak. There appears to be no reasonable mechanism which can produce such a spectrum.

(iii) Decay into $\pi^+\pi^-\gamma$. To produce the highly singular behavior shown in Fig. 3 it would be necessary for the γ ray to have an average energy of less than 1 MeV with the available energy extending to 209 MeV. We know of no physical process which would accomplish this.

We would conclude therefore that K_2^0 decays to two pions with a branching ratio $R = (K_2 \to \pi^+ + \pi^-)/(K_2^0 \to \text{all charged modes}) = (2.0 \pm 0.4) \times 10^{-3}$ where the error is the standard deviation. As emphasized above, any alternate explanation of the effect requires highly nonphysical behavior of the three-body decays of the K_2^0. The presence of a two-pion decay mode implies that the K_2^0 meson is not a pure eigenstate of CP. Expressed as $K_2^0 = 2^{-1/2}[(K_0 - \bar{K}_0) + \epsilon(K_0 + \bar{K}_0)]$ then $|\epsilon|^2 \cong R_T \tau_1 \tau_2$ where τ_1 and τ_2 are the K_1^0 and K_2^0 mean lives and R_T is the branching ratio including decay to two π^0. Using $R_T = \frac{3}{2}R$ and the branching ratio quoted above, $|\epsilon| \cong 2.3 \times 10^{-3}$.

We are grateful for the full cooperation of the staff of the Brookhaven National Laboratory. We wish to thank Alan Clark for one of the computer analysis programs. R. Turlay wishes to thank the Elementary Particles Laboratory at Princeton University for its hospitality.

---

*Work supported by the U. S. Office of Naval Research.

†This work made use of computer facilities supported in part by National Science Foundation grant.

‡A. P. Sloan Foundation Fellow.

§On leave from Laboratoire de Physique Corpusculaire à Haute Energie, Centre d'Etudes Nucléaires, Saclay, France.

[1]M. Bardon, K. Lande, L. M. Lederman, and W. Chinowsky, Ann. Phys. (N.Y.) 5, 156 (1958).

[2]D. Neagu, E. O. Okonov, N. I. Petrov, A. M. Rosanova, and V. A. Rusakov, Phys. Rev. Letters 6, 552 (1961).

[3]D. Luers, I. S. Mittra, W. J. Willis, and S. S. Yamamoto, Phys. Rev. 133, B1276 (1964).

[4]R. Adair, W. Chinowsky, R. Crittenden, L. Leipuner, B. Musgrave, and F. Shively, Phys. Rev. 132, 2285 (1963).

The discovery of charge-conjugation parity asymmetry*

Val L. Fitch

Department of Physics, Princeton University, Princeton, New Jersey 08540

Physics as a science has made incredible progress because of the delicate interplay between theory and experiment. Astonishing predictions based on theories devised to account for known phenomena have been confirmed by experiment. Experiments probing previously unexplored areas often reveal physical effects which are completely unanticipated by theoretical conjecture. The incorporation of the new effects into a theoretical framework then follows.

This year Professor Cronin and I are being honored for a purely experimental discovery, a discovery for which there were no precursive indications, either theoretical or experimental. It is a discovery for which after more than 16 years there is no satisfactory accounting. But showing as it does a lack of charge-conjugation parity symmetry and, correspondingly, a violation of time-reversal invariance, it touches on our understanding of nature at its deepest level.

The discovery of failure of *CP* symmetry was made in the system of *K* mesons. This observation is especially interesting because it was the study of these same particles that led to the overthrow of parity conservation, the notion that interactions and their mirror-reflected counterparts must be equal.

My own interest in *K* particles started in 1952–53 while I was at Columbia working with Jim Rainwater on μ-mesonic atoms. At that time the strange behavior of the particles newly discovered in cosmic rays[1] was a major topic of conversation in the corridors and over coffee. By strange behavior I am referring to the copious production but slow decay. Protons bombarded by pions would result in the production of Λ^0's at 10^{13} times the rate of their decay back to pions and protons. Pais came to Columbia and talked of his ideas on associated production to explain this anomaly (Pais, 1952). Gell-Mann visited and discussed the scheme which he and, independently, Nakano and Nishijima had devised to account for associated production (Gell-Mann, 1953; Nakano and Nishijima, 1953).

Their idea was implausible and daring in the face of available data. The scheme assigned the *K* mesons to two doublets, K^+K^0, and the antiparticles K^- and $\bar{K}^0$. The natural assignment would have been the same as for pions, a triplet of particles K^+, K^0, K^-. Nishijima also assigned quantum numbers, subsequently called strangeness, which were conserved in the strong interaction but not in the weak. The K^+K^0 were assigned +1, the $\bar{K}^0K^-$ as well as the Λ^0, −1.

Standing alone among the particles with positive strangeness were the K^+ and K^0 mesons, and I idly thought that if the situation was ever to be understood these objects might be the key. Most often experiments in physics are long and difficult. It takes some special tweaking of interest to make the commitment to a new area of research. The original motivation is, in the end, apt to appear naive. However, I did in fact join the Princeton Cosmic Ray Group headed by George Reynolds, and spent the Summer of 1954 on a mountain in Colorado learning about the ongoing experiments. During the same period the energy of the cosmotron at Brookhaven was being raised to 3 GeV. Associated production was clearly seen by Shutt and his group at Brookhaven (Fowler *et al.*, 1954), and *K* mesons produced in the cosmotron were identified in photographic emulsion. By the end of the summer I reluctantly decided the future was not in studying cosmic rays in the mountains I loved, but with the accelerators.

The following fall, with Bob Motley, a graduate student, we began to design an apparatus to detect *K* mesons using purely counter techniques at the cosmotron. As this work progressed the cascading interest in the tau-theta puzzle (Dalitz, 1953; Fabri, 1954)[2] led us naturally to explore the lifetime of the *K* particles as a function of their decay mode. We were successful with our detectors, and Motley and I published our results simultaneously with those from the Alvarez group at Berkeley which was using the bevatron as a source (Alvarez *et al.*, 1956; Harris *et al.*, 1955; Fitch and Motley, 1956, 1957). These results showed the degeneracy in the lifetime of the tau and theta mesons. Independently the masses of tau and theta had been shown to be the same to within 1% (Birge *et al.*, 1956). The situation then set the stage for the famous work of Lee and Yang (1956) followed by the experiments with the striking results showing maximal parity violation in the weak interactions (Wu *et al.*, 1957; Garwin *et al.*, 1957; Friedman and Telegdi, 1957). This remarkable story was told by Lee and Yang on this occasion in 1957.

At about this time there appeared a paper by Landau (1957) written before the results of the beta decay experiments were known. Addressing the tau-theta problem, he observed that a simple rejection of parity conservation would create difficult problems in physics. However, with what he called "combined inversion,"

*This lecture was delivered December 8, 1980, on the occasion of the presentation of the 1980 Nobel Prizes in Physics.

[1]For a review ca. 1953 see Rochester and Butler (1953).

[2]Among the strange particles, some were seen to decay to two and some to three pions. By using the analysis of Dalitz and Fabri, it was shown, with very few examples in hand, that the parity of the three-pion system was opposite to that of the two-pion system. If parity is conserved in the decay interaction then there must be distinguishable parents of opposite parity, the theta that decays to two and the tau that decays to three pions. The puzzle was in the question, if the particles are distinct entities, why should they have the same mass and lifetime? Now with parity violation both are recognized as *K* mesons, K_2 and K_3.

that is, space inversion and the simultaneous transformation of particle into antiparticle, the difficulties would be avoided. Indeed, this is a path that nature appeared to take. Subsequent experiments showed parity violation was compensated by a failure of charge conjugation. The weak interactions were therefore invariant under the combined operations of particle-antiparticle interchange and mirror reflection, charge-conjugation parity, *CP*.

One symmetry had been shown to be invalid but had been replaced by a still deeper one. This new symmetry was especially appealing because of the *CPT* theorem. This theorem, which is based on little more than special relativity and locality and which is the foundation of all quantum field theory, says that all interactions must be invariant under *C*, *P*, and *T*, time reversal, all combined. If *CP* is good so also is *T*, in complete accord with all experimental data. The subject was left in a highly satisfactory state. "Who would have dreamed in 1953 that studies of the decay properties of the *K* particles would lead to a new revolution in our understanding of invariance principles," wrote Sakurai in 1963 (Sakurai, 1964). But then in 1964 these same particles, in effect, dropped the other shoe.

It is difficult to give a better example of the mutually complementary roles of theory and experiment than in telling the story of the neutral *K* mesons. For a physicist the pleasures are special because there is scarcely a physical system which contains so many of the elements of modern physics. Two-state systems, of which this is an example, abound, but this one has special properties which give it a unique beauty. I hope that I can convey to you some of the reasons why this system has held such a fascination for us. The story begins with the isotopic spin, strangeness assignment of Gell-Mann and Nishijima. The assignment of the *K* mesons to two doublets makes the K^0 and $\bar{K}^0$ distinct entities. But both particles decay to two π mesons. If the physicist sees π^+ and π^- mesons in his detector, which is the source, the K^0 or $\bar{K}^0$? The problem was solved through the remarkable insight of Gell-Mann and Pais in their 1955 paper. In the spirit of quantum mechanics it is necessary that the source of the $\pi^+\pi^-$ mesons be some linear combination of K^0 and $\bar{K}^0$ states. They observed that a $\pi^+\pi^-$ final state is even under charge conjugation. By "even" we mean that the wave function does not change its algebraic sign upon interchange of particle and antiparticle. The evenness condition is obviously met by the combination $K^0+\bar{K}^0$. This they called the K_1^0.[3] If this is the case, there must be another state equally probable, the $K^0-\bar{K}^0$, the K_2^0 which is odd under charge conjugation, and, correspondingly, is forbidden to decay to $\pi^+\pi^-$. However, it can decay to many other states, three-body states such as $\pi^+\pi^-\pi^0$. It was expected that the decay to the three-body states would be substantially inhibited compared to the two-body state. The particle corresponding to the K_2^0 would have a longer lifetime than the K_1^0 by about 500. In addition, it was expected that the K_1^0 and K_2^0 would have somewhat different masses even though the masses of K^0 and $\bar{K}^0$ are strictly equal by the CPT theorem.

This long-lived neutral *K* meson, predicted by Gell-Mann and Pais, was then looked for and found by a Columbia group working at the Brookhaven cosmotron (Lande *et al.*, 1956). The theoretical model, based on the notion of charge-conjugation invariance in the weak interactions, had been confirmed. Then, suddenly, parity was found to be violated in the weak interactions along with charge conjugation! This dark cloud was almost immediately removed with the observation that one had only to replace *C* with *CP* and the story of the neutral *K* mesons would remain the same (Landau, 1957). With *CP* invariance the K_2^0 would continue to be absolutely forbidden to decay to two pions. The successful description of the neutral system of *K* mesons has been characterized by Feynman as "one of the greatest achievements of theoretical physics" (Feynman, 1962).

Additional features of the $K^0\bar{K}^0$ system become evident if we write the wave function including the lifetime and energy terms for the case of production of a K^0 at $t=0$:

$$\psi(t)=\frac{1}{\sqrt{2}}\left(|K_1^0\rangle e^{-t/2\tau_1+i\omega_1 t}+|K_2^0\rangle e^{-t/2\tau_2+i\omega_2 t}\right),$$

$$|K_1^0\rangle=\frac{1}{\sqrt{2}}\left(|K^0\rangle+|\bar{K}^0\rangle\right),$$

$$|K_2^0\rangle=\frac{1}{\sqrt{2}}\left(|K^0\rangle-|\bar{K}^0\rangle\right).$$

It is seen that after a time, long compared to the K_1^0 lifetime and short compared to the K_2^0 lifetime, the state that was originally a pure K^0 will become a K_2^0 which in turn in an equal mixture of K^0 and $\bar{K}^0$. To give a measure of the magnitudes involved we should point out that the K_1^0 meson, in a typical experimental situation, travels an average of a few centimeters before it decays, whereas the K_2^0 travels tens of meters. At a distance greater than about one meter from the point of production of a K^0 a nearly pure K_2^0 beam will be present.

Another important characteristic of the system becomes apparent when we consider the interaction of K_2^0's with matter. The K^0's and $\bar{K}^0$'s, by virtue of their opposite strangeness, have quite different interaction cross sections. Passing a beam of K_2^0's through a block of material will result in a mixture of K^0 and $\bar{K}^0$'s which, because of differential absorption of the two components, is no longer 50-50, but instead a mixture equivalent to a new combination of K_1^0's and K_2^0's. The newly produced short-lived K_1^0's decaying to $\pi^+\pi^-$ will appear behind the material. This phenomenon is called regeneration (Pais and Piccioni, 1955). In the case when the absorbing material is completely transparent to K^0's and opaque to $\bar{K}^0$'s, the intensity of the K_1^0's after the absorber will be $\frac{1}{4}$ the initial intensity of the K_2^0 incident on the absorber.

In the late 1950s Good (1957) observed that with a very small mass difference between the K_1^0 and K_2^0 the regeneration phenomena just discussed would result in a coherent process. By coherent we mean that the scattering process of K_2^0 to K_1^0 would not be from individual

[3]We have changed the notation to correspond to recent custom. Gell-Mann and Pais called them θ_1 and θ_2.

nuclei but from the whole block of scattering material! That is, the block of material would remain in its initial quantum mechanical state during the scattering process. In this case, as with ordinary light passing through glass, the regeneration material could be treated as having an index of refraction. The K_1^0's regenerated coherently would have precisely the same energy as the incident K_2^0's and an angular distribution identical to the incident beam but broadened by diffraction effects determined by the size of the regenerating material perpendicular to the beam. A characteristic wavelength for the K_2^0 mesons in a typical experiment is about 10^{-13} cm. The transverse dimensions are typically 10 cm. The corresponding diffraction pattern has a width of the order of 10^{-14} radians! In addition, the coherent addition of K_1^0 waves has been observed over distances greater than 10^{14} wavelengths. The unique feature of this coherently regenerated K_1^0 beam is that it can be distinguished from the original beam since it decays with a short lifetime to $\pi^+\pi^-$. To my knowledge, it is the only instance where a forward coherently scattered beam can be distinguished from the original beam.

It has become evident to physics students in the audience that the $K_1^0K_2^0$ story has an analogy in polarized light. The K_1^0 and K_2^0 correspond to the left and right circularly polarized light, and the K^0 and $\bar{K}^0$ states are equivalent to the x and y components of linear polarization. The passage of a K_2^0 beam through a block of condensed material is equivalent to the passage of left circular polarized light through a doubly refractive medium like calcite which has a different index refraction for the x and y components of polarization. The general picture of regeneration, coherent and incoherent, was confirmed in a definitive bubble-chamber experiment (Good *et al.*, 1961).

There are many associated phenomena still to be explored. For example, experiments coherently regenerating K_1^0's from the planes in crystals have yet to be done. At the particle momenta commonly available the Bragg angles are exceedingly small, and the extinction factor, the Debye-Waller factor, comes into play at correspondingly small angles, but the experiments could be done.

Unexpectedly, the $K^0\bar{K}^0$ system provides us with important and highly precise information about the gravitational interaction. It relates to the question of strong universality; that is, whether different objects, in this case particle and antiparticle, with the same inertial mass behave the same in a gravitational field. As observed by Good (1961), if the K^0 and its antiparticle, the $\bar{K}^0$, had an opposite gravitational potential energy, the $K^0\bar{K}^0$ system would mix so quickly that the long-lived particle would never be seen. By analyzing the system in more detail one can show that if the gravitational interaction of particle and antiparticle differ by a fraction κ, then κ must be less than 10^{-10} if we're dealing with the gravitational field of the Earth, 10^{-11} for the solar system, and 10^{-13} for the galaxy.

Voyages of discovery can be made in new uncharted waters but also in the familiar bays close to port, provided one has observing apparatus that can see familiar objects with detail greater than that previously possible. In 1963 we had the opportunity to investigate the neutral K-meson phenomena with resolution greater than that permitted before. The introduction of spark chambers as charged-particle detectors permitted precise track position determination, but also the chambers could be selectively triggered on appropriate classes of events.

Using such new devices with our colleagues, Jim Christenson and Rene Turlay, Jim Cronin and I initiated a systematic study of (1) the regeneration phenomena, (2) what we called *CP* invariance, and (3) neutral currents. We were interested in the regeneration phenomena in particular because of an anomaly that had just been reported by a group studying the passage of K_2^0's through a liquid hydrogen bubble chamber (Christenson *et al.*, 1964). Not many of our colleagues would have given us much credit for studying *CP* invariance, but we did anyway, and neutral currents, of long interest, were discussed by Professor Glashow on this occasion one year ago.

A plan view of the apparatus we used for these studies is shown in Fig. 1. It is a two-armed spectrometer,

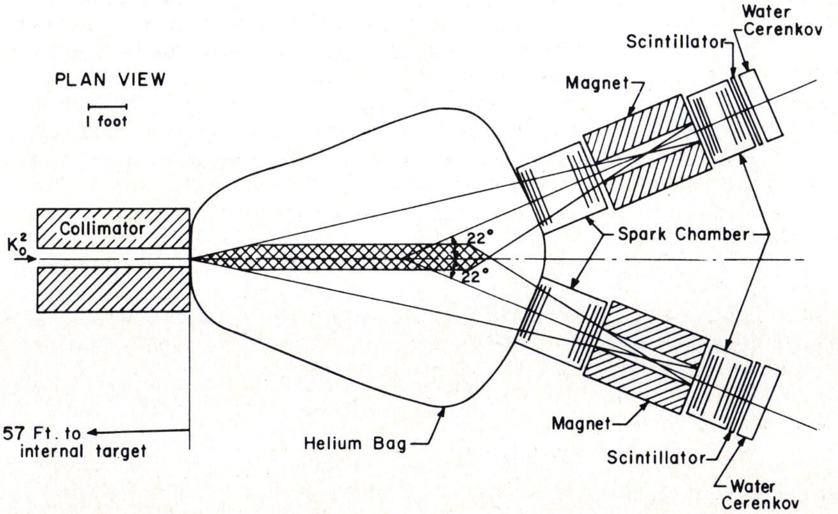

FIG. 1. Plan view of the apparatus as located at the alternating-gradient synchrotron.

each arm with spark chambers before and after a magnet for track delineation. Čerenkov and scintillation counters in both arms, operated in coincidence, provided the signals to trigger the spark chambers, which were recorded photographically. The apparatus was placed in a beam of neutral particles at the Brookhaven alternating gradient synchrotron at a distance such that K_1^0's would have decayed away leaving K_2^0's. The angle between the spectrometer arms was chosen to optimize the detection of K^0 mesons decaying to two π mesons. In the regeneration studies, blocks of various solid materials were placed in the neutral beam. In the studies of the free decay of $K_2^0 \rightarrow 2$ pions, the decay volume was filled with helium gas to minimize the interactions.

The decay to two pions is distinguished from the copious three-body decays in two ways. The sum of the momenta of the two detected particles must line up with the direction of the incident K_2^0's. In general this will not happen for three-body decay. In addition, the mass computed for the parent particle must match the mass of the K^0 meson. The original data are shown in Figs. 2 and 3. Figure 2 shows the data after measurement of the photographic records on a relatively coarse measuring machine. The presence of the peaking of events along the beam line stimulated more precise measurements, and these results are shown in Fig. 3. Clearly there are about 56 events in the forward peak in the proper mass interval where the background is 11. From this data we established that the branching ratio of K_2^0 to 2 pions relative to all the charge modes decay was 2×10^{-3}. Here was the first evidence for the decay completely forbidden by *CP* conservation (Christenson *et al.*, 1964). We were acutely sensitive to the importance of the result and, I must confess, did not initially believe it ourselves. We spent nearly half a year attempting to invent viable alternative explanations, but failed in every case.

The study of coherent regeneration was important for the *CP* measurement for several reasons. First, the results we found were entirely consistent with expectations; there were no anomalies. The measured coherent regeneration rates in tungsten, copper, carbon, and liquid hydrogen enabled us to show that coherent regeneration in the gaseous helium which filled the decay volume would produce a totally negligible contribution to the signal we observed. Second, the coherent regeneration of the K_1^0's, which subsequently decayed to $\pi^+\pi^-$ mesons, provided an invaluable calibration of the apparatus.

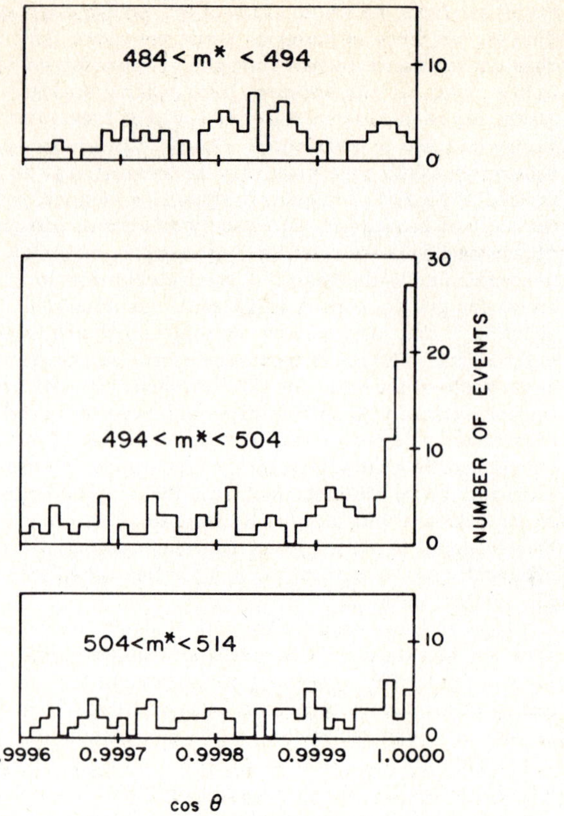

FIG. 3. Angular distribution of the events after measurement by a precise machine in three relevant mass regions.

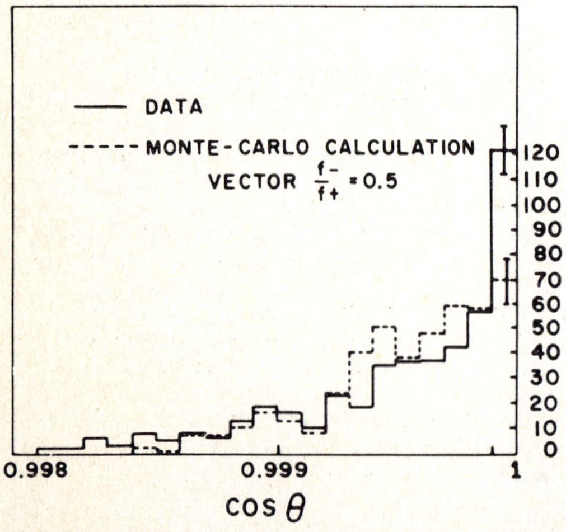

FIG. 2. Angular distribution of those events in the appropriate mass range as measured by a coarse measuring machine.

It is appropriate now to look at the neutral K system in a somewhat more quantitative way (Lee, Oehme, and Yang, 1957). Because of the mixing of the K^0 and $\bar{K}^0$ through the weak interaction, the time rate of change of a K^0 wave will depend not only on the K^0 amplitude, but also on the $\bar{K}^0$ amplitude, viz.,

$$-\frac{dK^0}{dt}=AK^0+p^2\bar{K}^0$$

and

$$-\frac{d\bar{K}^0}{dt}=B\bar{K}^0+q^2K^0 .$$

We have let the particle symbol stand for the amplitude of the corresponding wave. With invariance under *CPT*, particle and antiparticle masses and lifetimes must be precisely identical. In terms of the above equations, A must be equal to B. Now, *CP* violation can, in fact, occur in two ways, either through terms in the set of equations above, or in the amplitudes for the decay. Subsequent experiments show that most, if not all, of the violation is in the equations above, involving the so-called mass-decay matrix. Professor Cronin will discuss the ramifications of the effect being present also in the decay terms. Suffice it to say here that any departure of p^2 from q^2 will result in the decay of the $K_2^0 \to 2$ pions. With *CP* nonconservation the short- and long-lived particles are no longer the K_1^0 and K_2^0 previously defined, but rather

$$K_S^0 = \frac{1}{(p^2+q^2)^{1/2}} (p|K^0\rangle + q|\bar{K}^0\rangle)$$

and

$$K_L^0 = \frac{1}{(p^2+q^2)^{1/2}} (p|K^0\rangle - q|\bar{K}^0\rangle).$$

The fact that K_L^0 decays to two pions shows that the amplitude for particle to antiparticle transitions, in this case $K^0 \to \bar{K}^0$, does not quite equal the reverse, $\bar{K}^0 \to K^0$, and indeed we now know rather precisely that not only are the magnitudes somewhat different but that there is a small phase angle between the two amplitudes (see Fig. 4).

We indicated earlier that, through the *CPT* theorem, a violation of *CP* is equivalent to a violation of time-reversal invariance. As Professor Cronin will show, the *CPT* theorem has been shown to hold in the neutral *K* system independently, so in a self-contained way a violation of time-reversal invariance is demonstrated.

We are all familiar with the time asymmetry associated with entropy. Entropy in a closed system increases with time. This kind of time asymmetry results from the boundary conditions. But for the first time we have in the neutral *K* mesons a physical system that behaves asymmetrically in time as a result of an interaction, not a boundary condition.

Since the microscopic physical laws had always been throught to be invariant under time reversal, this discovery opens up a very wide range of profound questions. Professor Cronin will go into some of these questions in greater detail. I will mention two. Can this effect be used to decrease the entropy of an isolated system? We look out from the Earth and see a highly ordered universe. With entropy always increasing how can this be? Is *CP* violation an effect that can be used, in effect, to wind up the universe? The answers to these questions appear to be no (Ne'eman, 1972).

At the same time we look out from the Earth and see the remains of an earlier much hotter universe. In that earlier time one expects that matter and antimatter would condense out in equal amounts and eventually annihilate to gamma radiation. However no evidence of antimatter is seen. The gauge theories described on this occasion one year ago allow for the possibility of proton (and antiproton) decay. This process, coupled with *CP* violation, drives the universe towards a preponderance of matter over antimatter and can account for the observed ratio of the amount of matter to radiation (Sakharov, 1967).[4]

Lewis Thomas, whose essays on science grace our literature, has written, "You measure the quality of the work by the intensity of the astonishment." After 16 years, the world of physics is still astonished by *CP* and *T* noninvariance. I suspect that the Nobel Committee was motivated by considerations similar to those of Thomas in awarding to Professor Cronin and myself this highest of honors.

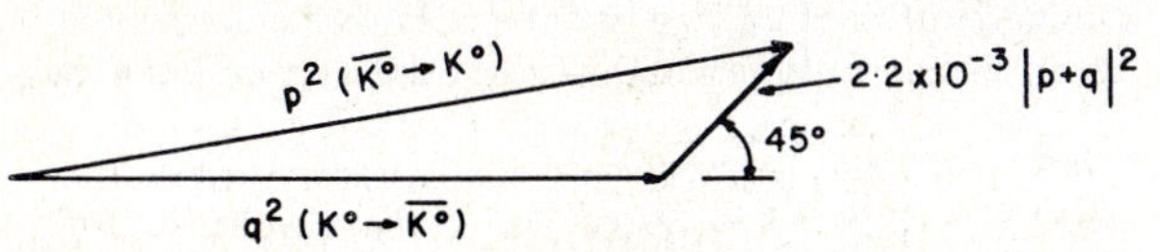

FIG. 4. Vector diagram showing schematically the difference in the amplitudes for $K^0 \to \bar{K}^0$ and $\bar{K}^0 \to K^0$.

REFERENCES

Alvarez, L. W., F. S. Crawford, M. L. Good, and M. L. Stevenson, 1956, Phys. Rev. **101**, 503.

Birge, R. W., D. H. Perkins, J. R. Peterson, D. H. Stork, and M. N. Whitehead, 1956, Nuovo Cimento **4**, 834.

Christenson, J., J. W. Cronin, V. L. Fitch, and R. Turlay, 1964, Phys. Rev. Lett. **13**, 138.

Dalitz, R. H., 1953, Philos. Mag. **44**, 1068.

Fabri, E., 1954, Nuovo Cimento **11**, 479.

Feynman, R. P., 1962, *The Theory of Fundamental Processes; a lecture note volume* (Benjamin, New York), p. 50.

Fitch, V., and R. Motley, 1956, Phys. Rev. **101**, 496.

Fitch, V., and R. Motley, 1957, Phys. Rev. **105**, 265.

Fowler, W. B., R. P. Schutt, A. M. Thorndike, and W. L. Whittemore, 1954, Phys. Rev. **93**, 861.

Friedman, J. I., and V. L. Telegdi, 1957, Phys. Rev. **105**, 1681.

Garwin, R., L. Lederman, and M. Weinrich, 1957, Phys. Rev. **105**, 1415.

Gell-Mann, M., 1953, Phys. Rev. **92**, 833.

Gell-Mann, M., and A. Pais, 1955, Phys. Rev. **97**, 1387.

Good, M. L., 1957, Phys. Rev. **106**, 591.

Good, M. L., 1961, Phys. Rev. **121**, 311.

Good, R. H., R. P. Matsen, F. Muller, O. Piccioni, W. M. Powell, H. S. White, W. B. Fowler, and R. W. Birge, 1961, Phys. Rev. **124**, 1223.

Harris, G., J. Orear, and S. Taylor, 1955, Phys. Rev. **100**, 932.

Landau, L., 1957, Nucl. Phys. **3**, 254.

Lande, K., E. T. Booth, J. Impeduglia, L. M. Lederman, and W. Chinowsky, 1956, Phys. Rev. **103**, 1901.

Lee, T. D., K. Oehme, and C. N. Yang, 1957, Phys. Rev. **106**, 340.

Lee, T. D., and C. N. Yang, 1956, Phys. Rev. **104**, 254.

Nakano, T., and K. Nishijima, 1953, Prog. Theor. Phys. **10**, 581.

Ne'eman, Y., 1972, Erice Summer School Lectures, June 16–July 6, 1972.

Pais, A., 1952, Phys. Rev. **86**, 663.

Pais, A., and O. Piccioni, 1955, Phys. Rev. **100**, 1487.

Rochester, G. D., and C. C. Butler, 1953, Rep. Prog. Phys. **16**, 364.

Sakharov, A. D., 1967, JETP Lett. **5**, 24.

Sakurai, J. J., 1964, *Invariance Principles and Elementary Particles* (Princeton University Press, Princeton, N.J.), p. 296.

Wilczek, F. W., 1980, Sci. Am. **243**, (Dec.), 82.

Wu, C. S., E. Ambler, R. W. Hayward, D. D. Hoppes, and R. P. Hudson, 1957, Phys. Rev. **105**, 1413.

---

[4]For a nontechnical discussion see Wilczek (1980).

CP symmetry violation—the search for its origin*

James W. Cronin

Department of Physics, University of Chicago, Chicago, Illinois 60637
and Enrico Fermi Institute, University of Chicago, Chicago, Illinois 60637

The greatest pleasure a scientist can experience is to encounter an unexpected discovery. I am always astonished when a simple apparatus, designed to ask the right question of nature, receives a clear response. Our experiment, carried out with James Christenson, Val Fitch, and René Turlay, gave convincing evidence that the long-lived neutral K meson (K_L) decayed into two charged pions, a decay mode forbidden by CP symmetry. The forbidden decay mode was found to be a small fraction $(2.0\pm0.4)\times10^{-3}$ of all charged decay modes. Professor Fitch has described our discovery of CP symmetry violation. He has discussed how it was preceded by brilliant theoretical insights and incisive experiments with K mesons. My lecture will review the knowledge that we have obtained about CP violation since its discovery.[1] The discovery triggered an intense international experimental effort. It also provoked many theoretical speculations which in turn stimulated a variety of experiments.

At present there is no satisfactory theoretical understanding of CP violation. Such understanding as we do have has come entirely from experimental studies. These studies have extended beyond the high-energy accelerator laboratories into nuclear physics laboratories and research reactor laboratories. The experiments which have sought to elucidate the tiny effect have involved both ingenuity and painstaking attention to detail.

Upon learning of the discovery in 1964, the natural reaction of our colleagues was to ask what was wrong with the experiment. Or, if they were convinced of the correctness of the measurements, they asked how the effect could be explained while still retaining CP symmetry. I remember vividly a special session organized at the 1964 International Conference on High Energy Physics at Dubna in the Soviet Union. There, for an afternoon, I had to defend our experiment before a large group of physicists who wanted to know every detail of the experiment—more details than could have been given in the formal conference session.

As the session neared a close, one of my Soviet colleagues suggested that, perhaps, the effect was due to regeneration of short-lived K mesons (K_S) in a fly unfortunately trapped in the helium bag. We did a quick "back of the envelope" estimate of the density of the fly necessary to produce the effect. The density required was far in excess of uranium.

More serious questions were raised at this session and by many other physicists who had thought deeply about our result. While we were confident that the experiment had been correctly carried out and interpreted, many sought reassurance through confirmation of the experiment by other groups. This confirmation came quickly from experiments at the Rutherford Laboratory (Galbraith *et al.*, 1965) in England, and at CERN (de Bouard *et al.*, 1965) in Geneva, Switzerland.

Another important issue was raised. In the original experiment, the decay to two pions was inferred kinematically, but no proof was given that these pions were identical to the ordinary pions or that the decay was not accompanied by a third light particle emitted at a very low energy. The direct proof that the effect was indeed a violation of CP symmetry was the demonstration of interference between the decay of the long-lived and short-lived K meson to two charged pions. This interference was first demonstrated in a simple and elegant experiment by my colleague Val Fitch with Roth, Russ, and Vernon (Fitch *et al.*, 1965).

Their experiment compared the rate of decay of a K_L beam into two charged pions in vacuum and in the presence of a diffuse beryllium regenerator. The density of the regenerator was adjusted so that the regeneration amplitude A_r was equal to the CP-violating amplitude η_{+-}. These amplitudes are defined by

$$\eta_{+-}=\frac{\text{amplitude}(K_L\to\pi^+\pi^-)}{\text{amplitude}(K_S\to\pi^+\pi^-)}$$

and

$$A_r=i\pi N\Lambda\left(\frac{f-\bar{f}}{k}\right)(i\delta+\tfrac{1}{2})^{-1}.$$

The yield of $K_L\to\pi^+\pi^-$ in the presence of the regenerator is proportional to

$$|A_r+\eta_{+-}|^2.$$

In the expression for A_r, δ is given by $(M_S-M_L)/\Gamma_S$, where M_S and M_L are the K_S and K_L masses, and Γ_S the decay rate of the K_S meson, Λ is the mean decay length of the K_S meson, k is the wave number of the incident K_L beam, and f and $\bar{f}$ are the forward scattering amplitudes for K and $\bar{K}$, respectively, on the nuclei of the regenerator. The regeneration amplitude is proportional to N, the number density of the material. The quantity $(f-\bar{f})/k$ was determined in an auxiliary experiment with a dense regenerator. Then a regenerator of appropriate density was constructed using the formula for A_r.[2] The actual regenerator was

*This lecture was delivered December 8, 1980, on the occasion of the presentation of the 1980 Nobel Prizes in Physics.

[1]This lecture cannot cover all the important details, refer to all the important work concerning CP violation, or do justice to the work that is referred to in the text. For a more complete discussion, the reader is referred to the most recent review by Kleinknecht (1976). Also, a good perspective of the progress in the field can be found by reading the appropriate sections of the proceedings of the biannual International Conferences on High Energy Physics, 1964–1974.

[2]The value of A_r depends on the K_s-K_L mass difference $|\delta|$ which was, at the time, measured to be 0.5 ± 0.1 by Christenson *et al.* (1965).

constructed of 0.5 mm sheets separated by 1 cm. Such an arrangement behaves as a homogeneous regenerator of ($\frac{1}{20}$) normal density if the separation of the sheets is small compared to the quantity $\delta\Lambda$.

In the earliest experiment Fitch and his colleagues found that with $|A_r|$ chosen to be equal to $|\eta_{+-}|$ the rate of $\pi^+\pi^-$ decays was about *four times* the rate without the regenerator. This result showed not only that there was interference, but also that the interference was fully constructive. Complete analysis of this experiment reported subsequently (Fitch *et al.*, 1967) gave the $\pi^+\pi^-$ yield as a function of density as shown in Fig. 1. The quantity α in the figure is the relative phase between the regeneration amplitude and the *CP*-violating amplitude.

The results of this experiment also permit the experimental distinction between a world composed of matter and a world composed of antimatter.[3] Imagine that this experiment were performed in the antiworld. The only difference would be that the regenerator material would be antimatter. If we assume *C* invariance for the strong interactions, the forward scattering amplitudes for K and $\bar{K}$ would be interchanged so that A_r would have the opposite sign. Thus, in the antiworld, an investigator performing the interference experiment would observe destructive interference similar to the dashed curve of Fig. 1, an unmistakable difference from the result found in our world. The interference experiment of Fitch and collaborators eliminated alternate explanations of the $K_L \to \pi^+\pi^-$ decay, since the effect was of such a nature that an experiment distinguishing a world of matter and antimatter was possible.

It was also suggested that the effect might be due to a long-range vector field of cosmological origin (Bell and Perring, 1964; Bernstein, Cabibbo, and Lee, 1964). Such a source of the effect would lead to a decay rate for $K_L \to \pi^+\pi^-$ which would be proportional to the square of the K_L energy in the laboratory. Our original experiment was carried out at a mean K_L energy of 1.1 GeV. The confirming experiments at the Rutherford Laboratory and CERN were carried out at mean K_L energies of 3.1 and 10.7 GeV, respectively. Since the three experiments found the same branching ratio for $K_L \to \pi^+\pi^-$, the possibility of a long-range vector field was eliminated.

Before continuing, it is necessary to state some of the phenomenology which describes the *CP* violation in the neutral *K* system. The basic notation was introduced by Wu and Yang (1964).[4] For this discussion *CPT* conservation is assumed. Later we shall refer to the evidence from *K*-meson decays which show that all data are consistent with a corresponding *T* violation. Any *CPT* violation is consistent with zero within the present sensitivity of the measurements.

There are two basic complex parameters which are required to discuss *CP* violation as observed in the two pion decays of K_L mesons. The first quantity ε is a measure of the *CP* impurity in the eigenstates $|K_S\rangle$ and $|K_L\rangle$. These eigenstates are given by

$$|K_S\rangle = \frac{1}{\sqrt{2}\,(1+|\varepsilon|^2)^{1/2}}\,[(1+\varepsilon)|K\rangle + (1-\varepsilon)|\bar{K}\rangle]$$

and

$$|K_L\rangle = \frac{1}{\sqrt{2}\,(1+|\varepsilon|^2)^{1/2}}\,[(1+\varepsilon)|K\rangle - (1-\varepsilon)|\bar{K}\rangle]$$

The quantity ε can be expressed in terms of the elements of the mass and decay matrices which couple and control the time evolution of time $|K\rangle$ and $|\bar{K}\rangle$ states. It is given by

$$\varepsilon = \frac{-\mathrm{Im}M_{12} + i\,\mathrm{Im}\Gamma_{12}/2}{i(M_S - M_L) + (\Gamma_S - \Gamma_L)/2}\,.$$

Limits on the size of $\mathrm{Im}\Gamma_{12}$ can be obtained from the observed decay rates of K_S and K_L to the various decay modes. If $\mathrm{Im}\Gamma_{12}$ were zero, then the phase of ε would be determined by the denominator, which is just the difference in eigenvalues of the matrix which couples K and $\bar{K}$. These quantities have been experimentally measured and give $\arg\varepsilon \sim 45°$.

The second quantity ε' is defined by

$$\varepsilon' = \frac{i}{\sqrt{2}}\,\mathrm{Im}\left(\frac{A_2}{A_0}\right)e^{i(\delta_2-\delta_0)}\,.$$

Here A_0 and A_2 are the amplitudes for a K meson to decay to standing-wave states of two pions in the isotopic spin-0 and spin-2 states, respectively. Time-reversal symmetry demands that A_0 and A_2 be relatively real (Wigner, 1932; see also Lee, Oehme, and Yang, 1957). The quantities δ_0 and δ_2 are the *s*-wave π-π scattering phase shifts for the states $I=0$ and $I=2$, respectively. The parameters ε and ε' are related to observable quantities defined by

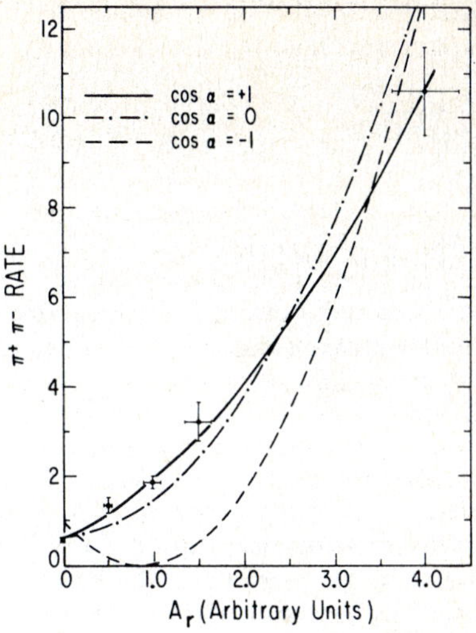

FIG. 1. Yield of $\pi^+\pi^-$ events as a function of the diffuse regenerator amplitude. The three curves correspond to the three stated values of the phase between the regeneration amplitude A_r and the *CP*-violating amplitude η_{+-}.

[3]This argument was presented in the literature by Wattenberg and Sakurai (1967).

[4]The phenomenology was first discussed by Lee, Oehme, and Yang (1957).

$$|\eta_{+-}|e^{i\phi_{+-}} = \frac{\mathrm{am}(K_L \to \pi^+\pi^-)}{\mathrm{am}(K_S \to \pi^+\pi^-)}\,,$$

$$|\eta_{00}|e^{i\phi_{00}} = \frac{\mathrm{am}(K_L \to \pi^0\pi^0)}{\mathrm{am}(K_S \to \pi^0\pi^0)}\,,$$

and

$$\delta_l = \frac{\Gamma(K_L \to \pi^- l^+ \nu_l) - \Gamma(K_L \to \pi^+ l^- \bar{\nu}_l)}{\Gamma(K_L \to \pi^- l^+ \nu_l) + \Gamma(K_L \to \pi^+ l^- \bar{\nu}_l)}\,.$$

These experimentally measured quantities are related to ε and ε' by the following expressions[5]:

$$\eta_{+-} = \varepsilon + \varepsilon'\,,$$

$$\eta_{00} = \varepsilon - 2\varepsilon'\,,$$

$$\delta_l = 2\,\mathrm{Re}\,\varepsilon\,.$$

The magnitude and phase of the quantity η_{+-} have been most precisely measured by studying the time dependence of $\pi^+\pi^-$ decays from a K beam which was prepared as a mixture of K_S and K_L. This experimental technique was suggested by Whatley (1962) long before the discovery of *CP* violation. If we let ρ be the amplitude for K_S at $t=0$, relative to the K_L amplitude, then the time dependence of $\pi^+\pi^-$ decays will be given by[6]

$$N_{+-}(t) = |\rho \exp[(-i\Delta M - \Gamma_S/2)t] + \eta_{+-}|^2\,.$$

The initial amplitude for the K_S component can be prepared by two different methods. In the first method we pass a K_L beam through a regenerator. Then ρ is the regeneration amplitude. Here the interference term is

$$2|\rho|\,|\eta_{+-}|e^{-\Gamma_S t/2}\cos(-\Delta M t + \phi_\rho - \phi_{+-})\,.$$

In the second method we produce a beam which is pure K (or $\bar{K}$) at $t=0$. In practice protons of ~20 GeV produce at small angles about three times as many K as $\bar{K}$. The $\bar{K}$ dilution is a detail which need not be of concern here. In this case $\rho = +1$, and the interference term is

$$2|\eta_{+-}|e^{-\Gamma_S t/2}\cos(-\Delta M t - \phi_{+-})\,.$$

The important *CP* parameters are $|\eta_{+-}|$ and ϕ_{+-}. We see, however, that a knowledge of the auxiliary parameters Γ_S and ΔM is also required. In the first method one measures $\phi_{+-} - \phi_\rho$ and one must also have a technique to independently measure ϕ_ρ. In both cases the $\pi^+\pi^-$ yield is most sensitive to the interference term when the two interfering amplitudes are of the same size. For the second method we require observation at $12\,K_S$ lifetimes. (We want $e^{-\Gamma_S t/2} \approx |\eta_{+-}| \approx 2\times 10^{-3}$.) As a consequence, a small error in ΔM can lead to a large uncertainty on ϕ_{+-}, and, more importantly, a systematic error in ΔM can lead to an incorrect value for ϕ_{+-}. A one-percent error in ΔM corresponds to an error in ϕ_{+-} of about 3°. The measurement of ΔM with satisfactory precision has required an effort as formidable as the interference experiments themselves.[7]

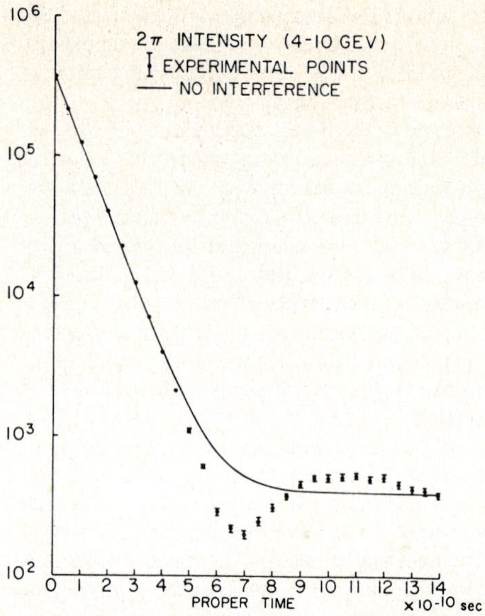

FIG. 2. Yield of $\pi^+\pi^-$ events as a function of proper time downstream from an 81 cm carbon regenerator placed in a K_L beam. Figure taken from thesis of T. Modis, Columbia University (1973); a published version of this work is given by Carithers *et al.* (1975).

Time and space do not permit a survey which does justice to the many groups at CERN, Brookhaven, Argonne, and SLAC who have made the meticulous measurements which have led to the following parameters[8]:

$$\eta_{+-} = [(2.27 \pm 0.02)\times 10^{-3}]\exp[i(44.7° \pm 1.2°)]\,,$$

$$\Delta M = M_S - M_L = -(0.535 \pm 0.002)\times 10^{10}\ \mathrm{sec}^{-1}\,,$$

$$\Gamma_S = (1.121 \pm 0.003)\times 10^{10}\ \mathrm{sec}^{-1}\,.$$

As an example of the quality of the measurements mentioned above, Fig. 2 shows a time distribution of $\pi^+\pi^-$ decays following the passage of a K_L beam of 4 to 10 GeV/c momentum through an 81 cm thick carbon regenerator. The destructive interference is clearly seen. If the experiment were carried out with a regenerator of anticarbon, then constructive interference would have been observed.

Measurements of the charge asymmetry δ_l for K_L

---

[5]Some approximations have been made. The first two expressions should read: $\eta_{+-} = \varepsilon + \varepsilon'/(1+\omega)$ and $\eta_{00} = \varepsilon - 2\varepsilon'/(1-2\omega)$, where $\omega = 1/\sqrt{2}\,\mathrm{Re}(A_2/A_0)e^{i(\delta_2-\delta_0)}$. The magnitude of $\omega \simeq 0.05$, so that its effect is not large. The charge asymmetry δ_L should be given by

$$\delta_L = 2\frac{(1-|x|^2)}{|1-x|^2}\,\mathrm{Re}\,\varepsilon\,,$$

where x is the ratio of the $\Delta Q = -\Delta S$ amplitude to the $\Delta Q = +\Delta S$ amplitude in the semileptonic decay. Evidence strongly favors $x \simeq 0$; see Niebergall *et al.* (1974).

[6]Here $\Delta M = M_S - M_L$ and we have neglected Γ_L compared to Γ_S. The interference experiments are always performed over a time scale such that $t \ll 1/\Gamma_L$ so that the decay of the K_L amplitude is negligible.

[7]Precise measurements of ΔM have been reported by Aronson *et al.* (1970); Cullen *et al.* (1970); Carnegie *et al.* (1971); Geweniger *et al.* (1974).

[8]The data for this compilation is most readily available from the Particle Data Group (1980).

decays began in 1966. This asymmetry is found in the abundant semileptonic decay modes $K_L \to \pi^{\pm} l^{\mp} \nu$, where l is either an electron or a muon. It basically measures the difference in amplitude of K and $\bar{K}$ in the eigenstate of the K_L. It does so by virtue of the $\Delta S = \Delta Q$ rule, which states that all semileptonic decays have the change in charge of the hadron equal to the change in strangeness. Thus K mesons decay to $\pi^- l^+ \nu$ and $\bar{K}$ mesons decay to $\pi^+ l^- \bar{\nu}$. The validity of the $\Delta S = \Delta Q$ rule was in doubt for many years, but it has finally been established that the $\Delta Q = -\Delta S$ transitions are no more than about 2% of the $\Delta Q = +\Delta S$ transitions (Niebergall *et al.*, 1974). The size of the charge asymmetry expected is $\sim\sqrt{2}\,|\eta_{+-}| \approx 3\times10^{-3}$. Millions of events are required to measure δ_l accurately, and excellent control of the symmetry of the apparatus and understanding of charge-dependent biases are needed to reduce systematic errors.

Again, we must omit a detailed review of all the asymmetry measurements. These have been carried out at CERN, Brookhaven, and SLAC. The net result of these measurements gives (Particle Data Group, 1980)

$$\delta_e = (3.33 \pm 0.14)\times10^{-3}$$

and

$$\delta_\mu = (3.19 \pm 0.24)\times10^{-3}.$$

We expect these two asymmetries to be equal since they both are a measure of $2\,\mathrm{Re}\,\varepsilon$. These asymmetries are measured for a pure K_L beam. For a beam which is pure K at $t=0$ the charge asymmetry shows a strong oscillation term with angular frequency ΔM. Figure 3 shows the time dependence of the charge asymmetry taken from the thesis of Lüth (Lüth, 1974). The small residual charge asymmetry of the K_L decays after the oscillations have died out is clearly resolved.

The charge asymmetry is a manifest violation of *CP*, and as such also permits an experimental distinction between a world and an antiworld. In our world we find that the positrons in the decay are slightly in excess. The positrons are leptons which have the same charge as our atomic nuclei. In the antiworld the experimenter will find that the excess leptons have opposite charge to his atomic nuclei; hence, he would report a different result for the same experiment.

Simple examination of the relations between the experimentally measurable parameters and the complex quantities ε and ε' show that measurements of $|\eta_{00}|$ and ϕ_{00} are essential to finding ε and ε'. The path to reliable results for $|\eta_{00}|$ and ϕ_{00} has been torturous. This statement is based on personal experience; six years of my professional life have been spent on the measurement of $|\eta_{00}|$.

Measurement of the parameters associated with $K_L \to \pi^0\pi^0$ is complicated by the fact that each π^0 decays rapidly (10^{-16} sec) into two photons. For typical K_L beams used in these experiments the photon energies are in the range of 0.25 to 5 GeV. It is difficult to measure accurately the direction and energy of such photons. In addition to that difficulty, the *CP*-conserving decay $K_L \to 3\pi^0$ occurs at a rate which is about 200 times as frequent, and presents a severe background.

Early results suggested that $|\eta_{00}|$ was about twice $|\eta_{+-}|$ with the consequence that ε' was a large number. By 1968, however, an improved experiment using spark chambers (Banner *et al.*, 1968) and a painstaking heavy-liquid bubble-chamber experiment from CERN (Budagov *et al.*, 1968) showed that $|\eta_{00}|$ was rather

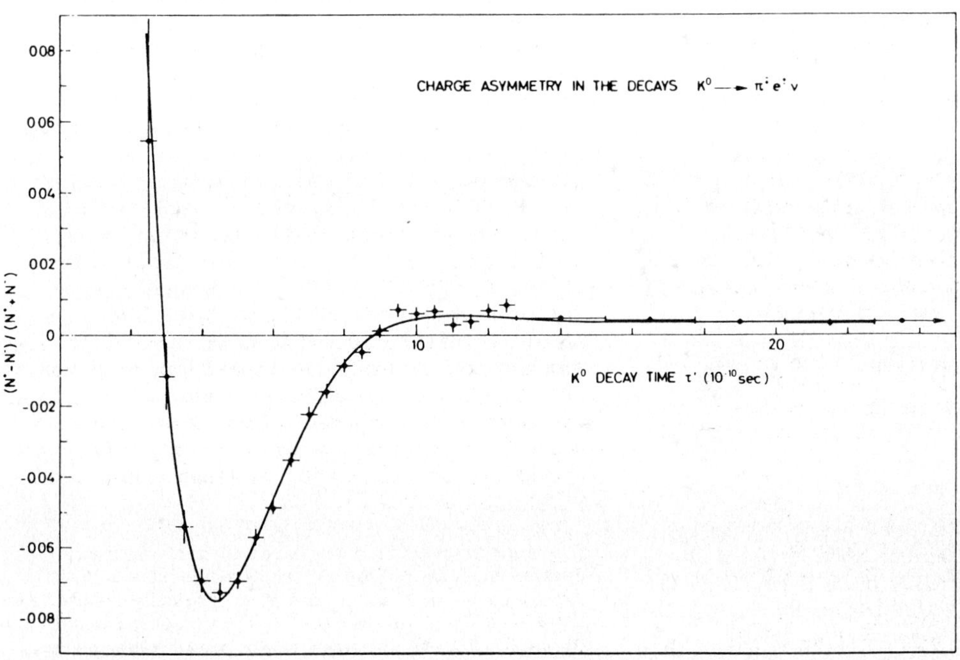

FIG. 3. Time dependence of the charge asymmetry of semileptonic decays. Figure taken from thesis of V. Luth, Heidelberg University (1974); a published version of this work can be found in Gjesdal *et al*. (1974).

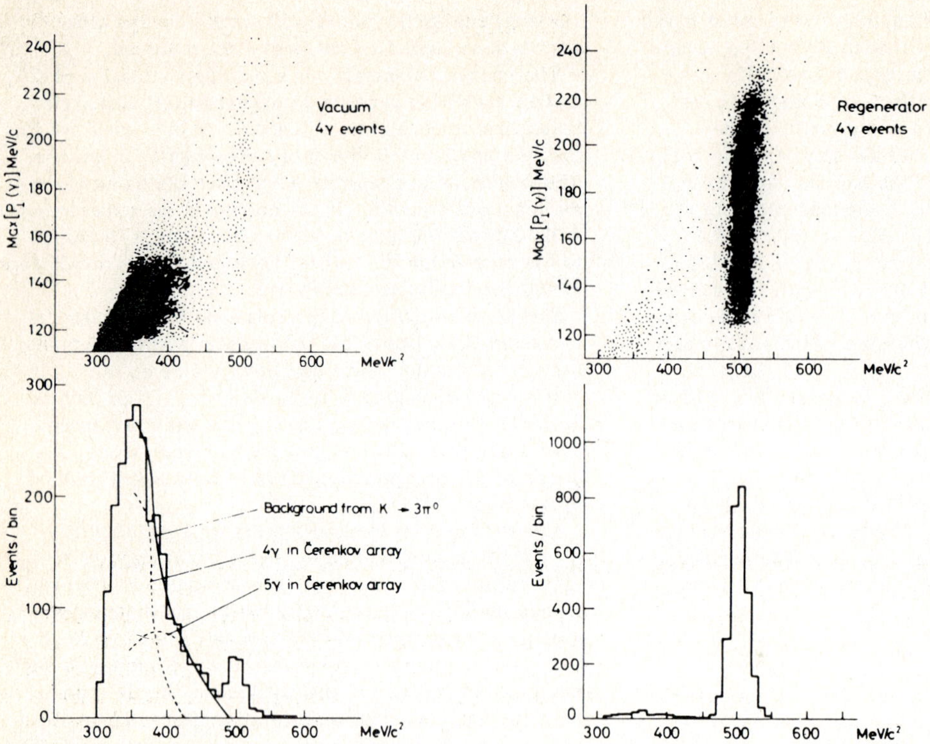

FIG. 4. Distributions of reconstructed $K_L \to \pi^0\pi^0$ events, and regenerated $K_S \to \pi^0\pi^0$ events.

close in value to $|\eta_{+-}|$. Figure 4 shows the results from the most accurate measurement of $|\eta_{00}|/|\eta_{+-}|$ (Holder *et al.*, 1972). Shown are reconstructed events from free K_L decays as well as a sample of $K_S \to \pi^0\pi^0$ from a regenerator used to determine the resolution of the apparatus. The serious background from the $3\pi^0$ decays is clearly seen. The result $|\eta_{00}|/|\eta_{+-}| = 1.00 \pm 0.06$ is based on only 167 events. The equality of $|\eta_{00}|$ and $|\eta_{+-}|$ means that the ratio of charged 2π decays to neutral 2π decays is the same for *CP*-violating K_L decays as for *CP*-conserving K_S decays. This result implies that ε' is very small providing ϕ_{00} is close to ϕ_{+-}.

The $K_L \to \pi^0\pi^0$ events cannot be collected at the rate of the $\pi^+\pi^-$ decays, nor can they be separated so cleanly from backgrounds. As a consequence, the precision with which we know the parameters $|\eta_{00}|$ and ϕ_{00} is much less than that for charged parameters. A weighted average of all the data presently available gives (Particle Data Group, 1980)

$$|\eta_{00}|/|\eta_{+-}| = 1.02 \pm 0.04$$

and

$$\phi_{00} - \phi_{+-} = 10° \pm 6° .$$

The results are quoted with reference to the charged decay mode parameters because the most accurate experiments have measured the quantity $|\eta_{00}|/|\eta_{+-}|$ directly. The result for ϕ_{00} is principally due to a recent experiment by Christenson *et al.* (1979).

The phase of the quantity ε' is given by the angle $\pi + \delta_2 - \delta_0$. Information concerning the pion-pion scattering phase shifts comes from several sources.[9] A compendium of these sources gives $\delta_2 - \delta_0 = -45° \pm 10°$. The phase of ε is naturally related to

$$\phi_n \equiv \arg([i(M_S - M_L) + (\Gamma_S - \Gamma_L)/2]^{-1}) = 43.7° \pm 0.2° .$$

This is the phase ε would have if there were no contributions from $\mathrm{Im}\Gamma_{12}$. The measured phase of η_{+-}(44.7° ± 1.2°) is within measurement precision equal to ϕ_n.

The measured parameters are plotted on the complex plane in Fig. 5(a). The size of the box for η_{+-} and η_{00} and the width of the bar for δ_I correspond to one standard deviation. The derived quantities ε and ε' are plotted in Fig. 5(b). Boxes corresponding to both one and two standard deviations are shown. Also plotted is the constraint coming from the π-π scattering phase shifts which defines the phase of ε' to be 45° ± 10°. With this constraint we find that ε, ε', η_{00}, and η_{+-} lie nearly on a common line. There is a mild disagreement between the π-π phase shift constraint and the result of Christenson *et al.* for ϕ_{00}.

A more general analysis of the neutral K system which includes the possibility of violation of *CPT* with T conservation as well as *CP* violation with *CPT* con-

[9]From extrapolation of the phase shift analysis of Ke_4 decays, one finds $\delta_2 - \delta_0 = 36° \pm 10°$ (Rosselet *et al.*, 1977). From analysis of K^+ and K_S decays to $\pi^+\pi^-$ and $\pi^0\pi^0$, one finds $\delta_2 - \delta_0 = \pm(53° \pm 6°)$ (Abbud, Lee, and Yang, 1967; Particle Data Group, 1980). From analysis of pion production by pions, one finds $\delta_2 - \delta_0 = 40° \pm 6°$ (see, for example, Baton *et al.*, 1970; Estabrooks and Martin, 1974).

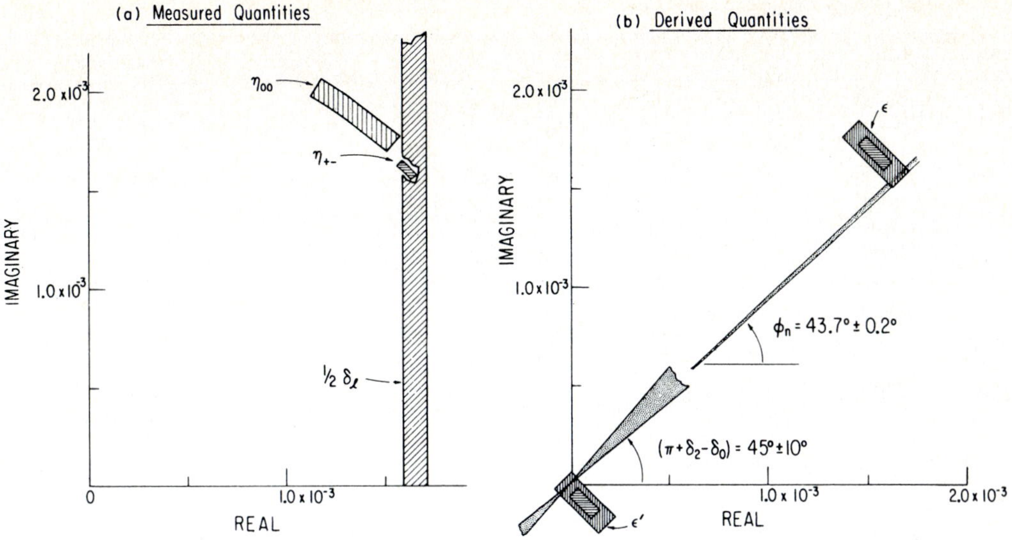

FIG. 5. Summary of *CP*-violating parameters in the neutral *K* system. (a) Measured quantities. (b) Derived quantities.

servation has been given by Bell and Steinberger (1965).[10] The analysis does depend on the assumption of unitarity, which requires that the *M* and Γ matrices remain Hermitian. The Bell-Steinberger analysis has been applied to the data with the conclusion that while a small *CPT* violation is possible, the predominant effect is one of *CP* violation. All experiments are consistent with exact *CPT* conservation (Schubert *et al.*, 1970), and hence imply a violation of time-reversal symmetry. The conservation or nonconservation of *CPT* remains, however, a question that must continue to be addressed by experiment. A brief discussion of the unitarity analysis is given in an appendix.

The essential point of this analysis rests on the measurement of the phase of η_{+-}. Limits on the contribution of $\mathrm{Im}\Gamma_{12}$ can be estimated from measured decay rates to all modes of decay of the neutral *K* mesons. The absence, within present experimental limits, of *CP* violation in the decay modes other than the 2π modes limits the contribution of $\mathrm{Im}\Gamma_{12}$ to ε to be $\leq 0.3\times10^{-3}$, a small value compared to $|\eta_{+-}|$. Thus the phase of ε and hence η_{+-} is expected to be close to ϕ_n. We can examine the other extreme, namely, that *CP* and *CPT* symmetry are both violated while time-reversal symmetry remains valid. Under these conditions we would find the natural phase ϕ_n to be ~135°, and would expect ϕ_{+-} to be close to 135°. The fact that this is not the case is the essence of the argument that *CPT* is not violated.

We note that the natural phase depends on the sign of the mass difference. We have assumed $\Delta M = (M_S - M_L) < 0$. If the sign of the mass difference were opposite, we would expect the phase of ε to be equal to 135° or −45° for *CP* violation with *CPT* symmetry. The phase of ε′ would remain the same, however, since it does not depend on ΔM in any way. Thus the conclusion that the phase of ε and ε′ are approximately the same is a consequence of the fact that the long-lived *K* is heavier than the short-lived *K*. The sign of the mass difference has been measured by several groups with complete agreement (Canter *et al.*, 1966; Meisner *et al.*, 1966; Mehlhop *et al.*, 1968).[11]

Independent of any particular theory, we would expect results which are similar to those observed. The constraint of unitarity and π-π scattering phase shifts force $\phi_{00}\approx\phi_{+-}$ for $\varepsilon' \ll \varepsilon$. Under these circumstances, a measurement of the ratio $(|\eta_{00}|/|\eta_{+-}|)^2$ is a direct measurement of the quantity ε′ by means of the relation $\varepsilon'/\varepsilon \approx [1-(|\eta_{00}|/|\eta_{+-}|)^2]/6$. Applying this relation to the present data we have $\varepsilon'/\varepsilon = -0.007\pm0.013$. New experiments at Fermilab and at Brookhaven will attempt to increase the sensitivity of the measurement by a factor of 10.

As we have shown, detailed analysis of the *CP* violation in the neutral *K*-meson system leads to the conclusion that time reversal is also violated. Table I gives a representative set of experiments which have searched for *T* violation, *CP* violation, and *C* violation (in nonweak interactions). None of these experiments has led to a positive result. Many of the experiments are approaching a sensitivity for the violation of 10^{-3}, but few have attained this value. A strength of 10^{-3} in amplitude or relative phase is what we might expect for the *CP* violation based on the results of *K* decay. For experiments involving decays with electromagnetic interactions in the final states, an apparent *T*-violation effect is usually expected at the 10^{-3} level. An example of this is the result for the ^{191}Ir decay in which a significant effect is found, but it is of the size expected on the basis of the final-state electromagnetic interaction.

Among the many measurements listed in Table I, we

---

[10] An analysis with a similar purpose has also been given by Sachs (1975). References to the literature concerning the analysis of the neutral *K*-meson decay data into a *T*-conserving part and a *T*-violating part are given by Sachs.

[11] The last reference uses a highly innovative technique, and will give pleasure to those who take the time to read it.

TABLE I. Searches for *CP*, *T*, and *C* violation.

Measurement	Result	Test	Reference
$\frac{\Gamma(K^+\to\pi^+\pi^+\pi^-)-\Gamma(K^-\to\pi^-\pi^-\pi^+)}{\text{average}}$	$(0.8\pm1.2)\times10^{-3}$	*CP*	Ford *et al.*, 1970
$\frac{\Gamma(K^+\to\pi^+\pi^0\pi^0)-\Gamma(K^-\to\pi^-\pi^0\pi^0)}{\text{average}}$	$(0.8\pm5.8)\times10^{-3}$	*CP*	Smith *et al.*, 1970
$\frac{a_{\tau^+}-a_{\tau^-}}{\text{average}}$, where $a_{\tau\pm}$ is the slope of the odd pion in the $K^\pm\to\pi^\pm\pi^\pm\pi^\mp$ Dalitz plot	$(-7.0\pm5.3)\times10^{-3}$	*CP*	Ford *et al.*, 1970
Muon polarization transverse to decay plane in $K_L\to\pi^-\mu^+\nu_\mu$	$(2.1\pm4.8)\times10^{-3}$	*T*	Schmidt *et al.*, 1979
Coefficient of *T* odd correlation $\langle \mathbf{J}\cdot\mathbf{P}_e\times\mathbf{P}_\nu\rangle$ in the β decay of polarized ^{19}Ne	$(-0.5\pm1.0)\times10^{-3}$	*T*	Baltrusaitis and Calaprice, 1977
Coefficient of *T* odd correlation $\langle \sigma_n\cdot\mathbf{P}_e\times\mathbf{P}_\nu\rangle$ in the β decay of the neutron	$(-1.1\pm1.7)\times10^{-3}$	*T*	Steinberg *et al.*, 1974
Asymmetry in distribution of $(T_{\pi^+}-T_{\pi^-})$ in the decay of $\eta\to\pi^+\pi^-\pi^0$	$(1.2\pm1.7)\times10^{-3}$	*C*	Layter *et al.*, 1972; Jane *et al.*, 1974
Electric dipole moment of the neutron	$(0.4\pm1.5)\times10^{-24}$ *e* cm $(0.4\pm0.75)\times10^{-24}$ *e* cm	*T*	Dress *et al.*, 1977 Altarev *et al.*, 1978
Angular correlation in γ decay of polarized iridium, $^{191}\text{Ir}^*\to{}^{191}\text{Ir}+\gamma$. Measure phase angle between E_2 and M_1 decay amplitudes.	$(4.7\pm0.3)\times10^{-3}$	*T*	Gimlett, 1979
Result expected on basis of electromagnetic interaction in final state	4.3×10^{-3}		Davis *et al.*, 1980
Detailed balance in nuclear reactions, e.g., $^{24}\text{Mg}+\alpha\rightleftarrows{}^{27}\text{Al}+p$. Measure: $\frac{\text{amplitude } T \text{ violating}}{\text{amplitude } T \text{ conserving}}$	$\leq3\times10^{-3}$	*T*	Weitkamp *et al.*, 1968; von Witsch *et al.*, 1968; Thornton *et al.*, 1971; Driller *et al.*, 1979

would like to single out the electric dipole moment of the neutron. The first measurement of this quantity was made in 1950 by Purcell, Ramsey, and Smith (Purcell and Ramsey, 1950; Smith, 1951) with the avowed purpose of testing the assumptions on which one presumed the electric dipole moment would be zero. Today, outside of the *K* system, the search for an electric dipole moment of the neutron is the most promising approach to the detection of *T* violation. At present the upper limit is $\sim10^{-24}$ *e* cm. New experiments using ultracold neutrons give promise of an increase in intensity by 100-fold within the next several years. The significance of a negative result for the electric dipole moment, or for any of the measurements in Table I, is difficult to assess without a theory of *CP* violation.[12]

Up to now our discussion has been entirely experimental. In the analysis of the *CP* violation in the neutral *K* system general principles of quantum mechanics have been used. The manifest charge asymmetry of the K_L semileptonic decays requires no assumptions at all for its interpretation. The literature abounds with theoretical speculations about *CP* violation. One of these speculations by Wolfenstein (1964) is frequently referred to. He hypothesizes a direct $\Delta S=2$ superweak interaction which is constructed to produce a *CP* violation. This direct interaction interferes with the second-order weak interaction to produce the *CP*-violating $\Delta S=2$ coupling between K and $\bar{K}$. Since the hypothesized superweak transition is first order, it need have only $\sim10^{-7}$ of the strength of the normal weak interaction. As such the only observable consequence is a *CP* violation in $K\to2\pi$ decay characterized by a single number, the value of $\text{Im}M_{12}$ in the mass matrix.

At present the data are in agreement with this hypothesis, which leads to predictions that $|\eta_{00}|=|\eta_{+-}|$, and $\phi_{00}=\phi_{+-}=\phi_n$. However, the relation $\phi_{00}=\phi_{+-}=\phi_n$ to a good approximation follows from the constraints of unitarity and the π-π scattering phase shifts with no further assumptions. On the other hand, the relation $|\eta_{00}|=|\eta_{+-}|$ has not been tested to very high accuracy, especially considering the difficulty of experiments which attempt to measure the properties of $K_L\to\pi^0\pi^0$. These experiments are more prone to systematic errors, and in truth $|\eta_{00}|$ and $|\eta_{+-}|$ could differ considerably more than appears to be allowed by the experiments. Thus, while the superweak hypothesis is in agreement with the present data, the data by no

[12]Weinberg (1976) has suggested a mechanism whereby the *CP* violation is due to Higgs mesons. The suggestion is attractive because the *CP* violation can be maximal and a neutron electric dipole moment of $\sim10^{-24}$ might be expected.

means make a compelling case for the superweak hypothesis.

In 1973, Kobayashi and Maskawa in a remarkable paper pointed out that with the (then) current understanding of weak interactions, *CP* violation could be accommodated only if there were three or more pairs of strongly interacting quarks. The paper was remarkable because at that time only three quarks were known to exist experimentally. Since then, strong evidence has been accumulated to support the existence of a charmed quark and a fifth bottom quark. It is presumed that the sixth quark, top, will be eventually found. With six quarks the weak hadronic current involving quarks can be characterized by three Cabibbo angles and a phase δ. This phase, if nonzero, would imply a *CP* violation in the weak interaction.

In principle, the magnitude of this phase δ which appears in the weak currents of quarks can be related to the *CP* violation observed in the laboratory. Unfortunately, all the experimental investigations are carried out with hadrons, which are presumed to be structures of bound quarks, while the parameter one wants to establish, δ, is expressed in terms of interactions between free quarks. The theoretical "engineering" required to relate the free quark properties to bound quark properties is difficult and, as a consequence, is not well developed. A balanced and sober view of this problem is given in a paper by Guberina and Peccei (1980). Even if the *CP* violation has its origin in the weak currents, it is not clear whether the experimental consequences with respect to *K* decay can be distinguished from the superweak hypothesis. If we are successful in establishing the fact that *CP* violation is the result of a phase in the weak currents between quarks, we will still have to understand why it has the particular value we find.

There are, however, on the horizon new systems which show some promise of giving additional information about *CP* violation. These are the new neutral mesons, D^0, B^0, B_s^0 (composed of $c\bar{u}$, $b\bar{d}$, and $b\bar{s}$ quarks), and their antiparticles $\bar{D}^0, \bar{B}^0, \bar{B}_s^0$. These mesons have the same general properties as *K* mesons. They are neutral particles that, with respect to strong interactions, are distinct from their own antiparticles, and yet are coupled to them by common weak decay modes. While we may not expect any stronger *CP* impurities on the eigenstates (the parameter analogous to ε), we might expect stronger effects in the decay amplitudes (the parameter analogous to ε'). We might expect this since the *CP* violation comes about through the weak interactions of the heavy quarks, c, b, t, which participate only virtually in *K* decay, but can be more influential in heavy neutral meson decay. At present, *D* mesons can be made rather copiously at the e^+e^- storage ring SPEAR at SLAC, (Lüth, 1979) and *B* mesons are beginning to be produced at the e^+e^- storage ring CESR at Cornell (Andrews *et al.*, 1980; Finocchiaro *et al.*, 1980).

It is conceivable that the effect of *CP* violation may become stronger with energy. Soon collisions of protons with antiprotons will be observed at CERN with a total center-of-mass energy greater than 500 GeV. It will be most interesting to look for *C* violations in the spectra of particles produced in those collisions. Also, improvements in the technology of detectors over the next several decades may permit sensitive searches for time-reversal-violating observables in high-energy neutrino interactions.

Recently, much attention has been given to the role that *CP* violation may play in the early stages of the evolution of the universe (Yoshimura, 1978; Dimopoulos and Susskind, 1978; Toussaint *et al.*, 1979; Ellis, Gaillard, and Nanopoulos, 1979; Weinberg, 1979). A mechanism has been proposed, with *CP* violation as one ingredient, which leads from matter-antimatter symmetry in the early universe to the small excess of matter observed in the universe at the present time. The first published account of this mechanism, of which I am aware, was made by Sakharov in 1967. He explicitly stated the three ingredients which form the foundation of the mechanism as it is presently discussed. These ingredients are: (1) baryon instability, (2) *CP* violation, and (3) appropriate lack of thermal equilibrium. The recent intense interest in this problem has arisen because baryon instability is a natural consequence of the present ideas of unification of the strong interactions with the successfully unified electromagnetic and weak interactions. This latter unification was discussed in the 1979 Nobel lectures of Glashow, Salam, and Weinberg (Glashow, 1980; Salam, 1980; Weinberg, 1980).

A very oversimplified explanation of the process which leads to a net baryon number can be given with the aid of Fig. 6(a). Quarks and leptons are linked by a very heavy boson *X* and its antiparticle $\bar{X}$. While the total decay rates of *X* and $\bar{X}$ may be equal, with *CP* violation the fractional partial rates r and $\bar{r}$ to $B=-\frac{1}{3}$ and $B=+\frac{1}{3}$ decay channels of *X* and $\bar{X}$ respec-

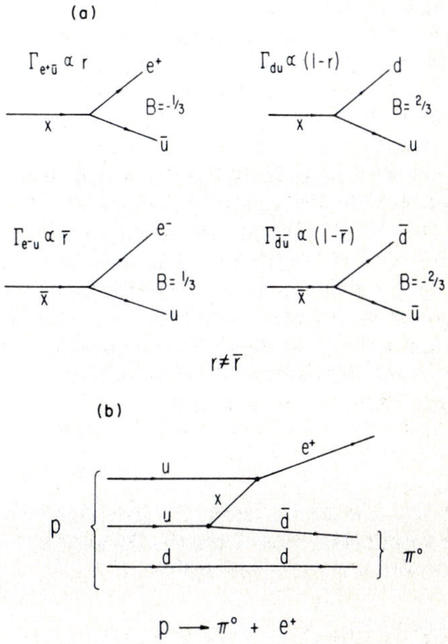

FIG. 6. (a) Simplified diagrams of baryon-number-nonconserving *X* boson decays. (b) A proton decay mediated by an *X* boson.

tively, can differ. At an early stage where the temperature is large compared to the mass of X, the density of X and $\bar{X}$ may be equal. On decay, however, the net evolution of baryon number is proportional to $(r-\bar{r})$. The excess can be quite small since the ratio of baryons to photons today is $\sim 10^{-9}$. Figure 6(b) shows how such an X boson can mediate the decay $p \to e^+ + \pi^0$. If nucleon decay is discovered it will give a strong support to these present speculations.

Whether the *CP* violation that we observe today is a "fossil remain" of these conjectured events in the early universe is a question that cannot be answered at present. This is to say, does the *CP* violation we observe today provide supporting evidence for these speculations? We simply do not know enough about *CP* violation. Our experimental knowledge is limited to its observation in only one extraordinarily sensitive system that nature has provided us. We need to know the theoretical basis for *CP* violation, and we need to know how to reliably extrapolate the behavior of *CP* violation to the very high energies involved.

At present our experimental understanding of *CP* violation can be summarized by the statement of a single number. If we state that the mass matrix which couples K and $\bar{K}$ has an imaginary off-diagonal term given by

$$\mathrm{Im} M_{12} = -1.16 \times 10^{-8}\ \mathrm{eV},$$

then all the experimental results related to *CP* violation can be accounted for. If this is all the information nature is willing to provide about *CP* violation it is going to be difficult to understand its origin. I have emphasized, however, that despite the enormous experimental effort, punctuated by some experiments of exceptional beauty, we have not reached a level of sensitivity for which a single parameter description should either surprise or discourage us.

We must continually remind ourselves that the *CP* violation, however small, is a very real effect. It has been used almost routinely as a calibration signal in several high-energy physics experiments. But more importantly, the effect is telling us that there is a fundamental asymmetry between matter and antimatter, and it is also telling us that at some tiny level interactions will show an asymmetry under the reversal of time. We must continue to seek the origin of the *CP* symmetry violation by all means at our disposal. We know that improvements in detector technology and quality of accelerators will permit even more sensitive experiments in the coming decades. We are hopeful, then, that at some epoch, perhaps distant, this cryptic message from nature will be deciphered.

APPENDIX

The evolution of a neutral K system characterized by time-dependent amplitudes a and $\bar{a}$ for the $|K\rangle$ and $|\bar{K}\rangle$ components, respectively, is given by

$$-\frac{d}{dt}\begin{pmatrix} a \\ \bar{a} \end{pmatrix} = (iM + \tfrac{1}{2}\Gamma)\begin{pmatrix} a \\ \bar{a} \end{pmatrix},$$

where M and Γ are each Hermitian matrices, and t is the time measured in the rest system of the K meson.

Expressed in terms of their elements the matrices are

$$\begin{pmatrix} M_{11} & M_{12} \\ M_{12}^* & M_{22} \end{pmatrix} \text{ and } \begin{pmatrix} \Gamma_{11} & \Gamma_{12} \\ \Gamma_{12}^* & \Gamma_{22} \end{pmatrix}.$$

The matrix $iM + \frac{1}{2}\Gamma$ has eigenvalues $\gamma_S = iM_S + \frac{1}{2}\Gamma_S$ and $\gamma_L = iM_L + \frac{1}{2}\Gamma_L$. We define small parameters

$$\varepsilon = (-\mathrm{Im}M_{12} + i\mathrm{Im}\Gamma_{12}/2)/(\gamma_S - \gamma_L)$$

and

$$\Delta = [i(M_{11} - M_{22}) + (\Gamma_{11} - \Gamma_{22})/2]/[2(\gamma_S - \gamma_L)].$$

We can then express the eigenvectors as

$$|K_S\rangle = \frac{1}{\sqrt{2}}\frac{1}{(1+|\varepsilon+\Delta|^2)^{1/2}}[(1+\varepsilon+\Delta)|K\rangle + (1-\varepsilon-\Delta)|\bar{K}\rangle]$$

and

$$|K_L\rangle = \frac{1}{\sqrt{2}}\frac{1}{(1+|\varepsilon-\Delta|^2)^{1/2}}[(1+\varepsilon-\Delta)|K\rangle - (1-\varepsilon+\Delta)|\bar{K}\rangle].$$

The parameter ε represents a *CP* violation with T nonconservation. The parameter Δ represents a *CP* violation with *CPT* nonconservation.

If we form a state $|K(t)\rangle\rangle$ which is an arbitrary superposition of $|K_S\rangle$ and $|K_L\rangle$ with amplitudes a_S and a_L at $t=0$, we can compute its norm $\langle K(t)|K(t)\rangle\rangle$ as a function of time. At $t=0$ by conservation of probability we have the relation

$$-\frac{d}{dt}\langle K(t)|K(t)\rangle\rangle\bigg|_{t=0} = \sum_f |a_S \mathrm{am}(K_S \to f) + a_L \mathrm{am}(K_L \to f)|^2,$$

where f represents the set of final states. Explicit evaluation of the expression gives

$$[-i(M_S - M_L) + (\Gamma_S + \Gamma_L)/2]\langle K_S|K_L\rangle = \sum_f [\mathrm{am}(K_S \to f)]^*[\mathrm{am}(K_L \to f)].$$

A number of definitions and a particular phase convention are used. We define $\tilde{\Delta} = \Delta - (A_0 - \bar{A}_0)/(A_0 + \bar{A}_0)$, where A_0 and $\bar{A}_0$ are the standing-wave amplitudes for K and $\bar{K}$, respectively, to decay to the $I=0$ state of two pions. A_0 and $\bar{A}_0$ are chosen real and define the phase convention used in the analysis. From the experimental parameters we define

$$\varepsilon_0 = \tfrac{2}{3}\eta_{+-} + \tfrac{1}{3}\eta_{00} \quad\text{and}\quad \varepsilon_2 = \tfrac{1}{3}\sqrt{2}\,(\eta_{+-} - \eta_{00}),$$

and

$$\alpha(f) = (1/\Gamma_S)[\mathrm{am}(K_S \to f)]^*[\mathrm{am}(K_L \to f)].$$

With these definitions we find to a good approximation that

$$(-i\Delta M/\Gamma_S + \tfrac{1}{2})(2\,\mathrm{Re}\varepsilon - 2i\,\mathrm{Im}\tilde{\Delta}) = \varepsilon_0 + \sum_f \alpha(f) \qquad \text{(A1)}$$

and

$$\varepsilon - \tilde{\Delta} = \varepsilon_0. \qquad \text{(A2)}$$

The sum over f, which now excludes the $I=0$ π-π state, consists of the following terms:

$$\alpha(\pi\pi, I=2) = A_2/A_0 e^{i(\delta_2 - \delta_0)}\varepsilon_2^*$$

$$\alpha(\pi^+\pi^-\pi^0) = [\Gamma(K_L \to \pi^-\pi^+\pi^0)/\Gamma_S]\eta_{+-0}^*,$$

$$\alpha(\pi^0\pi^0\pi^0) = [\Gamma(K_L \to \pi^0\pi^0\pi^0)/\Gamma_S]\eta^*_{000},$$

$$\alpha(\pi\mu\nu) = [\Gamma(K_L \to \pi e\nu)/\Gamma_S]2i\,\mathrm{Im}\,x_\mu,$$

and

$$\alpha(\pi e\nu) = [\Gamma(K_L \to \pi\mu\nu)/\Gamma_S]2i\,\mathrm{Im}\,x_\mu,$$

where

$$\eta_{+-0} = \mathrm{am}(K_S \to \pi^+\pi^-\pi^0)/\mathrm{am}(K_L \to \pi^+\pi^-\pi^0),$$

$$\eta_{000} = \mathrm{am}(K_S \to \pi^0\pi^0\pi^0)/\mathrm{am}(K_L \to \pi^0\pi^0\pi^0),$$

and x_l is the ratio, $\mathrm{am}(\Delta Q = -\Delta S)/\mathrm{am}(\Delta Q = \Delta S)$, for $K \to \pi l \nu_l$. The quantities η_{+-0} and η_{000} are *CP*-violating ratios. (The final state $\pi^+\pi^-\pi^0$ can be *CP* even or odd. Here we refer only to the odd state.) The measurements of η_{000} and η_{+-0} are not at present very accurate and are consistent with zero. If we use the experimental limits which exist (Particle Data Gr up, 1980), we find

$$\mathrm{Re}\,\alpha = \mathrm{Re}\sum_f \alpha(f) = (0.14 \pm 0.19)\times 10^{-3}$$

and

$$\mathrm{Im}\,\alpha = \mathrm{Im}\sum_f \alpha(f) = (-0.19 \pm 0.25)\times 10^{-3}.$$

The equations (A1) and (A2) take a very simple form if we resolve the components of ε and $\tilde{\Delta}$ parallel and perpendicular to the direction which makes an angle ϕ_n with the real axis, where

$$\phi_n = \tan^{-1}\left(-\frac{2(M_S - M_L)}{(\Gamma_S - \Gamma_L)}\right).$$

We then find

$$\varepsilon_\parallel = \varepsilon_{0_\parallel} + \cos\phi_n \mathrm{Re}\,\alpha,$$

$$\varepsilon_\perp = -\cos\phi_n \mathrm{Im}\,\alpha,$$

$$\tilde{\Delta}_\parallel = \cos\phi_n \mathrm{Re}\,\alpha,$$

and

$$\tilde{\Delta}_\perp = -\varepsilon_{0_\perp} - \cos\phi_n \mathrm{Im}\,\alpha.$$

The experimental values of $\varepsilon_{0_\parallel}$ and $\varepsilon_{0_\perp}$ are, respectively, $(2.27 \pm 0.03)\times 10^{-3}$ and $(0.16 \pm 0.09)\times 10^{-3}$. We then find

$$\varepsilon_\parallel = (2.37 \pm 0.19)\times 10^{-3},$$

$$\varepsilon_\perp = (0.14 \pm 0.18)\times 10^{-3},$$

$$\tilde{\Delta}_\parallel = (0.10 \pm 0.14)\times 10^{-3},$$

and

$$\tilde{\Delta}_\perp = (-0.02 \pm 0.20)\times 10^{-3}.$$

Within the present experimental limits we find that all the measurements are consistent with *T* violation and *CPT* conservation. In particular, we see that the limit on $\varepsilon_\perp$ is very small, so that we cannot expect ϕ_{+-} and ϕ_{00} to differ greatly from ϕ_n. Further, if the values of η_{000}, η_{+-0}, x_e, and x_μ were $<10^{-2}$, then we would find $|\varepsilon_\perp| \leq 10^{-5}$. Such an expectation is reasonable if the strength of the *CP* violation is roughly the same in all modes.

I would like to thank Professors S. Chandrasekhar, R. Oehme, R. G. Sachs, and B. Winstein for their advice and critical comments concerning this lecture. I would also like to thank Professors V. Telegdi, S. Treiman, and L. Wolfenstein for many valuable discussions concerning *CP* violation over the years.

REFERENCES

Abbud, F., B. W. Lee, and C. N. Yang, 1967, Phys. Rev. Lett. **18**, 980.

Altarev, I. S., Yu. V. Borisov, A. B. Brandin, A. I. Egorov, V. F. Ezhov, S. N. Ivanov, V. M. Lobashov, V. A. Nazarenko, G. D. Porsev, V. L. Rjabov, A. P. Serebrov, and R. R. Taldaev, 1978, in *Proceedings of the Third International Symposium on Neutron Capture Gamma-Ray Spectroscopy and Related Topics*, BNL and SUNY Stony Brook, edited by Robert E. Chrien and Walter R. Kane (Plenum, New York), p. 541.

Andrews, D., *et al.*, 1980, Phys. Rev. Lett. **45**, 219.

Aronson, S. H., R. D. Ehrlich, H. Hofer, D. A. Jensen, R. A. Swanson, V. L. Telegdi, H. Goldberg, J. Solomon, and D. Fryberger, 1970, Phys. Rev. Lett. **25**, 1057.

Baltrusaitis, R. M., and F. P. Calaprice, 1977, Phys. Rev. Lett. **38**, 464.

Banner, M., J. W. Cronin, J. K. Liu, and J. E. Pilcher, 1968, Phys. Rev. Lett. **21**, 1103.

Baton, J., G. Laurens, and J. Reignier, 1970, Phys. Lett. B **33**, 528.

Bell, J. S., and J. K. Perring, 1964, Phys. Rev. Lett. **13**, 348.

Bell, J. S., and J. Steinberger, 1965, in *Proceedings of the Oxford International Conference on Elementary Particles, 1965*, edited by R. G. Moorhouse, A. E. Taylor, and T. R. Walsh (Rutherford Laboratory, Chilton, England, 1966), p. 195.

Bernstein, J., N. Cabibbo, and T. D. Lee, 1964, Phys. Lett. **12**, 146.

Budagov, I. A., *et al.*, 1968, Phys. Lett. B **28**, 215.

Canter, J., Y. Cho, A. Engler, H. E. Fisk, R. W. Kraemer, C. M. Meltzer, D. G. Hill, D. K. Robinson, and M. Sakitt, 1966, Phys. Rev. Lett. **17**, 942.

Carithers, W. C., T. Modis, D. R. Nygren, T. P. Pun, E. L. Schwartz, H. Sticker, and J. H. Christenson, 1975, Phys. Rev. Lett. **34**, 1244.

Carnegie, R. K., R. Cester, V. L. Fitch, M. Strovink, and L. R. Sulak, 1971, Phys. Rev. D **4**, 1.

Christenson, J. H., J. H. Goldman, E. Hummel, S. D. Roth, T. W. L. Sanford, and J. Sculli, 1979, Phys. Rev. Lett. **43**, 1209.

Christenson, J. H., *et al.*, 1965, Phys. Rev. B **140**, 74.

Cullen, M., P. Darriulat, J. Deutsch, H. Foeth, C. Grosso, M. Holder, K. Kleinknecht, E. Radermacher, C. Rubbia, D. Shambroom, M. Sciré, A. Staude, and K. Tittel, 1970, Phys. Lett. B **32**, 523.

Davis, B. R., S. E. Koonin, and P. Vogel, 1980, Phys. Rev. C **22**, 1233.

de Bouard, X., D. Dekkers, B. Jordan, R. Mermod, T. R. Willitts, K. Winter, P. Scharff, L. Valentin, M. Vivargent, and M. Bott-Bodenhausen, 1965, Phys. Lett. **15**, 58.

Dimopoulos, S., and L. Susskind, 1978, Phys. Rev. D **18**, 4500.

Dress, W. B., P. D. Miller, J. M. Pendlebury, Paul Perrin, and Norman F. Ramsey, 1977, Phys. Rev. D **15**, 9.

Driller, H., E. Blanke, H. Genz, A. Richter, G. Schrieder, and J. M. Pearson, 1979, Nucl. Phys. A **317**, 300.

Ellis, J., M. K. Gaillard, and D. V. Nanopoulos, 1979, Phys. Lett. B **80**, 360.

Estabrooks, P., and A. D. Martin, 1974, Nucl. Phys. B **79**, 301.

Finocchiaro, G., *et al.*, 1980, Phys. Rev. Lett. **45**, 222.

Fitch, V. L., R. F. Roth, J. S. Russ, and W. Vernon, 1965, Phys. Rev. Lett. **15**, 73.
Fitch, V. L., R. F. Roth, J. Russ, and W. Vernon, 1967, Phys. Rev. **164**, 1711.
Ford, W. T., P. A. Piroué, R. S. Remmel, A. J. S. Smith, and P. A. Souder, 1970, Phys. Rev. Lett. **25**, 1370.
Galbraith, W., G. Manning, A. E. Taylor, B. D. Jones, J. Malos, A. Astbury, N. H. Lipman, and T. G. Walker, 1965, Phys. Rev. Lett. **14**, 383.
Geweniger, C., S. Gjesdal, G. Presser, P. Steffen, J. Steinberger, F. Vannucci, H. Wahl, F. Eisele, V. Lüth, G. Zech, and K. Kleinknecht, 1974, Phys. Lett. B **52**, 108.
Gimlett, J. L., 1979, Phys. Rev. Lett. **42**, 354.
Gjesdal, S., G. Presser, T. Kamae, P. Steffen, J. Steinberger, F. Vannucci, H. Wahl, F. Eisele, H. Filthuth, V. Lüth, G. Zech, and K. Kleinknecht, 1974, Phys. Lett. B **52**, 113.
Glashow, S. L., 1980, Rev. Mod. Phys. **52**, 539.
Guberina, B., and R. D. Peccei, 1980, Nucl. Phys. B **163**, 289.
Holder, M., E. Radermacher, A. Staude, P. Darriulat, J. Deutsch, M. Hansroul, S. Orito, J. Pilcher, C. Rubbia, P. Strolin, K. Tittel, A. Fainberg, C. Grosso-Pilcher, and M. Sciré, 1972, Phys. Lett. B **40**, 141.
Jane, M. R., B. D. Jones, N. H. Lipman, D. P. Owen, B. K. Penney, T. G. Walker, M. Gettner, P. Grannis, H. Uto, J. Anderson, E. H. Bellamy, M. G. Green, J. Kirkby, P. E. Osmon, J. A. Strong, D. H. Thomas, and C. M. Solomonides, 1974, Phys. Lett. B **48**, 260.
Kleinknecht, K., 1976, Annu. Rev. Nucl. Sci. **26**, 1.
Kobayashi, M., and K. Maskawa, 1973, Prog. Theor. Phys. **49**, 652.
Layter, J. G., J. A. Appel, A. Kotlewski, W. Lee, S. Stein, and J. J. Thaler, 1972, Phys. Rev. Lett. **29**, 316.
Lee, T. D., R. Oehme, and C. N. Yang, 1957, Phys. Rev. **106**, 340.
Lüth, V., 1974, Ph.D. thesis, Heidelberg University (unpublished). A published version of this work can be found in Gjesdal *et al*. (1974).
Lüth, V., 1979, in *Proceedings of the 1979 International Symposium on Lepton and Photon Interactions at High Energies*, edited by T. B. W. Kirk (Fermilab, Batavia, IL, 1980), p. 83.
Mehlhop, W. A. W., S. S. Murty, P. Bowles, T. H. Burnett, R. H. Good, C. H. Holland, O. Piccioni, and R. A. Swanson, 1968, Phys. Rev. **172**, 1613.
Meisner, G. W., B. B. Crawford, and F. S. Crawford, Jr., 1966, Phys. Rev. Lett. **17**, 492.
Modis, T., 1973, Ph.D. thesis, Columbia University (unpublished). A published version of this work is given by Carithers *et al*. (1975).
Niebergall, F., M. Regler, H. H. Stier, K. Winter, J. J. Aubert, X. de Bouard, V. Lepeltier, L. Massonet, H. Pessard, M. Vivargent, T. R. Willitts, M. Yvert, W. Bartl, G. Neuhofer, and M. Steuer, 1974, Phys. Lett. B **49**, 103.
Particle Data Group (Barash-Schmidt *et al*.), 1980, Rev. Mod Phys. **52**, S1.
Purcell, E. M., and N. F. Ramsey, 1950, Phys. Rev. **78**, 807.
Rosselet, L., P. Extermann, J. Fischer, O. Guisan, R. Mermod, R. Sachot, A. M. Diamant-Berger, P. Bloch, G. Bunce, B. Devaux, N. Do-Duc, G. Marel, and R. Turlay, 1977, Phys. Rev. D **15**, 574.
Sachs, R. G., 1975, Prog. Theor. Phys. **54**, 809.
Sakharov, A. D., 1967, Zh. Eksp. Teor. Fiz. Pis'ma Red. **5**, 32 [JETP Lett. **5**, 24 (1967)].
Salam, A., 1980, Rev. Mod. Phys. **52**, 525.
Schmidt, M. P., S. R. Blatt, M. K. Campbell, D. M. Grannan, M. J. Lauterbach, H. Kasha, R. K. Adair, W. M. Morse, L. B. Leipuner, and R. C. Larsen, 1979, Phys. Rev. Lett. **43**, 556.
Schubert, K. R., B. Wolff, J. C. Chollet, J.-M. Gaillard, M. R. Jane, T. J. Ratcliffe, and J.-P. Repellin, 1970, Phys. Lett. B **31**, 662.
Smith, J. H., 1951, Ph.D. thesis, Harvard University (unpublished).
Smith, K. M., P. S. L. Booth, H. R. Renshall, P. B. Jones, G. L. Salmon, W. S. C. Williams, P. J. Duke, W. M. Evans, R. E. Hill, W. R. Holley, D. P. Jones, and J. J. Thresher, 1970, Nucl. Phys. B **60**, 411.
Steinberg, R. I., P. Liaud, B. Vignon, and V. W. Hughes, 1974, Phys. Rev. Lett. **33**, 41.
Thornton, S. T., C. M. Jones, J. K. Bair, M. D. Mancusi, and H. B. Willard, 1971, Phys. Rev. C **3**, 1065.
Toussaint, D., S. B. Treiman, F. Wilczek, and A. Zee, 1979, Phys. Rev. D **19**, 1036.
von Witsch, W., A. Richter, and P. von Brentano, 1968, Phys. Rev. **169**, 923.
Wattenberg, A., and J. Sakurai, 1967, Phys. Rev. **161**, 1449.
Weinberg, S., 1976, Phys. Rev. Lett. **37**, 657.
Weinberg, S., 1979, Phys. Rev. Lett. **42**, 850.
Weinberg, S., 1980, Rev. Mod. Phys. **52**, 515.
Weitkamp, W. G., D. W. Storm, D. C. Shreve, W. J. Braithwaite, and D. Bodansky, 1968, Phys. Rev. **165**, 1233.
Whatley, M. C., 1962, Phys. Rev. Lett. **9**, 317.
Wigner, E. P., 1932, Göttinger Nachr. **31**, 546.
Wolfenstein, L., 1964, Phys. Rev. Lett. **13**, 569.
Wu, T. T., and C. N. Yang, 1964, Phys. Rev. Lett. **13**, 380.
Yoshimura, M., 1978, Phys. Rev. Lett. **41**, 381.

2. CP Violation in K^0 Decays

Introduction

The violation of CP invariance has been observed in three decay modes of the K^0 meson and in no other processes. These observations are summarized in two complex quantities η_{+-} and η_{00} and the semi-leptonic charge asymmetry δ. These are defined by

$$\eta_{+-} = \frac{A(K_L \to \pi^+\pi^-)}{A(K_S \to \pi^0\pi^0)} = |\eta_{+-}| \exp(i\phi_{+-}),$$

$$\eta_{00} = \frac{A(K_L \to \pi^0\pi^0)}{A(K_S \to \pi^0\pi^0)} = |\eta_{00}| \exp(i\phi_{00}),$$

$$\delta = \frac{\Gamma(K_L - \pi^- \ell^+ \nu) - \Gamma(K_L \to \pi^+ \ell^- \nu^-)}{\Gamma(K_L \to \pi^- \ell^+ \nu) + \Gamma(K_L \to \pi^+ \ell^- \nu^-)},$$

where ℓ is either an electron or muon.

An analysis, based on CPT invariance and unitarity by Wu and Yang [1], leads to the following parameterization,

$$\eta_{+-} = \epsilon + \epsilon', \tag{2.1a}$$

$$\eta_{00} = \epsilon - 2\epsilon', \tag{2.1b}$$

$$\delta = 2 \operatorname{Re} \epsilon, \tag{2.1c}$$

where the phases of ϵ and ϵ' are given by

$$\theta(\epsilon) = \tan^{-1}(2\,\Delta M/\Gamma_S) = 43.67 \pm 0.14°, \tag{2.2a}$$

$$\theta(\epsilon') = \delta_2 - \delta_0 + \pi/2 = 48 \pm 8°. \tag{2.2b}$$

Here δ_I are the final state pion–pion phase shifts, $\Delta M = m(K_L) - m(K_S)$ and Γ_S is the width of K_S. A number of approximations are used for the form given here. In particular we have left out small corrections of the order (Re A_2/Re A_0) ≈ 0.05, and assumed the $\Delta Q = \Delta S$ rule for semi-leptonic decays. The formula for the phase of ϵ holds exactly in the superweak theory and is sometimes called the superweak phase. It holds as an approximation in general to the extent that CP violation in decay channels other than the $I = 0$ 2π state can be neglected. These are the quantities y_ℓ and $y_{3\pi}$ in the paper of Wu and Yang.

Following the work of Wu and Yang, a number of more detailed analyses of the phenomenology were given [2,3]. In particular, Bell and Steinberger [2] showed how the phase information is related to a general unitarity relation. With CP violation for a decaying particle, the eigenstates appear to be non-orthogonal. Careful formulations for dealing with this, were given by Enz and Lewis [4] and by Sachs [5] among others. Enz and Lewis also present a good discussion of the $K^0-\overline{K}^0$ phenomenology inside a material medium.

The original experiment of Christenson et al. [11] measured the quantity $|\eta_{+-}|$. Interference experiments such as those discussed in the Bell–Steinberger paper have determined the phase of η_{+-}. The final $\pi^0\pi^0$ state is much harder to measure accurately. Only quite recently has a CERN experiment [6] given an accurate determination of $|\eta_{00}|$ via the ratio

$$|\eta_{00}/\eta_{+-}| = 0.990 \pm 0.003. \tag{2.3}$$

If eqs. (2.1) and (2.2) are accepted there are only two quantities to be measured: $|\epsilon|$ and $|\epsilon'|$. The result, eq. (2.3), can then be interpreted as giving

$$\epsilon'/\epsilon = (3.1 \pm 1) \times 10^{-3} \tag{2.4}$$

Note that the phases of ϵ and ϵ' are almost equal so that the ratio in eq. (2.4) is essentially real and the positive sign is significant. Given eq. (2.4), one finds from the measurement of η_{+-}

$$|\epsilon| = (2.27 \pm 0.02) \times 10^{-3}. \tag{2.5}$$

Measurements of the semi-leptonic decays [7] give values of δ consistent with eq. (2.1c) using eqs. (2.5) and (2.2a).

From the phenomenological point of view, the origin of CP violation can either be in the $K^0-\overline{K}^0$ mass mixing matrix or in specific decay amplitudes. As pointed out by Wu and Yang, the situation in which CP violation occurs in only one decay amplitude, such as K^0 goes to $\pi\pi$ in the $I=0$ state, cannot be distinguished from CP violation in the mass matrix. This is because of the phase ambiguity associated with a $U(1)$ symmetry (strangeness conservation) in the strong and electromagnetic interactions. If CP violation occurs only in the mass matrix then $\epsilon'=0$ and the phase of ϵ is given exactly by eq. (2.2a). The apparent observation of a non-zero value of ϵ' is the first indication that CP violation is not confined to the mass matrix. For this reason, it is very important that there be independent verification of the CERN result in eq. (2.3).

Once CPT invariance is assumed, it follows that time reversal must also be violated in K^0 decays. An experiment that could demonstrate time reversal in K^0 physics directly, is discussed by Kabir [8]. This experiment probably cannot be carried out, but there can be little doubt that the experiment would work as presented.

From an experimental view, it is important to test CPT invariance as much as possible. The fact that the phase of ϵ agrees closely with the theoretical prediction of eq. (2.2a) is the best test of CPT invariance [9]. This can be interpreted as giving a limit on $m(K^0)-m(\overline{K}^0)$ of less than 3×10^{-10} eV. A general analysis of the $K^0-\overline{K}^0$ system without assuming CPT invariance, but keeping the unitarity assumption, is given by Barmin et al. [10]. It is possible from this analysis to establish limits on a number of parameters that measure CPT violation.

References

[1] T.T. Wu and C.N. Yang, Phys. Rev. Lett. 13 (1964) 380. Paper 6, this volume.

[2] J.S. Bell and J. Steinberger, Proc. Oxford International Conference on Elementary Particle Physics (1965) 195. The first part of this article is Paper 7, this volume.

[3] L. Wolfenstein, Nuovo Cimento 52 (1966) 17.
[4] C.P. Enz and R.R. Lewis, Helv. Phys. Acta 38 (1965) 860. Paper 8, this volume.
[5] R.G. Sachs, Ann. Phys. (NY) 22 (1964) 239.
[6] H. Burkhardt et al. Phys. Lett. B 206 (1988) 169. Paper 9, this volume.
[7] C. Geweniger et al., Phys. Lett. B 48 (1974) 483. Paper 10, this volume.
[8] P.K. Kabir, Phys. Rev. D 2 (1970) 540. Paper 11, this volume.
[9] L. Wolfenstein, Nuovo Cimento A 64 (1969) 269.
[10] V. Barmin et al., Nucl. Phys. B 247 (1984) 293. Paper 12, this volume.
[11] J.H. Christenson, J.W. Cronin, V.L. Fitch and R. Turlay, Phys. Rev. Lett. 13 (1964) 138. Paper 3, this volume.

PHENOMENOLOGICAL ANALYSIS OF VIOLATION OF *CP* INVARIANCE IN DECAY OF K^0 AND $\bar{K}^0$†

Tai Tsun Wu* and C. N. Yang‡
Brookhaven National Laboratory, Upton, New York
(Received 18 August 1964)

1. It was recently discovered[1] that the long-lived component K_L^0 of K^0-$\bar{K}^0$ decays into the $\pi^+\pi^-$ mode. Now if *CP* invariance holds, the $CP = +1$ and $CP = -1$ components of K^0-$\bar{K}^0$ decay independently. The $\pi^+\pi^-$ mode in the *S*-wave state has $CP = 1$. Hence either the short-lived component K_S^0, or K_L^0, does not decay into $\pi^+ + \pi^-$, in contradiction to the new discovery.

Accepting the experimental result of reference 1, one is thus forced to the conclusion that *CP* invariance is violated in K^0-$\bar{K}^0$ decay, as explicitly stated in reference 1. Notice that this conclusion is independent of the details of the Weisskopf-Wigner formulation[2] of decay amplitudes, as applied to the K_0-$\bar{K}_0$ case by Lee, Oehme, and Yang,[3] whose notation we shall follow.[4] (In particular, small corrections to the exponential decay rule of the formalism cannot alter the conclusion that *CP* invariance is violated.)

In the present note we shall analyze the decay properties of K^0-$\bar{K}^0$, mostly from the phenomenological viewpoint. Possible further experiments will be discussed for their theoretical significance.

We shall assume *CPT* invariance, the validity of the Weisskopf-Wigner formulation,[2,3] and that for the strong and electromagnetic interactions, separate *C*, *P*, and *T* invariance hold.

In the next five sections, we shall also assume that electromagnetic interactions can be neglected, and that isotopic spin is conserved for the strong interactions. We shall come back to the electromagnetic effects in Sec. 7.

2. The experimental decay rates are tabulated in Table I. (We thank P. Franzini, J. Steinberger, and W. Willis for supplying the entries.)

To analyze the decay of K^0-$\bar{K}^0$ we consider the decay matrix

$$\Gamma = \Gamma_0 + \Gamma_2 + \Gamma_l + \Gamma_{3\pi}, \qquad (1)$$

and the mass matrix

$$M = M_0 + M_2 + M_l + M_{3\pi} + \cdots, \qquad (2)$$

as sums of contributions from the $\pi\pi(I=0)$, $\pi\pi(I=2)$, leptonic, and 3π modes. One has, in the notation of reference 3,

$$\Gamma_0 = \begin{pmatrix} A_0^2 & A_0^2 \\ A_0^2 & A_0^2 \end{pmatrix}, \qquad (3)$$

$$\Gamma_2 = \begin{pmatrix} A_2A_2^* & A_2^2 \\ A_2^{*2} & A_2A_2^* \end{pmatrix}, \qquad (4)$$

$$\Gamma_l = \begin{pmatrix} \alpha_l & x_l + iy_l \\ x_l - iy_l & \alpha_l \end{pmatrix}, \qquad (5)$$

and

$$\Gamma_{3\pi} = \begin{pmatrix} \alpha_{3\pi} & x_{3\pi} + iy_{3\pi} \\ x_{3\pi} - iy_{3\pi} & \alpha_{3\pi} \end{pmatrix}, \qquad (6)$$

Table I. Experimental decay rates in $10^6\ \text{sec}^{-1}$.

Mode	K_S^0	K_L^0
$\pi^+ + \pi^-$	$\frac{2}{3} \times 1.1 \times 10^4$	2.6×10^{-2}
$\pi^0 + \pi^0$	$\frac{1}{3} \times 1.1 \times 10^4$	Not known
Leptons	~11	~11
$\pi^+ + \pi^- + \pi^0$	$\lesssim 2$	~ 2
$3\pi^0$	$\lesssim 4^a$	~ 4
All modes	1.1×10^4	~18

[a]No available experimental information; the number given is based on the assumption that the $|\Delta| = \frac{1}{2}$ rule is approximately valid.

where A_0 and A_2 are the decay amplitudes of

$$K^0 \to \pi + \pi \ (I=0 \text{ standing wave})$$

and

$$K^0 \to \pi + \pi \ (I=2 \text{ standing wave}),$$

respectively. We have chosen the phase of K^0 so that

$$A_0 = \text{real} > 0. \qquad (7)$$

We emphasize that this choice, which is always possible, serves to define the phase of K and $\bar{K}$. The quantities p and q are given by

$$p^2 = A_0^2 + A_2^2 + x_l + iy_l + x_{3\pi} + iy_{3\pi} + iM_r - M_i,$$

and

$$q^2 = A_0^2 + A_2^{*2} + x_l - iy_l + x_{3\pi} - iy_{3\pi} + iM_r + M_i, \qquad (8)$$

with the real parts of p and q chosen ≥ 0. In Eq. (8) $M_r + iM_i = M_{12}$ is an off-diagonal element of M. The eigenstates and the eigenvalues of $\Gamma + iM$ were given by Eqs. (28) and (29) of reference 3. The decay amplitudes of K_S and K_L into π-π states are easily constructed, and are tabulated in Table II. The quantity F is

$$F = \exp[i(\delta_2 - \delta_0)], \qquad (9)$$

where δ_2 and δ_0 are the π-π S-wave scattering phase shifts for the $I=2$ and $I=0$ states at the energy of the rest mass of K^0. Obviously,

$$R[K_S^0 \to \pi + \pi] + R[K_L^0 \to \pi + \pi] = 2A_0^2 + 2A_2A_2^* \sim 1.1 \times 10^4,$$

$$R[K_S^0 \to \text{lep}] + R[K_L^0 \to \text{lep}] = 2\alpha_l \sim 22,$$

and

$$R[K_S^0 \to 3\pi] + R[K_L^0 \to 3\pi] = 2\alpha_{3\pi} \sim 12. \qquad (10)$$

The following quantities are of intrinsic experimental interest:

$$\eta_{+-} = a_{+-}^L / a_{+-}^S, \quad \eta_{00} = a_{00}^L / a_{00}^S. \qquad (11)$$

We shall also use

$$\epsilon = (p-q)/p. \qquad (12)$$

These quantities are useful because they are small parameters, as we shall see later.

3. The following remarks serve to orient further analysis:

(a) If $\text{Im}A_2 = 0$, $y_l = 0$, $y_{3\pi} = 0$, and $M_i = 0$, then $p = q$, and Table II shows that $K_L \not\to \pi + \pi$, in con-

Table II. Decay amplitudes of K_S and K_L into π-π states.

State	Amplitudes[a]
$I=0$ (standing wave)	$a_0 = A_0(pp^* + qq^*)^{-1/2}(p \pm q)$
$I=2$ (standing wave)	$a_2 = (pp^* + qq^*)^{-1/2}(A_2 p \pm A_2^* q)$
$\pi^+ + \pi^-$ (outgoing wave)	$a_{+-} = (pp^* + qq^*)^{-1/2}\{[(\frac{2}{3})^{1/2}A_0 + (\frac{1}{3})^{1/2}A_2 F]p \pm [(\frac{2}{3})^{1/2}A_0 + (\frac{1}{3})^{1/2}A_2^* F]q\}$
$\pi^0 + \pi^0$ (outgoing wave)	$a_{00} = (pp^* + qq^*)^{-1/2}\{[(\frac{1}{3})^{1/2}A_0 - (\frac{2}{3})^{1/2}A_2 F]p \pm [(\frac{1}{3})^{1/2}A_0 - (\frac{2}{3})^{1/2}A_2^* F]q\}$

[a]Upper sign for K_S, lower sign for K_L.

tradiction with the experimental result of reference 1.

(b) If $\Delta Q = \Delta S$ for the leptonic decay modes of K^0-$\bar{K}^0$, then Γ_l and M_l of Eqs. (1) and (2) are both multiples of the unit matrix. Therefore, $y_l = 0$. The leptonic mode does not in this case contribute to *CP* violation as observed in reference 1, even though the lepton mode itself could violate *CP* invariance. We shall, however, not make the assumption that $\Delta Q = \Delta S$ in this paper.

(c) Phenomenologically, it is not possible to distinguish between the four *M*'s on the right-hand side of Eq. (2). In other words, measurable quantities can only depend on *M*, but not on M_0, M_2, M_l, or $M_{3\pi}$ separately.

(d) If $\mathrm{Im}A_2 = 0$, then Table II gives directly

$$a_{+-}^{\mathrm{L}}/a_{+-}^{\mathrm{S}} = a_{00}^{\mathrm{L}}/a_{00}^{\mathrm{S}}. \tag{13}$$

(e) If $M_i = 0$, and $y_l + y_{3\pi} = 0$, then $p^2 - q^2 = A_2^2 - A_2^{*2}$. It will be clear in Sec. 4 that

$$|a_0^{\mathrm{L}}/a_2^{\mathrm{L}}| = O(\mathrm{Re}A_2/A_0), \tag{14}$$

provided that $|A_2|/A_0$ is small.

(f) If $\mathrm{Im}A_2 = 0$ and $M_i = 0$, then, given the experimental decay rates of K^0-$\bar{K}^0 \to$ leptons and 3π, and of $K_S \to \pi + \pi$, the rate $K_L \to \pi^+ + \pi^-$ is at most $1.75[1 + (M_\gamma/A_0^2)^2]^{-1} \times 10^{-2}$. This is too low to account for the experimental result of reference 1. We shall discuss this in more detail in Sec. 5.

Accordingly, roughly there are four ways to violate *CP* invariance in the decay of K^0-$\bar{K}^0$, namely, $\mathrm{Im}A_2 \neq 0$, $y_l \neq 0$, $y_{3\pi} \neq 0$, and/or $M_i \neq 0$. They correspond to *CP* (or *T*) noninvariance due to the interference between the dominant $\pi\pi(I=0)$ mode and (α) the $\pi\pi(I=2)$ mode ($\mathrm{Im}A_2/A_0 \neq 0$); ($\beta$) the lepton mode ($y_l \neq 0$), ($\gamma$) the 3π mode ($y_{3\pi} \neq 0$), and/or (δ) the off-energy shell contributions to the $K \leftrightarrow \bar{K}$ elements of the mass operator $M(M_i \neq 0)$. This possibility has been discussed by Sachs and Treiman.[5]

According to (f) above, (β) and (γ) together by themselves are too small to account for the magnitude of the observed effect. Thus the more important contribution to the observed *CP* violation has to come from (α) and/or (δ). [Theoretically it is, of course, to be expected that if any interference of the type (α), (β), or (γ) is present, then an interference of type (δ) is also present, in general.]

4. The $|\Delta I| = \frac{1}{2}$ rule is well verified in general, and for the $K \to \pi + \pi$ decay in particular. Thus, $|A_2/A_0| \ll 1$. Dropping A_2/A_0 in Table II, we obtain

$$\eta_{+-} \sim (p-q)/(p+q) = \epsilon/(2+\epsilon). \tag{15}$$

Thus the experimental small value of $|\eta_{+-}|$ shows that $|\epsilon| \ll 1$.

We proceed to expand various quantities to the lowest nonvanishing order of ϵ and A_2/A_0:

$$\eta_{+-} = \tfrac{1}{2}[\epsilon + (2)^{1/2} iF\,\mathrm{Im}A_2/A_0], \tag{16}$$

$$\eta_{00} = \tfrac{1}{2}[\epsilon - 2(2)^{1/2} iF\,\mathrm{Im}A_2/A_0], \tag{17}$$

and

$$R[K_S^0 \to \pi^+ + \pi^-] - 2R[K_S^0 \to \pi^0 + \pi^0] = 2\sqrt{2}[\mathrm{Re}A_2/A_0]\cos(\delta_2 - \delta_0) \times R[K_S^0 \to \pi + \pi(I=0)]. \tag{18}$$

Furthermore, since Γ_l is positive definite, $\alpha_l \geq |x_l|$, $\alpha_l \geq |y_l|$; or by Eq. (10),

$$|x_l| \lesssim 11, \quad |y_l| \lesssim 11. \tag{19}$$

Similarly,

$$|x_{3\pi}| \lesssim 6, \quad |y_{3\pi}| \lesssim 6. \tag{20}$$

Thus these elements are negligible compared with A_0^2. {In more general cases, we can establish a somewhat better bound for y. For any mode C,

$$y_C^2 \leq R[K_S^0 \to C]R[K_L^0 \to C].\}$$

Using Eq. (8) one sees that $|\epsilon| \ll 1$ implies $M_i \ll A_0^2$. Thus one has an approximate expression for the difference of the two eigenvalues of $\Gamma + iM$:

$$\lambda_+ - \lambda_- = 2A_0^2 + 2iM_r. \qquad (21)$$

Therefore,

$$M_r = m_S - m_L. \qquad (22)$$

Using these one obtains from Eq. (8)

$$\epsilon = \frac{-M_i + i(y_l + y_{3\pi})}{A_0^2 + i(m_S - m_L)}. \qquad (23)$$

Equations (16)–(23) form the basis of a phenomenological analysis.

5. If $\mathrm{Im}A_2 = 0$ and $M_i = 0$, then it follows from Eqs. (16) and (23) that

$$R[K_L^0 \to \pi^+ + \pi^-]$$
$$= 6(y_l + y_{3\pi})^2 \times 10^{-5}[1 + (M_r/A_0^2)^2]^{-1}. \qquad (24)$$

Statement 3(f) then follows from Eqs. (19) and (20). This is too small by a factor of 3 or 15 for[6] $|m_S - m_L| = 1/2\tau_1$ or $3/2\tau_1$. {If, moreover, we believe that the $\Delta Q = -\Delta S$ matrix element is at most 50% of that for $\Delta Q = \Delta S$ as a result of the Paris experiment,[7] then $(x_l^2 + y_l^2)^{1/2}/\alpha_l < 2(\frac{1}{2})/[1 + (\frac{1}{2})^2] = \frac{4}{5}$. Hence, $|y_l| < 9$. Thus we can strengthen the argument by about 20%.}

6. The amplitude ratios η_{+-} and η_{00} are experimentally measurable quantities. The experiment of reference 1 gives

$$|\eta_{+-}| = 2.1 \times 10^{-3}(1 \pm 0.1). \qquad (25)$$

A measurement[8] of $R[K_L^0 \to \pi^0 + \pi^0]$ would yield $|\eta_{00}|$. If one introduces further assumptions, this rate can be predicted. For example, according to Sec. 3, (d) and (e),

if $\mathrm{Im}A_2 = 0$, then $R[K_L^0 \to \pi^0 + \pi^0] = 1.3 \times 10^{-2}$;

if $M_i = y_l + y_{3\pi} = 0$, then

$$R[K_L^0 \to \pi^0 + \pi^0] = 5.2 \times 10^{-2}. \qquad (26)$$

Existing experiments yield only a rough upper bound:

$$R[K_L^0 \to \pi^0 + \pi^0] < (\sim 1), \quad |\eta_{00}| < (\sim 2 \times 10^{-2}).$$

Without additional assumptions, there are no theoretical arguments that $R[K_L^0 \to \pi^0 + \pi^0]$ cannot be as large as this experimental upper bound.

It is clear that in order to measure the phases of the amplitude ratios η_{+-} and η_{00}, interference between K_L and K_S decays into these modes must be studied. To obtain greater sensitivity, the intensity ratio of K_L and K_S must be such that their decay amplitudes into these modes are about comparable.[9] The relative phase θ between the K_L and K_S beams must also be known in order to determine the phase of η_{+-} (or of η_{00}). But to determine the difference of the phases of η_{+-} and η_{00} it is not necessary to know θ.

It is convenient to construct a diagram of the complex numbers η_{+-}, η_{00}, ϵ, and iF related through Eqs. (16) and (17), as shown in Fig. 1.

(a) If η_{+-} and η_{00} are completely measured, the quantities ϵ and $F\,\mathrm{Im}A_2/A_0$ are known. Thus $\delta_2 - \delta_0$ is measured up to $\pm n\pi$, and $\mathrm{Im}A_2/A_0$ is known up to a sign. If further $m_S - m_L$ is known, then through (23), M_i and $y_l + y_{3\pi}$ are determined.

(b) If $|\eta_{+-}|$, $|\eta_{00}|$, and the phase difference

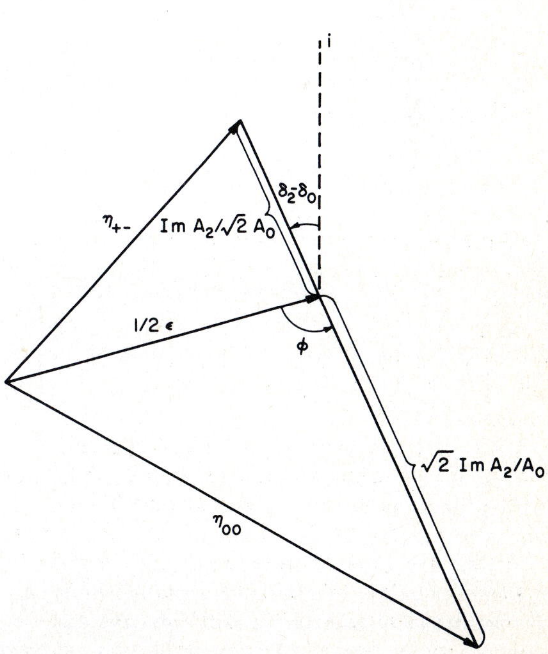

FIG. 1. Geometrical relation between η_{+-}, η_{00}, and other quantities.

of η_{+-} and η_{00} are known, then the triangle of Fig. 1 can be constructed, but its orientation relative to the real axis is known only if $\delta_2-\delta_0$ is independently obtained.

These experiments, however, do not yield any information on $\mathrm{Re}A_2/A_0$. It seems that the only experimental method of determining this ratio is to measure the rate difference on the left-hand side of Eq. (18). Existing experiment[10] gives

$$\mathrm{Re}A_2=(0\pm1)\sec(\delta_2-\delta_0). \quad (27)$$

7. In the above discussion, electromagnetic effects are completely neglected. Inclusion of these effects introduces (a) mass splits between $\pi^+\pi^-$ and $\pi^0\pi^0$ states (and related effects), and (b) additional channels like $\pi\pi\gamma$. To account for (a) one introduces two eigenstates of the S matrix for the strong and electromagnetic interactions of the $\pi\pi$ S-wave state at the K^0 mass. The resultant change comprises only small real corrections to the coefficients $(\frac{2}{3})^{1/2}$ and $(\frac{1}{3})^{1/2}$ and the phase shifts δ_2 and δ_0 in Table II.

As to (b), electromagnetic effects do not introduce CP noninvariance. Thus it is reasonable to expect additional channels, such as $\pi\pi\gamma$, not to introduce matrix elements which are imaginary in phase relative to A_0. In any case, experimentally the rates $K_{\mathrm{S,L}}\to\pi+\pi+\gamma$ are limited:

$$R[K_\mathrm{S}\to\pi^++\pi^-+\gamma]<(\sim1)$$

(from Kirsch et al.[11]) and

$$R[K_\mathrm{L}\to\pi+\pi+\gamma]<(\sim1)$$

(from total rate). It is reasonable to assume

$$R[K_\mathrm{S}\to\pi^0+\pi^0+\gamma]<(\sim1).$$

By an argument similar to that leading to Eq. (19), we have then

$$|y_{\pi\pi\gamma}|<(\sim1.5).$$

Thus electromagnetic effects are expected to be negligible in the discussion of CP noninvariances of the preceding sections.

8. We now make two supplementary remarks.

One may raise the question of the evidence of T invariance in neutron[12] and Λ[13] decays and perhaps other decays. To reconcile this evidence with the CP noninvariance of reference 1, it may be that the weak interaction consists of two terms:

$$H=H_\mathrm{strong}+H_\mathrm{el}+H_{W1}+H_{W2},$$

where (a) $H_\mathrm{strong}+H_\mathrm{el}+H_{W1}$ satisfies CP invariance, (b) $H_\mathrm{strong}+H_{W1}$ obeys the $|\Delta I|=\frac{1}{2}$ rule, and (c) H_{W2} is weaker than H_{W1}, and violates the $|\Delta I|=\frac{1}{2}$ rule.

Another question concerns the $\pi\pi(I=2)$ decay rates of K^+ and of K^0. From Eqs. (16) and (17),

$$|\mathrm{Im}A_2|=\tfrac{1}{3}\sqrt{2}A_0|\eta_{+-}-\eta_{00}|<(\sim\tfrac{2}{3}). \quad (28)$$

If the decay amplitude of $K^+\to\pi+\pi$ $(I=2)$ is A_2^+, the experimental rate of this process gives

$$|A_2^+|=3.9. \quad (29)$$

Comparing Eqs. (27), (28), and (29) one concludes that more accurate measurements of $R[K_\mathrm{S}\to\pi^++\pi^-]-2[K_\mathrm{S}\to\pi^0+\pi^0]$, and information about $|\eta_{00}|$ and $\delta_2-\delta_0$ can be used to analyze the $K\to\pi+\pi$ $(I=2)$ amplitude into $|\Delta I|=\frac{3}{2}$ and $|\Delta I|=\frac{5}{2}$ components.

It is a pleasure to acknowledge the hospitality we enjoyed at Brookhaven National Laboratory where this work was done. We thank J. W. Cronin, V. L. Fitch, P. Franzini, J. Steinberger, and W. Willis for many stimulating discussions.

For other recent discussions of CP noninvariance in K^0-$\bar{K}^0$ decay, see a recent article by Sachs.[14]

---

[1]J. H. Christenson, J. W. Cronin, V. L. Fitch, and R. Turlay, Phys. Rev. Letters 13, 138 (1964).

[2]V. F. Weisskopf and E. P. Wigner, Z. Physik 63, 54 (1930); 65, 18 (1930).

[3]T. D. Lee, R. Oehme, and C. N. Yang, Phys. Rev. 106, 340 (1957).

[4]To avoid confusion we shall call the long- and short-lived components K_L and K_S. If their wave functions are called Ψ_L and Ψ_S, then, in the notation of reference 3, $\Psi_\mathrm{L}=\Psi_-$ and $\Psi_\mathrm{S}=\Psi_+$. Throughout this paper we use the unit 10^6 sec^{-1} for decay rates.

[5]R. G. Sachs and S. B. Treiman, Phys. Rev. Letters 8, 137 (1962).

[6]W. F. Fry, International Conference on Fundamental Aspects of Weak Interactions (Brookhaven National Laboratory, Upton, New York, 1963), p. 3; J. H. Christenson, J. W. Cronin, V. L. Fitch, and R. Turlay, ibid., p. 74.

[7]B. Aubert, L. Behr, J. P. Lowys, P. Mittner, and C. Pascaud, Phys. Letters 10, 215 (1964). The large ratio of e^+/e^- in the first few lifetimes of K_S is a strong argument that $|x|$ is quite a bit less than 1.

[8]W. Willis has been considering, independently of us, the possibility of measuring this rate.

[9]J. Steinberger has reached this same conclusion independently of us.

[10]J. L. Brown, J. A. Kadyk, G. H. Trilling, B. P. Roe, D. Sinclair, and J. C. Van der Velde, Phys. Rev. 130, 769 (1963).
[11]L. Kirsch, R. J. Plano, J. Steinberger, and P. Franzini, Phys. Rev. Letters 13, 35 (1964).
[12]M. T. Burgy, V. E. Krohn, T. B. Novey, G. R. Ringo, and V. L. Telegdi, Phys. Rev. 120, 1829 (1960).
[13]J. W. Cronin and O. E. Overseth, Phys. Rev. 129, 1795 (1963).
[14]R. G. Sachs, Phys. Rev. Letters 13, 286 (1964).

Weak Interactions of Kaons

J. S. Bell and J Steinberger

CERN, Geneva

In this review the emphasis will be on developments associated with CP violation. We will be concerned both with phenomenology and with experiments and experimental results.

1. Phenomenology of Neutral Kaons

First we recall the essential ideas of the classical treatments[1-7] of the subject, applying in turn some elementary notions. As well as papers explicitly cited below, some other recent papers are listed[6-20] which are of phenomenological character, and which contain ideas whose value will increase with time.

1.1 Superposition

The particles K and $\bar{K}$ produced in strong interactions have definite hypercharges, +1 and -1 respectuvely. The weak interactions do not conserve hypercharge and so induce, as well as decays, transitions between K and $\bar{K}$. For this reason it is not K and $\bar{K}$ themselves, but certain superpositions thereof, which have simple exponential decay laws:

$$\begin{aligned} |L\rangle &= p\,|K\rangle + q\,|\bar{K}\rangle \\ |S\rangle &= r\,|K\rangle + s\,|\bar{K}\rangle \end{aligned} \qquad (1)$$

In a period of proper time τ (Lorentz invariance allows us to work always in the rest frame)

$$|L\rangle \to e^{-iM_L\tau}\,|L\rangle\,, \qquad |S\rangle \to e^{-iM_S\tau}\,|S\rangle$$

where M_L and M_S are complex:

$$M_L = \mathrm{Re}\, M_L - \frac{i}{2}\,\Gamma_L$$

$$M_S = \mathrm{Re}\, M_S - \frac{i}{2}\,\Gamma_S$$

where Γ_L and Γ_S are the decay rates of the long- and short-lived combinations, respectively.

We are free to specify arbitrary phases in $|L\rangle$ and $|S\rangle$, and between $|K\rangle$ and $|\bar{K}\rangle$, and therefore to adopt the following

Convention: p, q and r are real and positive (2)

The phase of $|S\rangle$ then is not conventional. Although we advocate this convention as a sensible one, we will not in general impose it on the equations. We suppose $|L\rangle$ and $|S\rangle$ to be normalized:

$$|p|^2 + |q|^2 = |r|^2 + |s|^2 = 1$$

Then the real quantity (q/p) and the complex (s/r) are constants of nature to be determined.

Inverting Eqs. (1)

$$\begin{aligned} |K\rangle &= (sp - qr)^{-1}\,\{\, s|L\rangle - q|S\rangle \,\} \\ |\bar{K}\rangle &= (sp - qr)^{-1}\,\{\, -r|L\rangle + p|S\rangle \,\} \end{aligned} \qquad (3)$$

Therefore in a period τ a state which is initially $|K>$ evolves thus

$$|K> \rightarrow (sp - qr)^{-1} \{e^{-iM_L\tau} s|L> - e^{-iM_S\tau} q|S>\} \quad (4)$$

$$= (sp - qr)^{-1} \begin{cases} (e^{-iM_L\tau} sp - e^{-iM_S\tau} qr) |K> \\ + (e^{-iM_L\tau} sq - e^{-iM_S\tau} qs) |\bar{K}> \end{cases} \quad (5)$$

Likewise

$$|\bar{K}> \rightarrow (sp - qr)^{-1} \{e^{-iM_S\tau} p|S> - e^{-iM_L\tau} r|L>\} \quad (6)$$

$$= (sp - qr)^{-1} \begin{cases} (e^{-iM_S\tau} pr - e^{-iM_L\tau} rp) |K> \\ + (e^{-iM_S\tau} ps - e^{-iM_L\tau} rq) |\bar{K}> \end{cases} \quad (7)$$

In a small interval $\delta\tau$

$$\begin{aligned} |K> &\rightarrow |K> - i\delta\tau \{M|K> + B|\bar{K}>\} \\ |\bar{K}> &\rightarrow |\bar{K}> - i\delta\tau \{A|K> + \bar{M}|\bar{K}>\} \end{aligned} \quad (8)$$

where

$$\begin{aligned} M &= (M_L sp - M_S qr) (sp - qr)^{-1} \\ \bar{M} &= (M_S sp - M_L qr) (sp - qr)^{-1} \\ A &= (M_S - M_L) rp (sp - qr)^{-1} \\ B &= (M_L - M_S) sq (sp - qr)^{-1} \end{aligned} \quad (9)$$

If the general state is expressed as

$$\Psi = \psi|K> + \bar{\psi}|\bar{K}>$$

it follows from (8) that

$$\frac{d}{d\tau}\begin{pmatrix}\psi \\ \bar{\psi}\end{pmatrix} = -i \begin{pmatrix} M & A \\ B & \bar{M}\end{pmatrix}\begin{pmatrix}\psi \\ \bar{\psi}\end{pmatrix} \quad (10)$$

The matrix on the right is called the mass matrix.

1.2 Unitarity

At any instant decays will occur to a specified final state F with a probability proportional to the square of the transition matrix element,

$$|(F|T|\Psi)|^2 = |(F|T|K)\psi + (F|T|\bar{K})\bar{\psi}|^2 \quad (11)$$

The total transition rate Γ is given by summing over all final states F consistent with energy and momentum conservation:

$$\begin{aligned} \Gamma = & |\psi|^2 \Sigma |(F|T|K)|^2 + |\bar{\psi}|^2 \Sigma |(F|T|\bar{K})|^2 \\ & + \psi^* \bar{\psi} \Sigma (F|T|K)^* (F|T|\bar{K}) + \bar{\psi}^* \psi \Sigma (F|T|\bar{K})^* (F|T|K) \end{aligned}$$

This must be compensated by a decrease in the norm of the K meson state; from the equations above

$$\begin{aligned} -\frac{d}{d\tau}(|\psi|^2 + |\bar{\psi}|^2) &= -2 \text{ Re } (\psi^* \frac{d\psi}{d\tau} + \bar{\psi}^* \frac{d\bar{\psi}}{d\tau}) \\ &= -2 \text{ Im } (\psi^* M\psi + \psi^* A\bar{\psi} + \bar{\psi}^* B\psi + \bar{\psi}^* \bar{M}\bar{\psi}) \quad (12) \\ &= -2 \text{ Im } M|\psi|^2 - 2 \text{ Im } \bar{M}|\bar{\psi}|^2 + i(A - B^*) \psi^*\bar{\psi} + i (B - A^*) \bar{\psi}^*\psi \end{aligned}$$

Comparing coefficients in (11) and (12), remembering that ψ and $\overline{\psi}$ are arbitrary

$$\begin{aligned} -2 \,\mathrm{Im}\, M &= \Sigma \,|(F|T|K)|^2 \\ -2 \,\mathrm{Im}\, \overline{M} &= \Sigma \,|(F|T|\overline{K})|^2 \\ i\,(B - A^*) &= \Sigma \,(F|T|\overline{K})^* \,(F|T|K) \end{aligned} \qquad (13)$$

The first two equations are familiar from more ordinary unstable particles, relating the imaginary part of the mass (decay rate) to transition probabilities. More generally, one sees here that the antihermitian part of the mass matrix is given by decay amplitues; for a Hermitian matrix we would have $A = B^*$. These consequences of probability conservation will be referred to as "unitarity relations", and play an important role.

We need other unitarity relations that can be obtained by manipulating the above, or more pleasantly by repeating the above reasoning with the form

$$\Psi = X\, e^{-iM_L \tau}\, |L> + Y\, e^{-iM_S \tau}\, |S> \qquad (14)$$

the norm is

$$N = |X|^2\, e^{-\Gamma_L \tau} + |Y|^2\, e^{-\Gamma_S \tau}$$

$$+ X^* Y\, e^{i(M_L^* - M_S)\tau} < L|S> + Y^* X\, e^{i(M_S^* - M_L)} < S|L>$$

At $\tau = 0$

$$-\frac{dN}{d\tau} = \Gamma_L |X|^2 + \Gamma_S |Y|^2 - i(M_L^* - M_S)X^*Y<L|S> - i(M_S^* - M_L)Y^*X<S|L>$$

This should equal the total transition rate

$$\Sigma\, |(F|T|L)\, X + (F|T|S) Y\, |^2$$

for arbitrary X and Y. Thus

$$\begin{aligned} \Gamma_L &= \Sigma\, |(F|T|L)|^2 \\ \Gamma_S &= \Sigma\, |(F|T|S)|^2 \\ -i(M_L^* - M_S)<L|S> &= \Sigma\, (F|T|L)^*\, (F|T|S) \end{aligned} \qquad (15)$$

Now either from the structure of the last three equations and the Schwarz inequality, or simply[11] from the fact that $(-dN/d\tau)$ should be positive for arbitrary X and Y,

$$|M_L^* - M_S|\, |<L|S>| \leqslant (\Gamma_L \Gamma_S)^{\frac{1}{2}} \qquad (16)$$

Substituting experimental values

$$\pm\, \mathrm{Re}\, (M_L - M_S) \approx \frac{1}{2}\, \Gamma_S$$

$$\Gamma_L \approx 1.6\; 10^{-3}\; \Gamma_S$$

one finds

$$|<L|S>| \lessapprox 0.06$$

Thus the states $|L>$ and $|S>$ are very nearly orthogonal. Note that if CP were a good quantum number $|L>$ and $|S>$ would be exactly orthogonal, having different eigenvalues of CP. Thus $<L|S>$ is one measure of CP violation in the $K\overline{K}$ system.

Let us investigate in more detail the expression

$$\Sigma\, (F|T|L)^* (F|T|S)$$

For any group G if final states

$$|\sum_G (F|T|L)^*(F|T|S)| \leqslant (\sum_G |(F|T|L)|^2)^{\frac{1}{2}}\, (\sum_G |(F|T|S)|^2)^{\frac{1}{2}}$$

$$\leqslant (\Gamma_L\, (G)\; \Gamma_S\, (G)\;)^{\frac{1}{2}}$$

where Γ_L (G) and Γ_S (G) are partial decay rates. The main decay modes are the leptonic, the three-pionic, and the two-pionic.

Experimentally

$$\Gamma_L \text{ (leptonic)} \approx \Gamma_S \text{ (leptonic)} \approx 12 \quad 10^{-6} \text{ sec}^{-1}$$
$$\approx 10^{-3} \quad \Gamma_S$$

Thus

$$\Big| \sum_{\substack{\text{Leptonic} \\ \text{modes}}} \Big| \leqslant 10^{-3} \; \Gamma_S$$

Now it is only if $|L\rangle$ and $|S\rangle$ can both decay into common states F, which would be forbidden by CP conservation, that Σ is not zero. The studies of leptonic modes (Paris, Padua, and Columbia) suggest that there is no gross violation of CP. So we can suppose

$$\Gamma_S^{-1} \Big| \sum_{\text{Leptonic}} \Big| \ll 10^{-3}$$

Experimentally

$$\Gamma_L \text{ (3-pionic)} \approx 7 \quad 10^{6} \text{ sec}^{-1} \approx 7 \quad 10^{-4} \; \Gamma_S$$

Not much is known about the corresponding modes of S, and they might even be as big as the above. Then

$$\Gamma_S^{-1} \Big| \sum_{3\pi} \Big| \leqslant 7 \quad 10^{-4}$$

However, if this approximate bound were reached, it would imply a gross violation of CP invariance in this mode. Tentatively, supposing that this does not occur

$$\Gamma_S^{-1} \Big| \sum_{3\pi} \Big| \ll 10^{-3}$$

There are also the radiative decay modes to be considered. Of course, in principle, every mode with charged final state particles is radiative, because of the soft photon effect. This should not upset the above estimates because for any X and Y in (14) the soft photon effect should mainly cause a slight degradation of energy without much affecting the total decay rate, and so should not much affect (15). With regard to hard photons, and with regard to hypothetical modes not yet observed, the most solid estimate would be based on comparison of the total decay rates, Γ_L and Γ_S, with the sums of known partial rates. With present accuracy this does not exclude significant contributions from unknown modes. However, it is also permissible to suppose that such modes account for less than or about 10^{-4} of Γ_L and Γ_S, in which case they can be ignored here.

We are left with the two pionic modes:

$$-i\,(M_L^* - M_S)\,\langle L|S\rangle = \sum_{2 \text{ pionic}} (F|T|L)^* \, (F|T|S) + \gamma \, \Gamma_S \qquad (17)$$

where γ, the sum of other contributions discussed above, is very small

$$|\gamma| \ll 10^{-3} \qquad (18)$$

From the experiments on $L \to \pi^+ \pi^-$

$$\Gamma \, (L \to \pi^+\pi^-) \approx (2 \quad 10^{-3})^2 \; \Gamma(S \to \pi^+\pi^-)$$

and it is probable[21] that

$$\Gamma(L \to \pi^0\pi^0) \leqslant \frac{1}{2} \, \Gamma(L \to \pi^0\pi^0\pi^0)$$
$$\leqslant \frac{1}{8} \, \Gamma_L \approx (1.4 \quad 10^{-2})^2 \; \Gamma_S$$

On this basis

$$|M_L^* - M_S| \, |\langle L|S\rangle| \leqslant 10^{-2} \; \Gamma_S$$
$$|\langle L|S\rangle| \leqslant 0.01 \qquad (19)$$

Weinberg and Glashow[21] showed that such an estimate can be obtained from the two-pionic modes alone, when PCT is assumed.

For future reference note the result

$$(M_L - M_S) \langle L|S\rangle = i \{ (0|T|L)^*(0|T|S) + (2|T|L)^*(2|T|S) + \gamma \ \Gamma_S \} \tag{20}$$

where

$$|\gamma| \ll 10^{-3}$$

and we have used the fact that $\Gamma_L \approx 1.6 \times 10^{-3}\, \Gamma_S$ to replace M_L^* by M_L. $|0\rangle$ and $|2\rangle$ denote the two-pionic states with I = 0 and 1 = 2, respectively.

1.3 CP

If CP had been a good symmetry, then the states $|L\rangle$ and $|S\rangle$ would necessarily be eigenstates of it, and so identical with the combinations

$$\frac{1}{\sqrt{2}}\,|K\rangle \pm \frac{1}{\sqrt{2}}\,|\overline{K}\rangle$$

which have CP = ∓ 1. The J = 0 state of two pions, either $\pi^+ \pi^-$ or $\pi^0 \pi^0$, would have CP = +1, so that only one of the two states could decay in this mode. Since the decay $|S\rangle \to 2\pi$ is copious we would identify $|S\rangle$ as the state with CP = +1:

$$|S\rangle = \frac{1}{\sqrt{2}}\,|K\rangle - \frac{1}{\sqrt{2}}\,|\overline{K}\rangle$$

$$|L\rangle = \frac{1}{\sqrt{2}}\,|K\rangle + \frac{1}{\sqrt{2}}\,|\overline{K}\rangle$$

Note that we use conventions such that

$$CP|K\rangle = -\,|\overline{K}\rangle \qquad CP|\overline{K}\rangle = -\,|K\rangle \tag{21}$$

Another common convention is that obtained from ours by the change of notation

$$\begin{aligned} |K\rangle_{\text{of us}} &= i\,\{\,|K\rangle_{\text{of them}}\,\} \\ |\overline{K}\rangle_{\text{of us}} &= -\,i\,\{\,|\overline{K}\rangle_{\text{of them}}\,\} \end{aligned} \tag{22}$$

However, it is now found that $|L\rangle$ also decays into two pions, although at a greatly reduced rate, so that CP symmetry does not hold exactly.

It is worth noting that this conclusion, that CP is not an exact symmetry, is arrived at only in a very indirect way. There is nothing manifestly unsymmetric between particles and antiparticles in the observation of long-lived neutral kaons decaying into pion pairs. It is only by an analysis involving the general ideas of quantum mechanics, and in particular the superposition principle, that CP violation is detected. One could therefore (Laurent and Roos[46]) give up the superposition principle instead of CP. To dispose of this possibility it will be necessary to demonstrate the absence of CP symmetry more directly. It would be sufficient for example to find some difference in the rates of decays involving positive and negative electrons from states which are initially pure $|K\rangle$ and $|\overline{K}\rangle$ respectively, or in the same states to find some difference in the time distributions of $(\pi^+ \pi^-)$ decays.

In the meantime there is the intriguing possibility that we have reached the limit of applicability not of CP but of ordinary quantum mechanics. This possibility will not be considered further here.

1.4 PCT

PCT symmetry is deeply rooted in local relativistic field theory. Although it is far from having the status of superposition, unitarity, or Lorentz invariance, we will tentatively regard PCT also as part of the experimenter's equipment. Of course any failure of the subsequent scheme to accommodate experimental data will be of great interest.

According to PCT the transition amplitudes are equal for the processes

$$I \to F \qquad\qquad \overline{F}' \to \overline{I}'$$

where the bar denotes interchange of particles and antiparticles, and the prime denotes reversal of spin directions.

Apply this first to the probability amplitudes for

$$K \rightarrow K \qquad\qquad \overline{K} \rightarrow \overline{K}$$

in a given period of time. They should be equal. Inspection of (5) + (7) shows that this requires

$$(e^{-iM_L\tau}\, sp - e^{-iM_S\tau}\, qr) = (e^{-iM_S\tau}\, ps - e^{-iM_L\tau}\, rq)$$

or

$$sp = -\, rq \tag{23}$$

In terms of the elements of the mass matrix, from (9), the requirement is simply

$$M = \overline{M} \tag{24}$$

We can learn something about a possible PCT violating difference, $M-\overline{M}$, from the observed difference M_L-M_S. From (9)

$$M - \overline{M} = (M_L - M_S)\,(sp + qr)\,(sp - qr)^{-1} \tag{25}$$

Now we have noted that $|L\rangle$ and $|S\rangle$ are very nearly orthogonal

$$|\langle L|S\rangle| = |pr^* + qs^*| \leqslant 0.01$$

Moreover, there is the identity

$$|sp - qr|^2 + |pr^* + qs^*|^2 = (|p|^2 + |q|^2)(|r|^2 + |s|^2) = 1 \tag{26}$$

Thus

$$sp - qr \approx 1 \tag{27}$$

very closely. Further

$$|sp + qr| \leqslant |s|\,|p| + |q|\,|r| \leqslant 2$$

Thus

$$|M - \overline{M}| \leqslant 2\,|M_L - M_S| \tag{28}$$

and experimentally M_L-M_S is very small. This inequality could be made much stronger if it were established experimentally, by looking at strong interactions induced by neutral kaons, that $|L\rangle$ and $|S\rangle$ are both more or less euqal mixtures of $|K\rangle$ and $|\overline{K}\rangle$:

$$|s| \approx |r| \approx \frac{1}{\sqrt{2}} \qquad\qquad |p| \approx |q| \approx \frac{1}{\sqrt{2}}$$

Then

$$|sp - qr| \approx 1$$

implies

$$|sp + qr| \ll 1$$

or

$$|M - \overline{M}| \ll |M_L - M_S|$$

From the PCT requirement

$$sp + qr = 0$$

we have

$$s/r = -\, q/p$$

Since $|p|^2+|q|^2 = |r|^2+|s|^2 = 1$, and p, q, and r were real and positive by convention, we now have

$$s = -q \qquad r = p$$

We rewrite the essential formulae with this simplification:

$$|L\rangle = p|K\rangle + q|\overline{K}\rangle \tag{29}$$

$$|S\rangle = p|K\rangle - q|\overline{K}\rangle \tag{30}$$

$$2p|K\rangle = |L\rangle + |S\rangle \tag{31}$$

$$2q|\overline{K}\rangle = |L\rangle - |S\rangle \tag{32}$$

$$|K\rangle \rightarrow \frac{1}{2}(e^{-iM_L\tau} + e^{-iM_S\tau})|K\rangle + \frac{1}{2}\frac{q}{p}(e^{-iM_L\tau} - e^{-iM_S\tau})|\overline{K}\rangle \tag{33}$$

$$|\overline{K}\rangle \rightarrow \frac{1}{2}(e^{-iM_L\tau} + e^{-iM_S\tau})|\overline{K}\rangle + \frac{1}{2}\frac{p}{q}(e^{-iM_L\tau} - e^{-iM_S\tau})|K\rangle \tag{34}$$

$$\langle L|S\rangle = |p|^2 - |q|^2 \leqslant 0.01 \tag{35}$$

$$M = \overline{M} = \frac{1}{2}(M_L + M_S) \tag{36}$$

$$A = \frac{1}{2}\frac{p}{q}(M_L - M_S) \tag{37}$$

$$B = \frac{1}{2}\frac{q}{p}(M_L - M_S) \tag{38}$$

2. Two-Pion Decay

2.1 Isospin analysis of S, L → 2π

We wish to make an isospin analysis of $L \rightarrow 2\pi$ and $S \rightarrow 2\pi$, following Wu and Yang[9] and Wolfenstein[10]. There are four amplitudes involved:

$$(0|T|L), \quad (0|T|S), \quad (2|T|L), \quad (2|T|S)$$

We could hope to determine the relative magnitudes and phases of these amplitudes. The over-all phase of course could not be determined, and the absolute magnitude, when the various ratios are known, is given by, say, Γ_S. Thus we define the ratios to be determined

$$\epsilon = \frac{(0|T|L)}{(0|T|S)} \qquad \epsilon' = \frac{(2|T|L)}{(0|T|S)} \qquad \omega = \frac{(2|T|S)}{(0|T|S)} \tag{39}$$

- six real numbers.

Ideally we would have available the experimental information

$$\left.\begin{aligned} R &= \frac{\Gamma_S(\pi^0\pi^0)}{\Gamma_S(\pi^+\pi^-)} = \frac{1}{2}\left|\frac{1-\sqrt{2}\,\omega}{1+\omega/\sqrt{2}}\right|^2 \\ \eta_{+-} &= \frac{(\pi^+\pi^-|T|L)}{(\pi^+\pi^-|T|S)} = \frac{\epsilon + \epsilon'/\sqrt{2}}{1+\omega/\sqrt{2}} \\ \eta_{00} &+ \frac{(\pi^0\pi^0|T|L)}{(\pi^0\pi^0|T|S)} = \frac{\epsilon - \sqrt{2}\,\epsilon'}{1 - \sqrt{2}\,\omega} \end{aligned}\right\} \tag{40}$$

which is only five numbers. However, in addition, we have the unitarity condition, from (20),

$$\begin{aligned} &\mathrm{Im}[\,i(M_L - M_S)^{-1}\{(0|T|L)^*(0|T|S) + (2|T|L)^*(2|T|S) + \gamma\,\Gamma_S\}] \\ &= \mathrm{Im}\,\langle L|S\rangle = 0 \end{aligned} \tag{41}$$

where only the last step [from (35)] depends on PCT. Then in so far as γ ($\ll 10^{-3}$) can be neglected, as will be discussed in more detail below, we can make a zero constraint fit.

Unfortunately, the phase of η_{00} may not be measured for some time, requiring an interference experiment on the two neutral pion decay. Therefore it is expedient to make the assumption

$$|\omega| \ll 1$$

as suggested by the $\Delta I = \frac{1}{2}$ (non-leptonic) rule. This cannot be justified from the data alone because R is insensitive to Imω. However, it is consistent with the data, and the low rate of $K^+ \to \pi^+\pi^0$ decay increases confidence in the $\Delta I = \frac{1}{2}$ rule - although in principle that might be due to an accidental cancellation of $\Delta I = \frac{3}{2}$ and $\Delta I = \frac{5}{2}$ parts.

With $|\omega|$ small, R determines Reω, Imω is undetermined, and for the rest we have the equations

$$\left.\begin{aligned} \eta_{+-} &= \epsilon + \epsilon'/\sqrt{2} \\ |\eta_{00}| &= |\epsilon - \sqrt{2}\,\epsilon'| \\ i(M_L - M_S)^{-1}[\epsilon + \epsilon'\omega^* + \gamma^*]^* &= \text{real} \end{aligned}\right\} \tag{42}$$

If it turns out, as will be indicated by the branching ratio of $(K_L \to \pi^0\pi^0)/(K_L \to \pi^+\pi^-)$, that L decay is not dominated by I = 2, then

$$|\omega^* \epsilon'| \ll |\epsilon|$$
$$|\epsilon| \geqslant 2\ 10^{-3}$$

Since $|\gamma| \ll 10^{-3}$, the last equation simplifies to

$$i\,(M_L - M_S)^{-1}\,\epsilon^* = \text{real} \tag{43}$$

and we have a no-constraint situation. There are in general two solutions, as can be seen by simple graphical considerations[9,10].

2.2 PCT and Decay Amplitudes

For decay into a given final state $|F\rangle$, PCT implies

$$(F|T|\overline{K}) = (K|T|\overline{F}')$$

When final state interactions can be neglected the matrix T is Hermitian:

$$(K|T|\overline{F}') = (\overline{F}'|T|K)^*$$

and we have the useful relation

$$(F|T|\overline{K}) = (\overline{F}'|T|K)^* \tag{44}$$

between decay amplitudes of K and $\overline{K}$. Time reversal invariance T would imply a similar relation without the bar, and so lead to reality conditions - again when final state interactions are negligible. The problem of generalizing this relation when final state interactions are not negligible is solved by Watson's theorem, with the result

$$(F|T|\overline{K}) = e^{2i\delta}\,(\overline{F}'|T|K)^* \tag{45}$$

where δ is the scattering phase shift for the states $|F\rangle$ and $|\overline{F}'\rangle$.

In view of this result we can write[9,10] for the I = 0 2π decay amplitude of $|K\rangle$:

$$\begin{aligned} (0|T|K) &= e^{i\delta_0}\, i\, A_0 \\ (0|T|\overline{K}) &= -e^{i\delta_0}\, i\, A_0^* \end{aligned} \tag{46}$$

Likewise

$$\begin{aligned} (2|T|K) &= e^{i\delta_2}\, i\, A_2 \\ (2|T|\overline{K}) &= -e^{i\delta_2}\, i\, A_2^* \end{aligned} \tag{47}$$

where δ_0 and δ_2 are S wave scattering phases in the I = 0 and 2 states at the energy of the kaon mass.

In terms of the A's

$$\epsilon = \frac{(0|T|L)}{(0|T|S)} = \frac{1 - (q\, A_0^*)/(p\, A_0)}{1 + (q\, A_0^*)/(p\, A_0)}$$

$$\epsilon' = \frac{(2|T|L)}{(0|T|S)} = \frac{1 - (q\, A_2^*)/(p\, A_2)}{1 + (q\, A_0^*)/(p\, A_0)} \frac{A_2}{A_0}\, e^{i(\delta_2-\delta_0)} \tag{48}$$

$$\omega = \frac{(2|T|S)}{(0|T|S)} = \frac{1 + (q\, A_2^*)/(p\, A_2)}{1 + (q\, A_0^*)/(p\, A_0)} \frac{A_2}{A_0}\, e^{i(\delta_2-\delta_0)}$$

Now

$$|\epsilon|^2 \leqslant \Gamma_L/\Gamma_S \approx 1.6\ 10^{-3}$$

Thus ϵ is a small quantity

$$|\epsilon| \leqslant 4\ 10^{-2}$$

Thus to a good approximation

$$(q\, A_0^*)/(p\, A_0) \approx 1$$

and

$$\epsilon = \frac{1}{2}\left(1 - \frac{q\, A_0^*}{p\, A_0}\right)$$

$$\epsilon' = \frac{1}{2}\left(\frac{A_2}{A_0} - \frac{A_2^*}{A_0^*}\right) e^{i(\delta_2-\delta_0)} \tag{49}$$

$$\omega = \frac{1}{2}\left(\frac{A_2}{A_0} + \frac{A_2^*}{A_0^*}\right) e^{i(\delta_2-\delta_0)}$$

So when the phase of ϵ' is determined we know $\delta_2 - \delta_0$ up to a multiple of π :

$$\epsilon' = \pm\, |\epsilon'|\ e^{i(\delta_2-\delta_0 + \frac{1}{2}\pi)} \tag{50}$$

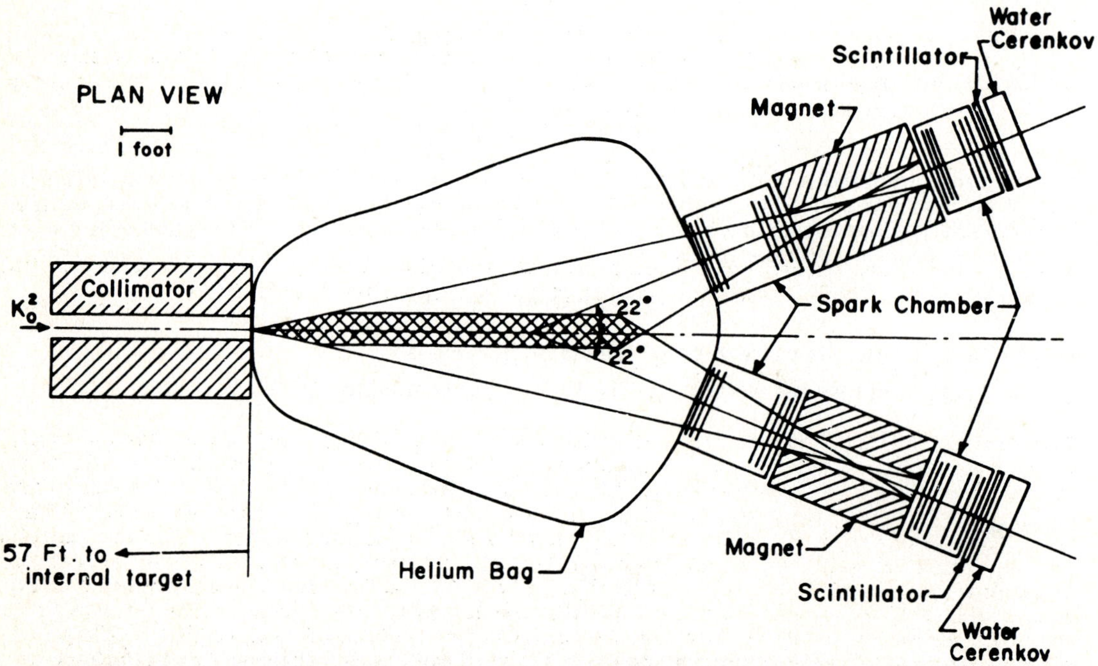

Figure 1 Arrangement of $\pi^+\pi^-$ detector in experiment of Christenson et al.[22]

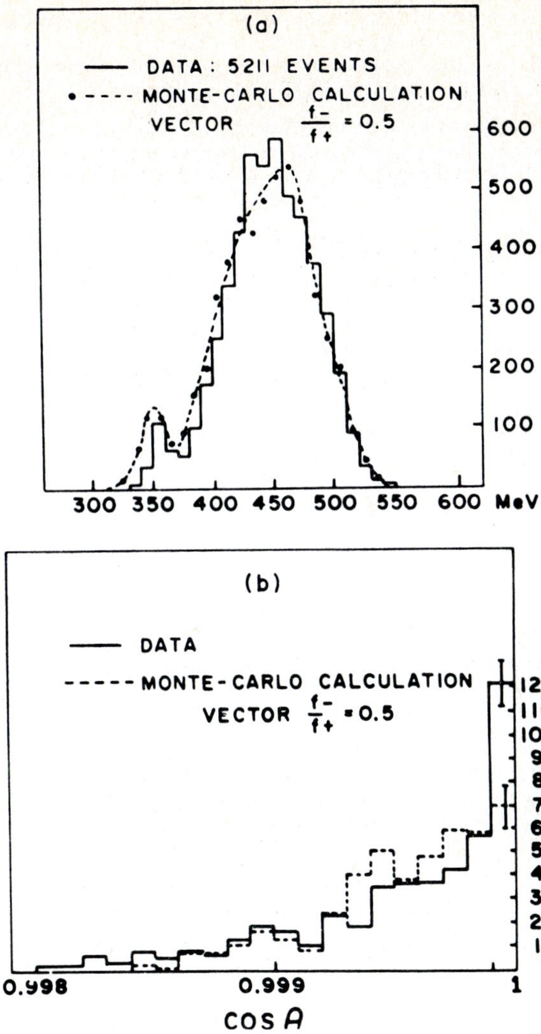

Figure 2 a) Invariant mass distribution in the experiment of Christenson et al.[22];
b) Angular distribution of events in the mass interval $490 < m < 510$ in the same experiment. The calculated distributions are the results of a Monte Carlo calculation on the expectation for leptonic decays.

So then is the phase of ω known

$$\omega = \pm |\omega| e^{i(\delta_2 - \delta_0)} \tag{51}$$

Then $|\omega|$ can be determined from the branching ratio R and the assumpyion that $|\omega| \ll 1$ verified. If necessary, one can then remove this assumption by an iterative or numerical procedure.

2.3 Experimental Status of CP Violation in $K^0 \rightarrow \pi^+ \pi^-$ Decay

A) $K_L \rightarrow \pi^+ + \pi^-$

Observations of this decay by Christenson, Cronin, Fitch and Turlay[22] and almost simultaenously by Abashian et al.[23] constituted the demonstration of CP violation. The apparatus of Christenson et al is shown in Fig. 1. A neutral beam is produced at 30° to an internal target at the Brookhaven A.G.S. It traverses a region filled with an atmosphere of helium gas at a distance of about 15 m from the target. The He region is viewed by two telescopes consisting of spark chambers, bending magnet, spark chambers and scintillation and Cerenkov counters, set at about twenty degrees on either side of the beam. A coincidence between the counters in the telescope triggers the spark chambers and the event is photographed and subsequently measured. The momenta of the two tracks are then reconstructed in space and the invariant mass $m = \{[(p_+^2+m_\pi^2)^{\frac{1}{2}} + (p^2+m_\pi^2)^{\frac{1}{2}}]^2 - (\vec{p}_+ + \vec{p}_-)^2\}^{\frac{1}{2}}$ is computed. The observed distributions (Fig. 2, upper part) is that expected for the leptonic decay of K_L, which accounts therefore for the bulk of the observed events. The events in the region

490 < m < 510 MeV are then plotted as a function of angle with respect to the beam direction (Fig. 2, bottom). Again the distribution is that expected for the leptonic decay, except for some 50 events which are colinear with the central beam, within measurement error. These are then due to the $\pi^+\pi^-$ decay of a long-lived particle of mass 500, presumably the K_L.

The experiment has been repeated both at the Rutherford Laboratory[24] and at CERN[25], using beams of higher average momentum. The results confirm the earlier experiments. The relative rate ($\Gamma_{L,\pi^+\pi^-}/\Gamma_{L,\text{all charged}}$) is given in Table I for these experiments.

Table I

Rate for the decay $K_L \rightarrow \pi^+ + \pi^-$

Experiment	Relative Rate ($\Gamma_{L,\pi^+\pi^-}/\Gamma_{L,\text{all charged}}$)
Christenson et al.[22]	$(1.87 \pm .18) \times 10^{-3}$ *
Abashian et al.[23]	$(2 - 3) \times 10^{-3}$
Galbraith et al.[24]	$(2.08 \pm .35) \times 10^{-3}$
de Bouard et al.[25]	$(2.24 \pm .23) \times 10^{-3}$ †
Combined	$(2.04 \pm .13) \times 10^{-3}$

The corresponding absolute value of the relative amplitude for K_L and K_S decay into π^+ and π^- is $(1.85 \pm .12) \times 10^{-3}$.

B) Interference of K_S and K_L in 2π decay

A coherent beam of K_S and K_L will show interference phenomena in common decay modes. The observation of such interference in the two pion decay would be in itself valuable, since it would be a clear demonstration that it is the K_L, and not some other particle, which decays into two pions, not two other particles. The interference would be most clear in the time dependence of $K^0 \rightarrow 2\pi$ decay, in that characteristic oscillatory behaviour should be observed.

Consider a beam for which the K_S amplitude is ρ and the K_L amplitude is 1 at time t = 0. The $\pi^+\pi^-$ decay per unit time is then

$$\frac{d}{d\tau}N_{+-} = \Gamma_{S+-} \left| \rho e^{-iM_S\tau} + \eta_{+-} e^{-iM_L\tau} \right|^2$$

$$= \Gamma_{S+-}\left\{|\rho|^2 e^{-\Gamma_S\tau} + |\eta_{+-}|^2 e^{-\Gamma_L\tau} + 2|\rho|\,|\eta_{+-}|\, e^{-(\Gamma_S+\Gamma_L)\frac{\tau}{2}} \cos\chi\right\} \tag{52}$$

$$\chi = \varphi_\rho - \varphi_\eta - \Delta m\tau \qquad \Delta m = \mathrm{Re}\,(M_S - M_L)$$

τ is the proper time of decay. It is related to the path length x before decay: $\tau = x\,.\,m_K/P_K$. The interference is most prominent at the time $\Gamma_S\tau \approx 2\ln|\rho/\eta_{+-}|$. There are two slightly different experiments possible. In one of these the K_S amplitude is generated from a pure K_L beam by placing an absorber in the beam. This method has advantages of experimental simplicity, but the disadvantage that the regeneration phase φ_ρ must be obtained before $\varphi_{\eta_{+-}}$ can be determined from the experimental result. The measurement of φ_ρ is not entirely easy. In Fig. 3 typical time distributions are shown, corresponding to several values for the phase $\varphi_\rho - \varphi_{\eta_{+-}}$, and to a particular regenerator intensity and mass difference.

In the second method the detector views a decay region close to the production region of K^0 mesons. In this case ρ is known to be +1. This type of experiment may be technically more difficult, but the phase of $\varphi_{\eta_{+-}}$ can be found directly, providing Δm is known. Both experiments are in progress at CERN.

A somewhat different experiment has been suggested by Cabibbo[26] and by Lewis and Enz[16]. In this method the K_S mesons are generated by an initially pure K_L beam in a material of low density. The decays are observed inside the regenerator, and the decay rate is measured as a function of the density of the material (only the average density is changed, not the nuclear species). If one looks for the decays in a region which is sufficiently far from the front face of the regenerator, then transient effects can be neglected and the ratio of K_S to K_L amplitudes is independent of position:

$$\frac{A_S}{A_L} = \rho_0 N = \frac{\pi\,(\bar{f}(o) - f(o))}{(\Delta m - \frac{1}{2}\,i\,\Gamma_S)\,M_K}\,N \tag{53}$$

---

* As reported at this conference by J. Cronin.

† As reported at this conference by T.R. Willits.

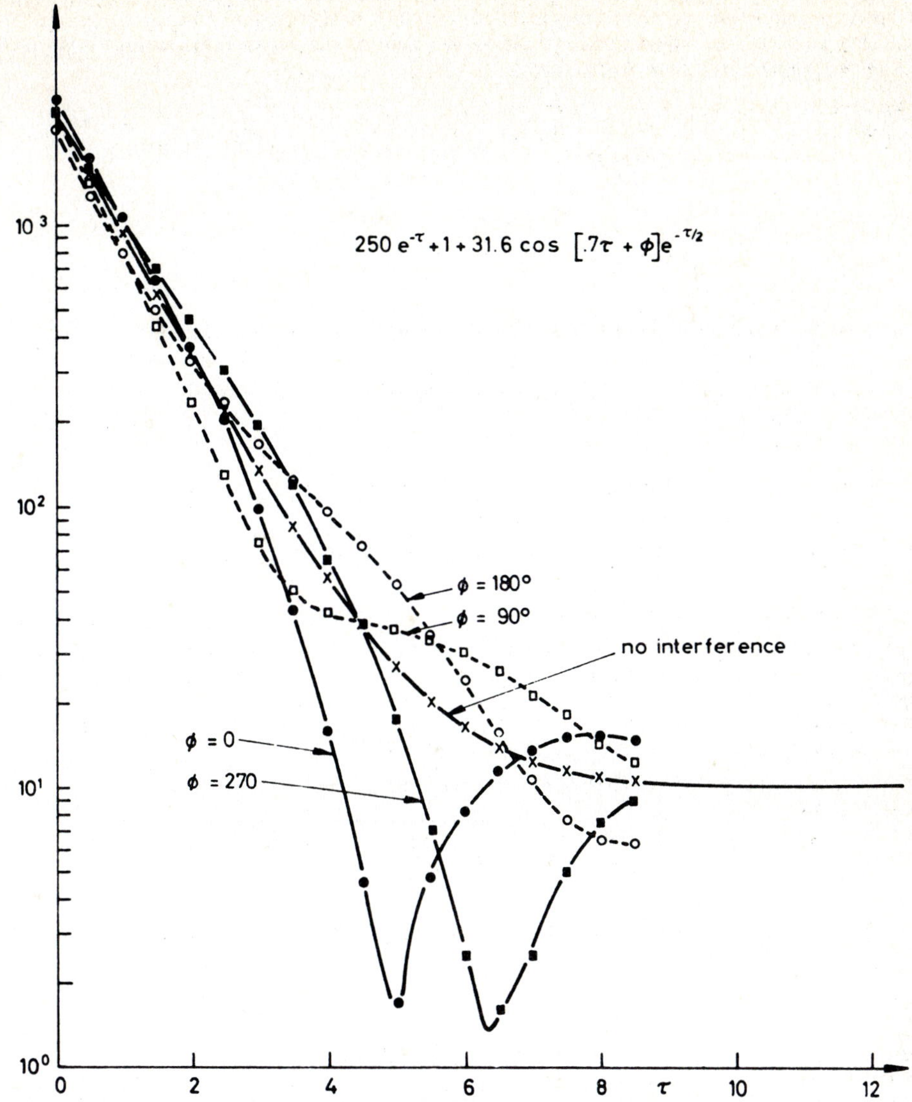

Figure 3 Expected time distributions for K_L, $K_S \rightarrow \pi^+\pi^-$ interference (52) for several values of the relative phase $\varphi_\rho - \varphi_{\eta_{+-}}$, assuming $\Delta m = .7\,\Gamma_S$ and a regeneration $|\rho|^2 = 1.2 \times 10^{-3}$.

where N is the number of nuclei per unit volume in the regenerator. The two pion decay rate as a function of the density N is then a parabola

$$I_{\pi^+\pi^-}(N) = \left\{ \begin{array}{l} |\eta_{+-}|^2 + N^2\,|\rho_0|^2 \\ \\ \quad + 2\,N|\rho_0|\;|\eta_{+-}|\;\cos\,(\varphi_{\rho_0} - \varphi_{\eta_{+-}}) \end{array} \right\} \tag{54}$$

Demonstration of a non-zero slope at N = 0 would be demonstration of interference, and the relative phase of ρ_0 and η_{+-} is given by the actual slope. Again φ_{ρ_0} must be independently determined to arrive at $\varphi_{\eta_{+-}}$. The advantage of this method is in the fact that uncertainties in Δm do not enter into the determination of $(\varphi_{\eta_{+-}} - \varphi_{\rho_0})$.

A variation of this experiment, in which this last feature was however relinquished, has been reported by Fitch, Roth, Russ and Vernon[27]. The apparatus is similar to that of Christenson et al.[22]. The experiment consists of three parts:

Part I Measurement of the rate $K^0 \rightarrow \pi^+ + \pi^-$ without regenerator. This is essentially a repetition of the earlier experiment.

Part II Measurement of the rate $K^0 \rightarrow \pi^+ + \pi^-$ with a dense regenerator (Be), thick enough so that the regeneration amplitude is much greater than η_{+-}, but thin compared to the length of the sensitive region. In this part of the experiment the absolute value of the regeneration amplitude of Be found by means of the expression (see Appendix):

$$R = \frac{I(K_S)}{I(K_L)} = \left| \frac{\pi N (\bar{f}(o) - f(o)) (1 - e^{-(i\Delta m + \frac{1}{2}\Gamma_S)\tau})}{(\Delta m - \frac{1}{2} i \Gamma_S) M_K} \right|^2$$

$$\approx \frac{4\pi^2}{M_K^{\,2}} \; | f(o) - \bar{f}(o) |^2 \; N^2 \tau^2 \;\; \text{(for small } \tau) \qquad (55)$$

where τ is related to the thickness x of the regenerator: $\tau = x m_K / p_K$.

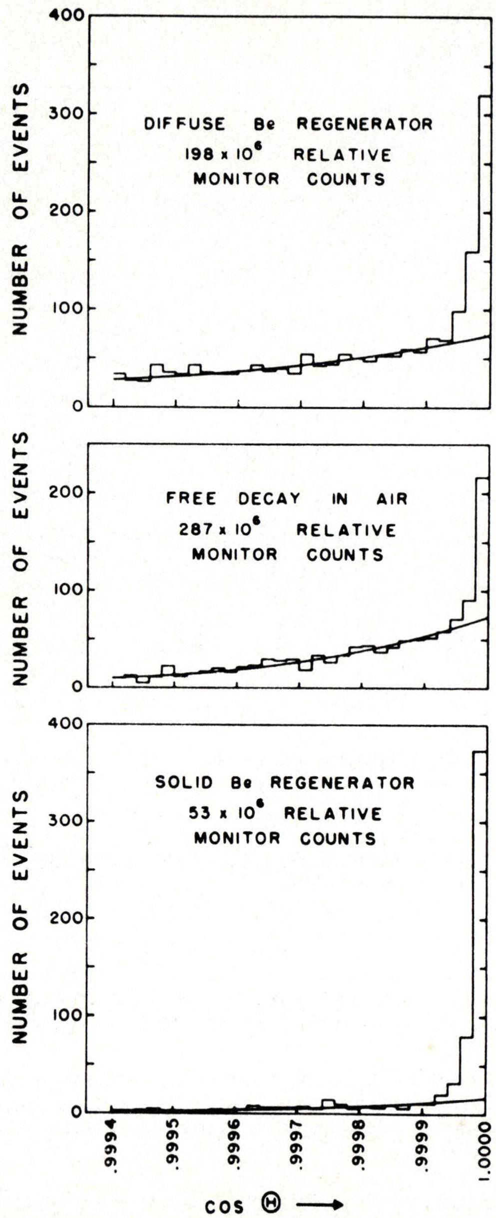

Figure 4 Results for the three parts of the experiment of Fitch et al.[27] in the form of angular distributions of the events in the mass region $485 < m < 505$.

Part III The sensitive region is now entirely filled with a sandwich of $\frac{1}{2}$ mm Be sheets spaced 1 cm apart. The average density is chosen on the basis of Part II such that the regeneration amplitude is comparable to η_{+-} in the regenerator. The intensity within the converter is then given by (54).

It should be noted that because the geometry of Part II differs from that of Part III, the results are not independent of uncertainty in the mass difference.

The experimental results of the three parts are shown in Fig. 4 in the form of angular distribution. The forward peaks are attributed to $K \rightarrow \pi^+ + \pi^-$, and the resultant 3 rates yield the mass dependent interference term of Fig. 5 on the basis of (53), (54) and (55).

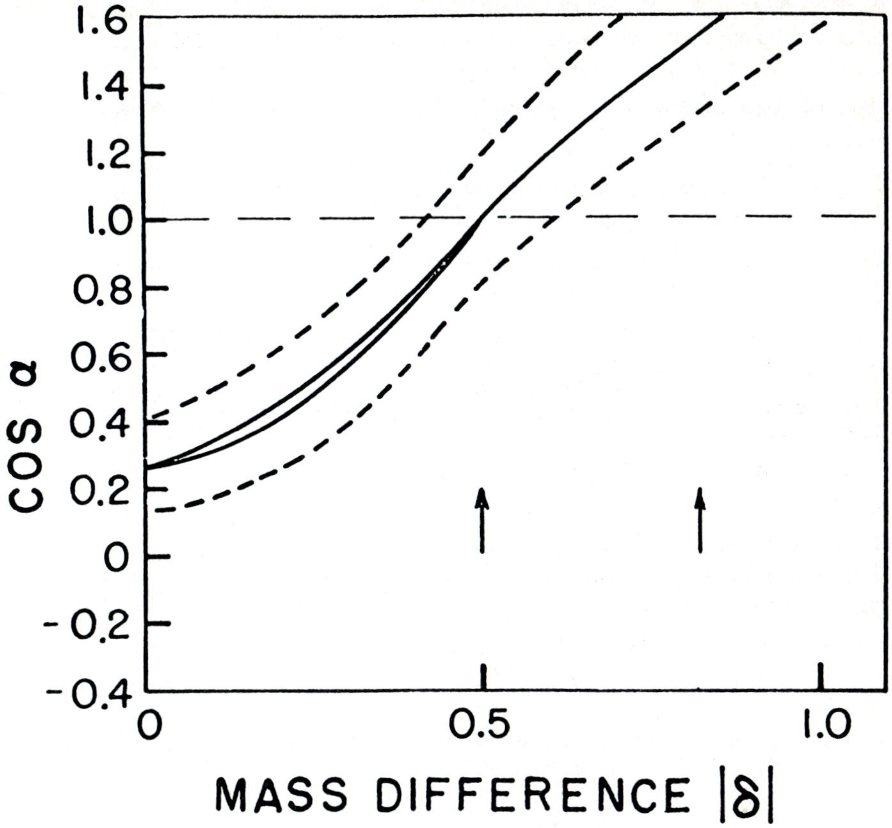

Figure 5 Cos($\varphi_{\rho_0} - \varphi_{\eta_{+-}}$) as a function of assumed mass difference. Experiment of Fitch et al.[27]; $\alpha = \varphi_{\rho_0} - \varphi_{\eta_{+-}}$.

At present there is still substantial experimental uncertainty in Δm (recent measurements range from $|\Delta m| = .15$ to $|\Delta m| = .8$). For $\Delta m = .5$, the observed intensity is twice that which is expected without interference, so that the result corresponds to constructive intereference:

$$\varphi_{\rho_0} \approx \varphi_{\eta_{+-}}$$

No measurements exist on φ_{ζ_0}. However[27], arguments which are not entirely compelling can be presented to the point that f(0) and $\bar{f}(0)$ are essentially imaginary. If this is granted then

$$\varphi_{\rho_0} \approx \tan^{-1} \frac{2\Delta m}{\Gamma_S}$$

and the experimental result can be interpreted in terms of the phase of η_{+-} :

$$\varphi_{\eta_{+-}} \approx \tan^{-1} \frac{2\Delta m}{\Gamma_S}$$

It is interesting to notice that this is the result expected in the simplest models of CP violation, namely all those models in which there is no especially strong violation in any of the K^0 decay channels, especially in the I = 2 channel. In that case the phase of η_{+-} is just $\tan^{-1}(2\Delta m/\Gamma_S)$, for in (43), (42) ϵ' should be small.

1. M. Gell-Mann and A. Pais, Phys. Rev. **97**, 1387 (1955).

2. R.G. Sachs and S.B. Treiman, Phys. Rev. **103**, 1545 (1956).

3. T.D. Lee, R. Oehme and C.N. Yang, Phys. Rev. **105**, 1671 (1957).

4. M.L. Good, Phys. Rev. **106**, 591 (1957).

5. S. Weinberg, Phys. Rev. **110**, 782 (1958).

6. R.G. Sachs and S.B. Treiman, Phys. Rev. Letters 8, 137 (1962).

7. R.G. Sachs, Ann. Phys. 22, 239 (1963).

8. M.C. Whatley, Phys. Rev. Letters 9, 317 (1962).

9. T.T. Wu and C.N. Yang, Phys. Rev. Letters **13**, 501 (1964).

10. L. Wolfenstein, CERN preprint TH. 583, July 13 (1965).

11. T.D. Lee and L. Wolfenstein, Phys. Rev. **138**, B1490 (1965).

12. G. Lüders and B. Zumino, Phys. Rev. **106**, 385 (1957).

13. S. Okubo, Phys. Rev. **109**, 984 (1958).

14. S.M. Berman and M.J.G. Veltman, Phys. Letters **12**, 275 (1964).

15. S.W. MacDowell, Ann. Phys. 18, 171 (1962).

16. R.R. Lewis and C.P. Enz, Phys. Letters **16**, 73 (1965); R.R. Lewis, Phys. Rev. Letters **14**, 749 (1965).

17. J.M. Kaplan, Phys. Rev. **139**, B1065 (1965).

18. M. Kenny and R.G. Sachs, Phys. Rev. (to appear).

19. E. Okonov et al., J.I.N.R. P2083 (1965).

20. P.K. Kabir and R.R. Lewis, Phys. Rev. Letters **15**, 306 (1965), and Rutherford Laboratory preprint RPP/A/5 (1965).

21. S.L. Glashow and S. Weinberg, Phys. Rev. Letters **14**, 835 (1965).

22. J.H. Christenson, J.W. Cronin, V.L. Fitch and R. Turlay, Phys. Rev. Letters **13**, 138 (1964).

23. A. Abashian, R.J. Abrams, D.W. Carpenter, G.P. Fisher, B.M.K. Nefkens and J.H. Smith, Phys. Rev. Letters **13**, 243 (1964).

24. W. Galbraith, G. Manning, A.E. Taylor, B.D. Jones, J. Malos, A. Astbury, N.H. Lipman and G.T. Walker, Phys. Rev. Letters **14**, 383 (1965).

25. X. de Bouard, D. Dekkers, B. Jordan, R. Mermod, T.R. Willitts, K. Winter, P. Scharff, L. Valentin, M. Vivargent and M. Bott-Bodenhausen, Phys. Letters **15**, 58 (1965).

26. N. Cabibbo, Lectures at the International School of Physics, Erice, (1964), p. 285, Academic Press.

27. V.L. Fitch, R.F. Roth, J.S. Russ and W. Vernon, Phys. Rev. Letters **15**, 73 (1965).

28. The τ^+ slope is from values quoted by Taylor et al. (to appear in Phys. Rev.) for 3587 events, and by Ferro-Luzzi et al. [Nuovo Cimento 22, 1087 (1961)] for 899 events. The τ' slope is that given by Kalmus et al. [Phys. Rev. Letters **13**, 99 (1964)]. The K^L slope (S^L_{0+-}) is from the mean value $\alpha = -8.1 \pm .7$ quoted by Chuvilo (Dubna Conference) and the recent measurements $\alpha = -7.6 \pm 1.7$ [Adair et al., Phys. Letters **12**, 67 (1964)], $\alpha = -7.2 \pm .7$ [Astbury et al., Phys. Letters 18, 175 (1965)] and $\alpha = -7.5 \pm 1.1$ (Aubert et al., Paris-Milano-Padua, at this conference), where

$$S = (\alpha\, Q/3M_K)(1 + \alpha\, Q/3M_K)^{-1}$$

29 The phase space numbers 1, 1.246, 1.226, 1.494 are M_K^{-1} x the relative areas of the allowed regions in the various Dalitz plots $K^+ \to \pi^- \pi^+ \pi^+$, $K^+ \to \pi^+ \pi^0 \pi^0$, $K^L \to \pi^0 \pi^+ \pi^-$, $K^L \to \pi^0 \pi^0 \pi^0$. We are indebted to M.J.G. Veltman for these numbers, and for informing us that numbers previously quoted in the literature are not quite correct.

30. The first and second ratios are from the latest (August 1965) compilation of Rosenfeld et al., and also Γ_{+00} in the last ratio. Γ^L_{0+-} has been taken from recent experiments of Astbury et al. [Phys. Letters 18, 175 (1965)], $(2.51 \pm 0.38)\ 10^6\ sec^{-1}$, and of Aubert et al. (Paris-Milan-Padua) as given at this conference, $(2.57 \pm 0.21)\ 10^6\ sec^{-1}$.

31. N. Cabibbo, Proceedings of Erice School (1964), p. 284.

32. M.K. Gaillard, Nuovo Cimento **35**, 1225 (1965).

33. J.A. Anderson et al., Phys. Rev. Letters **14**, 475 (1965).

34. The τ^{+} slope is that described above[28]. The τ^{-} slope is given by L.T. Smith et al., on the basis of 2306 events [Phys. Letters **2**, 204 (1962)].

35. G.E. Kalmus, A. Kernan, R.T. Pu and W.M. Powell, Phys. Rev. Letters **13**, 35 (1964).

36. N. Cabibbo and A. Maksymowicz, Phys. Letters **9**, 352 (1964).

37. U. Camerini, R. Hantman, R.H. March, D. Murphree, G. Gidal, G.E. Kalmus, W.M. Powell, R.T. Pu and C.L. Sandler, Phys. Rev. Letters **14**, 989 (1965).

38. The use of the same formulae for muons depends on supposing the second form factor f_{-} to be negligible.

39. B. Aubert, L. Behr, F.L. Canavan, L.M. Chounet, J.P. Lowys, P. Mittner and C. Pascaud, Phys. Letters **27**, 59 (1965).

40. M. Baldo-Ceolin, E. Calimani, S. Giampolillo, C. Filippi-Filosofo, H. Huzita, F.Mattioli and G. Miari, Nuovo Cimento **38**, 684 (1965).

41. P. Franzini, L. Kirsch, P. Schmidt and J. Steinberger, Phys. Rev. (in press).

42. H. Courant et al., Phys. Rev. **136**, B1791 (1964).

43. U. Nauenberg, P. Schmidt, J. Steinberger, S. Marateck, R.J. Plano, N. Blumenfeld and L. Seidlitz, Phys. Rev. Letters **12**, 679 (1964).

44. C.T. Murphy, Phys. Rev. **134**, B188 (1964).

45. R.W. Birge et al., UCRL preprint 16083, submitted to Phys. Rev.

46. B. Laurent and M. Roos, Phys. Letters **13**, 269 (1964), **15**, 104 (1965) and CERN preprint TH. 580, June (1965).

47. A preliminary comparison of two-pion decay rates of K^{+} and K^{-} has been reported [A. Callahan and D. Cline, Phys. Rev. Letters **15**, 129 (1965)] and seems to show no significant discrepancy.

48. R.G. Sachs, Phys. Rev. Letters **13**, 286 (1964) and references therein.

On the Phenomenological Description of *CP* Violation for *K*-mesons and its Consequences

by **C. P. Enz** and **R. R. Lewis** *)**)

Institut de Physique Théorique, Université de Genève

(16. IX. 1965)

Résumé. Récemment une violation de CP a été mise en évidence par la découverte du mode de décomposition $\pi^+ + \pi^-$ du méson K à longue vie. Ce cas de violation de CP dans le système K^0, $\overline{K^0}$ est décrit phénoménologiquement par deux paramètres complexes, l'un mesurant la non-conservation de T, l'autre la non-conservation de CPT dans la matrice d'énergie. Le but de ce travail est de discuter en détail de ces paramètres et d'examiner les possibilités expérimentales de les déterminer. Des propositions d'expériences sont faites dans les domaines suivants: (i) interactions fortes (mesure de la phase de l'ocillation dans l'intensité), (ii) désintégrations (comparaison de l'oscillation dans le taux de désintégration pour un K^0 initial et un $\overline{K^0}$ initial), (iii) régénération cohérente dans la matière (mesure du taux de désintégration en fonction de la densité), iv) production de paires $K\overline{K}$ (détection de décompositions ou de réactions provenant de paires initiales $K\overline{K}$ à un moment cinétique pair, produites, par exemple, par des résonances).

Introduction

The recent discovery[1]) of the existence of a long-lived $\pi^+\pi^-$-decay mode with the mass and lifetime of the K_2^0, has seriously challenged the previous belief in exact CP-invariance. Since the original formulation of the theory of neutral kaons[2]) was based on this invariance, it has been necessary to reexamine the status of symmetry breaking interactions, especially for the K^0, $\overline{K^0}$ system[3]). Assuming that there are only two particles involved[4]), the phenomenological description of their behavior rests entirely on the energy matrix; the parameters necessary to specify this matrix constitute the observable properties of the system. It is the main purpose of this paper to present a careful discussion of these parameters, and to examine the experimental possibilities of determining them. Some of our proposals have been briefly presented elsewhere[5]).

If CP-invariance is assumed, there are only three observable quantities to determine: the widths and the mass difference of the two CP eigenstates, K_1^0 and K_2^0. If this symmetry is broken in such a way that TCP-invariance is preserved, as is the case in most field theoretic models[6]), then the energy matrix contains one additional complex parameter (r) which determines the admixture of K^0 and $\overline{K^0}$ in the 'mass eigenstates'. At this phenomenological level, it is also possible to describe a violation of TCP-invariance, whether by an intrinsic interaction or by a static external field[7]).

---

*) Permanent address: University of Michigan, Ann Arbor, Michigan, USA.

**) Supported in part by the Swiss National Fund.

Another independent complex parameter (ϱ) is thereby introduced. Determination of these parameters is therefore equivalent to establishing the validity of T, CP, and TCP for this system. This is not necessarily the same goal as determining the specific dynamical origin of the symmetry-breaking, since these parameters are defined solely by the total energy matrix, rather than by the decay amplitudes for any particular channel[8]). However, the determination of r and ϱ is an important step in discovering the origin of the CP-violation. One of the contributions of this work is a general discussion of the limitations imposed on the measurability of the energy matrix, and of the physical interpretation of the 'mass eigenstates'.

The main results presented are several proposals for experiments which could determine the parameters r and ϱ. We show that by using strong interactions to produce and detect kaons of definite strangeness (see *strong interactions of K-mesons*) one can measure the phase of the small oscillatory term, either in the total intensity (K^0 *and* $\overline{K}^0$) of an initially pure beam (K^0 *or* $\overline{K}^0$), or in the partial intensity (K^0 *or* $\overline{K}^0$) of an initially even mixture (K^0 *and* $\overline{K}^0$). The phase difference between the cases of TCP-conservation and of T-conservation is π, and the oscillations determine Re $(1 - r)$ in the first case and Im $(1 - \varrho)$ in the second. Similar expressions are derived for the decay rates for single channels, in particular for 2π-decay (see *decays of K-mesons*). Here a comparison of the oscillations in the decay rates for initial K^0 and for initial $\overline{K}^0$ may also yield a test of TCP-versus T-conservation. In the case of TCP-conservation the oscillations have exactly opposite phase but slightly different amplitude, in the case of T-conservation they have exactly equal amplitude but a phase difference which is slightly different from π. In the case of coherent regeneration in matter (see *coherent effects in media*) we show that by measuring the decay rate for a particular channel as a function of density the phase of the decay amplitude may be determined. Finally we examine the possibility of obtaining information on r and ϱ from an investigation of $K\overline{K}$ states (see *kaon pair states*). We find that useful information with respect to TCP may be obtained from initial pair states of *even* angular momentum. One possibility here might be the detection of decay or reaction products of pairs from $K\overline{K}$ decays of resonances of even spin.

Form of the Energy Matrix

The propagation of a single free kaon can be described[2]) by the introduction of the two by two energy matrix of a meson at rest, $\boldsymbol{\Lambda}$, with the equation of motion

$$i \frac{\partial}{\partial t} \psi(t) = \boldsymbol{\Lambda}\, \psi(t)\,, \tag{1}$$

where t is the proper time*). In the absence of weak interactions, the conservation of strangeness and invariance under TCP (or CP or T) implies that the two particles

*) An alternative to this single particle formalism is the field theoretic approach, in which the K-meson propagator plays the central role rather than the energy matrix; see R. G. Sachs, Reference 15. We consider the use of single particle states more familiar and elementary. But we also recognize that some nettlesome questions are more readily understood using a formalism in which the kaons are only virtual. Example: How can a long-lived K_2^0 with well defined momentum, scatter from a crystal with zero momentum transfer and emerge as a K_1^0 with identical momentum but different mass? (coherent regeneration).

are stable and degenerate and that they have definite strangeness $S = \pm 1$. These states $| K^0 \rangle$ and $| \overline{K}^0 \rangle$ provide a standard representation for the subsequent calculations. The effect of the weak interactions, which may break each of these symmetries, is to destroy the degeneracy and the stability of these states. This can be described by the introduction of both a (hermitian) mass matrix $\boldsymbol{M}$ and a (hermitian) damping matrix $\boldsymbol{\Gamma}$, which are the hermitian and anti-hermitian parts of the energy matrix, $\boldsymbol{\Lambda} \equiv \boldsymbol{M} - i/2\,\boldsymbol{\Gamma}$. A suitable formalism for studying the form of $\boldsymbol{\Lambda}$ is provided by the non-relativistic perturbation theoretic result of WEISSKOPF-WIGNER[9]).

$$\Lambda_{SS'} = M\,\delta_{SS'} + \langle S \,|\, \mathcal{H}_W \,|\, S' \rangle + \sum_n \frac{\langle S \,|\, \mathcal{H}_W \,|\, n \rangle \langle n \,|\, \mathcal{H}_W \,|\, S' \rangle}{M - E_n + i\,\varepsilon} + \cdots$$

$$= \langle S \,|\, M + \mathcal{H}_W + \mathcal{H}_W \left[P \frac{1}{M - \mathcal{H}_S} - i\,\pi\,\delta(M - \mathcal{H}_S) \right] \mathcal{H}_W + \cdots \,|\, S' \rangle, \qquad (2)$$

where $\mathcal{H}_W$ and $\mathcal{H}_S$ are the hamiltonians of the weak and strong interactions. The intermediate states $| n \rangle$ are understood to be eigenstates of the strong interactions, thus including the effect of final state interactions. If there is an external field present, it is to be included in $\mathcal{H}_W$.

It will be important for the subsequent discussion that the choice of the many arbitrary phase factors be made unambiguously (and intelligently!). We will choose the phases of the states so that the representation is *real*,

$$T \,|\, K^0 \rangle = |\, K^0 \rangle, \quad T \,|\, \overline{K}^0 \rangle = |\, \overline{K}^0 \rangle, \qquad (3a)$$

which satisfies $T^2 = 1$. This does *not* fix the arbitrary phase in the definition of T for the kaon *field*. In addition we will define CP so that

$$CP \,|\, K^0 \rangle = |\, \overline{K}^0 \rangle, \quad CP \,|\, \overline{K}^0 \rangle = |\, K^0 \rangle, \qquad (3b)$$

satisfying $(CP)^2 = 1$ and $[CP, T] = 0$. For the kaon *field* this commutativity restricts the phases of T and CP in such a way that their product is ± 1. If we represent $| K^0 \rangle$ by $\left| {1 \atop 0} \right|$ and $| \overline{K}^0 \rangle$ by $\left| {0 \atop 1} \right|$, then we must represent the strangeness operator by σ_3, CP by σ_1, and T by complex conjugation. In this standard representation, the form of $\boldsymbol{\Lambda}$ can easily be discovered once the symmetries of $\mathcal{H}_W$ and $\mathcal{H}_S$ are given. Of course, other choices of phases of the representation can be made[8]); but we must not lose sight of the fact that this implies other phases of the off-diagonal matrix elements of $\boldsymbol{\Lambda}$. The specific advantage of choosing a real representation is that an operator which commutes with T has *real* matrix elements, while an operator which anticommutes with T has *pure imaginary* matrix elements.

If CP is strictly conserved, then $\boldsymbol{\Lambda}$ must have the form

$$\boldsymbol{\Lambda}^{(0)} = \begin{vmatrix} \Lambda^{(0)}_{1,1} & \Lambda^{(0)}_{1,-1} \\ \Lambda^{(0)}_{1,-1} & \Lambda^{(0)}_{1,1} \end{vmatrix} \qquad (4)$$

with four real parameters, which define the two complex eigenvalues

$$\lambda_1 = \Lambda^{(0)}_{1,1} + \Lambda^{(0)}_{1,-1} = m_1 - i\,\gamma_1/2\,, \quad \lambda_2 = \Lambda^{(0)}_{1,1} - \Lambda^{(0)}_{1,-1} = m_2 - i\,\gamma_2/2\,. \qquad (5)$$

The corresponding CP eigenstates

$$| K_1^0 \rangle = (| K^0 \rangle + | \overline{K}^0 \rangle)/\sqrt{2}, \quad | K_2^0 \rangle = (|K^0 \rangle - | \overline{K}^0 \rangle)/\sqrt{2} \tag{6}$$

are independent of the energy matrix elements. Since $\boldsymbol{M}^{(0)}$ and $\boldsymbol{\Gamma}^{(0)}$ commute, the transformation of $\boldsymbol{\Lambda}^{(0)}$ to diagonal form can be made by a *unitary* matrix

$$\boldsymbol{U} = \frac{1}{\sqrt{2}} \begin{vmatrix} 1 & 1 \\ 1 & -1 \end{vmatrix} . \tag{7}$$

The experimental determination of the parameters $m_1 - m_2, \gamma_1$, and γ_2 has required nearly a decade of effort; it is well established[10]) that the K_1^0 has a larger width ($\gamma_1 = 1{,}09 \pm 0{,}02 \cdot 10^{10}$ sec^{-1}) than the K_2^0 ($\gamma_2 = 1{,}78 \pm 0{,}21 \cdot 10^7$ sec^{-1}), and that the mass difference is comparable to γ_1. The precise value of $| m_1 - m_2 |$ is still subject to some discrepancy[11]) between different determinations, but appears to settling down to a value[12]) $| m_1 - m_2 | = (0{,}5 \pm 0{,}1) \cdot \gamma_1$; the sign is reported to be negative ($m_2 > m_1$)[13]). We can conclude, subject to small corrections due to symmetry-breaking contributions, that $\Gamma_{1,1}$, $\Gamma_{1,-1}$, and $-M_{1,-1}$ are all positive and of comparable magnitude.

Since CP-invariance implied symmetry of $\boldsymbol{\Lambda}$ both along and across the diagonal ($\boldsymbol{\Lambda}^{(0)} = \Lambda_{1,1}^{(0)} \cdot 1 + \Lambda_{1,-1}^{(0)} \cdot \sigma_1$), symmetry-breaking interactions will add terms either antisymmetric along the diagonal ($\sim \sigma_3$) or across the diagonal ($\sim \sigma_2$). The first of these violates TCP-invariance and preserves T-invariance; the second violates T and preserves TCP. In each case, two additional real parameters are added to $\boldsymbol{\Lambda}$. By straight-forward solution of the secular equations for the general case

$$\boldsymbol{\Lambda} = \begin{vmatrix} \Lambda_{1,1} & \Lambda_{1,-1} \\ \Lambda_{-1,1} & \Lambda_{-1,-1} \end{vmatrix} \tag{8}$$

we find for the eigenvalues and eigenvectors

$$\begin{aligned} \lambda_S &= \Lambda_{1,1} + r\varrho\Lambda_{1,-1}, \quad & | K_S \rangle &= (| K^0 \rangle + r\varrho | \overline{K}^0 \rangle)/\sqrt{2} \\ \lambda_L &= \Lambda_{1,1} - \frac{r}{\varrho}\Lambda_{1,-1}, \quad & | K_L \rangle &= (\varrho | K^0 \rangle - r | \overline{K}^0 \rangle)/\sqrt{2}\,. \end{aligned} \tag{9}$$

where we have defined

$$r = \sqrt{\Lambda_{-1,1}/\Lambda_{1,-1}}, \quad \varrho = \eta + \sqrt{1 + \eta^2}, \quad \eta = \frac{\Lambda_{-1,-1} - \Lambda_{1,1}}{2\sqrt{\Lambda_{1,-1}\Lambda_{-1,1}}} . \tag{10}$$

If we write these eigenvectors as

$$| K_S \rangle = \boldsymbol{V} | K^0 \rangle, \quad | K_L \rangle = \boldsymbol{V} | \overline{K}^0 \rangle \tag{11}$$

then

$$\boldsymbol{V}(r, \varrho) = \frac{1}{\sqrt{2}} \begin{vmatrix} 1 & \varrho \\ r\varrho & -r \end{vmatrix} \tag{12}$$

is *not* a unitary matrix. We note that neither $\boldsymbol{M}$ nor $\boldsymbol{\Gamma}$ are diagonalized by $\boldsymbol{V}$, but only the combination $\boldsymbol{\Lambda} = \boldsymbol{M} - i/2\,\boldsymbol{\Gamma}$. The complex parameter r measures the strength of the T-violating part of Λ, and ϱ measures the strength of the TCP-violating part. The square roots are defined so that r, ϱ are close to $+1$ for small CP-violations. It follows that the state $| K_S \rangle$ will have a smaller mass and larger width than $| K_L \rangle$; therefore for S and L one should read short- and longlived. We continue to use $| K_1^0 \rangle$ and $| K_2^0 \rangle$ to denote the CP-eigenstates.

It will be useful to have the inversion of the relations between λ_S, λ_L, r, ϱ and the energy matrix elements, at least for small CP-violations. If we write Λ as

$$\Lambda = \begin{vmatrix} \Lambda_{1,1}^{(0)} + \delta\Lambda_{1,1} & \Lambda_{1,-1}^{(0)} + \delta\Lambda_{1,-1} \\ \Lambda_{1,-1}^{(0)} - \delta\Lambda_{1,-1} & \Lambda_{1,1}^{(0)} - \delta\Lambda_{1,1} \end{vmatrix} \tag{13}$$

then to first order in $\delta\Lambda$ we obtain

$$\begin{aligned} \lambda_S &\cong \Lambda_{1,1}^{(0)} + \Lambda_{1,-1}^{(0)} = \lambda_1 \,, \quad r \cong 1 - \delta\,\Lambda_{1,-1}/\Lambda_{1,-1}^{(0)} \\ \lambda_L &\cong \Lambda_{1,1}^{(0)} - \Lambda_{1,-1}^{(0)} = \lambda_2 \,, \quad \varrho \cong 1 - \delta\,\Lambda_{1,1}/\Lambda_{1,-1}^{(0)} \,, \end{aligned} \tag{14}$$

or inverting,

$$\delta\Lambda_{1,-1} \cong \frac{1}{2}\,(1 - r)\,(\lambda_S - \lambda_L) \,, \quad \delta\Lambda_{1,1} \cong \frac{1}{2}\,(1 - \varrho)\,(\lambda_S - \lambda_L) \,. \tag{15}$$

Notice that, since the degeneracy is already removed by the symmetry-preserving terms, there is no first order change in the eigenvalues due to the symmetry-breaking terms.

The energy matrix elements, and therefore λ_S, λ_L, ϱ, r, are not completely independent, but are restricted by the conditions that the eigenvalues of $\boldsymbol{M}$ and $\boldsymbol{\Gamma}$ be positive, as follows from Equation (2)*). This is known[14]) to place rather severe restrictions on the off-diagonal elements of $V^+ V$, and therefore on the strength of the symmetry-breaking terms. With large symmetry-breaking terms one cannot obtain the observed ratio of the two widths, $\gamma_1/\gamma_2 \simeq 610$. The condition det $\boldsymbol{\Gamma} \geqslant 0$ leads to the inequality

$$\frac{|\,\varrho - \varrho^* \,|r|^2\,|^2}{(1 + |\varrho\, r|^2)\,(|\varrho|^2 + |r|^2)} \leqslant \frac{4\,\mathrm{Im}\,\lambda_S\,\mathrm{Im}\,\lambda_L}{|\lambda_S - \lambda_L^*|^2} \tag{16}$$

which provides a quantitative estimate of how large the deviation of r and ϱ from unity can become**). We find, for small $(1 - r)$ and $(1 - \varrho)$,

$$[\mathrm{Re}\,(1 - r)]^2 + [\mathrm{Im}\,(1 - \varrho)]^2 < 4\,\frac{\gamma_2}{\gamma_1} \cong 6{,}5 \cdot 10^{-3} \,. \tag{17}$$

Before proceeding to the discussion of experiments, we will consider some questions of physical interpretation and measurability. Although we could continue to work exclusively with the states K^0 and $\overline{K}{}^0$, and directly solve the original equation of motion for a given energy matrix, it is very useful instead to introduce the "mass

*) $\sum_{SS'} \Gamma_{SS'}\, x_S^*\, x_{S'} = 2\pi \cdot \langle \psi \,|\, \mathcal{H}_W\, \delta(M - \mathcal{H}_S)\, \mathcal{H}_W \,|\, \psi \rangle \geqslant 0$, where $|\psi\rangle \equiv \sum_S |\,S\rangle\, x_S$ and x_S is arbitrary.

**) There are other limits set by existing experiments, such as the fact that, to within the accuracy of the branching ratios, the fraction of kaons which decay after $\gamma_1 t \sim 5$ is 1/2. This implies $(|\varrho|^4 + |\varrho|^2)/|1 + \varrho^2|^2 \sim 1/2$ to within about 20%, setting limits on the real part of ϱ as well.

eigenstates' and 'mass eigenvalues'. But since the transformation to these states with $\boldsymbol{V}$ is *not* unitary, there are important differences in the physical interpretation of the eigensolution. For example, we note that the real and imaginary parts of the 'mass eigenvalues' are *not* the eigenvalues of $\boldsymbol{M}$ and $\boldsymbol{\Gamma}$. It is Re $(\lambda_S - \lambda_L)$ which is measured in the experiments designed to determine the 'mass difference'. The situation is not unique to K-meson physics, but presumably will occur whenever the mass and damping matrices of an unstable system fail to commute. For the neutral kaons, this pertains to the case in which CP-invariance is broken.

In dealing with the 'mass eigenstates' $| K_S \rangle$ and $| K_L \rangle$, it is useful to define some kind of adjoint states $\langle \tilde{K}_S |$ and $\langle \tilde{K}_L |$, and an inner product. If we use the ordinary hermitian adjoint, then it follows directly from the non-unitary character of $\boldsymbol{V}$ that the states are not orthogonal, $\langle K_S | K_L \rangle \neq 0$. This clearly makes it impossible to interpret $\langle K_S | \psi \rangle$ as the probability amplitude of obtaining the short-lived state in the state $| \psi \rangle$, since there would be a non-vanishing probability amplitude even for $| \psi \rangle = | K_L \rangle$. The customary completeness and orthonormality relations can be guaranteed by introducing[15]) instead the 'inverse states', defined by

$$\langle \tilde{K}_S | \equiv \langle K^0 | \boldsymbol{V}^{-1}, \quad \langle \tilde{K}_L | \equiv \langle \overline{K}^0 | \boldsymbol{V}^{-1}, \tag{18}$$

and given explicitly by

$$\langle \tilde{K}_S | = (\langle K^0 | r + \langle \overline{K}^0 | \varrho) \frac{\sqrt{2}}{r(1+\varrho^2)}, \quad \langle \tilde{K}_L | = (\langle K^0 | \varrho r - \langle \overline{K}^0 |) \frac{\sqrt{2}}{r(1+\varrho^2)}. \tag{19}$$

The major difference from the usual situation in quantum mechanics, is that the orthonormality conditions

$$\langle \tilde{K}_S | K_S \rangle = \langle \tilde{K}_L | K_L \rangle = 1, \quad \langle \tilde{K}_S | K_L \rangle = \langle \tilde{K}_L | K_S \rangle = 0 \tag{20}$$

leave undetermined not only the *phase* of the states, but also the *scale*. If we change basis by

$$| K_S \rangle \to c_S | K_S \rangle, \quad | K_L \rangle \to c_L | K_L \rangle \tag{21a}$$

then the 'inverse states' change according to

$$\langle \tilde{K}_S | \to \langle \tilde{K}_S | c_S^{-1}, \quad \langle \tilde{K}_L | \to \langle \tilde{K}_L | c_L^{-1} \tag{21b}$$

(not c^* but c^{-1}!), and the orthonormality is preserved for *any* value of c_S, c_L. As a result, a transition amplitude $\langle \tilde{K}_S | T | K_L \rangle$ is undefined to a complex factor (c_L/c_S), rather than simply to a phase factor; we can not consider $|\langle \tilde{K}_S | T | K_L \rangle|^2$ an observable[15]).

The resolution of these difficulties lies in the recognition that the 'mass eigenstates' have only an intermediate role, and cannot be prepared or observed (i.e., used as 'in' and 'out' states). If we insist that they appear only in the expansion of another initial state, thus occuring only in the combinations $| K_S \rangle \langle \tilde{K}_S |$ and $| K_L \rangle \langle \tilde{K}_L |$, then the normalization constants c_S and c_L do not appear and can be chosen arbitrarily. This criterion is implicit in the use of eigenstates of the K-meson propagator in the field theoretic treatment of Sachs.

To illustrate this point, we will give an example: the rate of $\pi^+\pi^-$-decay of a long-lived kaon beam. For an initial K^0, making use of the identity

$$e^{-i\Lambda t} = | K_S > e^{-i\lambda_S t} < \tilde{K}_S | + | K_L > e^{-i\lambda_L t} < \tilde{K}_L | \tag{22}$$

and dropping the short-lived component, we have

$$R_{(K_L \to \pi^+\pi^-)} = c \, |< 2\pi | \mathcal{H}_W | K_L > e^{-i\lambda_L t} < \tilde{K}_L | K^0 >|^2. \tag{23}$$

For an initial $\overline{K}^0$, the result is different

$$\overline{R}_{(K_L \to \pi^+\pi^-)} = c \, |< 2\pi | \mathcal{H}_W | K_L > e^{-i\lambda_L t} < \tilde{K}_L | \overline{K}^0 >|^2. \tag{24}$$

We do not consider it possible to define the rate of decay $|< 2\pi | \mathcal{H}_W | K_L >|^2$ unambiguously, irrespective of its initial state; the extra factors $< \tilde{K}_L | K^0 >$ and $< \tilde{K}_L | \overline{K}^0 >$ must be included to make the expression well-defined. In practice, these extra factors usually have a trivial effect for small CP violations, if we choose the normalization of the 'mass eigenstates' so that $| K_S >$ and $| K_L >$ differ from $| K_1^0 >$ and $| K_2^0 >$ only in first order of the symmetry-breaking terms. We have made this choice in Equation 9.

Next we consider the measurability of the complex parameters r and ϱ. In order to determine them, we generally must have another interaction of kaons with known matrix elements which can be used as a comparison. Consider first ϱ, which depends on the difference $\Lambda_{1,1} - \Lambda_{-1,-1}$. We expect that ϱ can be determined by utilizing another interaction which is different for K^0 and $\overline{K}^0$: the strong interactions involved in charge exchange scattering, hyperon production, etc. We shall see explicitly in the next section that such interactions permit the measurement of both the phase and magnitude of ϱ.

The measurement of r, which depends on the off-diagonal matrix elements $\Lambda_{1,-1} - \Lambda_{-1,1}$, is more subtle. Because of the conservation of strangeness, there is no *strong* interaction which couples K^0 to $\overline{K}^0$; this will be shown explicitly to imply that, using strong interactions alone, the magnitude of r can be measured, but not the phase. If weak interactions are used, a further difficulty intrudes: the phase of a weak interaction coupling K^0 to $\overline{K}^0$ is known only if T-invariance is valid. Therefore, if we have some a priori consideration which implies T-invariance of the decay amplitudes for a particular decay channel, then this channel can be used to measure the phase of r. Without such a priori considerations, which are usually associated with special models of the weak interaction, we cannot measure the phase of r.

Finally we comment that all this discussion concerns the problem of determining the energy matrix *in our standard representation*. A change of the relative phase of the K^0 and $\overline{K}^0$ states will change the phase of the off-diagonal elements of $\boldsymbol{\Lambda}$, thereby admixing the mass and damping matrices,

$$\Lambda_{1,-1} \to \Lambda_{1,-1} e^{i\phi} = \left(M_{1,-1} \cos\phi + \frac{1}{2} \Gamma_{1,-1} \sin\phi\right) + i\left(M_{1,-1} \sin\phi - \frac{1}{2} \Gamma_{1,-1} \cos\phi\right). \tag{25}$$

Thus, it is not possible to distinguish the effects of a symmetry-breaking term in $\boldsymbol{M}$ from one in $\boldsymbol{\Gamma}$, except with reference to a particular representation.

Strong Interactions of K-Mesons

The first test of any model of neutral kaons is the study of the subsequent strong interactions of a particle initially known to be either a K^0 or a $\overline{K}^0$ [16]; that is, reactions in which a kaon is both produced and later absorbed in strong interactions. There are clearly four relevant amplitudes, for the propagation of a particle of initial strangeness S_i having, after proper time t, a final strangeness S_f. We define

$$A(S_f, S_i) = \langle S_f \mid e^{-i\Lambda t} \mid S_i \rangle$$
$$= \langle S_f \mid K_S \rangle e^{-i\lambda_S t} \langle \tilde{K}_S \mid S_i \rangle + \langle S_f \mid K_L \rangle e^{-i\lambda_L t} \langle \tilde{K}_L \mid S_i \rangle . \qquad (26)$$

By explicit calculation, we find

$$A(+1, +1) = \frac{1}{1+\varrho^2} (e^{-i\lambda_S t} + \varrho^2 e^{-i\lambda_L t})$$
$$A(+1, -1) = \frac{\varrho}{r(1+\varrho^2)} (e^{-i\lambda_S t} - e^{-i\lambda_L t})$$
$$A(-1, +1) = \frac{\varrho r}{1+\varrho^2} (e^{-i\lambda_S t} - e^{-i\lambda_L t})$$
$$A(-1, -1) = \frac{1}{1+\varrho^2} (\varrho^2 e^{-i\lambda_S t} + e^{-i\lambda_L t}) . \qquad (27)$$

Several deductions can be drawn from these results. First we notice, as previously remarked, that r occurs only as a factor, and that therefore the phase of r cannot be determined in this way. Second, we notice that, in addition to the well-known oscillations in the intensity of the K^0 and $\overline{K}^0$, there are also small oscillations in the total number of mesons present (and therefore in the total rates of decay). For an initial K^0, we find

$$\mathcal{A}(t) = | A(+1, +1) |^2 + | A(-1, +1) |^2$$
$$= \frac{1}{|1+\varrho^2|^2} \Big\{ (1 + |\varrho r|^2) e^{-\gamma_S t} + [(\varrho^2 - |\varrho r|^2) e^{i\Delta m t}$$
$$+ \text{c. c.}] e^{-(\gamma_S + \gamma_L) t/2} + (|\varrho|^4 + |\varrho r|^2) e^{-\lambda_L t} \Big\}, \qquad (28)$$

whereas for an initial $\overline{K}^0$, we find

$$\overline{\mathcal{A}}(t) = | A(+1, -1) |^2 + | A(-1, -1) |^2$$
$$= \frac{1}{|1+\varrho^2|^2} \left\{ \left(\left| \frac{\varrho}{r} \right|^2 + |\varrho|^4 \right) e^{-\gamma_S t} + \left[\left(\varrho^{*2} - \left| \frac{\varrho}{r} \right|^2 \right) e^{i\Delta m t} \right. \right.$$
$$\left. \left. + \text{c. c.} \right] e^{-(\gamma_S + \gamma_L) t/2} + \left(\left| \frac{\varrho}{r} \right|^2 + 1 \right) e^{-\gamma_L t} \right\}, \qquad (29)$$

where $\Delta m = \text{Re}\,(\lambda_S - \lambda_L)$, $\gamma_{S,L} = -2\,\text{Im}\,\lambda_{S,L}$. It has been noted earlier[14] that the occurrence of oscillatory terms in the total intensity is a test of CP-violation. A further conclusion concerns the phase of these small oscillations, which, for either K^0 or $\overline{K}^0$, is

$$[\text{Re}\,(1 - r) \cos(\Delta m t) + \text{Im}\,(1 - \varrho) \sin(\Delta m t)] e^{-(\gamma_S + \gamma_L) t/2}$$

correct to first order in the symmetry-breaking terms. Thus TCP-conservation implies an oscillation $\sim \cos(\Delta m t)$ and T-conservation implies an oscillation $\sim \sin(\Delta m t)$. In view of the continuing breakdown of symmetry in the weak interactions, it seems very important to provide a test of TCP as well as CP and T; the above effect is one possible way. The amplitude is known from Equation 17 to be less than 8%, and is expected from the experiment of CHRISTENSON, CRONIN, FITCH, and TURLAY to be about a factor 10 smaller than this limit. To observe this effect by counting both the K^0 and the $\overline{K}^0$ intensity, one faces the experimental difficulties of obtaining adequate statistical significance as well as determining accurately the relative detection efficiency for K^0 and $\overline{K}^0$. As an alternative, one could attempt to measure the total rate of decay (the time derivative of this curve), by summing all the decay channels. The most sensitive region is then about $10\,\gamma_1^{-1}$, where the rates of decay of the two modes are comparable. In this case, the measurement of the absolute phase of the oscillatory term requires a precise knowledge of the mass difference.

It is important to realize that the same information can be obtained in another way, by observing the strong interactions of a K^0 (or a $\overline{K}^0$) in a beam which is known initially to have equal numbers of K^0 and $\overline{K}^0$ (incoherent mixture). Thus, for measurement of a K^0 after time t, we define

$$\mathcal{P}(t) = |A(+1, +1)|^2 + |A(+1, -1)|^2$$
$$= \frac{1}{|1+\varrho^2|^2} \left\{ \left(1 + \left|\frac{\varrho}{r}\right|^2\right) e^{-\gamma_S t} + \left[\left(\varrho^2 - \left|\frac{\varrho}{r}\right|^2\right) e^{i\Delta m t} + \text{c. c.}\right] \right.$$
$$\left. \times\, e^{-(\gamma_S + \gamma_L)t/2} + \left(|\varrho|^4 + \left|\frac{\varrho}{r}\right|^2\right) e^{-\gamma_L t} \right\}, \tag{30}$$

and for a $\overline{K}^0$,

$$\overline{\mathcal{P}}(t) = |A(-1, +1)|^2 + |A(-1, -1)|^2$$
$$= \frac{1}{|1+\varrho^2|^2} \left\{ (|\varrho\, r|^2 + |\varrho|^4)\, e^{-\gamma_S t} + [(\varrho^{*2} - |\varrho\, r|^2)\, e^{i\Delta m t} \right.$$
$$\left. + \text{c. c.}]\, e^{-(\gamma_S + \gamma_L)t/2} + (1 + |\varrho\, r|^2)\, e^{-\gamma_L t} \right\} \tag{31}$$

which have oscillatory terms

$$[\operatorname{Re}(1 - r) \cos(\Delta m t) - \operatorname{Im}(1 - \varrho) \sin(\Delta m t)]\, e^{-(\gamma_S + \gamma_L)t/2}\,.$$

In this case one need only determine that equal numbers of K^0 and $\overline{K}^0$ production events are counted in the sample, and the actual detection efficiency is not needed, except insofar as it depends on the momentum spread in the beam. It may be feasible to measure these curves, especially at higher energies where the increased laboratory lifetime enhances the relative probability of interaction versus decay.

The close similarity in form of $\mathcal{A}(t)$ and $\mathcal{P}(t)$ is an example of the close analogy to the theory of polarized spin 1/2 beams. The first expression, for $\mathcal{A}(t)$, would correspond to determining the asymmetry of an initially polarized beam. The second expression, for $\mathcal{P}(t)$, would correspond to determining the polarization of an initially unpolarized beam. The close relationship between these two is familiar in the theory of polarization.

We should make two further comments. Although we have stressed the measurement of the oscillatory terms, the hyperbolic terms also depend on the symmetry-breaking terms, containing Re $(1 - r)$ and Re $(1 - \varrho)$. These terms do not allow a test of T versus TCP however. Finally we repeat a comment of SACHS[15]), that there do exist oscillatory rates, such as $|A(+1, -1)|^2$ and $|A(-1, +1)|^2$, whose time dependence is independent of r and ϱ and which can therefore be used to determine the mass difference without corrections due to CP-violation.

Decays of K-Mesons

In the previous sections we have emphasized the measurement of the total energy matrix, whereas the measurement of K-decays normally gives information about the individual channels. To calculate the rate of decay into one particular state, it is necessary to introduce the decay amplitudes for K^0 and $\overline{K}^0$ into that state, and therefore the partial damping matrix. If we denote the amplitudes by

$$D_S^{(j)} = \langle j \mid \mathcal{H}_W \mid S \rangle ,$$

then the jth partial damping matrix is defined by

$$\Gamma_{SS'}^{(j)} = D_S^{(j)*} D_{S'}^{(j)} , \tag{32}$$

where the appropriate kinematical factors have been absorbed into the definition of the amplitudes. The general form of the partial damping matrices is thus

$$\Gamma^{(j)} = \begin{vmatrix} \alpha_j & x_j + i y_j \\ x_j - i y_j & \beta_j \end{vmatrix} \tag{33}$$

with $\alpha \beta = x^2 + y^2$. If the jth channel is CP-self-conjugate, as for the various 2π and 3π channels, then TCP-invariance implies $\alpha_j = \beta_j$. For any channel, if we continue to define the states as real, T-invariance implies $y = 0$*).

Using this general form and the definition (26) we can express the rate of decay of the beam into the jth channel, as

$$R_j(t) \equiv |\langle j \mid \mathcal{H}_W e^{-iAt} \mid +1 \rangle|^2 = |\sum_S D_S^{(j)} A(S, +1)|^2$$

$$= \frac{1}{|1+\varrho^2|^2 \alpha_j} \left| [\alpha_j + \varrho r(x_j + i y_j)] e^{-i\lambda_S t} + \varrho [\alpha_j \varrho - r(x_j + i y_j)] e^{-i\lambda_L t} \right|^2 \tag{34}$$

for a single initial K^0 meson. If instead the initial state was a single $\overline{K}^0$, we obtain

$$\overline{R}_j(t) \equiv |\langle j \mid \mathcal{H}_W e^{-iAt} \mid -1 \rangle|^2 = |\sum_S D_S^{(j)} A(S, -1)|^2$$

$$= \frac{1}{|1+\varrho^2|^2 \alpha_j |r|^2} \left| \varrho [\alpha_j + \varrho r (x_j + i y_j)] e^{-i\lambda_S t} - [\alpha_j \varrho - r (x_j + i y_j)] e^{-i\lambda_L t} \right|^2 . \tag{35}$$

*) In Reference 8, the relative phase of K^0 and $\overline{K}^0$ is chosen to make the dominant decay amplitude (2π, $I = 0$ channel) real, and so in that channel, $y = 0$ by convention. We prefer to use real states, for which the phase of a decay amplitude is determined by whether the interaction preserves T-invariance. The presence of final state interactions will give additional contributions to the phase of D_S.

For example, the rate of decay of the long-lived mode in the 2π-channel is, for a K^0,

$$R_{2\pi}(t) \xrightarrow[\gamma_S t \gg 1]{} \frac{|\varrho|^2 e^{-\gamma_L t}}{\alpha_{2\pi}|1+\varrho^2|^2} \, |\alpha_{2\pi} \varrho - r(x_{2\pi} + i\, y_{2\pi})|^2 \tag{36}$$

or, correct to first order in the symmetry-breaking terms, for either K^0 or $\overline{K}^0$ ($\alpha = x + \delta\alpha$),

$$R_{2\pi}(t) \cong x_{2\pi}\, e^{-\gamma_L t} \left| \frac{\delta\alpha_{2\pi} - i\, y_{2\pi}}{2\, x_{2\pi}} + \frac{\delta\Lambda_{1,-1} - \delta\Lambda_{1,1}}{\lambda_S - \lambda_L} \right|^2 . \tag{37}$$

This general result illustrates the fact that there are, at the phenomenological level, several independent origins possible for the CP violation.

The limitations on the measurability of the phase of r can be illustrated with the above formulas. If, in a particular model, there is a reason to assume CP-invariance for a given channel, then the phase of the oscillatory term in the decay rate in that channel depends on the phase of r. From a completely empirical view, making no assumption about the origin of the CP violation, we cannot determine that phase.

Finally, let us note that, if we compare the rates of decay of beams which are initially known to be either K^0 or $\overline{K}^0$, it is possible to obtain information about the total energy matrix without knowing the partial damping matrix. From Equations (34) and (35) we see that TCP-conservation implies that the oscillatory terms have exactly opposite *phase*, but amplitude ratio $|r|^{-4}$ while T-conservation implies that their *amplitudes* are the same whereas their phase difference is $-\varrho^*/\varrho$. On the other hand, the ratio of the rates for the long-lived component is

$$R_j(t)/\overline{R}_j(t) \xrightarrow[\gamma_S t \gg 1]{} |\varrho\, r|^2 \tag{38}$$

which can be used to determine Re $(2 - r - \varrho)$, but not to test T or TCP. Such experiments require the identification of the production vertex of the kaon. For the measurement of the oscillatory terms in the 2π-channel, the most sensitive region is around $\gamma_1 t \sim 12$, where the rates for the two modes are known to be comparable.

Coherent Effects in Media

The strong interactions to which we referred in the earlier section (i.e., charge exchange, hyperon production) are incoherent reactions, in the sense that an observable change of state of the medium results from the interaction. It was first pointed out by GOOD[17]) that it is also possible to experiment with those mesons which scatter in the medium without changing its state. These particles define a 'coherent wave' propagating through the medium according to the familiar laws of optics. Assuming only that these scattering processes conserve strangeness, we can write the form of the index of refraction for homogeneous matter as

$$n^2 = \mathbf{1} + 4\pi \frac{N}{k^2} \begin{vmatrix} f & 0 \\ 0 & \overline{f} \end{vmatrix} , \tag{39}$$

where N is the number density of scatterers and $f, \overline{f}$ are the forward elastic scattering amplitudes of the K^0 and the $\overline{K}^0$, respectively. The necessary modification of our mass matrix formulation is readily found from the dispersion law for kaons. If the kaon has velocity v, then relativistic covariance implies that the frequency and wave vector are given in terms of the mass by

$$\boldsymbol{\omega} = \gamma \boldsymbol{\Lambda}, \quad \boldsymbol{k} = \gamma v \boldsymbol{\Lambda}, \tag{40}$$

satisfying $\boldsymbol{\omega}^2 - \boldsymbol{k}^2 = \boldsymbol{\Lambda}^2$, with $\gamma = (1 - v^2)^{-1/2}$. The coherent scattering in matter alters the wave vector but not the frequency,

$$\boldsymbol{\omega}' = \boldsymbol{\omega}, \quad \boldsymbol{k}' = n\,\boldsymbol{k} \tag{41}$$

giving

$$\boldsymbol{\Lambda}'^2 = \boldsymbol{\omega}'^2 - \boldsymbol{k}'^2 = \boldsymbol{\Lambda}^2 - 4\pi N \begin{vmatrix} f & 0 \\ 0 & \bar{f} \end{vmatrix}. \tag{42}$$

Since $(n - 1)$ is very small, we retain only terms linear in the density. The final result is simply an additional contribution to the mass matrix

$$\boldsymbol{\Lambda}' = \boldsymbol{\Lambda} + \boldsymbol{\Lambda}_{SC} = \boldsymbol{\Lambda} - \frac{2\pi N}{M} \begin{vmatrix} f & 0 \\ 0 & \bar{f} \end{vmatrix}. \tag{43}$$

The wave equation corresponding to the dispersion law Equation (42) is the two-component Klein-Gordon equation, with mass $\boldsymbol{\Lambda}'$.

Since we have already discussed the eigenvalues and eigenvectors of the general energy matrix, the effect of adding $\boldsymbol{\Lambda}_{SC}$ to $\boldsymbol{\Lambda}$ is simply to make the parameters λ_S', λ_L', r', ϱ' depend on the density. To first order in the density, we find from our linearized connection formulas Equations (13), (14)

$$\lambda_S' = \lambda_S - \pi N (f + \bar{f})/M, \quad \lambda_L' = \lambda_L - \pi N (f + \bar{f})/M, \quad r' = r, \quad \varrho' = \varrho + 2\delta, \tag{44}$$

where $\delta = \pi N (f - \bar{f})/[M(\lambda_S - \lambda_L)]$. The primed quantities all refer to the beam inside matter and the unprimed quantities to the vacuum. The coherent scattering can be considered another source of CP-violation (since matter is not CP-invariant) which preserves T-invariance, and therefore appears to violate TCP-invariance. Since the degeneracy is already removed, the effect of the scattering is a small pertubation on the vacuum eigenstates, governed by the dimensionless parameter δ which is about 10^{-2} for condensed matter[18]). There is also a CP-conserving term $\pi N (f + \bar{f})/M$ in each of the eigenvalues, which decreases the lifetimes of both modes due to scattering out of the beam, but which does not alter the mass difference.

The theory of coherent regeneration of kaons in matter can be based on the calculation of the S-matrix elements according to

$$S_{fi} = \langle f \,|\, e^{-i(\boldsymbol{\Lambda} + \boldsymbol{\Lambda}_{SC})t} \,|\, i \rangle, \tag{45}$$

where t is the proper time elapsed in the matter. This result neglects the specularly reflected wave, whose amplitude is of first order in $(n - 1)$; it is a special case of the 'eikonal approximation' to the forward scattering amplitude of the medium. If we choose $|K_S\rangle$ and $|K_L\rangle$ as 'initial' states, $\langle \tilde{K}_S|$ and $\langle \tilde{K}_L|$ as 'final' states, introducing explicitly the matrix $\boldsymbol{V}(r, \varrho)$ which diagonalizes $\boldsymbol{\Lambda}$, we obtain

$$S_{fi} = \langle \tilde{K}_f \,|\, \boldsymbol{V} e^{-i\boldsymbol{\Lambda}''t} \boldsymbol{V}^{-1} \,|\, K_i \rangle, \tag{46}$$

where

$$\Lambda'' = V^{-1}(\Lambda + \Lambda_{SC})\, V = \begin{vmatrix} \lambda_S & 0 \\ 0 & \lambda_L \end{vmatrix} - \frac{2\pi N}{M(1+\varrho^2)} \begin{vmatrix} f + \varrho^2 \bar{f}, & \varrho(f - \bar{f}) \\ \varrho(f - \bar{f}), & \varrho^2 f + \bar{f} \end{vmatrix} . \qquad (47)$$

We remark that the symmetry of Λ'' is not due to any physical property, but is simply a consequence of our choice of normalization of $| K_S \rangle$ and $| K_L \rangle$.

The above result establishes the very weak dependence of regeneration phenomena on the violation of CP, first noted by GOOD[19]). In the above representation, Λ'' is independent of r, and only weakly dependent on ϱ, through terms of order $(1 - \varrho)^2$ and $\delta\,(1 - \varrho)$. Neglecting these terms, we can then calculate the regenerated amplitudes of $| K_S \rangle$ and $| K_L \rangle$ independently of the symmetry breaking terms. Of course, we must not lose sight of the fact that the 'mass eigenstates' themselves contain r and ϱ. Explicit evaluation of Equation (46) can easily be carried out and shown to lead to the formulas first derived by GOOD; we have no need for the explicit regeneration amplitudes here.

There are several significant observations to be drawn from these formulas:

1. The original motivation for this paper was to study the 'Stokes reversibility' of kaons in matter, as a possible test of T-invariance. This reversibility concerns the equality of transmission of particles travelling in opposite directions through an inhomogeneous medium. If we consider a double regenerator composed of two homogeneous pieces of different material, denoted by subscripts 1 and 2, then the transmission of K_L mesons from one side is

$$\langle \tilde{K}_L | S_1 S_2 | K_L \rangle = \langle \tilde{K}_L | S_1 | K_L \rangle \langle \tilde{K}_L | S_2 | K_L \rangle + \langle \tilde{K}_L | S_1 | K_S \rangle \langle \tilde{K}_S | S_2 | K_L \rangle \qquad (48)$$

and the transmission from the other side is obtained by interchanging 1 and 2. However, since Equation (48) is explicitly invariant under the transformation (21) we can always choose Λ'' symmetric, as in Equation (47). It follows that the expression is symmetric in 1 and 2. Therefore, 'Stokes reversibility' is implied even in the absence of T-invariance; it is simply a consequence of the 'eikonal approximation'. This is probably well-known in optics, but was unknown to us! One can also show in a similar way that this property holds for K_S, K^0, and $\overline{K}^0$ mesons as well. There do not appear to be any interesting experiments of this type for testing T and TCP.

2. There is a very sensitive test of the violation of CP by measurement of the decay rates of the coherent beam in matter. Since the effect of both the weak interactions and the coherent scattering is to admix the states K^0 and $\overline{K}^0$, it is possible to compensate the one with the other, and therefore to determine the parameters of the decay amplitudes in terms of the known (or separately measurable) parameters in Λ_{SC}. Using Equations (37) and (44), the rate of 2π-decay for the long-lived mode in matter is

$$R'_{2\pi}(t) \cong x_{2\pi}\, e^{-\gamma'_L t} \left| \delta + \left(\frac{\delta\, \alpha_{2\pi} - i\, y_{2\pi}}{2\, x_{2\pi}} + \frac{\delta\, \Lambda_{1,-1} - \delta\, \Lambda_{1,1}}{\lambda_S - \lambda_L} \right) \right|^2 . \qquad (49)$$

By measuring this rate as a function of the density, one can determine the phase of the quantity

$$\frac{\delta\, \alpha_{2\pi} - i\, y_{2\pi}}{2\, x_{2\pi}} + \frac{\delta\, \Lambda_{1,-1} - \delta\, \Lambda_{1,1}}{\lambda_S - \lambda_L}$$

relative to the phase of δ. From the known rate of the long-lived mode in vacuum, we estimate that the critical density necessary for strong interference is about 1/10 of the normal density of condensed materials, which can be obtained by use of a stratified medium[5] *). Such an experiment will give important evidence concerning the source of the CP-violating terms, although it cannot differentiate between a violation in the partial damping matrix for that channel ($\delta\alpha$, $y \neq 0$) and a violation in the total energy matrix ($\delta\Lambda_{1,1}$, $\delta\Lambda_{1,-1} \neq 0$).

Kaon Pair States

It is very natural to consider the self-conjugate system of a kaon pair ($K^0 \overline{K}^0$) for a study of invariance under CP, T, and TCP. The wide variety of production mechanisms (such as meson resonance decay, antiproton annihilation, πp and $p p$ reactions, etc.) and the many possibilities for observing decays and/or interactions of the kaons, make this subject very rich and varied. The subtle correlations in the two particle state due to the Einstein-Podolsky effect must also be correctly treated[20]).

We begin the discussion by noting that of the sixteen states of two kaons[21]), only the two zero strangeness states of neutral kaons, $K^0(k_1)\,\overline{K}^0(k_2)$ and $\overline{K}^0(k_1)\,K^0(k_2)$, need be considered since it is assumed that strangeness, charge, and the momenta are well specified in the production process.

The two particles are distinguished by their momenta. The states of interest are antisymmetric and symmetric combinations which we write in the center of mass system,

$$|\psi_\mp\rangle \equiv \frac{1}{\sqrt{2}}\,|K^0(k)\,\overline{K}^0(-k) \mp \overline{K}^0(k)\,K^0(-k)\rangle\,. \tag{50}$$

A further decomposition into partial waves could be made, but the only result of importance here is that, as a consequence of Bose statistics, the state $|\psi_-\rangle$ contains only odd L, and $|\psi_+\rangle$ contains only even L. Only propagation in the vacuum is considered. The coherent effects of a scattering medium can be discussed as in the previous section, but it is complicated by the relative motion of the matter and the meson pair reference frames[22]).

The temporal evolution of the pair states will be assumed to be given by an energy matrix separable and symmetric in the two particles, $\mathcal{H}(1,2) = \Lambda(1) + \Lambda(2)$. This neglects the weak final state interaction between the two particles, but includes the weak self-interaction. Since weak interactions result in a very short range force, this procedure is correct for pair states with an appreciable relative momentum. With this assumption, and with the definition (40) the time dependence of $|\psi_\mp\rangle$ is given by

$$|\psi_\mp(t)\rangle = e^{-(\boldsymbol{\omega}(1)+\boldsymbol{\omega}(2))t}\,|\psi_\mp\rangle\,, \tag{51}$$

where t is the time after production of the pair. With the help of Equations (22) and (19) we find for the time evolution of the one-kaon states

$$e^{-i\boldsymbol{\omega}t}\,|K^0(k)\rangle = \frac{\sqrt{2}}{1+\varrho^2}\,\left(e^{-i\gamma\lambda_S t}\,|K_S(k)\rangle + \varrho\,e^{-i\gamma\lambda_L t}\,|K_L(k)\rangle\right)$$

$$e^{-i\boldsymbol{\omega}t}\,|\overline{K}^0(k)\rangle = \frac{\sqrt{2}}{r(1+\varrho^2)}\,\left(\varrho\,e^{-i\gamma\lambda_S t}\,|K_S(k)\rangle - e^{-i\gamma\lambda_L t}\,|K_L(k)\rangle\right)\,. \tag{52}$$

*) A successful experiment of this type has recently been reported by V. L. Fitch, R. F. Roth, J. S. Russ, W. Vernon, see Reference 12.

This gives for Equation (51)

$$| \psi_-(t) \rangle = \frac{\sqrt{2}}{r(1+\varrho^2)} e^{-i\gamma(\lambda_S+\lambda_L)t} | K_L(k) K_S(-k) - K_S(k) K_L(-k) \rangle . \tag{53}$$

Note that in the center of mass system $\gamma = (1 - v^2)^{-1/2}$ is the same for both mesons. If we express this result in K_1, K_2 or in K^0, $\overline{K}^0$ then

$$| \psi_-(t) \rangle = \frac{1}{\sqrt{2}} e^{-i\gamma(\lambda_S+\lambda_L)t} | K_2(k) K_1(-k) - K_1(k) K_2(-k) \rangle$$

$$= e^{-i\gamma(\lambda_S+\lambda_L)t} | \psi_- \rangle \tag{54}$$

which is *independent of* r, ϱ, and therefore identical with the results of a CP-invariant model. As a consequence we see that there cannot be simultaneous decay of both mesons into $\pi^+ \pi^-$ from odd L.

The corresponding result for $| \psi_+ \rangle$ is more complicated; we find from (52)

$$| \psi_+(t) \rangle = \frac{\sqrt{2}}{r(1+\varrho^2)^2} \Big\{ 2\varrho (e^{-i\gamma 2\lambda_S t} | K_S(k) K_S(-k) \rangle - e^{-i\gamma 2\lambda_L t} | K_L(k) K_L(-k) \rangle) - (1-\varrho^2) e^{-i\gamma(\lambda_S+\lambda_L)t} \times | K_S(k) K_L(-k) + K_L(k) K_S(-k) \rangle \Big\} . \tag{55}$$

Note that TCP conservation implies the absence of the last term in Equation (55). Written interms of K^0, $\overline{K}^0$ $| \psi_+(t) \rangle$ contains the states $| K^0(k) K^0(-k) \rangle$ and $| \overline{K}^0(k) \overline{K}^0(-k) \rangle$ with strangeness ± 2, in addition to $| \psi_+ \rangle$*). As is seen by applying Bose statistics these three states all contain only even L; they are the only pair states of neutral kaons with this property. The explicit expression is

$$| \psi_+(t) \rangle = \frac{e^{-i\gamma(\lambda_S \lambda + L)t}}{(1+\varrho^2)^2} \Big\{ [(1-\varrho^2)^2 + 4\varrho^2 \cos \Delta\omega t] \cdot | \psi_+ \rangle + \sqrt{2} \frac{\varrho}{r} (1 - e^{i\Delta\omega t}) \times [(\varrho^2 + e^{-i\Delta\omega t}) \cdot | K^0(k) K^0(-k) \rangle + r^2 (1 + \varrho^2 e^{-i\Delta\omega t}) \cdot | \overline{K}^0(k) \overline{K}^0(-k) \rangle] \Big\} , \tag{56}$$

where $\Delta\omega \equiv \gamma (\lambda_S - \lambda_L)$.

Quantities of more direct physical significance are the decay rates of the states $| \psi_\mp \rangle$ whereby one kaon decays into channel j_1 at time t_1 and the other into j_2 at t_2,

$$R_\mp(j_1 t_1 ; j_2 t_2) \equiv | \langle j_1 j_2 | \mathcal{H}_W(1) \cdot \mathcal{H}_W(2) \, e^{-i(\boldsymbol{\omega}(1) t_1 + \boldsymbol{\omega}(2) t_2)} | \psi_\mp \rangle |^2 . \tag{57}$$

With Equations (34), (35), (40), and (50) we find

$$R_-(j_1 t_1 ; j_2 t_2) = \frac{| e^{-i\gamma(\lambda_S t_1 + \lambda_L t_2)} |^2}{2 | r |^2 | 1 + \varrho^2 |^2 \alpha_{j_1} \alpha_{j_2}} | a_{j_1} b_{j_2} - b_{j_1} a_{j_2} e^{i\Delta\omega(t_1 - t_2)} |^2 \tag{58}$$

*) The following elegant comment is due to Dr. P. K. KABIR: These results can be simply understood in terms of the analogy of kaons with spin 1/2 particles, $K^0 \sim$ spin up, $\overline{K}^0 \sim$ spin down. Then $| \psi_- \rangle$, $| \psi_+ \rangle$ are the singlet and the triplet $m = 0$ states, respectively. Since we have assumed separability, our energy matrix expresses the precession of each spin in a fictitious 'magnetic field'. But the singlet state is unaffected by a magnetic field, since the total spin is conserved, whereas the triplet state undergoes a precession, admixing the other two triplet states $K^0 K^0$ and $\overline{K}^0 \overline{K}^0$!

and

$$R_+(j_1\,t_1\,;\,j_2\,t_2) = \frac{|\,e^{-i\gamma(\lambda_S t_1 + \lambda_L t_2}\,|^2}{2\,|\,r\,|^2\,|\,1+\varrho^2\,|^4\,\alpha_{j_1}\,\alpha_{j_2}}\,\big|\,2\,\varrho\,(a_{j_1}\,a_{j_2}\,e^{-i\Delta\omega t_2} - b_{j_1}\,b_{j_2}\,e^{+i\Delta\omega t_1})$$

$$-\,(1-\varrho^2)\,(a_{j_1}\,b_{j_2} + b_{j_1}\,a_{j_2}\,e^{+i\Delta\omega(t_1-t_2)})\,\big|^2 \tag{59}$$

where $a_j \equiv \alpha_j + \varrho\,r\,(x_j + i\,y_j)$, $b_j \equiv \varrho\,\alpha_j - r\,(x_j + i\,y_j)$. The cross term $\sim (1-\varrho^2)$ (const. $+\,e^{i\Delta\omega(t_1-t_2)}$) in R_+ measures a possible violation of *TCP*. No such test can be made with R_-. For simultaneous decay of the two mesons into the same channel

$$R_-(j\,t;\,j\,t) = 0 \tag{58'}$$

and

$$R_+(j\,t;\,j\,t) = 2\,R_j(t)\cdot\overline{R}_j(t)\,. \tag{59'}$$

Equation (58′) generalizes an earlier observation: Symmetrical decay of a pair state with odd L is impossible, independent of *CP* violation. In particular the resonance decay $\phi \to K^0\,\overline{K}^0$ only yields asymmetrical events.

Instead of decays, strong interactions of the pair may be detected. The reaction rate for producing a K^0 at t_1 and an other K^0 at t_2 from $|\,\psi_+\,\rangle$, for example, is

$$|\,\langle +1,+1\,|\,e^{-i(\boldsymbol{\omega}(1)t_1+\boldsymbol{\omega}(2)t_2}\,|\,\psi_+\,\rangle|^2 = \frac{|\,e^{-i\gamma(\lambda_S t_1+\lambda_L t_2)}\,|^2}{2\,|\,r\,|^2\,|\,1+\varrho^2\,|^4}$$

$$\times\,|\,2\,(e^{-i\Delta\omega t_2} - \varrho^2\,e^{+i\Delta\omega t_1}) - (1-\varrho^2)\,(1+e^{+i\Delta\omega(t_1-t_2)})\,|^2\,. \tag{60}$$

Similar expressions hold for production of $\overline{K}^0\,K^0$ or $\overline{K}^0\,\overline{K}^0$. Again they all contain a *TCP*-violating term $\sim (1-\varrho^2)$ (const. $+\,e^{i\Delta\omega(t_1-t_2)}$), whereas no such terms occur with $|\,\psi_-\,\rangle$.

As a final comment we assert that in spite of the correlations of K^0 and $\overline{K}^0$ in $|\,\psi_\mp\,\rangle$, we can use any incoherent mixture of these pair states as a source of mesons with equal mixture of K^0 and $\overline{K}^0$, as is required for measurement of Equations (30) or (31). It is only necessary to use *one* of the two particles produced in the pair, regardless of the state of the other particle. (However, it is vital to be certain that it is a $K^0\,\overline{K}^0$ production event!). To prove this, introduce the density matrix for such an initial state

$$\varrho = |\,\psi_-\,\rangle\,P_-\,\langle\,\psi_-\,| + |\,\psi_+\,\rangle\,P_+\,\langle\,\psi_+\,| = \frac{1}{4}\,(1-\sigma(1)\cdot\sigma(2))\cdot P_-$$

$$+\,\frac{1}{4}\,(1+\sigma(1)\cdot\sigma(2) - 2\,\sigma_z(1)\,\sigma_z(2))\cdot P_+ \tag{61}$$

in our standard representation. Here P_- and P_+ are the probabilities of producing $|\,\psi_-\,\rangle$ and $|\,\psi_+\,\rangle$. Now performing a trace over matrices (1), corresponding to summing over all decay products of meson (1), gives

$$\mathrm{tr}_1\,\varrho = \frac{1}{4}\,(P_- + P_+)\,. \tag{62}$$

which describes a mixture with equal weights for K^0 and $\overline{K}^0$, as desired. Summing instead over matrices (2) gives the same result, since ϱ is symmetric in (1) and (2).

Acknowledgments

The authors are pleased to acknowledge fruitful conversations on these subjects with their colleagues, especially with Dr. L. HORWITZ and Prof. L. WOLFENSTEIN. One of us (R.R.L.) is indepted to the Swiss National Fund for a grant supporting this work, and to Prof. J. M. JAUCH for the hospitality of the Institut de Physique Théorique.

Bibliography

[1]) J. H. CHRISTENSON, J. W. CRONIN, V. L. FITCH, R. TURLAY, Phys. Rev. Letters *13*, 138 (1964); X. DE BOUARD et al., Phys. Letters *15*, 58 (1965); W. GALBRAITH et al., Phys. Rev. Letters *14*, 383 (1965).

[2]) M. GELL-MANN and A. PAIS, Phys. Rev. *97*, 1387 (1955).

[3]) For some of the earliest discussions, see T. D. LEE, R. OEHME, and C. N. YANG, Phys. Rev. *106*, 340 (1957); STEVEN WEINBERG, Phys. Rev. *110*, 782 (1958); R. G. SACHS and S. B. TREIMAN, Phys. Rev. Letters *8*, 137 (1962); R. G. SACHS, Phys. Rev. *129*, 2280 (1963).

[4]) Several authors have suggested that there are more than two neutral kaons, J. L. URETSKY, Phys. Letters *14*, 154 (1965); K. NISHIJIMA and H. SAFFOURI, Phys. Rev. Letters *14*, 205 (1965); for a discussion of the consequences, see also L. B. OKUN and I. YA. POMERANCHUK, Phys. Letters *16*, 338 (1965); P. K. KABIR and R. R. LEWIS, Phys. Rev. Letters *15*, 306 (1965).

[5]) R. R. LEWIS and C. P. ENZ, Phys. Letters *16*, 73 (1965); R. R. LEWIS, Phys. Rev. Letters *14*, 749 (1965).

[6]) R. G. SACHS, Phys. Rev. Letters *13*, 286 (1965); N. CABIBBO, Phys. Letters *12*, 137 (1964); T. N. TRUONG, Phys. Rev. Letters *13*, 358 (1964); L. WOLFENSTEIN, Phys. Rev. Letters *13*, 562 (1964); and many others.

[7]) J. BERNSTEIN, N. CABIBBO, and T. D. LEE, Phys. Letters *12*, 146 (1964); J. S. BELL and J. K. PERRING, Phys. Rev. Letters *13*, 348 (1964); F. GUERSEY and A. PAIS (unpublished).

[8]) T. T. WU and C. N. YANG, Phys. Rev. Letters *13*, 380 (1964); T. D. LEE and L. WOLFENSTEIN, Phys. Rev. *138*, B1490 (1965); L. WOLFENSTEIN, CERN report TH.583 (1965).

[9]) V. F. WEISSKOPF and E. P. WIGNER, Zs. für Physik *63*, 54 (1930); KO AIZU, Nuovo Cimento *VI*, 1040 (1957).

[10]) A. H. ROSENFELD et al., Revs. Mod. Phys. *36*, 977 (1964).

[11]) T. FUJII, J. V. JOVANOVITCH, F. TURKOT, G. T. ZORN, Phys. Rev. Letters *13*, 253 (1964).

[12]) V. L. FITCH, R. F. ROTH, J. S. RUSS, W. VERNON, Phys. Rev. Letters *15*, 73 (1965); J. H. CHRISTENSON, J. W. CRONIN, V. L. FITCH, R. TURLAY, Phys. Rev. (to be published).

[13]) F. S. CRAWFORD, B. B. CRAWFORD, R. L. GOLDEN, G. W. MEISNER, Bull. American Phys. Soc. *9*, 443 (1964).

[14]) T. D. LEE, R. OEHME, and C. N. YANG, see Reference 3.

[15]) R. G. SACHS, see Reference 3 and Annals of Phys. *22*, 239 (1964).

[16]) A. PAIS and O. PICCIONI, Phys. Rev. *100*, 1487 (1955).

[17]) M. L. GOOD, Phys. Rev. *106*, 591 (1957); *110*, 550 (1958).

[18]) R. H. GOOD and E. PAULI, Phys. Rev. Letters 8, 223 (1962).

[19]) See Reference 17 and also MALCOLM C. WHATLEY, Phys. Rev. Letters *9*, 317 (1962).

[20]) T. B. DAY, Phys. Rev. *121*, 1204 (1961).

[21]) M. GOLDHABER, T. D. LEE, and C. N. YANG, Phys. Rev. *112*, 1796 (1958).

[22]) V. L. LYUBOSHITZ, E. O. OKONOV, and M. I. PODGORETSKII, Sov. Phys. JETP *20*, 1257 (1965). Several equations contain errors of sign.

FIRST EVIDENCE FOR DIRECT *CP* VIOLATION

CERN–Dortmund–Edinburgh–Mainz–Orsay–Pisa–Siegen Collaboration

H. BURKHARDT [1], P. CLARKE, D. COWARD [2,3], D. CUNDY, N. DOBLE, L. GATIGNON, V. GIBSON, R. HAGELBERG, G. KESSELER, J. VAN DER LANS, I. MANNELLI [4], T. MICZAIKA [5], A.C. SCHAFFER [6], J. STEINBERGER, H. TAUREG, H. WAHL, C. YOUNGMAN [7]
CERN, CH-1211 Geneva 23, Switzerland

G. DIETRICH, W. HEINEN [8]
Institut für Physik, Universität Dortmund, D-4600 Dortmund 50, Fed. Rep. Germany [9]

R. BLACK, D.J. CANDLIN, J. MUIR, K.J. PEACH, B. PIJLGROMS [10], I.P. SHIPSEY [11], W. STEPHENSON
Physics Department, University of Edinburgh, Edinburgh EH9 3JZ, UK

H. BLÜMER, M. KASEMANN, K. KLEINKNECHT, B. PANZER, B. RENK
Institut für Physik, Universität Mainz, D-6500 Mainz, Fed. Rep. Germany [9]

E. AUGÉ, R.L. CHASE, M. CORTI, D. FOURNIER, P. HEUSSE, L. ICONOMIDOU-FAYARD, A.M. LUTZ, H.G. SANDER [1]
Laboratoire de l'Accélérateur Linéaire, Université de Paris-Sud, F-91405 Orsay, France [12]

A. BIGI, M. CALVETTI [13], R. CAROSI, R. CASALI, C. CERRI, G. GARGANI, E. MASSA, A. NAPPI, G.M. PIERAZZINI
Dipartimento di Fisica e Sezione INFN, I-56100 Pisa, Italy

C. BECKER [5], D. HEYLAND [5], M. HOLDER, G. QUAST, M. ROST, W. WEIHS and G. ZECH
Fachbereich Physik, Universität Siegen, D-5900 Siegen 21, Fed. Rep. Germany [14]

Received 31 March 1988

[1] Present address: Fachbereich Physik, Universität Siegen, D-5900 Siegen 21, Fed. Rep. Germany.
[2] On leave from SLAC, Stanford, CA 94305, USA.
[3] Work supported in part by the US Department of Energy contract DE-AC03-765F00515.
[4] Present address: Scuola Normale Superiore e Sezione INFN, I-56100 Pisa, Italy.
[5] Present address: DVFLR, D-5000 Cologne, Fed. Rep. Germany.
[6] Present address: LAL, Université Paris-Sud, F-91405 Orsay, France.
[7] Present address: II. Institut für experimentelle Physik, Universität Hamburg, D-2000 Hamburg, Fed. Rep. Germany.
[8] Present address: IP-Systems, D-7500 Karlsruhe, Fed. Rep. Germany.
[9] Funded by the German Federal Minister for Research and Technology (BMFT) under contract 05 4MZ18.
[10] Present address: FWI, University of Amsterdam, 1018 XE Amsterdam, The Netherlands.
[11] Present address: Syracuse University, Syracuse, NY 13244-1130, USA.
[12] Funded by Institut National de Physique des Particules et de Physique Nucléaire, France.
[13] Present address: Dipartimento di Fisica e Sezione INFN, I-06100 Perugia, Italy.
[14] Funded by the German Federal Minister for Research and Technology (BMFT) under contract 054 Si74.

The double ratio R of the relative decay rates of the short- and long-lived neutral kaons into two charged and two neutral pions was measured to be $0.980\pm0.004\pm0.005$. The deviation of R from unity implies CP violation in the transition of the CP-odd K_2 into two pions with $\epsilon'/\epsilon=(3.3\pm1.1)\times10^{-3}$.

Since its first observation in the decay of the long-lived neutral kaon into two pions [1], CP violation remains one of the enigmas in particle physics. While CP violation is manifest in neutral kaon decays, the search for CP-violating effects elsewhere has been unsuccessful. In the phenomenology of CP violation in the neutral kaon system [2] it is convenient to define the CP eigenstates $K_1=(K^0+\bar{K}^0)/\sqrt{2}$ and $K_2=(K^0-\bar{K}^0)/\sqrt{2}$ with $K_1=+CP\,K_1$ and $K_2=-CP\,K_2$. The short- and long-lived K^0 are the mass eigenstates which can be written as $K_S\approx K_1+\epsilon K_2$ and $K_L\approx K_2+\epsilon K_1$. The parameter ϵ describes CP violation induced by kaon state mixing. Direct CP violation may also occur in the decay of the K_2 into two pions with a relative amplitude of [2]

$$\epsilon'=\mathrm{i}/\sqrt{2}\,\mathrm{Im}(A_2/A_0)\,\exp[\mathrm{i}(\delta_2-\delta_0)]\,,$$

where A_0 and A_2 are the amplitudes for the decay into isospin 0 and 2 two-pion states; δ_0 and δ_2 are the corresponding $\pi\pi$ scattering phase shifts at the mass of the K^0. With these definitions the ratios of K_L and K_S decay amplitudes into $2\pi^0$ and $\pi^+\pi^-$ respectively are

$$\eta_{00}\equiv\langle 2\pi^0|T|K_L\rangle/\langle 2\pi^0|T|K_S\rangle$$
$$\equiv|\eta_{00}|\,\exp(\mathrm{i}\Phi_{00})=\epsilon-2\epsilon'$$

and

$$\eta_{+-}\equiv\langle\pi^+\pi^-|T|K_L\rangle/\langle\pi^+\pi^-|T|K_S\rangle$$
$$\equiv|\eta_{+-}|\,\exp(\mathrm{i}\Phi_{+-})=\epsilon+\epsilon'\,.$$

The relevant experimental measurements are the magnitudes and phases of these two parameters, and the real part of ϵ determined from the charge asymmetry in semileptonic K_L decays: $|\eta_{+-}|=(2.27\pm0.02)\times10^{-3}$ [3]; $\Phi_{+-}=44.6°\pm1.2°$ [4]; $|\eta_{00}/\eta_{+-}|=1.00\pm0.01$ [5]; $\Phi_{00}=55°\pm5°$ [6]; and $\mathrm{Re}\,\epsilon=(1.62\pm0.09)\times10^{-3}$ [7]. All experimental results are compatible with $\epsilon=2.27\times10^{-3}\exp(\mathrm{i}\,43.7°)$ and the superweak model [8] in which state mixing is the only source of CP violation and $\epsilon'=0$ [#1].

[#1] We disregard here the two-standard-deviation discrepancy in Φ_{00}. A considerably more sensitive measurement of $\Phi_{00}-\Phi_{+-}$ is being carried out by this group.

In the theory of six weakly interacting quarks [9], direct CP violation as well as state mixing are introduced by transitions via heavy-quark intermediate states. Based on this, a small but non-zero value of ϵ'' is predicted [10]. From $\pi\pi$ scattering, its phase is determined to be $61°\pm3°$ [11]. This angle is close to the phase of ϵ, so that to a good approximation $\mathrm{Re}\,\epsilon'/\epsilon=1/6(1-|\eta_{00}/\eta_{+-}|^2)$. This relation is used to determine ϵ'/ϵ from the double ratio of K_S and K_L decay rates into charged and neutral pions:

$$R=\left|\frac{\eta_{00}}{\eta_{+-}}\right|^2=\frac{\Gamma(K_L\to2\pi^0)/\Gamma(K_L\to\pi^+\pi^-)}{\Gamma(K_S\to2\pi^0)/\Gamma(K_S\to\pi^+\pi^-)}\,.$$

This experiment has been performed at the CERN Super Proton Synchrotron. It is based on the concurrent detection of $2\pi^0$ and $\pi^+\pi^-$ decays. Collinear K_S and K_L beams are employed alternately, changing frequently from one to the other to reduce time-dependent effects. Details of the apparatus and beams have been given elsewhere [12]. Kaons with energies around 100 GeV are produced by 450 GeV protons incident upon one of two targets at an angle of 3.6 mrad with respect to the kaon beam line. The K_L are derived from $\sim10^{11}$ protons per pulse and are selected by two-stage collimation at distances of 48 m and 120 m, respectively, from the first production target. Alternatively, $\sim10^7$ protons per pulse are brought onto the second target, from which the K_S are selected by collimation after 7 m. The retractable K_S target station, sweeping magnet and collimator system are mounted on a train which can be moved through 48 m of the K_L decay region. The K_S data are taken with the beam train displaced in 1.2 m steps so that both the K_S and K_L decay distributions become effectively uniform in the fiducial region (the average K_S decay length is 6 m).

The detector is based on calorimetry and is designed for good stability and high efficiency, large acceptance and fast data-acquisition. A schematic layout of the apparatus is shown in fig. 1. The principal features are summarized as follows:

- both K_S and K_L beams are transported in vacuum;
- an anticounter in the K_S beam, preceded by a 7 mm

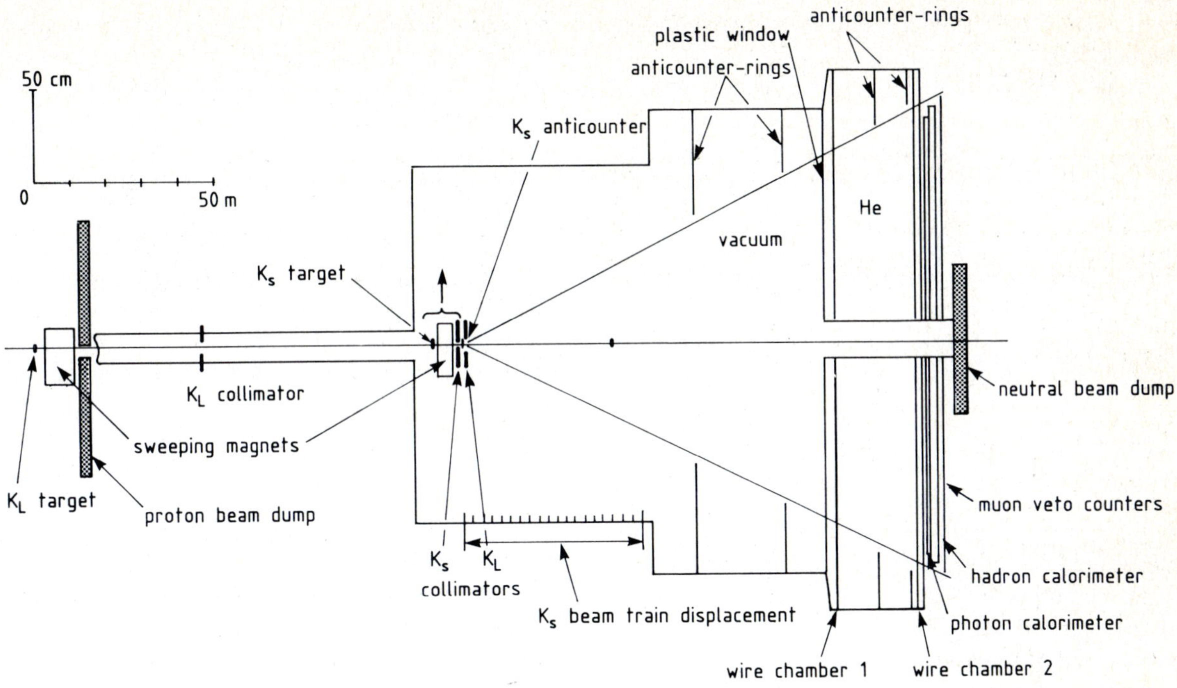

Fig. 1. Schematic layout of apparatus and beams.

lead converter, vetos decays in the collimator, defines the upstream edge of the decay region, and permits the relative calibration of the $2\pi^0$ and $\pi^+\pi^-$ energy scales to a precision better than $\pm 10^{-3}$;
– four ring-shaped anticounters surrounding the decay region detect large-angle photons and thus reduce unwanted three-body decays;
– two wire chambers spaced 25 m apart, with ± 0.5 mm resolution in each projection, track charged pions;
– a hodoscope of scintillation counters triggers on $\pi^+\pi^-$ decays by a coincidence of hits in opposite quadrants;
– a liquid-argon/lead sandwich calorimeter with strip readout measures photons with ± 0.5 mm position and $\pm 7.5\%/\sqrt{E}$ (GeV) energy resolution;
– a plane of scintillation counters, installed in the liquid argon after 13 radiation lengths of material, triggers on $2\pi^0$ decays;
– an iron/scintillator sandwich calorimeter measures, in conjuction with the liquid-argon calorimeter, the energy of charged pions with $\pm 65\%/\sqrt{E}$ (GeV) resolution;
– two planes of scintillators, after a total of 3 m of iron equivalent, reject $K^0 \to \pi\mu\nu$ decays.

The decay region is evacuated and the volume between the chambers is filled with helium. A thin composite Kevlar window of 3×10^{-3} radiation length separates the decay region from the wire chamber section. A tube of 20 cm diameter, through the centre of the window and the detectors, allows the neutral beam to continue in vacuum as far as the final beam dump.

Single counting rates are typically 10^5 Hz, originating predominantly from K^0 decays and beam-associated muons. The trigger on two-body K^0 decays is done in three steps. A pretrigger signal is generated from a coincidence of hits in opposite quadrants of the scintillator hodoscope, or from a left–right coincidence of the liquid-argon scintillators, with a veto from the ring and muon anticounters. A trigger signal is accepted, subject to further conditions on calorimeter energies, the number of hits in the first chamber, and the number of peaks in the liquid-argon calorimeter. After digitization of pulse heights and chamber information, three-body decays are rejected using online processors. The pretrigger rate is about 10 kHz, and typically 1000 events are recorded per burst in the K_L beam.

The $K^0 \to 2\pi^0 \to 4\gamma$ decays are reconstructed from the measured positions and energies of the photons. Details of the reconstruction method may be found in ref. [13]. Events with extra photons of more than 2.5 GeV are rejected. For accepted events, the photon energies have to be above 5 GeV with at least 5 cm separation between shower centres, and the centre of gravity of the energies of all photons has to lie within the beam region. The K^0 energy is measured with typically 1% accuracy. The distance of the decay vertex from the calorimeter is calculated, using the K^0 mass as a constraint, with similar precision. Constraints on the masses of two-photon pairs are used to reduce the background, which is primarily due to $K_L \to 3\pi^0 \to 6\gamma$ decays with undetected photons. This background is uniformly distributed in a two-dimensional scatter plot of photon-pair masses. The π^0 mass resolution is ~2 MeV. Signal and background events are counted in equal-area χ^2 contours around the region defined for accepted events (see fig. 2). The signal region is taken as $\chi^2 < 9$. Background is subtracted by linear extrapolation into the signal region. It is about 4% in the K_L beam but depends strongly on the longitudinal vertex position because of the apparent vertex shift due to the missing energy in $3\pi^0$ decays with undetected photons. It is negligible in the K_S beam.

The $K^0 \to \pi^+\pi^-$ decays are reconstructed from space points defined by at least three hits out of the four planes in each of the two wire chambers. Events with extra space points in the first chamber are rejected for both charged and neutral decays. The longitudinal vertex resolution is better than 1 m. The K^0 energy is calculated with 1% precision from the kaon mass, the opening angle between the two tracks, and the ratio of track energies as measured in the calorimeter. This ratio is limited to a maximum of 2.5, in order to achieve this resolution and also to reduce the contribution of $\Lambda \to p\pi$ decays to a negligible level. Events with isolated photons, such as $K^0 \to \pi^+\pi^-\pi^0$ decays and events with accidental photons, are rejected. The $K^0 \to \pi e \nu$ events are identified and re-

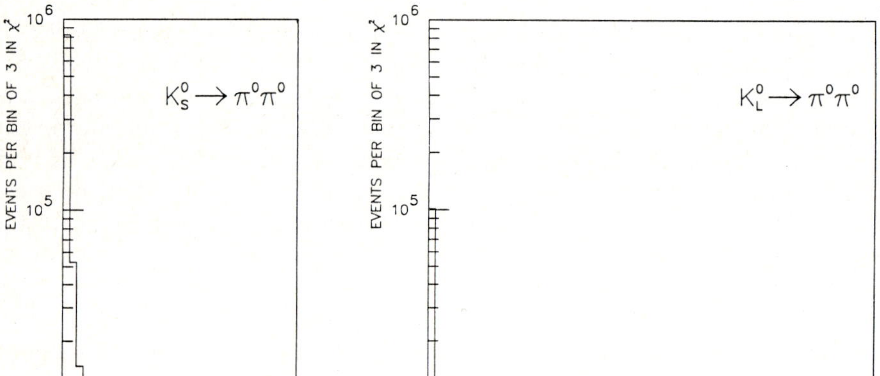

Fig. 2. Number of accepted 4γ events as a function of χ^2, for $K_S \to 2\pi^0$ and $K_L \to 2\pi^0$ data, and a Monte Carlo calculation for background originating from $K_L \to 3\pi^0$ decays. The signal region is taken as $\chi^2 < 9$.

jected by comparing, for each track, the energy deposited in the front half of the photon calorimeter with the energy deposited in the hadron calorimeter. About half of the detected $\pi^+\pi^-$ events are lost equally from K_S and K_L because of all these requirements. Possible variations of the rejection rate due to systematic changes in detector response are monitored by the observed $\pi^+\pi^-$ mass. The response of the hadron calorimeter is evaluated to be constant within $\pm 0.5\%$, leading to an uncertainty ($<0.1\%$) in the measured ratio of $K_S \to \pi^+\pi^-$ and $K_L \to \pi^+\pi^-$ event rates. After cuts on the $\pi^+\pi^-$ mass and on the reconstructed centre of gravity with respect to the beam axis, a residual background of the three-body decays must be subtracted.

In a two-body decay the decay plane should contain the production target, but because of measurement errors and multiple scattering, a certain distribution of the perpendicular distance d_T of this plane to the target is expected, and can be measured with $K_S \to \pi^+\pi^-$ decays. In the three-body decays of K_L, because of the non-coplanarity of the decay, a much broader d_T distribution is expected. This is illustrated in fig. 3, where the d_T distributions of accepted two-track events in the K_S and K_L beams are shown separately. The d_T distribution in K_S decays is scaled geometrically in order to compare directly with K_L.

The signal region is taken to be $d_T < 5$ cm, and the control region for the background extrapolation is taken as 7 cm $< d_T <$ 12 cm. The fraction of events in the background region is $(3.6 \pm 0.1) \times 10^{-3}$ of the signal. This background consists mainly of $K_L \to \pi e \nu$ decays and has contributions from $K \to \pi\mu\nu$ where the muon loses its energy by bremsstrahlung in the hadron calorimeter, from $K \to \pi^+\pi^+\pi^0$ where one photon overlaps the shower of one of the charged pions, and a small amount of K_S production in the final K_L collimator. The $K^0 \to \pi^+\pi^-\pi^0$ background is subtracted directly by counting events with identified photons as a function of the distance between the photon and the nearest track. The remaining $K^0 \to \pi e \nu$ candidates are identified by the well-defined electron shower width in the photon calorimeter and longitudinal energy deposition, and the $K^0 \to \pi\mu\nu$ candidates by the shower width in the hadron calorimeter for events in

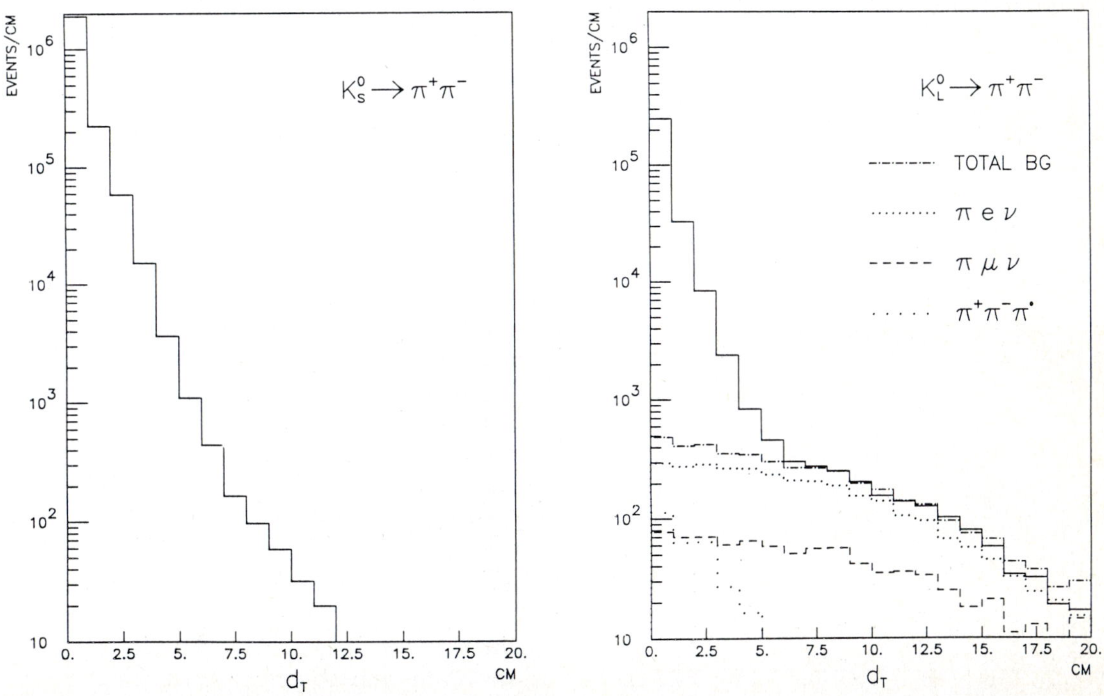

Fig. 3. Event distribution for charged decays as a function of distance d_T between the decay plane and the production target, for K_S and K_L decays and for various background components.

Table 1
Background composition for $K_L \to \pi^+\pi^-$ decays.

Background	Control region	Signal region
$K^0 \to \pi^+\pi^-\pi^0$	$(0.1 \pm 0.1) \times 10^{-3}$	$(1.0 \pm 1.0) \times 10^{-3}$
$K^0 \to \pi e \nu$	$(2.8 \pm 0.2) \times 10^{-3}$	$(4.4 \pm 0.3) \times 10^{-3}$
$K^0 \to \pi \mu \nu$	$(0.5 \pm 0.2) \times 10^{-3}$	$(0.7 \pm 0.3) \times 10^{-3}$
regenerated K_S	$(0.2 \pm 0.1) \times 10^{-3}$	$(0.4 \pm 0.2) \times 10^{-3}$
total	$(3.6 \pm 0.1) \times 10^{-3}$	$(6.5 \pm 2.0) \times 10^{-3}$

which both charged particles deposit less than 5 GeV in the photon calorimeter. The shape of the d_T distributions for the background events is determined from data samples for $K^0 \to \pi e \nu$ and residual $K^0 \to \pi^+\pi^-\pi^0$ decays, and by Monte Carlo for $K^0 \to \pi \mu \nu$ decays. The inelastic K_S regeneration on the K_L beam collimator has been determined from events with the vertex close to the collimator. The average background subtracted by extrapolation in d_T is $(6.5 \pm 2.0) \times 10^{-3}$, including systematic uncertainties (see table 1). In the case of K_S the background is negligible.

The total available statistics is $\sim 10^6$ K_L and $\sim 10^7$ K_S two-pion decays. The energy spectra of accepted $2\pi^0$ and $\pi^+\pi^-$ events are shown in fig. 4. After reconstruction, the relative energy scales of neutral and charged decays are adjusted to be the same within $\pm 10^{-3}$ by fits of the vertex distributions to the position of the anticounter in the K_S beam (see fig. 5). In this analysis, the data were selected in the energy range 70–170 GeV and with vertices between 10.5 and 48.9 m from the position of the final K_L beam collimator. A breakdown of event statistics is given in table 2 second column. The double ratio is evaluated in 10×32 bins in energy and vertex position, for each of 16 self-contained data sets of K_S and K_L. The weighted average, corrected for acceptance and resolution, is $R = 0.977 \pm 0.004$ (statistical error). In principle, the detection efficiencies for the two decay modes cancel. A Monte Carlo calculation has been used to determine the acceptance ratio. It includes the effect of the known difference in K_S and K_L beam divergences (0.7%) and the scattering of the K_S beam in the anticounter and collimator (0.3%, as measured from events without centre-of-gravity and d_T cuts, see table 2, fourth column), and the effects due to finite bin size and to energy and vertex resolution. The net total Monte Carlo correction amounts to 0.3% on R for the weighted average of all bins.

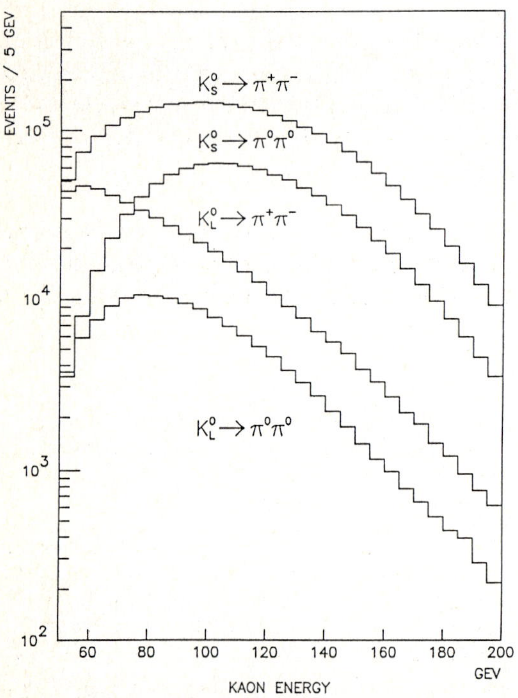

Fig. 4. Energy spectra for K_S and K_L decays into two pions.

The trigger system and analysis procedures are designed such that no significant bias should result in the events retained for analysis. Event losses due to inefficiencies of the pretrigger hodoscope counters and of the trigger system itself are measured using a sample of events with relaxed trigger conditions. Since those results are consistent with the expectation of no bias (see table 2, fifth and sixth columns), no correction has been applied. Gains and losses of good events due to accidentals are measured by overlaying a sample of events with events taken with a random trigger, at a rate proportional to the neutral beam intensity. These are primarily due to the cuts on extra space points in the first wire chamber on the number of photons. An asymmetry between charged and neutral decays is observed (table 2, last column), and a correction is applied $(0.34 \pm 0.1\%)$. This asymmetry is mainly due to a loss of charged decays in the K_S beam caused by additional background in the first chamber. Another correction of 0.06% accounts for the difference in efficiency, for charged and neutral decays, of the anticounter in the K_S beam.

The various systematic uncertainties are listed in table 3. Of these, the dominant ones are due to a pos-

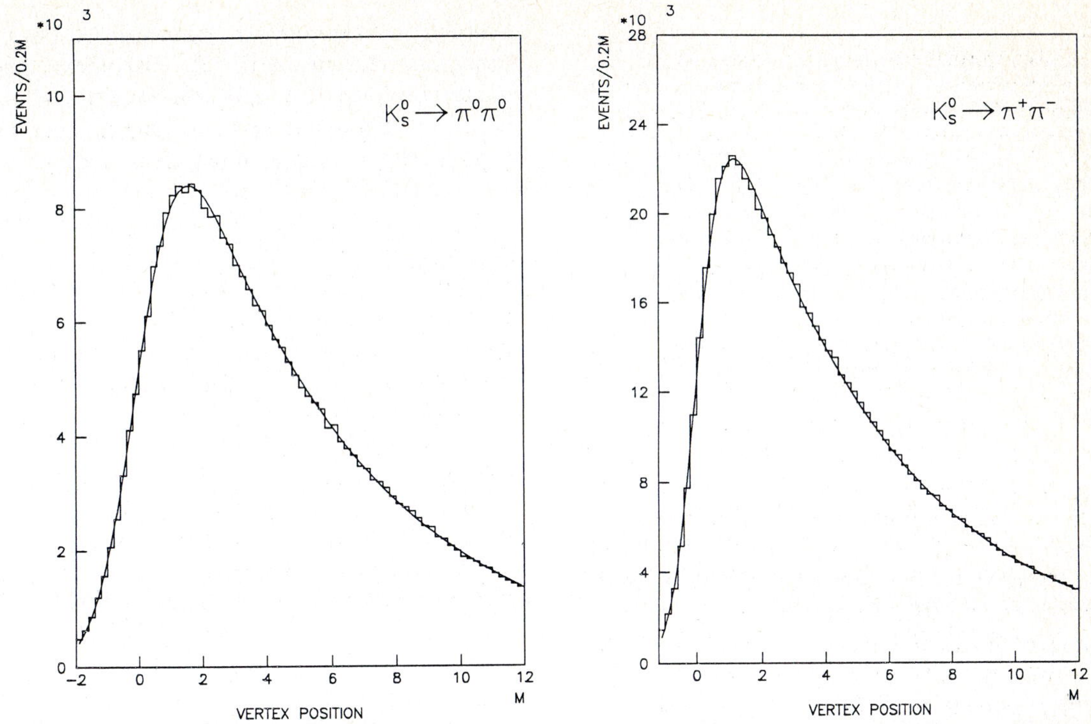

Fig. 5. $K_S \to 2\pi^0$ and $K_S \to 2\pi^+\pi^-$ event distributions as a function of distance from the anticounter in the K_S beam. The continuous lines show the best fits to the data.

sible energy scale difference between charged and neutral decays, the subtraction of backgrounds in the K_L beam, and the net losses of good events by accidentals. After all corrections we obtain the final result, $R = 0.980 \pm 0.004 \pm 0.005$, with statistical and systematic uncertainties given separately. With combined errors this corresponds to $\mathrm{Re}\, \epsilon'/\epsilon = (3.3 \pm 1.1) \times 10^{-3}$. This is the first time that evidence of *CP*-violating effects is seen in the decay of the *CP*-odd K_2 into two pions, as implied by a non-zero value of ϵ'. It is at the level predicted recently by several evaluations of the standard model for a t-quark mass in the range 50–100 GeV [14] and does not agree with the superweak model [8].

We are indebted to our colleagues L. Bertanza, F. Eisele and P. Steffen for their participation in the early stages of the experiment. We wish to thank the tech-

Table 2
Event statistics and corrections.

	Signal events ($\times 1000$)	Background (%)	Scattering (%)	Inefficiencies: pretrigger (%)	Inefficiencies: trigger (%)	Accidental losses (%)
$K_L \to 2\pi^0$	109	4.0	<0.1	0.06 ± 0.06	0.20 ± 0.10	2.6 ± 0.07
$K_L \to \pi^+\pi^-$	295	0.6		0.37 ± 0.07	0.05 ± 0.06	2.6 ± 0.05
$K_S \to 2\pi^0$	932	<0.1	0.3	0.04 ± 0.02	0.12 ± 0.03	2.5 ± 0.05
$K_S \to \pi^+\pi^-$	2300	<0.1		0.48 ± 0.03	0.01 ± 0.01	2.8 ± 0.05
effect on R			0.3	-0.12 ± 0.10	-0.03 ± 0.12	-0.34 ± 0.10

Table 3
Systematic uncertainties on the double ratio R (in %).

background subtraction for $K_L \to 2\pi^0$	0.2
background subtraction for $K_L \to \pi^+\pi^-$	0.2
$2\pi^0/\pi^+\pi^-$ difference in energy scale	0.3
regeneration in the K_L beam	<0.1
scattering in the K_S beam	0.1
K_S anticounter inefficiency	<0.1
difference in K_S/K_L beam divergence	0.1
calorimeter instability	<0.1
Monte Carlo acceptance	0.1
gains and losses by accidentals	0.2
pretrigger and trigger inefficiency	0.1
total systematic uncertainty	±0.5%

nical staff of the participating laboratories and universities for their dedicated effort in establishing the beam, the detectors, and the data-acquisition system for the experiment. We are indebted, in particular, to H.W. Atherton, C. Bizeau, F. Blin, M. Clément, G. Di Tore, G. Dubail, G. Dubois, G.P. Ferri, G. Fersuella, D. Guyon, R. Harfield, D. Jacobs, G. Juban, Ch. Lasseur, G. Laverrière, P. Le Cossec, K.D. Lohmann, A. Lovell, R. Maleyran, M. Marin, M.L. Mathieu, R. McLaren, L. Pregernig, P. Ponting, E.M. Rimmer, P. Schilly and B. Tomat (CERN); H.J. Büttner, U. Dretzler, K. Noffke and K. Wydinski (Institute of Physics, University of Dortmund); P. McInnes and A. Main (Physics Department, University of Edinburgh); the UK Science and Engineering Research Council, and the Rutherford Appleton Laboratory; C. Arnault, A. Bellemain, R. Bernier, A. Bozzone, J.P. Coulon, J.C. Drulot, J.P. Marolleau, E. Plaige, J.P. Richer and A. Roudier (Linear Accelerator Laboratory, Orsay); the IN2P3 Computing Centre at Lyons; C. Avanzini, R. Fantechi, S. Galeotti, G. Gennaro, F. Morsani, G. Pagani, D. Passuello, R. Ruberti, P. Salvadori and L. Zaccarelli (Department of Physics and INFN, University of Pisa); G. Iksal, M. Roschangar and R. Seibert (Physics Department, University of Siegen).

References

[1] J.H. Christenson et al., Phys. Rev. Lett. 13 (1964) 138.
[2] T.T. Wu and C.N. Yang, Phys. Rev. Lett. 13 (1964) 380.
[3] C. Geweniger et al., Phys. Lett. B 48 (1974) 487;
R. Messner et al., Phys. Rev. Lett. 30 (1973) 876;
D.P. Coupal et al., Phys. Rev. Lett. 55 (1985) 566.
[4] C. Geweniger et al., Phys. Lett. B 52 (1974) 119.
[5] J.K. Black et al., Phys. Rev. Lett. 54 (1985) 1628;
R.H. Bernstein et al., Phys. Rev. Lett. 54 (1985) 1631;
M. Woods et al., Chicago University preprint EFI 88-03, submitted to Phys. Rev. Lett. (1988).
[6] J.H. Christenson et al., Phys. Rev. Lett. 43 (1979) 1209.
[7] C. Geweniger et al., Phys. Lett. B 48 (1974) 483.
[8] L. Wolfenstein, Phys. Rev. Lett. 13 (1964) 562.
[9] M. Kobayashi and K. Maskawa, Prog. Theor. Phys. 49 (1973) 652.
[10] J. Ellis, M.K. Gaillard and D.V. Nanopoulos, Nucl. Phys. B 109 (1976) 213;
F.J. Gilman and M.B. Wise, Phys. Lett. B 83 (1979) 83; Phys. Rev. D 20 (1979) 2392;
B. Guberina and R. Peccei, Nucl. Phys. B 163 (1980) 289;
L. Wolfenstein, Annu. Rev. Nucl. Sci. 36 (1986) 137, and references therein.
[11] N.N. Biswas et al., Phys. Rev. Lett. 47 (1981) 1378.
[12] H. Burkhardt et al., The beam and detector for a high precision measurement of *CP* violation in neutral kaon decays, preprint CERN-EP/87-166 (1987), Nucl. Instrum. Methods, to be published.
[13] H. Burkhardt et al., Phys. Lett. B 199 (1987) 139.
[14] A.J. Buras and J.-M. Gerard, ϵ'/ϵ in the standard model, Munich preprint MPI-PAE/PTh 84-87 (1987), and references therein.

MEASUREMENT OF THE CHARGE ASYMMETRY IN THE DECAYS $K_L^o \to \pi^{\pm} e^{\mp} \nu$ AND $K_L^o \to \pi^{\pm} \mu^{\mp} \nu$

C. GEWENIGER*[1], S. GJESDAL*[2], T. KAMAE*[3], G. PRESSER*[1], P. STEFFEN, J. STEINBERGER, F. VANNUCCI*[4] and H. WAHL
CERN, Geneva, Switzerland

F. EISELE*[1], H. FILTHUTH, K. KLEINKNECHT*[1], V. LÜTH and G. ZECH*[5]
Institut für Hochenergiephysik, Universität Heidelberg, Germany

The charge asymmetry in semi-leptonic K_L^o decays has been measured in a high statistics experiment using multiwire proportional chambers. The asymmetry $\delta = (N^+ - N^-)/(N^+ + N^-)$, where N^+ and N^- are the partial decay rates for $K_L^o \to \pi^- \ell^+ \nu$ and $K_L^o \to \pi^+ \ell^- \bar{\nu}$, respectively, is found to be $\delta_L^e = (3.41 \pm 0.18) \times 10^{-3}$ for the K_{e3} mode, and $\delta_L^{\mu} = (3.13 \pm 0.29) \times 10^{-3}$ for the $K_{\mu 3}$ mode. Assuming *CPT* invariance and the absence of $\Delta S = -\Delta Q$ transitions, these results lead to a value of the real part of the *CP*-violation parameter ϵ, Re $\epsilon = (1.67 \pm 0.08) \times 10^{-3}$.

The charge asymmetry in the semi-leptonic decay modes of the long-lived neutral kaon is, apart from the two-pion decay of the long-lived kaon, the only manifestation of *CP* non-invariance observed up to now. A measurement of this effect is important to the understanding of the phenomenon of *CP* violation, as it can be related to ϵ, the *CP* mixing amplitude in the neutral K-meson state

$$\delta_L = \frac{N^+ - N^-}{N^+ + N^-} = 2 \, \mathrm{Re}\, \epsilon \, \frac{1 - |x|^2}{|1 - x|^2} \, .$$

Here *CPT* invariance has been assumed.

$N^{\pm}$ are the number of observed K_{l3} decays with positive and negative lepton, respectively. The parameter x is defined as the ratio of $\Delta S = -\Delta Q$ to $\Delta S = \Delta Q$ amplitudes and vanishes if the $\Delta S = \Delta Q$ selection rule is valid. The value of x could, in general, be different for K_{e3} and $K_{\mu 3}$ decays and may depend on the Dalitz plot variables. Recent measurements indicate, however, that x is small and not significantly different from zero [1].

*[1] Now at Universität Dortmund, Abteilung Physik, Germany.
*[2] Now at Distriktshøgskolen Stavanger, Norway.
*[3] Now at Institute for Nuclear Study, Tokyo University, Japan.
*[4] On leave from Institut de Physique Nucléaire, Orsay, France.
*[5] Now at Gesamthochschule Siegen, Germay.

The experiment was performed in a short neutral beam at the CERN Proton Synchrotron, providing neutral kaons over the momentum range 3–15 GeV/*c*. The main elements of the set-up [2] are shown in fig. 1. The neutral hadrons are produced by an external proton beam of 24 GeV/*c* hitting a 4.5 cm long platinum target, 4 × 4 mm^2 in cross-section. The secondaries are selected at an average angle of 75 mrad by a tapered uranium collimator, 2 m long, imbedded in a magnetic field of 20 kG. It is followed by a 9 m long decay volume filled with helium. The vector momenta of the charged decay products of the neutral kaons are measured in a spectrometer consisting of three multiwire proportional chambers and a bending magnet.* All chambers are divided into a left and a right half, each equipped with a horizontal and a vertical signal plane. The wire spacing is 2 mm. A 6 m long threshold Čerenkov counter filled with hydrogen gas at atmospheric pressure is used to label electrons. Muons are identified by a coincidence signal in two counter hodoscopes behind an absorber of 800 g/cm^2 of light concrete.

The use of large size multiwire proportional chambers introduces several advantageous features. Their good time resolution and zero dead-time combined with a fast selective read-out allow considerably higher

* A fourth chamber was introduced in a later stage of the experiment, but not used in the event reconstruction.

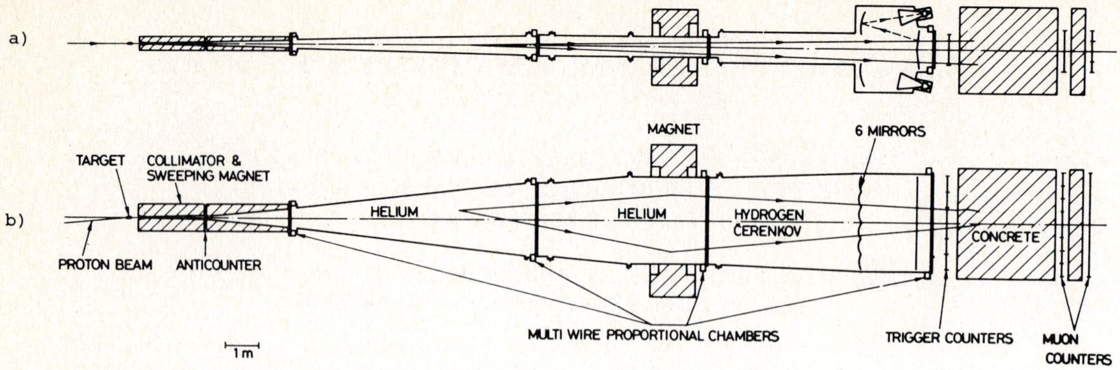

Fig. 1. Experimental lay-out. a) side view, b) top view

data acquisition rates than in earlier experiments. Furthermore, since they are d.c. operated, the proportional chambers do not require additional scintillation counters, permitting a substantial reduction of matter in the path of detected particles. The Čerenkov counter is a novel design with this in mind. The only optical element traversed by the particles is an aluminized mylar foil. The total amount of matter traversed by the decay particles in the 15 m long detector is 0.4 g/cm^2, mostly helium.

The read-out system [3] selects events with two signals in each of the wire planes so that practically all events recorded on magnetic tape can be uniquely reconstructed as a V^o. With an average rate of more than 1000 events per machine cycle of 350 m sec more than 10^9 events were registered. The total sample is divided into runs of equal length corresponding to about 0.5 million events each. The polarity of the momentum-analysing magnet was reversed at the end of each run.

After removing events with no signal in the lepton counters about 140 million events remained. The following selection procedure, designed to reduce possible contamination by other K^o decay modes, hyperon decays, or beam interactions in the decay volume, has been applied:

1) The geometrical reconstruction of the vertex is constrained by cuts on the skewness and the kinks of the tracks in the vertical projection.

2) The decay vertex is confined to a fiducial volume.

3) Electron pairs and events with poor vertex reconstruction are removed by the requirement $\alpha_x p_z >$ 0.1 GeV/c, where α_x denotes the horizontal opening angle of the V^o, and p_z its momentum component along the beam axis.

4) A unique and unambiguous assignment of the lepton track is required.

5) In the K$_{e3}$ analysis an upper limit of 8 GeV/c for the reconstructed momenta of both secondaries is set to eliminate pions radiating in the Čerenkov counter. In the K$_{\mu 3}$ data a cut is applied only on the μ momentum, $p_\mu > 1.6$ GeV/c. The threshold momentum for penetration of the absorber is 1.45 GeV/c.

6) The K$_{3\pi}$ contamination is separated from K$_{e3}$ decays by the requirement $p_o^{/\!/2} < -0.003$ (GeV/c)^2. The variable $p_o^{/\!/2}$ denotes the square of the π^o momentum in the frame in which the longitudinal momentum of the charged pair vanishes.

7) Events that are not compatible with the kinematics of the leptonic decays are eliminated by limits on the neutrino momentum in the K^o rest system,

$$p_\nu > 0$$

$$p_\nu^{/\!/2} = p_\nu^2 - p_\perp^2 > -0.003 \text{ (GeV/}c)^2 .$$

8) The effects of K$_S$–K$_L$ interference as well as decays of the Λ hyperon and K$_{2\pi}$ decays are kept to an acceptable level by a requirement on the proper time τ' *,

$$K_{e3}: \tau' > 12.75 \times 10^{-10} \text{ sec}$$

$$K_{\mu 3}: \tau' > 14.75 \times 10^{-10} \text{ sec.}$$

* τ' is derived from the smaller of the two kinematical solutions for the kaon laboratory momentum.

Table 1
Summary of the systematic corrections and results
(All quantities are measured in ppm)

Corrections	K_{e3}	$K_{\mu 3}$
Pion absorption in hydrogen	+ 61 ± 4	+ 61 ± 4
Positron annihilation	+ 28 ± 1	–
Delta-ray production	– 100 ± 20	–
Pion decay and penetration	– 169 ± 41	– 163 ± 47
$K_{3\pi}$ contamination	0	– 167 ± 85
Wrong charge assignment	0	0 ± 19
Neutron interactions	+ 3 ± 15	+ 255 ± 87
K_S–K_L interference	– 62 ± 12	+ 4 ± 4
Λ contamination	+ 5 ± 2	– 74 ± 40
Regeneration in helium	– 17 ± 5	– 17 ± 5
Č pulse height variation	– 1 ± 2	–
Accidentals	+ 8 ± 5	+ 1 ± 2
Total correction	– 244 ± 50	– 100 ± 140
Uncorrected asymmetry	+ 3652 ± 171	+ 3232 ± 257
Corrected charge asymmetry	+ 3408 ± 178	+ 3132 ± 293

In total, 34 million K_{e3} and 15 million $K_{\mu 3}$ events satisfied these criteria, giving a raw asymmetry of

$$\delta_L^e = (3.65 \pm 0.17) \times 10^{-3}$$

$$\delta_L^\mu = (3.23 \pm 0.26) \times 10^{-3}.$$

These results have to be corrected for effects due to the presence of matter in the detector and other instrumental biases.

The geometric asymmetry of the apparatus has a negligible effect, since the essential measurement is not one of left-right asymmetry, but an asymmetry with respect to the magnet polarity. Potentially important biases are field reversal effects on the detection efficiency of the wire chambers and the phototubes. The Čerenkov counter tubes are at a distance of 6 m from the magnet. The change in pulse height was found to be negligibly small.

The differential absorption of π^+ and π^- in the hydrogen filled Čerenkov counter has been evaluated from the measured π^-p and π^+p total cross-sections [4].

The corrections due to positron annihilation and production of delta electrons have been calculated according to the theoretically known cross-sections.

Pion decay (1.8%) and the penetration of the concrete absorber by pions (0.6%) can lead to an additional μ signal. Such events are rejected from the K_{l3} samples, because of two lepton signals. This omission of events introduces a systematic bias that was measured using K_{e3} decays with a coincidence signal in the μ hodoscopes.

Misidentification of the muon occurs with a probability of 1.1%, whenever the muon escapes detection, mainly for geometrical reasons, and simultaneously the pion gives a μ signal, due to decay or penetration. This is equivalent to a wrong charge assignment.

The $K_{3\pi}$ contamination in the K_{e3} sample is negligible, because of the rejection of pions in the Čerenkov counter by a factor of the order of 10^3. However, $K_{3\pi}$ decay contributes a 4.7% background to the $K_{\mu 3}$ data. The bulk of this is due to $\pi \to \mu\nu$ decay, the remainder is due to pion penetration of the muon shield. The correction required has been studied for pions in K_{e3} decays.

Interactions of beam particles, chiefly neutrons, have been investigated by placing copper plates of various thicknesses at several positions in the decay volume. The number of interactions in helium and their charge asymmetry was estimated on the basis of these data. Whereas the contamination is exceedingly small in the K_{e3} sample, a 10% correction is obtained for $K_{\mu3}$. This takes into account neutron stars as well as the production of Λ hyperons and neutral kaons.

Specific for a measurement in a rather short beam is a correction for $K_S - K_L$ interference at small decay times. In addition, the contamination due to Λ particles produced at the target has been determined and a correction is applied.

A list of all systematic corrections is given in table 1. The corrected values for the charge asymmetry

$$K_{e3}: \quad \delta_e = (3.41 \pm 0.18) \times 10^{-3}$$

$$K_{\mu3}: \quad \delta_\mu = (3.13 \pm 0.29) \times 10^{-3}$$

are in good agreement with each other, and consequently there is no indication that the $\Delta S = \Delta Q$ factor $(1-|x|^2)/|1-x|^2$ is significantly different for the two semi-leptonic decay modes of the K°_L. Furthermore these results are well consistent with earlier work [5–12] and the errors have been appreciably reduced, in particular for $K_{\mu3}$.

Assuming the validity of the $\Delta S = \Delta Q$ selection rule, the combined result leads to a value for the real part of the *CP* mixing parameter ϵ,

$$\mathrm{Re}\ \epsilon = (1.67 \pm 0.08) \times 10^{-3}.$$

We are grateful for the programming assistance of Dr. E.M. Rimmer and the technical help provided by Messrs. H. Dieperink, J. Olsfors, P. Schilly and Dr. M. Vysočansky. The Heidelberg members of the Collaboration acknowledge the financial support of the Bundesministerium für Forschung und Technologie.

References

[1] S. Wojcicki, Proc. 16th Int. Conf. on High-energy physics, Chicago-Batavia, 1972 (NAL, Batavia, 1972), vol. 2, p. 209.
[2] J.H. Dieperink, et al., Proc. Int. Conf. on Instrumentation for High-energy physics, Dubna, 1970 (Dubna, 1971), p. 251.
[3] H. Cunitz, W. Sippach and J.H. Dieperink, Nuclear Instrum. Methods 91 (1971) 211.
[4] A. Citron, et al., Phys. Rev. 144 (1966) 1101.
A.A. Carter, et al., Phys. Rev. 168 (1968) 1457.
[5] H. Saal, Ph. D. thesis, Columbia University, Internal report Nevis 169 (1969).
[6] J. Marx, Ph. D. thesis, Columbia University, Internal report Nevis 179 (1970).
J. Marx, et al., Phys. Lett. 32B (1970) 219.
[7] V.A. Ashford, et al., Phys. Lett. 38B (1972) 47.
[8] V.L. Fitch, et al., Phys. Rev. Lett. 31 (1973) 1524.
[9] H.H. Williams, et al., Phys. Rev. Lett. 31 (1973) 1521. This is a combined measurement of both leptonic decay modes, K_{e3} and $K_{\mu3}$.
[10] M.A. Paciotti, Ph. D. thesis, University of California, Berkeley, UCRL-19446 (1969).
[11] R.L. McCarthy, et al., Phys. Rev. D7 (1973) 687.
[12] R.L. Piccioni, Ph.D. thesis, Stanford University, SLAC-155 (1972).
R.L. Piccioni, et al., Phys. Rev. Lett. 29B (1972) 1412.

What Is Not Invariant under Time Reversal?

P. K. KABIR
Rutherford High-Energy Laboratory, Chilton, Didcot, Berkshire, England
(Received 25 February 1970)

The departure from T invariance in neutral K-meson decays is shown to be directly demonstrable through suitable measurements. The predicted asymmetries include a *CP- and T*-noninvariant effect larger than any which has been measured so far.

UNDER the assumption of TCP invariance, the observation of $K_2^0 \to 2\pi$ decays[1] demonstrates T noninvariance to the same extent that it demonstrates the failure of CP invariance. Tests of T invariance in K^0 decay, independent of the assumption of TCP invariance, have also been discussed,[2–4] and it was shown that available data cannot be easily reconciled[5,6] with the hypothesis of T invariance. In particular, if the phase of η_{00}, like the phase of η_{+-}, lies in the first quadrant,[7] the violation of T invariance can be directly demonstrated. However, as far as I am aware, none of the published analyses relates this observation to the lack of reversibility in any reaction. The purpose of this paper is to show the primitive connection between measurements of neutral K-meson decay and the breakdown of microscopic reversibility.[8]

We first demonstrate the failure of T invariance, i.e., reciprocity,[9] under the assumption of TCP invariance. Under this assumption, the short- and long-lived neutral kaon states can be written in the form[10,2]

$$|K_{1,2}^0\rangle = (1+|r|^2)^{-1/2}(|K^0\rangle \pm r|\bar{K}^0\rangle), \qquad (1)$$

where the complex parameter r is related to the more commonly used[2] parameter ϵ through

$$r = (1-\epsilon)/(1+\epsilon).$$

Inversion of Eq. (1) yields

$$|K^0\rangle = \tfrac{1}{2}(1+|r|^2)^{1/2}(|K_1^0\rangle + |K_2^0\rangle), \qquad (2a)$$

$$|\bar{K}^0\rangle = \tfrac{1}{2}r^{-1}(1+|r|^2)^{1/2}(|K_1^0\rangle - |K_2^0\rangle). \qquad (2b)$$

After a time τ has elapsed, the amplitude of a K_1^0 state changes by a factor $\theta_1 \equiv e^{-(\frac{1}{2}\gamma_1 + im_1)\tau}$, while that of a K_2^0 state is multiplied by $\theta_2 \equiv e^{-(\frac{1}{2}\gamma_2 + im_2)\tau}$. Therefore, according to Eqs. (1) and (2), a state created initially as $|K^0\rangle$ will be found after a lapse of time τ to have transformed into

$$|K^0\rangle \to \tfrac{1}{2}[(\theta_1+\theta_2)|K^0\rangle + r(\theta_1-\theta_2)|\bar{K}^0\rangle], \qquad (3a)$$

while an initial $\bar{K}^0$ state will change during the same time into

$$|\bar{K}^0\rangle \to \tfrac{1}{2}[r^{-1}(\theta_1-\theta_2)|K^0\rangle + (\theta_1+\theta_2)|\bar{K}^0\rangle]. \qquad (3b)$$

Consequently, the probability to find, after a time τ, that a K^0 state has changed into a $\bar{K}^0$ is $|r|^4$ times the probability that an initial $\bar{K}^0$ state is found to have transformed[11] into a K^0. The parameter $|r|^2 \approx 1-4\,\mathrm{Re}\epsilon$ has already been measured, under the assumption of TCP invariance, through the leptonic charge asymmetry[12,13] in K_2^0 decays. If we accept the conclusion of Refs. 12 and 13 that $\mathrm{Re}\epsilon > 0$, we deduce that a $\bar{K}^0$ is more likely to be transformed into a K^0 than a K^0 into a $\bar{K}^0$. The probability for a neutral kaon produced with negative strangeness to reappear spontaneously with positive strangeness after a time τ should exceed that for an $S=+1$ neutral kaon to transform spontaneously into one with $S=-1$ by a fraction

$$|r|^{-4} - 1 \approx 8\,\mathrm{Re}\epsilon, \qquad (4)$$

which is independent of time.[14] Observation of this effect would directly demonstrate CP noninvariance and T noninvariance simultaneously. The asymmetry of Eq. (4), which according to the values of $\mathrm{Re}\epsilon$ reported in Refs. 12 and 13 has a value between 1.1 and 1.6%, is the largest CP-noninvariant effect which can be predicted with assurance at the present time.

As a concrete example of nonreversibility resulting from this effect, let us consider a *Gedanken* experiment in which a K^- beam collides head on with a proton beam. We consider the production of K^+'s in the K^-

[1] J. H. Christenson, J. W. Cronin, V. L. Fitch, and R. Turlay, Phys. Rev. Letters **13**, 138 (1964).

[2] T. D. Lee and C. S. Wu, Ann. Rev. Nucl. Sci. **16**, 471 (1966).

[3] M. Gourdin, Nucl. Phys. **B3**, 207 (1967).

[4] P. K. Kabir, Nature **220**, 1310 (1968).

[5] R. C. Casella, Phys. Rev. Letters **21**, 1128 (1968); **22**, 554 (1969).

[6] J. Ashkin and P. K. Kabir, Phys. Rev. D **1**, 868 (1970).

[7] J. Chollet *et al.*, Phys. Letters **31B**, 658 (1970).

[8] Some of the conclusions of this note are implicitly contained in the paper by C. P. Enz and R. R. Lewis, Helv. Phys. Acta. **38**, 860 (1965).

[9] As defined, for example by J. M. Blatt and V. F. Weisskopf, *Theoretical Nuclear Physics* (Wiley, New York, 1952), p. 529.

[10] T. D. Lee, R. Oehme, and C. N. Yang, Phys. Rev. **106**, 340 (1957).

[11] If K^0 and $\bar{K}^0$ states are distinguished, as at present, by their opposite values of strangeness, these probabilities will not be measurable with arbitrary accuracy, even in principle, because strangeness is not conserved absolutely. However, since strangeness-nonconserving interactions are expected to be of order 10^{-6} times weaker than strangeness-conserving interactions (see Ref. 2, for example), this effect is neglibible for our discussion.

[12] D. Dorfan *et al.*, Phys. Rev. Letters **19**, 987 (1967).

[13] J. Steinberger, in *Proceedings of the Lund International Conference on Elementary Particles, 1969*, edited by G. von Dardel (Berlingska, Lund, 1969).

[14] The constancy of the ratio of the two probabilities does not depend on TCP invariance.

beam direction, through the two-stage process

$$K^- + p \to n + \bar{K}^0 \searrow K^0 + p \to n + K^+. \quad (5)$$

Even if we assume that the charge-exchange processes are perfectly reversible—if they were not, one would have evidence for nonreversibility already—the rate of the inverse reaction

$$K^+ + n \to p + K^0 \searrow \bar{K}^0 + n \to p + K^- \quad (6)$$

will *not* be deducible directly from that of (5) simply because the probability for a K^0 to change into a $\bar{K}^0$ is not equal to that for a $\bar{K}^0$ to undergo the inverse transformation.

A corollary of our earlier theorem is that an incoherent equal mixture of K^0 and $\bar{K}^0$ particles will always be found to have acquired a net positive strangeness at a later time. This is because TCP invariance requires that the fraction of K^0's which survive as K^0 be exactly equal to that of $\bar{K}^0$'s which remain $\bar{K}^0$, while we have seen that the number of $\bar{K}^0$'s transforming into K^0 is expected to exceed that of K^0's undergoing the inverse transmutation. If the $\Delta S = \Delta Q$ rule is valid, this excess of positive strangeness, which is a manifestly CP-noninvariant effect, will be reflected in a corresponding excess of positively charged leptons in the leptonic decays of this beam. The mean strangeness at time τ is given by

$$S(\tau) = \frac{\langle K_2^0 | K_1^0 \rangle [e^{-\gamma_1 \tau} + e^{-\gamma_2 \tau} - 2e^{-(\gamma_1+\gamma_2)\tau/2} \cos(m_2 - m_1)\tau]}{e^{-\gamma_1 \tau} + e^{-\gamma_2 \tau} - 2e^{-(\gamma_1+\gamma_2)\tau/2} \langle K_2^0 | K_1^0 \rangle^2 \cos(m_2 - m_1)\tau}. \quad (7)$$

It increases from zero at $\tau = 0$ to a maximum of $\sim 1.1 \times \langle K_2^0 | K_1^0 \rangle$ at $T \sim 5\tau_1$, and finally reaches the asymptotic value of $\langle K_2^0 | K_1^0 \rangle$.

We now show that similar effects can be predicted even without assuming TCP invariance. Making no symmetry assumption whatever, the short- and long-lived kaon states can be written as[15]

$$|K_1^0\rangle = \cos\alpha_1 \, e^{i\beta} |K^0\rangle + \sin\alpha_1 \, e^{-i\beta} |\bar{K}^0\rangle, \quad (8a)$$

$$|K_2^0\rangle = \cos\alpha_2 \, e^{-i\beta} |K^0\rangle - \sin\alpha_2 \, e^{i\beta} |\bar{K}^0\rangle, \quad (8b)$$

where β, $(\frac{1}{4}\pi - \alpha_1)$, and $(\frac{1}{4}\pi - \alpha_2)$ were shown to be small angles not exceeding 10^{-2} in magnitude. From equations analogous to (3a) and (3b), we find the probability for a K^0 state to transform into $\bar{K}^0$ after time τ to be[16,17]

$$P_{\bar{K}K}(\tau) = (1 - |\langle K_2^0 | K_1^0 \rangle|^2)^{-1} \times |\theta_1 - \theta_2|^2 \sin^2\alpha_1 \sin^2\alpha_2, \quad (9a)$$

while the probability for a $\bar{K}^0$ to be transmuted into K^0 in the same time is

$$P_{K\bar{K}}(\tau) = (1 - |\langle K_2^0 | K_1^0 \rangle|^2)^{-1} \times |\theta_1 - \theta_2|^2 \cos^2\alpha_1 \cos^2\alpha_2. \quad (9b)$$

Thus

$$\frac{P_{K\bar{K}}(\tau) - P_{\bar{K}K}(\tau)}{P_{K\bar{K}}(\tau) + P_{\bar{K}K}(\tau)} = \frac{2\cos(\alpha_1 + \alpha_2)\cos(\alpha_1 - \alpha_2)}{1 + \cos 2\alpha_1 \cos 2\alpha_2} \approx 2 \,\mathrm{Re}\langle K_2^0 | K_1^0 \rangle \quad (10)$$

to lowest order in β, $(\frac{1}{4}\pi - \alpha_1)$, and $(\frac{1}{4}\pi - \alpha_2)$. Equation (10) is equivalent to our earlier result (4) if TCP invariance holds. Even if TCP invariance is not assumed, we can estimate the right-hand side of Eq. (10) by using the unitarity relation,[16] which yields

$$\tfrac{1}{2}(1 + 4\delta^2)^{1/2} \,\mathrm{Re}\langle K_2^0 | K_1^0 \rangle = \mathrm{Re}[e^{-i\phi_W} \sum_j \eta_j \gamma_1^j / (\gamma_1 + \gamma_2)], \quad (11)$$

where $\delta = (m_2 - m_1)/(\gamma_1 + \gamma_2)$ and $\phi_W = \tan^{-1}(2\delta) \approx 43°$. The contributions to the sum from 2π channels are expected to be much larger than those of other channels, for which we can set an upper bound by using Schwarz's inequality and known limits on the corresponding partial decay rates; a conservative estimate[4] based on available data is 6×10^{-4}. If we neglect γ_2/γ_1 in comparison with unity and set $C = \gamma_1^{+-}/\gamma_1^{00}$, the 2π contribution to (11) is given by[18]

$$[C|\eta_{+-}|\cos(\phi_{+-} - \phi_W) + |\eta_{00}|\cos(\phi_{00} - \phi_W)]/(C+1). \quad (12)$$

C is known[19,20] to be slightly greater than 2 and $|\eta_{+-}|$ is known with fair precision[21] to be

$$|\eta_{+-}| = (1.90 \pm 0.05) \times 10^{-13},$$

[15] P. Eberhard, Phys. Rev. Letters **16**, 150 (1966).

[16] J. S. Bell and J. Steinberger, in *Proceedings of the Oxford International Conference on Elementary Particles, 1965* (Rutherford High-Energy Laboratory, Chilton, Berkshire, England, 1966).

[17] P. K. Kabir, in *Springer Tracts in Modern Physics* (Springer, Berlin, 1970), Vol. 52, p. 91.

[18] If we assume no symmetry, the composition of K_1^0 and K_2^0 states may be slightly different from that assuming TCP invariance. Then the same experimental results would be represented by η's which differ slightly from those deduced assuming Eq. (1). From the known limits on β, $(\frac{1}{4}\pi - \alpha_1)$, and $(\frac{1}{4}\pi - \alpha_2)$, we can be confident that this would not change (12) by more than 1%.

[19] B. Gobbi *et al.*, Phys. Rev. Letters **22**, 682 (1969).

[20] J. G. Morfin and D. Sinclair, Phys. Rev. Letters **23**, 660 (1969).

[21] J. W. Cronin, in *Proceedings of the Fourteenth International Conference on High-Energy Physics, Vienna, 1968*, edited by J. Prentki and J. Steinberger (CERN, Geneva, 1968), p. 281.

and ϕ_{+-} coincides with ϕ_W within the experimental error of a few degrees.[22] The $\pi^+\pi^-$ contribution to (12) is therefore 1.3×10^{-3}. There is far greater uncertainty about η_{00}, but it is consistent with all reported measurements to say that it has a magnitude comparable to $|\eta_{+-}|$. The only reported[7] measurement of ϕ_{00} yielded a value in the first quadrant, which assures a positive contribution to (12). We can therefore predict a finite nonvanishing T asymmetry [Eq. (10)] of several parts in a thousand, independent of any symmetry assumptions. The expected asymmetry could vanish only in the unlikely circumstance that $|\eta_{00}|$ were significantly larger than $|\eta_{+-}|$ and had a phase ϕ_{00} differing from ϕ_W by considerably more than 90°.

I wish to thank Professor K. E. Eriksson and NORDITA for hospitality at the Chalmers Institute of Technology, where the basic argument was first presented at a Seminar.

[22] D. A. Jensen *et al.*, Phys. Rev. Letters **23**, 615 (1969).

Nuclear Physics B247 (1984) 293-312

CPT SYMMETRY AND NEUTRAL KAONS

V.V. BARMIN, V.G. BARYLOV, I.V. CHUVILO, G.V. DAVIDENKO, V.S. DEMIDOV, A.G. DOLGOLENKO, A.G. MESHKOVSKY, L.B. OKUN and V.A. SHEBANOV

Institute for Theoretical and Experimental Physics, Moscow, USSR

M. BALDO-CEOLIN, E. CALIMANI, S. CIAMPOLILLO, F. MATTIOLI, G. MIARI and A. SCONZA

Dipartimento di Fisica dell' Università di Padova, Padova, Italy, and Istituto Nazionale di Fisica Nucleare, Sezione di Padova, Padova, Italy

Received 2 July 1984

We discuss the accuracy of *CPT* invariance tests in the neutral kaon decays. The possible *CPT* non-invariance is parametrized by amplitudes of two types: (i) *CP* violating, but *T* conserving, and (ii) *T* violating, but *CP* conserving. It is shown that the experimental data are consistent with *CPT* symmetry essentially within two standard deviations.

The source of this not quite satisfactory agreement is the experimental value of the phase ϕ_{00} of the ratio $A(K_L^0 \to \pi^0\pi^0)/A(K_S^0 \to \pi^0\pi^0)$, which differs by two standard deviations from the phase ϕ_{+-} of the ratio $A(K_L^0 \to \pi^+\pi^-)/A(K_S^0 \to \pi^+\pi^-)$ decay amplitude. We argue that to get better agreement with *CPT* invariance, the value of the phase ϕ_{00} has to be equal to the value of the phase ϕ_{+-} within the errors of the latter. New, more accurate measurements of the phase ϕ_{00} would be very interesting to test *CPT* invariance, especially taking into account that neutral kaons allow the test with much better accuracy than any other particle.

1. Introduction

The experimental test of *CPT* invariance in the neutral kaon decays is not a new problem [1–10]. We return to the subject for two reasons.

First, a new experimental upper limit for the amplitude of the *CP*-violating $K_S^0 \to 3\pi^0$ decay [11] was obtained recently. This result allows one to test *CPT* invariance with a higher accuracy using the Bell–Steinberger relation. A short letter with the results of this analysis was published recently [12].

Second, as will be shown below, the subject deserves discussion by itself independently of the new data [11], and, to a certain extent, independently of the Bell–Steinberger relation. The point is that the phenomenological analysis in the literature [1–10] is slightly incomplete. Although from the theoretical-technical point of view this analysis boils down to multiplication and addition of complex numbers, nevertheless, as it will be seen below, twenty years were not enough to settle all the details of *CPT*-violation phenomenology. Surely this paper will not be the final one.

We will concentrate our attention on $K_{L,S}^0$ decays into states of two pions with $I = 0, 2$. In sects. 2 and 3 the amplitudes describing these decays are introduced and the phase of the K and $\bar{K}$ mesons wave functions is chosen. Here, on the whole we follow the papers [7, 10]. The only difference is that alongside with amplitudes which are CP violating ($CP=-1$) and T conserving ($T=+1$), we explicitly discuss amplitudes with $CP=+1$ and $T=-1$. It is stressed that the present experimental limits on the latter amplitudes are much less stringent than on the former.

The possible CPT violation in the channel with $I=2$ is analyzed in sects. 4 and 5. We analyze the channel with $I=0$ in sects. 6–9, and in connection with that we discuss the Bell–Steinberger relation (sect. 8) and the contributions of three-pion and semileptonic channels to the right-hand side of the relation (sect. 9). Our main conclusions are presented in sect. 10.

2. The isotopic amplitudes of $K_{L,S} \to 2\pi$ decays

Let us define as usual

$$K^0_{L,S} = (K^0_{2,1} + \varepsilon_{L,S} K^0_{1,2}) \frac{1}{\sqrt{1+|\varepsilon_{L,S}|^2}}, \tag{1}$$

where $K^0_{1,2}$ are the states with $CP=\pm 1$:

$$K^0_{1,2} = \sqrt{\tfrac{1}{2}}\,(K^0 \pm \bar{K}^0), \tag{2}$$

and ε_S and ε_L are measures of the "wrong" CP admixture in the K_S and K_L states respectively. As it will be seen later the parameters ε_S and ε_L are small; if their difference is not zero it will be a signal of CPT violation.

Let us introduce the parameters η_{+-} and η_{00} of $K^0_L \to \pi^+\pi^-$ and $K^0_L \to \pi^0\pi^0$ decays respectively:

$$\eta_{+-} = |\eta_{+-}| e^{i\phi_{+-}} = \frac{A(K^0_L \to \pi^+\pi^-)}{A(K^0_S \to \pi^+\pi^-)}, \tag{3}$$

$$\eta_{00} = |\eta_{00}| e^{i\phi_{00}} = \frac{A(K^0_L \to \pi^0\pi^0)}{A(K^0_S \to \pi^0\pi^0)}. \tag{4}$$

According to ref. [13]

$$|\eta_{+-}| = (2.274 \pm 0.022) \times 10^{-3}, \qquad \phi_{+-} = 44.6° \pm 1.2°, \tag{5}$$

$$|\eta_{00}| = (2.33 \pm 0.08) \times 10^{-3}, \qquad \phi_{00} = 54.5° \pm 5.3°. \tag{6}$$

The most accurate measurement of the $\phi_{00} - \phi_{+-}$ phase difference was done in experiment [14]:

$$\phi_{00} - \phi_{+-} = 12.6° \pm 6.2°. \tag{7}$$

The fitted value according to the tables in [13] is

$$\phi_{00} - \phi_{+-} = 9.8° \pm 5.4°. \tag{8}$$

Let us introduce as usual the parameters ε_0 and ε_2 describing the $K^0_L \to 2\pi$ decays with $I=0, 2$:

$$\varepsilon_0 = \frac{A(K^0_L \to 2\pi,\ I=0)}{A(K^0_S \to 2\pi,\ I=0)}, \tag{9}$$

$$\varepsilon_2 = \frac{1}{\sqrt{2}} \frac{A(K_L^0 \to 2\pi, I=2)}{A(K_S^0 \to 2\pi, I=0)}, \tag{10}$$

and the parameter ω:

$$\omega = \frac{1}{\sqrt{2}} \frac{A(K_S^0 \to 2\pi, I=2)}{A(K_S^0 \to 2\pi, I=0)}. \tag{11}$$

Using the Clebsch–Gordan coefficients let us express the $\pi^+\pi^-$ and $\pi^0\pi^0$ states through the states with isospin $I = 0, 2$:

$$|\pi^+\pi^-\rangle = \sqrt{\tfrac{2}{3}}|2\pi, I=0\rangle + \sqrt{\tfrac{1}{3}}|2\pi, I=2\rangle, \tag{12}$$

$$|\pi^0\pi^0\rangle = \sqrt{\tfrac{1}{3}}|2\pi, I=0\rangle - \sqrt{\tfrac{2}{3}}|2\pi, I=2\rangle, \tag{13}$$

and obtain the relations between amplitudes:

$$A(K_{L,S}^0 \to 2\pi, I=0) = \sqrt{\tfrac{2}{3}}A(K_{L,S}^0 \to \pi^+\pi^-) + \sqrt{\tfrac{1}{3}}A(K_{L,S}^0 \to \pi^0\pi^0), \tag{14}$$

$$A(K_{L,S}^0 \to 2\pi, I=2) = \sqrt{\tfrac{1}{3}}A(K_{L,S}^0 \to \pi^+\pi^-) - \sqrt{\tfrac{2}{3}}A(K_{L,S}^0 \to \pi^0\pi^0). \tag{15}$$

Further, one finds

$$\begin{aligned} A(K_L^0 \to 2\pi, I=0) &= \sqrt{\tfrac{2}{3}}\eta_{+-}A(K_S^0 \to \pi^+\pi^-) + \sqrt{\tfrac{1}{3}}\eta_{00}A(K_S^0 \to \pi^0\pi^0) \\ &= \tfrac{2}{3}\eta_{+-}A(K_S^0 \to 2\pi, I=0) + \tfrac{1}{3}\eta_{00}A(K_S^0 \to 2\pi, I=0) \\ &\quad + \tfrac{1}{3}\sqrt{2}\eta_{+-}A(K_S^0 \to 2\pi, I=2) \\ &\quad - \tfrac{1}{3}\sqrt{2}\eta_{00}A(K_S^0 \to 2\pi, I=2). \end{aligned} \tag{16}$$

Then

$$\begin{aligned} \varepsilon_0 &= \tfrac{1}{3}(2\eta_{+-} + \eta_{00}) + \tfrac{2}{3}(\eta_{+-} - \eta_{00})\omega \\ &= \tfrac{2}{3}\eta_{+-}(1+\omega) + \tfrac{1}{3}\eta_{00}(1-2\omega). \end{aligned} \tag{17}$$

By analogy one gets

$$\begin{aligned} A(K_L^0 \to 2\pi, I=2) &= \sqrt{\tfrac{1}{3}}\eta_{+-}A(K_S^0 \to \pi^+\pi^-) - \sqrt{\tfrac{2}{3}}\eta_{00}A(K_S^0 \to \pi^0\pi^0) \\ &= \tfrac{1}{3}\sqrt{2}\eta_{+-}A(K_S^0 \to 2\pi, I=0) - \tfrac{1}{3}\sqrt{2}\eta_{00}A(K_S^0 \to 2\pi, I=0) \\ &\quad + \tfrac{1}{3}\eta_{+-}A(K_S^0 \to 2\pi, I=2) + \tfrac{2}{3}\eta_{00}A(K_S^0 \to 2\pi, I=2), \end{aligned} \tag{18}$$

$$\varepsilon_2 = \tfrac{1}{3}(\eta_{+-} - \eta_{00}) + \tfrac{1}{3}(\eta_{+-} + 2\eta_{00})\omega. \tag{19}$$

All the expressions of this paragraph are well known. We have listed them for the sake of completeness and to facilitate the subsequent analysis.

3. *CPT*-conserving and *CPT*-violating amplitudes

In order to discuss the accuracy of the experimental tests of *CPT* invariance it is necessary to introduce explicitly *CPT*-violating amplitudes alongside with

CPT-conserving ones. Let us define

$$A(K^0 \to 2\pi, I) = (A_I + B_I)\, e^{i\delta_I}\,, \tag{20}$$

$$A(\bar{K}^0 \to 2\pi, I) = (A_I^* - B_I^*)\, e^{i\delta_I}\,. \tag{21}$$

Here A_I are *CPT*-invariant amplitudes, B_I are *CPT*-non-invariant amplitudes, δ_I are the $I = 0$ and $I = 2$ $\pi\pi$ scattering phase shifts. It follows from the definition of K_1^0 and K_2^0 (see eq. (2)) that

$$A(K_1^0 \to 2\pi, I) = \sqrt{2}(\mathrm{Re}\, A_I + i\, \mathrm{Im}\, B_I)\, e^{i\delta_I}\,, \tag{22}$$

$$A(K_2^0 \to 2\pi, I) = \sqrt{2}(i\, \mathrm{Im}\, A_I + \mathrm{Re}\, B_I)\, e^{i\delta_I}\,. \tag{23}$$

It is easy to see that the amplitudes

Re A_I are *CPT* invariant, *CP* invariant and *T* invariant;
Im A_I are *CPT* invariant, *CP* non-invariant and *T* non-invariant;
Re B_I are *CPT* non-invariant, *CP* non-invariant and *T* invariant;
Im B_I are *CPT* non-invariant, *CP* invariant and *T* non-invariant.

In the literature the discussion of experimental tests of *CPT* invariance is usually confined to the possible effects of amplitudes of the type Re B_I. Such amplitudes give contributions of opposite sign to the decays of particles and antiparticles. As regards the amplitudes of the type Im B_I, they give the same contribution to the decays of the particles and antiparticles, but this contribution has a "wrong" phase (it contradicts the unitarity of the *S*-matrix, the hermiticity of the hamiltonian).

Let us refer to the case when $|B_I| \lesssim |\mathrm{Re}\, A_I|$ as superstrong *CPT* violation, when $|B_I| \lesssim |\mathrm{Im}\, A_I|$ as strong *CPT* violation and when $|B_I| \ll \mathrm{Im}\, A_I$ as a weak *CPT* violation.

As it will be shown below by considering as an example the parameter ω, the present experimental limits on the amplitudes Im B_I are much less stringent than on the amplitudes Re B_I. The limits on the amplitudes Im B_I do not exclude the possibility of a superstrong *CPT* violation. However, we consider this possibility as absolutely unrealistic and we will assume that $|\mathrm{Im}\, B_I| \ll |\mathrm{Re}\, A_I|$. We will be mainly interested in the limits on the *CPT*-violating amplitudes Re B_I and on the possible violation of *CPT* invariance in the parameters ε_L and ε_S.

Let us express the parameters ε_0, ε_2 and ω through the amplitudes A_I and B_I. We will neglect terms of order $\varepsilon_{L,S}$ when they are small corrections, and of course will take them into account when their contribution is essential.

Using eqs. (1), (2), (9)–(11), (22) and (23) one obtains

$$\varepsilon_0 = \frac{A(K_L^0 \to 2\pi, I=0)}{A(K_S^0 \to 2\pi, I=0)} \simeq \frac{A(K_2^0 \to 2\pi, I=0) + \varepsilon_L A(K_1^0 \to 2\pi, I=0)}{A(K_1^0 \to 2\pi, I=0)} = \varepsilon_L + a\,, \tag{24}$$

where

$$a = \frac{A(K_2^0 \to 2\pi, I=0)}{A(K_1^0 \to 2\pi, I=0)} = \frac{\text{Re } B_0 + i \text{ Im } A_0}{\text{Re } A_0 + i \text{ Im } B_0}, \tag{25}$$

$$\varepsilon_2 = \frac{1}{\sqrt{2}} \frac{A(K_L^0 \to 2\pi, I=2)}{A(K_S^0 \to 2\pi, I=0)} \simeq \frac{A(K_2^0 \to 2\pi, I=2) + \varepsilon_L A(K_1^0 \to 2\pi, I=2)}{\sqrt{2} A(K_1^0 \to 2\pi, I=0)}$$

$$= \varepsilon' + \varepsilon_L \omega, \tag{26}$$

where

$$\varepsilon' = \frac{1}{\sqrt{2}} \frac{A(K_2^0 \to 2\pi, I=2)}{A(K_1^0 \to 2\pi, I=0)} = \frac{1}{\sqrt{2}} \frac{\text{Re } B_2 + i \text{ Im } A_2}{\text{Re } A_0 + i \text{ Im } B_0} e^{i(\delta_2 - \delta_0)}, \tag{27}$$

and, finally

$$\omega = \frac{1}{\sqrt{2}} \frac{A(K_S^0 \to 2\pi, I=2)}{A(K_S^0 \to 2\pi, I=0)} \simeq \frac{1}{\sqrt{2}} \frac{A(K_1^0 \to 2\pi, I=2)}{A(K_1^0 \to 2\pi, I=0)}$$

$$= \frac{1}{\sqrt{2}} \frac{\text{Re } A_2 + i \text{ Im } B_2}{\text{Re } A_0 + i \text{ Im } B_0} e^{i(\delta_2 - \delta_0)}. \tag{28}$$

One can find from eqs. (24)–(28) that *CPT* invariance implies

$$\arg a = \pm \tfrac{1}{2}\pi, \tag{29}$$

$$\arg \varepsilon' = \delta_2 - \delta_0 \pm \tfrac{1}{2}\pi, \tag{30}$$

$$\arg \omega = \delta_2 - \delta_0 \ (\text{or } \delta_2 - \delta_0 + \pi). \tag{31}$$

A disagreement of experimental data with any of these equalities would be a signal of *CPT* violation.

When discussing the phases of the various parameters one has not to forget that the phase of the state K may be chosen arbitrarily (the phase of the state $\bar{K}$ has to be the same as that of K but with opposite sign). Following the papers [7, 10], we will choose the phases of K and $\bar{K}$ so that

$$\text{Im } a = 0. \tag{32}$$

The convention (32) is a generalization of the convention of Wu and Yang [15]. As is well known the convention of Wu and Yang applies to the case of *CPT* invariance and has the form Im $A_0 = 0$, which means that under this convention the direct transitions $K_2^0 \to 2\pi$, $I = 0$ are forbidden. Comparing eqs. (32) and (29) we see that *CPT* invariance implies that

$$a = 0. \tag{33}$$

4. On the parameter ω

According to eq. (31), *CPT* invariance requires the phase of the parameter ω to be equal to $\delta_2 - \delta_0$ or $\delta_2 - \delta_0 + \pi$. A summary of data on the $\delta_2 - \delta_0$ phase difference can be found in papers [10, 16]. In particular, from the analysis of pion production by pions, one finds $\delta_2 - \delta_0 = -40° \pm 6°$. From K_{e4} decays, one finds $\delta_2 - \delta_0 = -36° \pm 10°$. According to ref. [16] a weighted world average is $\delta_2 - \delta_0 = -41.4° \pm 8.1°$ and the fitted one is $\delta_2 - \delta_0 = -45.3° \pm 4.6°$. On the other hand the comparison of $K_S^0 \to \pi^+\pi^-$, $K_S^0 \to \pi^0\pi^0$ and $K^+ \to \pi^+\pi^0$ decay rates allows one to find $|\omega|$ and Re ω:

$$|\omega|^2 \simeq \frac{2}{3} \frac{\Gamma(K^+ \to \pi^+\pi^0)}{\Gamma(K_S^0 \to 2\pi)}, \tag{34}$$

$$\mathrm{Re}\,\omega \simeq \frac{1}{12}\left(\frac{\rho_{00}}{\rho_{+-}} \frac{\Gamma(K_S^0 \to \pi^+\pi^-)}{\Gamma(K_S^0 \to \pi^0\pi^0)} - 2\right), \tag{35}$$

where the ρ's are the phase space factors*.

From these relations one obtains

$$|\omega| = 31.9 \cdot 10^{-3}, \tag{36}$$

$$\arg \omega = \pm 55.6°. \tag{37}$$

From the above value of $|\omega|$ and assuming *CPT* invariance one would obtain

$$\frac{\mathrm{Re}\,A_2}{\mathrm{Re}\,A_0} = 4.5 \cdot 10^{-2}. \tag{38}$$

If the minus sign is chosen in eq. (37) (the plus sign would mean catastrophic *CPT* violation), one finds a rough compatibility with *CPT* invariance:

$$\arg \omega \simeq \delta_2 - \delta_0 \simeq -45°. \tag{39}$$

Note, however, that on the basis of existing data it is impossible to exclude a *CPT*-violating phase $\arg \omega - \delta_2 + \delta_0$ of the order of 10°–20°. In the following we suppose that there is no such superstrong *CPT* violation with $\mathrm{Im}\,B_2 \leqslant \frac{1}{5}\,\mathrm{Re}\,A_2$ and/or $\mathrm{Im}\,B_0 \leqslant \frac{1}{5}\,\mathrm{Re}\,A_0$ (see (28)) and assume that $\arg \omega = \delta_2 - \delta_0$. Then, from the branching ratio of K_S^0 into $\pi^+\pi^-$ and $\pi^0\pi^0$ channels (see (35)) we obtain

$$|\omega| = (26 \pm 4) \cdot 10^{-3}, \tag{40}$$

and from (28)

$$\mathrm{Re}\,A_2/\mathrm{Re}\,A_0 = (3.7 \pm 0.6) \cdot 10^{-2}. \tag{41}$$

Some possible small deviations of $\arg \omega$ from $\delta_2 - \delta_0$ will not change the results of our subsequent analysis.

* To obtain the equation for $|\omega|^2$ one has to assume that there are no transitions with $\Delta I = \frac{5}{2}$, see for example ref. [17], and for the discussion of electromagnetic corrections and further references see ref. [18].

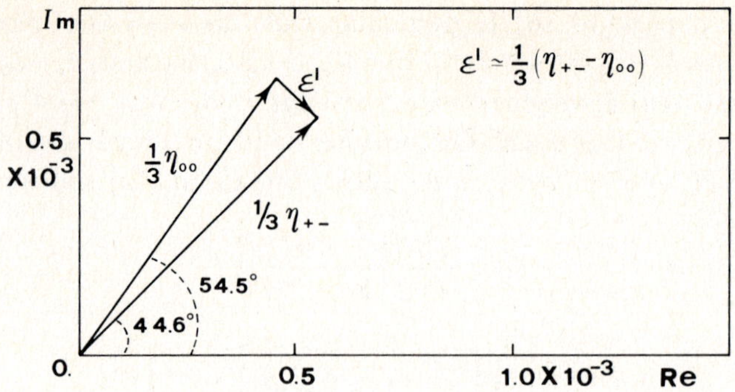

Fig. 1. The parameter ε' computed according to the approximate expression (43).

5. On the parameter ε'

Let us consider now the parameter ε' which describes the direct decay $K_2^0 \to 2\pi$, $I = 2$ (see eqs. (26) and (27)). By using eq. (19) it is convenient to rewrite the equation $\varepsilon_2 = \varepsilon' + \varepsilon_L \omega$ in the form

$$\varepsilon' = \tfrac{1}{3}(\eta_{+-} - \eta_{00}) + (\tfrac{1}{3}\eta_{+-} + \tfrac{2}{3}\eta_{00} - \varepsilon_L)\omega . \tag{42}$$

Taking into account that $|\omega| \ll 1$ and that according to the experimental data the possible difference between η_{+-} and η_{00} and ε_L is small (this point will be discussed in detail in sects. 8 and 9) one can neglect the second term in eq. (42). Then we obtain the very simple relation

$$\varepsilon' \simeq \tfrac{1}{3}(\eta_{+-} - \eta_{00}) . \tag{43}$$

Consider now the projections of the vector ε': $\varepsilon'_\perp$ and $\varepsilon'_\parallel$, the former being orthogonal and the latter parallel to the vector η_{+-} in the complex plane (see fig. 1). Using the known values for η_{+-} and η_{00} (eqs. (5), (6)), one can easily find that

$$\varepsilon'_\perp \simeq \tfrac{1}{3}|\eta_{+-}|(\phi_{+-} - \phi_{00}) = (-0.13 \pm 0.07) \cdot 10^{-3} , \tag{44}$$

$$\varepsilon'_\parallel \simeq \tfrac{1}{3}(|\eta_{+-}| - |\eta_{00}|) = (-0.02 \pm 0.03) \cdot 10^{-3} . \tag{45}$$

In fig. 2 are shown the one- and two-standard deviations contours for the parameter ε', obtained without any approximation from the exact relation (42). It is seen from (44), (45) and figs. 1 and 2 that in the complex plane the vector ε' has an appreciable component orthogonal to the vector η_{+-} whose phase is equal to about $\frac{1}{4}\pi$. So experimentally $\varepsilon'_\perp$ looks dominant while *CPT* invariance requires (see eq. (30)) the component $\varepsilon'_\parallel$ to be dominant.

The value $\varepsilon' = 0$ (*CPT* and *CP* conserved in the transition $K_2^0 \to 2\pi$, $I = 2$) turns out to be compatible with the experimental data only within the 15% confidence

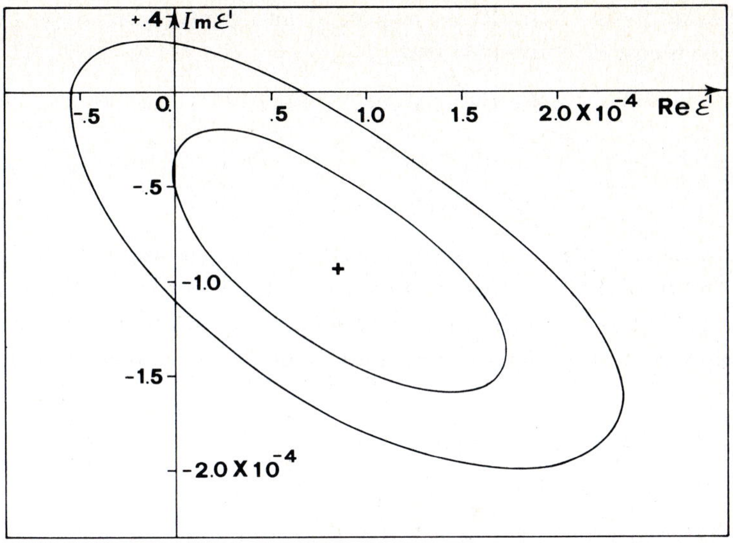

Fig. 2. One- and two-standard deviation contours for the exact computation of ε'.

level. Furthermore, the *CPT* invariance hypothesis in the same transition is compatible with the data at the 18% level or less*.

With the present world average values of ϕ_{+-} and ϕ_{00}, if the error of ϕ_{00} were smaller, we would face a strong *CPT* violation: according to eq. (27) a dominating $\varepsilon'_{\perp}$ implies that

$$|\mathrm{Re}\, B_2| \gg |\mathrm{Im}\, A_2| , \tag{46}$$

and/or

$$|\mathrm{Im}\, B_0| \gg |\mathrm{Re}\, A_0| , \tag{47}$$

Case (47) requires such a superstrong *CPT* violation that it seems absolutely unacceptable. So we will discuss here case (46) assuming that $|\mathrm{Im}\, B_0| \ll |\mathrm{Re}\, A_0|$.

* If, assuming *CPT* invariance, we try to estimate a limit for the ratio ε'/η_{+-} we obtain at the 90% confidence level

$$-0.021 \leq \frac{\varepsilon'}{\eta_{+-}} \leq 0.004 .$$

From this confidence interval for ε' using (27) we can obtain the corresponding interval for $\mathrm{Im}\, A_2/\mathrm{Re}\, A_0$:

$$-6.7 \cdot 10^{-5} \leq \mathrm{Im}\, A_2/\mathrm{Re}\, A_0 \leq 1.3 \cdot 10^{-5} ,$$

and using (41) we can estimate in the *CPT*-invariance hypothesis a 90% confidence interval for $\mathrm{Im}\, A_2/\mathrm{Re}\, A_2$:

$$-1.7 \cdot 10^{-3} \leq \mathrm{Im}\, A_2/\mathrm{Re}\, A_2 \leq 0.3 \cdot 10^{-3} ,$$

For the sake of simplicity let us neglect $\varepsilon'_{\parallel}$ and consider $\varepsilon'_{\perp}$ only. In this case it follows from eqs. (27) and (44) that

$$\mathrm{Re}\, B_2/\mathrm{Re}\, A_0 \simeq (1.8 \pm 1) \cdot 10^{-4}, \tag{48}$$

and from (41)

$$\mathrm{Re}\, B_2/\mathrm{Re}\, A_2 = (4.8 \pm 2.8) \cdot 10^{-3}. \tag{49}$$

It is interesting to compare (49) with the limit on the violation of *CPT* invariance which follows from the comparison of experimental data for the decays of K^+ and K^- mesons. If there is no special compensation between $\Delta I = \frac{5}{2}$ and $\Delta I = \frac{3}{2}$ transitions (we have no reason for such a compensation) then for $K^{\pm} \to \pi^{\pm}\pi^0$ decays we would get

$$\frac{\Gamma(K^+ \to \pi^+\pi^0) - \Gamma(K^- \to \pi^-\pi^0)}{\Gamma(K^+ \to \pi^+\pi^0)} \simeq 4\frac{\mathrm{Re}\, B_2}{\mathrm{Re}\, A_2} \simeq (19 \pm 11) \cdot 10^{-3}, \tag{50}$$

which is compatible with the experimental data [13]:

$$2\frac{\Gamma(K^+ \to \pi^+\pi^0) - \Gamma(K^- \to \pi^-\pi^0)}{\Gamma(K^+ \to \pi^+\pi^0) + \Gamma(K^- \to \pi^-\pi^0)} = (8 \pm 12) \cdot 10^{-3}. \tag{51}$$

We can also compare (50) with the experimental data on the lifetimes of K^+ and K^-. Taking into account that BR $(K^+ \to \pi^+\pi^0) = 21\%$ we would get from (50)

$$\frac{\tau^+ - \tau^-}{\tau^+} \simeq (4 \pm 2.3) \cdot 10^{-3}. \tag{52}$$

This has to be compared with the world average [13]

$$\frac{\tau^+ - \tau^-}{\tau^+} \simeq (1.1 \pm 0.9) \cdot 10^{-3}. \tag{53}$$

Thus, considering the parameter ε' we again come to the conclusion that compatibility with *CPT* invariance is not too good. The most probable cause of this is the difference of the ϕ_{+-} and ϕ_{00} phases.

It would be interesting to improve the experimental accuracy in ε', especially in $\varepsilon'_{\perp}$ and at the same time to improve the accuracy in the comparison of the lifetimes of K^+ and K^-.

6. On the parameter a and the parameters ε_0 and δ

Let us begin with the parameter ε_0. From eqs. (17) and (24) it follows:

$$\varepsilon_0 = \varepsilon_L + a = \tfrac{1}{3}(2\eta_{+-} + \eta_{00}) + \tfrac{2}{3}(\eta_{+-} - \eta_{00})\omega, \tag{54}$$

or

$$\mathrm{Re}\, a = \mathrm{Re}\, \eta_{+-} - \mathrm{Re}\, \varepsilon_L - \tfrac{1}{3}\, \mathrm{Re}\, (\eta_{+-} - \eta_{00}) + \tfrac{2}{3}\, \mathrm{Re}\, [(\eta_{+-} - \eta_{00})\omega]. \tag{55}$$

We have singled out Re a just because according to (29) the inequality Re $a \pm 0$ signals *CPT* violation. As for Re ε_L, it is well known (see for example [17]) that this parameter determines the value of the leptonic charge asymmetry in the K_L^0 decays:

$$\delta = \frac{\Gamma(K_L^0 \to \ell^+ \pi^- \nu) - \Gamma(K_L^0 \to \ell^- \pi^+ \bar{\nu})}{\Gamma(K_L^0 \to \ell^+ \pi^- \nu) + \Gamma(K_L^0 \to \ell^- \pi^+ \bar{\nu})}$$

$$= \frac{|1+\varepsilon_L|^2 - |1-\varepsilon_L|^2}{|1+\varepsilon_L|^2 + |1-\varepsilon_L|^2} \simeq 2 \text{ Re } \varepsilon_L \tag{56}$$

(here $\ell = e, \mu$).

The experimental data on δ(see [13]) give

$$\delta = (3.30 \pm 0.12) \cdot 10^{-3}, \tag{57}$$

and therefore

$$\text{Re } \varepsilon_L = (1.65 \pm 0.06) \cdot 10^{-3}. \tag{58}$$

In eq. (56) we have neglected the possible violation of the $\Delta Q = \Delta S$ rule, we will take it into account later on, see eq. (62).

From the experimental data (5) and (6) it follows:

$$\text{Re } \eta_{+-} = (1.62 \pm 0.04) \cdot 10^{-3}, \tag{59}$$

$$\text{Re } \eta_{00} = (1.37 \pm 0.17) \cdot 10^{-3}. \tag{60}$$

Thus, neglecting the last term in eq. (55) we have

$$\text{Re } a = (-0.12 \pm 0.08) \cdot 10^{-3}. \tag{61}$$

It is seen that the value Re a is compatible with zero within 1.5 standard deviations.

Let us now compute Re ε_L taking into account the possible admixture of the $\Delta Q = -\Delta S$ amplitude. In this case (see, for example [13])

$$\text{Re } \varepsilon_L = \frac{\delta}{2} \frac{|1-x|^2}{1-|x|^2}, \tag{62}$$

where

$$x = \frac{A(\bar{K}^0 \to \ell^+ \pi^- \nu)}{A(K^0 \to \ell^+ \pi^- \nu)}. \tag{63}$$

On the basis of experimental data on δ (57) and using the experimental value of x [13]

$$x = [(9 \pm 20) - i(4 \pm 26)] \cdot 10^{-3}, \tag{64}$$

we would obtain

$$\text{Re } \varepsilon_L = (1.64 \pm 0.08) \cdot 10^{-3}, \tag{65}$$

$$\text{Re } a = (-0.10 \pm 0.11) \cdot 10^{-3}. \tag{66}$$

Thus the comparison of ε_0 and δ shows no signs of *CPT* violation. As we will see below the agreement becomes worse when we compare ε_0 with ε_L and ε_S.

7. On the parameters ε_L, ε_S, ε and Δ

We introduce the parameters ε and Δ following refs. [7, 10]:

$$\varepsilon_L = \varepsilon - \Delta\,, \tag{67}$$

$$\varepsilon_S = \varepsilon + \Delta\,. \tag{68}$$

While *CPT* violation in the $|\Delta S| = 1$ transitions is parametrized by the amplitudes B_I, the *CPT* violation in the $\Delta S = 0$ and $|\Delta S| = 2$ transitions for the neutral kaons system is parametrized by ε_L and ε_S. The *CPT* violating transitions with $\Delta S = 0$ give a non-zero value of Δ, while those with $|\Delta S| = 2$ give a "wrong" phase of ε.

Let us consider the *CP*-odd mixing $K_2^0 \leftrightarrow K_1^0$ due to which the states K_L^0 and K_S^0 arise. Using the perturbation theory it is easy to write

$$\varepsilon_L = \frac{V_{12}}{(m_2 - m_1) - \frac{1}{2}i(\Gamma_2 - \Gamma_1)} \simeq \frac{V_{12}}{(m_L - m_S) - \frac{1}{2}i(\Gamma_L - \Gamma_S)}\,, \tag{69}$$

$$\varepsilon_S = \frac{V_{21}}{(m_1 - m_2) - \frac{1}{2}i(\Gamma_1 - \Gamma_2)} \simeq \frac{V_{21}}{(m_S - m_L) - \frac{1}{2}i(\Gamma_S - \Gamma_L)}\,. \tag{70}$$

The matrix elements V_{12} and V_{21} have the form

$$\begin{aligned} V_{12} &= \langle K_1^0|H|K_2^0\rangle = \tfrac{1}{2}\langle K^0 + \bar{K}^0|H|K^0 - \bar{K}^0\rangle \\ &= \tfrac{1}{2}[\langle K^0|H|K^0\rangle - \langle \bar{K}^0|H|\bar{K}^0\rangle - \langle K^0|H|\bar{K}^0\rangle + \langle \bar{K}^0|H|K^0\rangle]\,, \end{aligned} \tag{71}$$

$$\begin{aligned} V_{21} &= \langle K_2^0|H|K_1^0\rangle = \tfrac{1}{2}\langle K^0 - \bar{K}^0|H|K^0 + \bar{K}^0\rangle \\ &= \tfrac{1}{2}[\langle K^0|H|K^0\rangle - \langle \bar{K}^0|H|\bar{K}^0\rangle + \langle K^0|H|\bar{K}^0\rangle - \langle \bar{K}^0|H|K^0\rangle]\,. \end{aligned} \tag{72}$$

So that

$$\begin{aligned} \varepsilon &= \tfrac{1}{2}(\varepsilon_S + \varepsilon_L) = \frac{1}{2}\,\frac{-V_{21} + V_{12}}{(m_L - m_S) + \frac{1}{2}i(\Gamma_S - \Gamma_L)} \\ &= \frac{1}{2}\,\frac{\langle \bar{K}^0|H|K^0\rangle - \langle K^0|H|\bar{K}^0\rangle}{(m_L - m_S) + \frac{1}{2}i(\Gamma_S - \Gamma_L)}\,, \end{aligned} \tag{73}$$

$$\begin{aligned} \Delta &= \tfrac{1}{2}(\varepsilon_S - \varepsilon_L) = \frac{1}{2}\,\frac{-V_{21} - V_{12}}{(m_L - m_S) + \frac{1}{2}i(\Gamma_S - \Gamma_L)} \\ &= \frac{1}{2}\,\frac{\langle \bar{K}^0|H|\bar{K}^0\rangle - \langle K^0|H|K^0\rangle}{(m_L - m_S) + \frac{1}{2}i(\Gamma_S - \Gamma_L)}\,. \end{aligned} \tag{74}$$

If *CPT* is not violated ($CPT = +1$), one has

$$\langle K^0|H|K^0\rangle - \langle \bar{K}^0|H|\bar{K}^0\rangle = 0\,.$$

This means that

$$V_{12} = -V_{21},$$

and it follows from eqs. (69), (70), (67), (68) and (74) that

$$\varepsilon_L = \varepsilon_S = \varepsilon, \tag{75}$$

$$\Delta = 0. \tag{76}$$

Furthermore, if $CPT = +1$ and $CP = -1$, then $T = -1$ and, therefore, the term $\langle \bar{K}^0|H|K^0\rangle - \langle K^0|H|\bar{K}^0\rangle$ has to be purely imaginary if all the intermediate states in the $K_2^0 \leftrightarrow K_1^0$ transitions are off-mass shell. This determines (see eq. (73)) the phase of ε, which in this case is the same as in the case of the superweak interaction [19]:

$$\arg \varepsilon = \phi_{SW}, \tag{77}$$

where

$$\phi_{SW} = \text{arc tg}\, \frac{2(m_L - m_S)}{\Gamma_S - \Gamma_L} = 43.72° \pm 0.14°. \tag{78}$$

Note that the experimental value of the phase ϕ_{+-} is compatible within one standard deviation with ϕ_{SW}, while the experimental value of the phase ϕ_{00} is compatible with ϕ_{SW} within two standard deviations. The nearness of all three phases indicates that the deviation from CPT symmetry, if it takes place at all, has to be small.

In order to perform a more accurate comparison of the experimental data with the requirements of CPT symmetry it is necessary to take into account not only off-mass shell transitions $K_2^0 \leftrightarrow K_1^0$ but also on-mass shell transitions $K_2^0 \leftrightarrow K_1^0$ through the common channels 2π, 3π, $\pi e\nu$, $\pi\mu\nu$, $\pi\pi\gamma$, etc. Due to these channels the term $\langle \bar{K}^0|H|K^0\rangle - \langle K^0|H|\bar{K}^0\rangle$ acquires a non-zero absorptive part, which is real, and $\arg \varepsilon$ differs from ϕ_{SW} even if CPT is conserved. To take into account this difference between $\arg \varepsilon$ and ϕ_{SW} it is necessary to consider the unitarity relation which for the neutral kaon system is known as the Bell–Steinberger relation. The other reason for considering the Bell–Steinberger relation is that the CP-violating effects are observed up to now in the K_L^0 mesons decays only. It is much more difficult to observe these effects with K_S^0 mesons. Thus the value of ε_S remains inaccessible to experiment. On the other hand ε_S enters the Bell–Steinberger relation and so it is possible to obtain some knowledge about the value of ε_S and hence about Δ by analyzing this relation.

8. On the parameter $\tilde{\Delta}$ and the Bell–Steinberger relation

By equating the decrement of kaons to the increment of their decay products it is easy to derive the Bell–Steinberger relation [1]

$$[-i(m_S - m_L) + \tfrac{1}{2}(\Gamma_S + \Gamma_L)]\langle K_S^0|K_L^0\rangle = \sum_f A^*(K_S^0 \to f) A(K_L^0 \to f), \tag{79}$$

where f numerates the decay channels: $2\pi, I=0$; $2\pi, I=2$; 3π; $\pi e\nu$; $\pi\mu\nu$; etc. Picking out the main channel 2π, $I=0$, which practically determines Γ_S it is easy to rewrite eq. (79) in the form

$$(1+i\,\mathrm{tg}\,\phi_{SW})\tfrac{1}{2}(\varepsilon_S^*+\varepsilon_L)\simeq\varepsilon_0(1-2|\omega|^2)+\alpha\,. \tag{80}$$

In eq. (80) we have neglected Γ_L with respect to Γ_S and also neglected terms of order $|\varepsilon_L|^2$ and $|\varepsilon_S|^2$ and finally introduced the quantity

$$\alpha=\frac{1}{\Gamma_S}\sum_f{}' A^*(K_S^0\to f)A(K_L^0\to f)\,, \tag{81}$$

where the symbol $\sum'$ indicates summation over all states f, except 2π, $I=0$. The correction $(1-2|\omega|^2)$ is due to the equation*

$$\Gamma_S\simeq|A(K_S^0\to 2\pi, I=0)|^2\cdot(1+2|\omega|^2)\,. \tag{82}$$

Let us for the sake of argument neglect for the moment all decay channels except 2π, $I=0, 2$. In this case relation (80) may be rewritten in the form

$$(1+i\,\mathrm{tg}\,\phi_{SW})(\mathrm{Re}\,\varepsilon-i\,\mathrm{Im}\,\Delta)=\varepsilon_0(1-2|\omega|^2)+\alpha_{2\pi,I=2}\,, \tag{83}$$

or using eqs. (81), (82), (24) and (26)

$$(1+i\,\mathrm{tg}\,\phi_{SW})(\mathrm{Re}\,\varepsilon-i\,\mathrm{Im}\,\Delta)=\varepsilon_0+2\omega^*\varepsilon'\,. \tag{84}$$

Let us now take into account (see eqs. (24) and (67)) that

$$\varepsilon_0=\varepsilon_L+a=\varepsilon-\Delta+a\,, \tag{85}$$

and, following ref. [7], introduce the parameter $\tilde{\Delta}$:

$$\tilde{\Delta}=\Delta-a\,. \tag{86}$$

On the generalized Wu–Yang gauge (32) $\mathrm{Im}\,a=0$ and hence

$$\mathrm{Im}\,\tilde{\Delta}=\mathrm{Im}\,\Delta\,. \tag{87}$$

As a result we have two equations for ε and $\tilde{\Delta}$:

$$(1+i\,\mathrm{tg}\,\phi_{SW})(\mathrm{Re}\,\varepsilon-i\,\mathrm{Im}\,\tilde{\Delta})=\varepsilon-\tilde{\Delta}+2\omega^*\varepsilon'\,, \tag{88}$$

$$\varepsilon-\tilde{\Delta}=\varepsilon_0\,. \tag{89}$$

It is convenient, following (2), (10), to introduce the projections of the vectors $\tilde{\Delta}$ and ε: $\tilde{\Delta}_\parallel$, $\varepsilon_\parallel$, $\tilde{\Delta}_\perp$ and $\varepsilon_\perp$, where indexes $_\parallel$ and $_\perp$ refer to the direction determined by the angle ϕ_{SW} on the complex plane:

$$\tilde{\Delta}_\parallel=\cos\phi_{SW}\,\mathrm{Re}\,\tilde{\Delta}+\sin\phi_{SW}\,\mathrm{Im}\,\tilde{\Delta}\,, \tag{90}$$

* As the correction $2|\omega|^2$ is of the same order of magnitude as Γ_L/Γ_S it would be more consistent to omit it. We keep it here in order to consider the contribution of channels 2π, $I=0$ and 2π, $I=2$ as accurately as possible. However, as we will see, the contribution of the latter channel is all the same negligible.

$$\varepsilon_{\parallel} = \cos\phi_{SW}\,\mathrm{Re}\,\varepsilon + \sin\phi_{SW}\,\mathrm{Im}\,\varepsilon\,, \tag{91}$$

$$\tilde{\Delta}_{\perp} = -\sin\phi_{SW}\,\mathrm{Re}\,\tilde{\Delta} + \cos\phi_{SW}\,\mathrm{Im}\,\tilde{\Delta}\,, \tag{92}$$

$$\varepsilon_{\perp} = -\sin\phi_{SW}\,\mathrm{Re}\,\varepsilon + \cos\phi_{SW}\,\mathrm{Im}\,\varepsilon\,. \tag{93}$$

In the general case

$$\tilde{\Delta}_{\parallel} = \mathrm{Re}\,\alpha\,\cos\phi_{SW}\,, \tag{94}$$

$$\varepsilon_{\parallel} = \varepsilon_{0\parallel} + \mathrm{Re}\,\alpha\,\cos\phi_{SW}\,, \tag{95}$$

$$\tilde{\Delta}_{\perp} = -\varepsilon_{0\perp} - \mathrm{Im}\,\alpha\,\cos\phi_{SW}\,, \tag{96}$$

$$\varepsilon_{\perp} = -\mathrm{Im}\,\alpha\,\cos\phi_{SW}\,. \tag{97}$$

In the case (88) we obtain

$$\tilde{\Delta}_{\parallel} = \cos\phi_{SW}\,\mathrm{Re}\,(2\omega^*\varepsilon')\,, \tag{98}$$

$$\varepsilon_{\perp} = -\cos\phi_{SW}\,\mathrm{Im}\,(2\omega^*\varepsilon')\,, \tag{99}$$

and from eq. (89)

$$\tilde{\Delta}_{\perp} = \varepsilon_{\perp} - \varepsilon_{0\perp}\,, \tag{100}$$

$$\varepsilon_{\parallel} = \varepsilon_{0\parallel} + \tilde{\Delta}_{\parallel}\,. \tag{101}$$

Experimentally

$$\alpha_{2\pi, I=2} = [(0.007 \pm 0.003) + i(0.002 \pm 0.003)] \cdot 10^{-3}\,, \tag{102}$$

$$2\,\mathrm{Im}\,(\omega^*\varepsilon') < 2\,\mathrm{Re}\,(\omega^*\varepsilon') \simeq 10^{-5}\,. \tag{103}$$

So we conclude that

$$\tilde{\Delta}_{\parallel} \ll \tilde{\Delta}_{\perp} \simeq -\varepsilon_{0\perp} = (-0.16 \pm 0.07) \cdot 10^{-3}\,, \tag{104}$$

$$\varepsilon_{\perp} \ll \varepsilon_{\parallel} \simeq \varepsilon_{0\parallel} = (2.27 \pm 0.06) \cdot 10^{-3}\,. \tag{105}$$

As it is seen from fig. 3 the non-vanishing value of the *CPT*-violating parameter $\tilde{\Delta}$ is caused by the non-vanishing difference of the experimental values of the phases ϕ_{00} and ϕ_{SW}.

Thus the cause of the poor agreement with *CPT* symmetry for both parameters ε' and $\tilde{\Delta}$ is the experimental value of the phase ϕ_{00}. However, it should be emphasized that in spite of the common root of the evil, the poor agreement with *CPT* in the two cases is of a different nature. To illustrate this point let us consider some numerical examples. For instance it is easy to see that with the same mean value of $\phi_{00} - \phi_{SW} \simeq 10°$ we would have nonetheless $\varepsilon_{0\perp} = 0$ and, hence, $\tilde{\Delta}_{\perp} = 0$, if the angle ϕ_{+-} were such that

$$\phi_{+-} - \phi_{SW} = -\tfrac{1}{2}(\phi_{00} - \phi_{SW}) \simeq -5°\,. \tag{106}$$

(Experimentally, $\phi_{+-} - \phi_{SW} = 0.9 \pm 1.2°$).

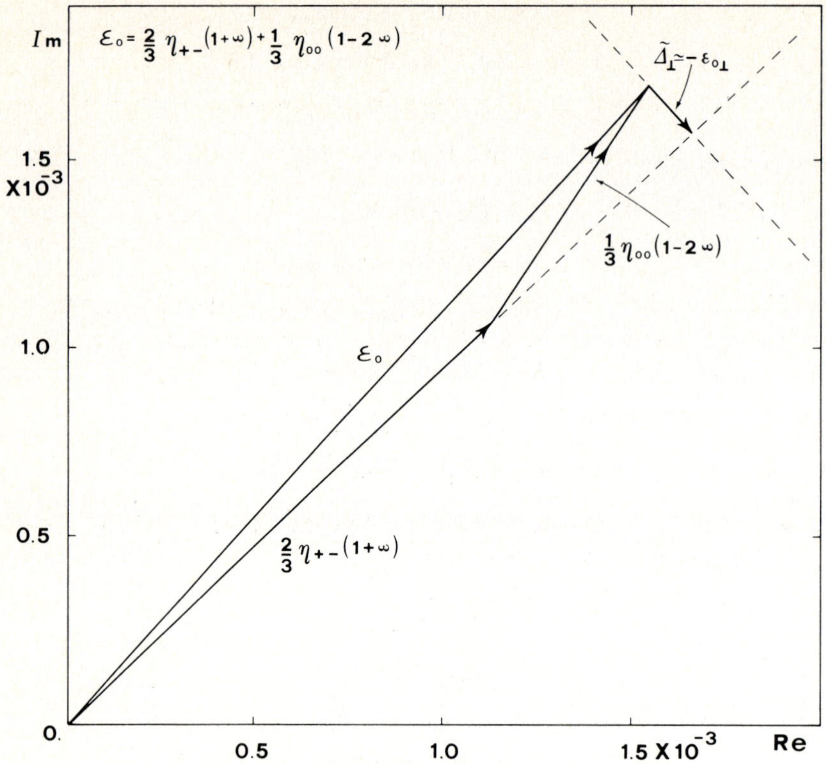

Fig. 3. Vector diagram for $\tilde{\Delta}_\perp \simeq -\varepsilon_{0\perp}$ (eq. (104)).

Note that the value of ε' would become even larger since in this case the difference between η_{+-} and η_{00} would increase.

On the other hand, even a large value of the parameter $|\varepsilon'|$ would not signal *CPT* violation were the $\pi\pi$ scattering phase shift $\delta_2-\delta_0$ close to $+\frac{1}{4}\pi$ and not to $-\frac{1}{4}\pi$ (as it is experimentally). At the same time the non-vanishing value of $\tilde{\Delta}$ signals *CPT* violation irrespective of the value of the $\delta_2-\delta_0$ phase shift.

9. On the parameter α

The conclusions of the preceding paragraph, eqs. (104), (105), are valid when we take into account the contributions to α of the 2π channel only. Would the role of α become more important if we took into account the contributions of other channels ($\pi^+\pi^-\pi^0$, $\pi^0\pi^0\pi^0$, $\pi e\nu$, $\pi\mu\nu$), and would then the agreement with *CPT* symmetry become better?

Let us begin with the contribution of the $K^0\to 3\pi$ decays and introduce the parameters:

$$\eta_{000}=\frac{A(K^0_S\to\pi^0\pi^0\pi^0)}{A(K^0_L\to\pi^0\pi^0\pi^0)}, \tag{107}$$

$$\eta_{+-\circ} = \frac{A(K_S^0 \to \pi^+\pi^-\pi^0)}{A(K_L^0 \to \pi^+\pi^-\pi^0)}\,. \tag{108}$$

The most accurate experiment on the search for $K_S^0 \to \pi^0\pi^0\pi^0$ decays gives [11]

$$\eta_{000} = (-0.08 \pm 0.18) + i(-0.05 \pm 0.27)\,. \tag{109}$$

According to a summary contained in ref. [20]

$$\eta_{+-0} = (0.05 \pm 0.07) + i(0.26 \pm 0.13)\,. \tag{110}$$

Assuming that there is no strong *CPT* violation in the $K^0 \to 3\pi$ decays and neglecting ε_S and the small phases due to $\pi\pi$ scattering, the parameters η_{000} and η_{+-0} have to be pure imaginary. Experimentally we have in this hypothesis

$$\eta_{000} = i(-0.02 \pm 0.18)\,, \tag{111}$$

$$\eta_{+-0} = i(0.01 \pm 0.11)\,. \tag{112}$$

Neglecting the charge exchange $\pi^+\pi^- \to \pi^0\pi^0$ the contributions of $3\pi^0$ and $\pi^+\pi^-\pi^0$ channels to α may be obtained separately:

$$\alpha_{000} = \frac{\Gamma_L}{\Gamma_S} B_{000}\eta^*_{000}\,, \tag{113}$$

$$\alpha_{+-0} = \frac{\Gamma_L}{\Gamma_S} B_{+-0}\eta^*_{+-0}\,. \tag{114}$$

Here $\Gamma_L/\Gamma_S = 1.72 \cdot 10^{-3}$ and $B_{000} = (21.5 \pm 1)\%$ and $B_{+-0} = (12.4 \pm 0.2)\%$ are the branching ratios for the decays $K_L^0 \to \pi^0\pi^0\pi^0$ and $K_L^0 \to \pi^+\pi^-\pi^0$ respectively. Let us consider the sum $\alpha_{3\pi} = \alpha_{000} + \alpha_{+-0}$. Without imposing *CPT* symmetry we get

$$\alpha_{3\pi}^{\text{exp}} = [(-0.02 \pm 0.07) + i(-0.04 \pm 0.10)] \cdot 10^{-3}\,, \tag{115}$$

while imposing *CPT* symmetry one finds

$$\alpha_{3\pi} = i(-0.01 \pm 0.07) \cdot 10^{-3}\,. \tag{116}$$

As for the theoretically expected value of $\alpha_{3\pi}$ it is natural to believe that *CP* violation in the three-pion decays is not larger than in the two-pion case, and therefore

$$\eta_{000} \simeq \eta_{+-0} \simeq \eta_{+-}\,, \tag{117}$$

so that

$$\alpha_{3\pi}^{\text{th}} = [(0.9 \pm 0.01) - i(0.9 \pm 0.01)] \cdot 10^{-6}\,. \tag{118}$$

Let us consider now the K_{e3} and $K_{\mu 3}$ decays. The contributions of both these decays to α can be expressed using the parameters δ and x:

$$\alpha_{e3} = \frac{\Gamma(K_L^0 \to \pi e \nu) + \Gamma(K_L^0 \to \pi\mu\nu)}{\Gamma_S}[\delta(1 + 2\,\text{Re}\,x) - 2i\,\text{Im}\,x]\,. \tag{119}$$

For the definitions of the parameters δ and x and their values see eqs. (56), (57), (63), (64). Using the decay rates (see ref. [13])

$$\Gamma(\mathrm{K}_L \to \pi e \nu) = (7.47 \pm 0.11) \cdot 10^6\ \mathrm{sec}^{-1}, \tag{120}$$

$$\Gamma(\mathrm{K}_L \to \pi \mu \nu) = (5.23 \pm 0.09) \cdot 10^6\ \mathrm{sec}^{-1}, \tag{121}$$

$$1/\Gamma_S = (0.8923 \pm 0.0022) \cdot 10^{-10}\ \mathrm{sec}, \tag{122}$$

we get

$$\alpha_{\ell 3}^{\mathrm{exp}} = [(0.004 \pm 0.0002) + i(0.01 \pm 0.06)] \cdot 10^{-3}. \tag{123}$$

Note that according to theoretical expectations the imaginary part of $\alpha_{\ell 3}$ has to be zero, since it is impossible to violate the $\Delta Q = \Delta S$ rule in the framework of the standard quark currents. In this case

$$\alpha_{\ell 3}^{\mathrm{th}} = (3.7 \pm 0.2) \cdot 10^{-6}. \tag{124}$$

Summing up the experimental contributions to α, i.e. (102), (115) and (123), we get

$$\alpha_{\mathrm{exp}} = [(-0.006 \pm 0.068) - i(0.026 \pm 0.120)] \cdot 10^{-3}, \tag{125}$$

while the theoretically expected value is

$$\alpha_{\mathrm{th}} = [(0.010 \pm 0.003) + i(0.001 \pm 0.003)] \cdot 10^{-3}. \tag{126}$$

It is interesting to compare (125) and (126) with the value

$$\alpha = [(0.14 \pm 0.19) - i(0.19 \pm 0.25)] \cdot 10^{-3}, \tag{127}$$

obtained by Cronin on the basis of previous data on $\mathrm{K}_S^\circ \to 3\pi^\circ$ and used in his Nobel lecture [10]. The comparison shows that the mean value of α and its error were considerably higher.

The results of the computation of the parameter $\tilde{\Delta}$ for each of the three values of α (eqs. (125), (126) and (127)) are shown in figs. 4 and 5.

It is seen that the agreement with *CPT* symmetry obtained with (127) looks nearly perfect, the mean value of $\tilde{\Delta}$ being compatible with zero within one standard deviation. The agreement with *CPT* symmetry looks satisfactory if we use the present experimental value (125) of α. As for the case (126) the vanishing of $\tilde{\Delta}$ is unfavoured by two standard deviations (see fig. 5).

Thus, answering the question put at the beginning of this section we can say that with the present value of ϕ_{00} one can improve the compatibility with *CPT* symmetry only at the price of an anomalously strong *CP* violation in 3π decays and/or the violation of $\Delta Q = \Delta S$ rule and *CP* symmetry in $\mathrm{K}_{\ell 3}$ decays.

10. Conclusions

The compatibility of experimental data with the requirements of *CPT* symmetry within two standard deviations does not look quite satisfactory. Taking into account

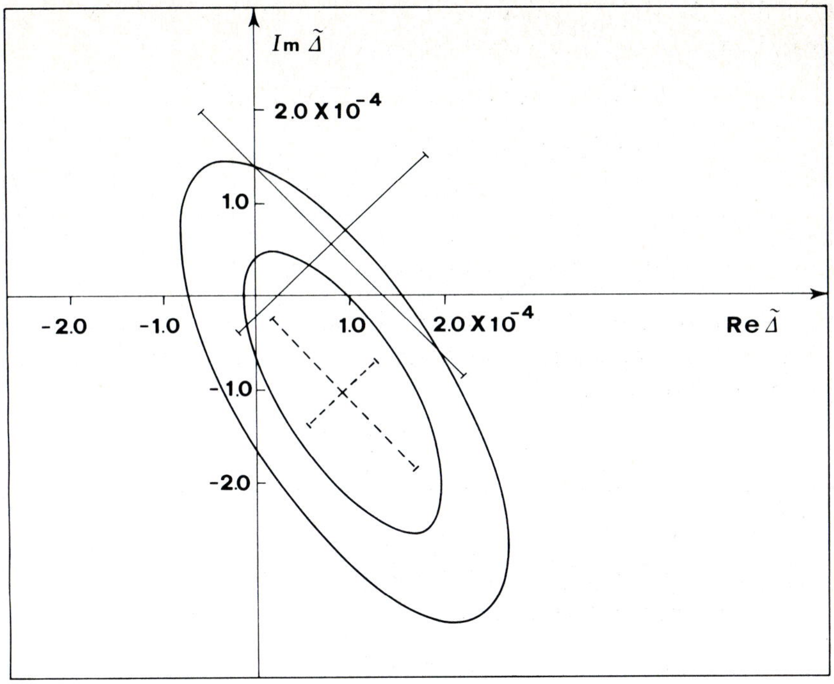

Fig. 4. One- and two-standard deviation contours for $\tilde{\Delta}$ in the case (125). The solid cross is Cronin's result: $\tilde{\Delta}_{\parallel}(-0.02\pm0.20)\cdot10^{-3}$, $\tilde{\Delta}_{\perp}=(-0.10\pm0.14)\cdot10^{-3}$ (case (127)). The dashed cross is our result expressed in the parallel and perpendicular projections (considered as independent variables): $\tilde{\Delta}_{\parallel}=(-0.14\pm0.11)\cdot10^{-3}$, $\tilde{\Delta}_{\perp}=(-0.005\pm0.05)\cdot10^{-3}$.

the fundamental role of *CPT* symmetry in modern physics, there is little doubt (see however refs. [21, 22]) that new more accurate experimental data will agree with *CPT*. It follows from the above analysis that in the first place it is the phase ϕ_{00} which it is necessary to remeasure with higher accuracy.

At present two experiments at Batavia [23] and Brookhaven [24] which measure the ratio $|\eta_{00}|^2/|\eta_{+-}|^2$ with accuracy of 2% and 1% respectively are nearing their completion. Two other experiments, measuring this ratio with accuracy of 0.5% are planned to be completed in FNAL and CERN in 1986 [25, 26]. The main goal of these experiments is to test the Kobayashi–Maskawa model [27] predictions [28] according to which $|\varepsilon'/\varepsilon|\sim2\cdot10^{-3}$ and the phases of the parameters ε' and ε practically coincide so that $\varepsilon'_{\perp}=0$ and

$$\left|\frac{\varepsilon'}{\varepsilon}\right|\simeq\frac{1}{6}\left(1-\frac{|\eta_{00}|^2}{|\eta_{+-}|^2}\right). \tag{128}$$

Eq. (128) is valid also for the models [29] where *CP* violation originates in the Higgs sector. In these models one expects $|\varepsilon'/\varepsilon|\sim2\cdot10^{-2}$.

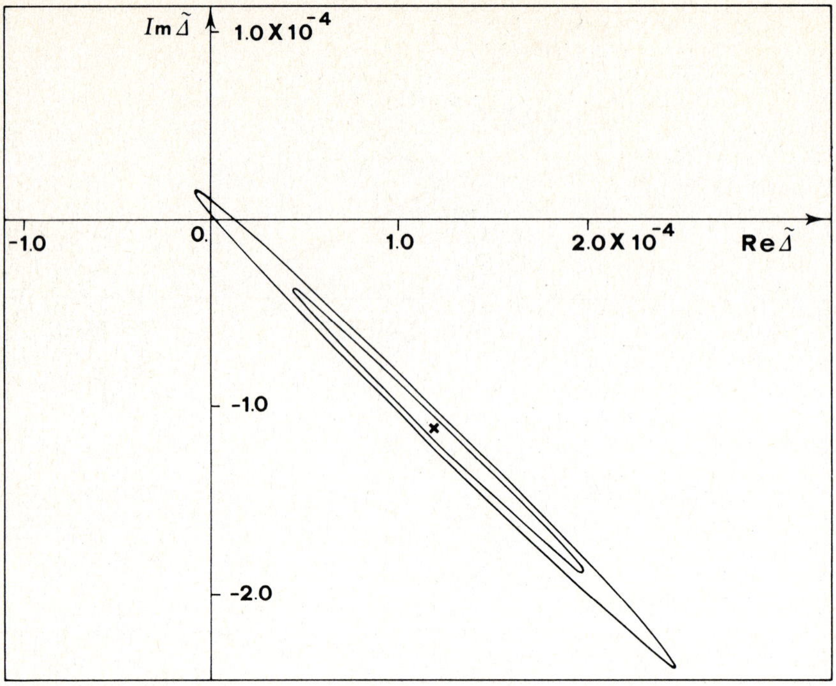

Fig. 5. One- and two-standard deviation contours for $\tilde{\Delta}_{\perp} = (-0.16 \pm 0.07) \cdot 10^{-3}$, $\tilde{\Delta}_{\parallel} = (0.007 \pm 0.002) \cdot 10^{-3}$.

Unfortunately we do not know of any proposal to measure not the module of η_{00} but its phase. It seems to us that it would be very interesting to measure ϕ_{00} with an accuracy of the order of one degree, i.e. with the accuracy with which the phase ϕ_{+-} is known. Our analysis shows that the ϕ_{00} and ϕ_{+-} phases must coincide within this accuracy due to *CPT* invariance. As a next step it would be expedient to improve the accuracy of ϕ_{+-} and to look for a non-zero $\phi_{+-} - \phi_{SW}$ phase difference. The accurate measurement of ϕ_{00} is especially interesting because an accuracy in $\tilde{\Delta}$ of the order of 10^{-4} corresponds to the accuracy of $\Delta m/m \sim 10^{-18}$ for the masses of K and $\bar{K}$ (see eq. (74)). This is roughly 14 orders of magnitude better than the accuracy achieved in testing the equality of masses of a particle and the corresponding antiparticle in any other case.

We thank J.W. Cronin, N.N. Nikolaev and M.B. Voloshin for helpful discussions.

References

[1] J.S. Bell and J. Steinberger, Proc. Oxford Int. Conf. on elementary particles, 1965, p. 195

[2] L.B. Okun and C. Rubbia, Proc. Heidelberg Int. Conf. on elementary particles, ed. H. Filtuth (North Holland, 1968) p. 301

[3] L.I. Lapidus, Usp. Fiz. Nauk 95 (1968) 657
[4] L.B. Okun, Comments Nucl. Part. Phys. 2 (1968) 116
[5] L.B. Okun, Proc. IV Winter School on Theory of nucleus and high energy physics, Physics-Technical Institute, Leningrad, Part I (1969) p. 210 (in Russian)
[6] S.M. Bilenkij, Particles and nucleus, vol. I, no. 1 (Dubna, 1970) p. 227 (in Russian)
[7] K.R. Shubert et al., Phys. Lett. 31B (1970) 662
[8] S. Gjesdal et al., Phys. Lett. 52B (1974) 119
[9] K. Kleinknecht, Ann. Rev. Nucl. Sci. 26 (1976) 36
[10] J.W. Cronin, Rev. Mod. Phys. 53 (1981) 373
[11] V.V. Barmin et al., Phys. Lett. 128B (1983) 129
[12] M. Baldo–Ceolin et al., JETP Lett. 38 (1983) 459
[13] Particle Data Group, Phys. Lett. 111B (1982)
[14] J.A. Christenson et al., Phys. Rev. Lett. 43 (1979) 1209
[15] T.T. Wu and C.N. Yang, Phys. Rev. Lett. 13 (1964) 380
[16] T.J. Devlin and J.O. Dickey, Rev. Mod. Phys. 51 (1979) 237
[17] L.B. Okun, Leptons and quarks (North-Holland, 1982)
[18] B. Gobbi et al., Phys. Rev. Lett. 22 (1969) 682
[19] L. Wolfenstein, Phys. Rev. Lett. 13 (1964) 569
[20] M. Baldo–Ceolin et al., Nuovo Cim. 25A (1975) 688
[21] I.I. Bigi, Z. Phys. C12 (1982) 235
[22] R.M. Wald, Phys. Rev. D21 (1980) 2742
[23] FNAL experiment E617. Spokesman: B. Winstein
[24] BNL experiment N749. Spokesmen: L.B. Leipuner, R.K. Adair
[25] G. Gollin et al., A measurement of the magnitude of $|\eta_{00}|^2/|\eta_{+-}|^2$ in the neutral kaon system to a precision of .001. FNAL proposal for an upgraded E617 experiment at Tevatron (February 1983)
[26] CERN experiment NA31. Spokesman: H. Wahl
[27] M. Kobayashi and K. Maskawa, Prog. Theor. Phys. 49 (1973) 652
[28] F. Gilman and M. Wise, Phys. Rev. D20 (1979) 2392; Phys. Lett. 83B (1979) 83;
L. Wolfenstein, Nucl. Phys. B160 (1979) 501;
V.V. Prokhorov, Yad. Fiz. 31 (1980) 1019;
M.I. Vysotsky, Yad. Fiz. 31 (1980) 1535;
B. Guberina and R. Peccei, Nucl. Phys. B163 (1980) 289;
J. Hagelin, Nucl. Phys. B193 (1981) 123;
M.B. Voloshin, preprint ITEP N22 (1981);
S. Pakvasa, CERN preprint TH 3464 (1983);
A.J. Buras, W. Slominski and H. Steger, Nucl. Phys. B238 (1984) 529;
F.J. Gilman and J.S. Hagelin, Phys. Lett. 126B (1983) 111
[29] S. Weinberg, Phys. Rev. Lett. 37 (1976) 657;
A.A. Anselem and D.I. Dyakonov, Nucl. Phys. B145 (1978) 271;
A.A. Anselem and N.G. Uraltsev, Yad. Fiz. 30 (1979) 465;
A. Sande, Phys. Rev. D23 (1982) 2467

3. Models of CP Violation

Introduction

At the time CP violation was discovered experimentally, there was no fundamental problem devising theories that were not invariant under CP or T. On the other hand, there was no obvious clue as to how to proceed. It soon became clear that there were three general classes of theories distinguished by the strangeness property of the CP-violating interaction.

(1) *Superweak*. The interaction includes terms that allow the strangeness change $\Delta S = 2$. A strength of the order of 10^{-9} times the normal weak interactions is then needed. This interaction contributes to the $K^0 - \overline{K}^0$ mass matrix in lowest order.

(2) *Milliweak*. $\Delta S = 1$ is allowed, but not $\Delta S = 2$, like the normal weak interactions. It is called "milliweak" because effectively, at least for the K^0 system, its effect is down by the factor 2×10^{-3}. The standard Kobayashi–Maskawa (KM) model discussed in chapter 4 is classified as milliweak.

(3) *Millistrong*. CP violation shows up as a small correction to the $\Delta S = 0$ strong parity-conserving interaction.

This classification was presented by Prentki and Veltman [1] and also by Lee and Wolfenstein [2]. The superweak theory was first discussed by Wolfenstein [3]. The specific illustration given was a current–current interaction between a normal weak current and an ultra-weak $\Delta Q = -\Delta S$ current. It was emphasized that this was a general class of theories with the distinguishing feature that it was very difficult to find CP or T violation in any other experiment outside of K^0 physics. The other extreme, the millistrong, was first discussed by Prentki and Veltman [1] and Okun' [4]. In this case, CP or T violation should be seen in every process, once one probed at the level of the order 10^{-3}.

An interesting alternative idea was that CP was not really violated in the interactions in nature, but rather the effect that was observed was due to interactions with the surrounding earth which was made of matter and not anti-matter. This proposal, made by Bernstein et al. [5] and independently by Bell and Perring [6] was equivalent to making K^0 and $\overline{K}^0$ have different effective masses and thus to an effective CPT violation. As discussed in the formalism presented in chapter 2, this predicts an incorrect phase for η_{+-}. Also, such a theory predicts a dependence of η_{+-} on the velocity relative to the earth [7]. In a somewhat artificial way, it is possible, to imagine that the superweak effect is due to the surroundings and that it might disappear if the experiment were done in outer space or in an empty universe, but such an alternative is designed to be phenomenologically equivalent to the superweak model.

With the success of modern gauge theories the problem of constructing models with CP violation became better defined. The strong interactions described by QCD had no place for CP violation of the millistrong variety and therefore millistrong theories became very unattractive. On the other hand, in the electroweak gauge theory (and extensions of it) CP violation may be introduced in the part of the Lagrangian associated with the Higgs bosons. This may be done by:

(1) explicit introduction of complex coefficients in the Yukawa potential,

(2) explicit introduction of complex coefficients in the Higgs potential, or

(3) by constructing a Higgs potential which is CP-invariant but which results in the

spontaneous violation of CP invariance as a result of non-trivial complex vacuum expectation values.

In the original minimal $SU(2) \times U(1)$ model with only two generations of quarks and one Higgs boson doublet, none of these possibilities exist. As first shown by Lee [8], spontaneous CP violation can be accomplished using two Higgs doublets. With only two Higgs doublets, spontaneous CP violation is accompanied by flavor-changing neutral Higgs boson exchanges. It thus becomes possible to have $\Delta S = 2$ interactions at tree level and one obtains a type of superweak model. However, a detailed analysis of this particular model [9] shows that it has interesting CP-violating effects beyond a generic superweak model.

Weinberg [10] proposed a model with three Higgs doublets with a discrete symmetry that eliminated flavor-changing neutral boson exchange. The origin of CP violation could be either a complex coupling among the three bosons in the Higgs potential or spontaneous CP violation. The latter possibility was developed by Branco [11], who showed that in this case the only source of CP violation was due to virtual Higgs boson processes even in the case of more than two generations. For the K meson system the major contributions to CP violation are diagrams with virtual charged Higgs bosons. A simple analysis of these diagrams led Sanda [12] and Deshpande [13] to conclude that the theory predicted a value ϵ'/ϵ of -0.05, clearly in disagreement with experiment. A re-analysis by Donoghue and Holstein [14], following earlier work of Pham [15], showed that this result was not reliable. Their best estimate was $\epsilon'/\epsilon = -0.006$, also in disagreement with the latest experiments, but this result is very uncertain even as to sign, so that the model may not be ruled out. The calculation of the neutron electric dipole model D_n in this model depends upon parameters that are not determined from fitting the CP violation in the K^0 system. Under reasonable assumptions, one expects [16] that neutral Higgs boson contributions yield a value considerably larger than the present limit on D_n of about 10^{-25} e cm. A characteristic feature of the Weinberg model is a T-violating asymmetry in $K \to \pi\mu\nu$, but the expected value is considerably lower than present experimental limits. The status of the Weinberg model has been reviewed by Cheng [17].

Mohapatra and Pati [18] considered an extension of the standard model gauge group to $SU(2)_L \times SU(2)_R \times U(1)$. This extension is interesting in that it allows parity conservation to be restored at high energy. In this model, CP violation may occur as a result of complex Yukawa couplings even when there are only two generations. Quantitative conclusions concerning this model are difficult, particularly when one includes all three generations. In general there are a large number of CP-violating phases in the quark mixing matrices and there is also CP violation of the superweak variety because the model necessarily contains a flavor-changing neutral Higgs boson. An important aspect of $SU(2)_L \times SU(2)_R \times U(1)$ models is that box diagrams containing the gauge boson $W_R^{\pm}$ associated with the right-handed currents can make a large contribution to the $K^0 - \bar{K}^0$ mass matrix [19]. Indeed the mass of W_R must be at least twenty times $m(W_L)$ to avoid too large a value of Δm. If $m(W_R)$ is not much above this limit, it becomes quite likely that a major contribution to CP violation could come from contributions to the mass matrix involving W_R.

Calculations of ϵ'/ϵ in the simplest versions of the $SU(2)_L \times SU(2)_R \times U(1)$ give values [20] quite consistent with the present experiment. However, the model is distinguished from the standard model (see chapter 4) by the prediction of an experimentally detectable electric dipole moment D_n of the neutron. Note particularly eq. (5.39) of the paper by Ecker and Grimus [20], where we can set ϵ'_{SM} equal to zero and substitute the experimental value of ϵ', which indicates a value of D_n of the order of, or greater than, 10^{-26} e cm.

Many other models have been proposed. Of particular interest may be those [21] based on supersymmetry. However, it is now possible to explain existing data on CP violation without extending the standard $SU(2) \times U(1)$ model. Thus the alternative models discussed in this chapter are of interest only if new data renders the standard model insufficient to explain CP violation, or if other evidence leads to a belief in certain extensions of the standard model.

References

[1] J. Prentki and M. Veltman, Phys. Lett. 15 (1965) 88. Paper 13, this volume.
[2] T.D. Lee and L. Wolfenstein, Phys. Rev. B 138 (1965) 1490.
[3] L. Wolfenstein, Phys. Rev. Lett. 13 (1964) 562. Paper 14, this volume.
[4] L.B. Okun', Sov. J. Nucl. Phys. 1 (1965) 670. Paper 15, this volume.
[5] J. Bernstein, N. Cabibbo and T.D. Lee, Phys. Lett. 13 (1964) 146. Paper 16, this volume.
[6] J.S. Bell and J.K. Perring, Phys. Rev. Lett. 13 (1964) 348.
[7] O. Nachtmann, Acta Phys. Austriaca Suppl. 6 (1969) 485.
[8] T.D. Lee, Phys. Rep. 9 (1974) 145.
The first part of this paper is Paper 17, this volume.
[9] J. Liu and L. Wolfenstein, Nucl. Phys. B 289 (1987) 1.
[10] S. Weinberg, Phys. Rev. Lett. 34 (1976) 657. Paper 18, this volume.
[11] G.C. Branco, Phys. Rev. D 22 (1980) 2901. Paper 19, this volume.
[12] A. Sanda, Phys. Rev. D 23 (1981) 2647.
[13] N. Deshpande, Phys. Rev. D 23 (1981) 2654.
[14] J.F. Donoghue and B.R. Holstein, Phys. Rev. D 32 (1985) 1152.
[15] Y. Dupont and T.N. Pham, Phys. Rev. D 28 (1983) 2169.
[16] A.A. Anselm, V.E. Bunakov, V.P. Gudkov and N.G. Uratsev, Phys. Lett. B 152 (1985) 116.
[17] H.Y. Cheng, Phys. Rev. D 34 (1986) 1397. Paper 20, this volume.
[18] R.N. Mohapatra and J.C. Pati, Phys. Rev. D 11 (1975) 566. Paper 21, this volume.
[19] G. Beall, M. Bander and A. Soni, Phys. Rev. Lett. 48 (1982) 848.
[20] G. Ecker and W. Grimus, Nucl. Phys. B 258 (1985) 328. Paper 22, this volume.
[21] J.M. Gerard, W. Grimus, A. Raychaudri and G. Zoupanos, Phys. Lett. B 140 (1984) 349;
M. Dugan, B. Grinstein and L. Hall, Nucl. Phys. B 255 (1985) 413.

POSSIBILITY OF *CP* VIOLATION IN SEMI-STRONG INTERACTIONS

J. PRENTKI and M. VELTMAN
CERN-Geneva

Received 27 January 1965

The recently observed [1] decay $K_{02} \to 2\pi$ can be interpreted in two essentially different manners. Either, a) one has to do with an apparent *CP* violation due to the presence of some external field [2], in which case a velocity dependence of the decay rate would show up, or b) one has to do with some real *CP* violation due to some interaction as yet unspecified. In the latter case, the following possibilities emerge:

i) the decay is entirely due to a new type of very weak interaction;
ii) the *CP* violation is inherent in weak interactions;
iii) the *CP* violation is due to perturbation by some interaction on an otherwise normal weak process.

The difference between these points of view is mainly a difference in suspected order of magnitude of the coupling constant of the *CP* violating interaction. In case i) one would expect a coupling constant small with respect to that of the weak interactions, and no prediction on any further properties, like $\Delta I = \frac{1}{2}$ can be made [3]. In case ii) the usual weak interaction selection rules are supposedly holding also for this mode of decay [4], so the K_{02} rate in $\pi^+\pi^-$ or $\pi^0\pi^0$ is according to the $\Delta I = \frac{1}{2}$ rule. As has been pointed out [5], no firm conclusions can be drawn on the magnitude of the coupling constant of the *CP* violating part, and it could be as large as the weak interaction coupling constant itself.

In this note we intend to entertain possibility iii). First we note that the coupling constant involved, by the same argument as in the case of ii), is undetermined. However, it should be at least of the order 1/500, in view of the observed branching ratio. This implies that this interaction should have a strength at least comparable with electromagnetic interactions. But the fundamental electromagnetic interaction is *CP* invariant thus it cannot be the electromagnetic interaction itself.

Obviously, by the introduction of an ad hoc new interaction with some "good" properties, one can explain the $K_2 \to 2\pi$ decay. What we want to point out in this paper is that already an interaction exists which could lead to some predictions which could be experimentally verified. As we will see some difficulties exist but we consider these difficulties as being not extremely serious. The interactions in question are the so-called semi-strong interactions, i.e., SU_3 violating interactions.

If the interplay of semi-strong and weak interactions is responsible for the *PC* violation in weak interactions, the semi-strong interactions must violate *PC*. However, *P* in this case must be conserved to a very high degree of accuracy as is seen from the measurement of the electric dipole moment of the neutron. Moreover these interactions conserve the isospin quite accurately as is seen from the isospin multiplet structure of the elementary particles, from the charge independence of nuclear forces, as seen for example from the reaction $d+d \to He+\pi_0$. From *PCT* one has then to deduce that medium strong interactions violate *C* and *T* respectively. Thus we will inspect now the evidence for *C* and *T* invariance of this interaction. To be more specific, we will also make the usual assumption that it transforms like the eighth component of an octet (although it is conceivable that the *C* violating part of the SU_3 violating interactions behaves differently). Let us first consider the question of order of magnitude of the coupling constant. The only information on this point is given by the mass differences between partners in an SU_3 multiplet. We may compare those mass differences with the electromagnetic mass differences. Let us consider the baryon multiplet. The charged-uncharged baryon mass differences are of second order in e, and are of the order of 1 (for p-n) to 5 or 6 (for Σ and Ξ) MeV. Typical mass differences within the octet are of order 75 - 150 MeV, and if we assume that the SU_3 breaking interactions do not have a completely 90° phase difference with the strong interactions, these differences are of the first order in f ($\equiv$ the coupling constant for the SU_3 breaking interaction). Thus f may be estimated to be 5-10 times the electromagnetic coupling constant. Roughly speaking one expects therefore 10% perturbation effects on processes involving strong interacting particles. It requires some stretching of the imagination (although we are in better position than in many models already proposed) to explain the small $K_{02} \to 2\pi$ branching ratio, but we will not enter a discussion on this point. As we take the SU_3 breaking interaction to be I spin conserving, the CP violating decay should also obey the $\Delta I = \frac{1}{2}$ rule of weak interactions.

Let us now consider the pion-nucleon interaction Lagrangian. According to the theorem by Soloviev [6], generalized to SU_3 invariant interactions by Pais [7], P and CPT are enough to guarantee C and T invariance for interactions like

$$ig(\bar{N}\gamma^5 \boldsymbol{\tau} N)\boldsymbol{\pi}$$

because the addition of the Hermitian conjugate cancels the imaginary part of g. This argument is valid also in SU_3 context, when one considers the couplings $g_1(\bar{B}BM)$ and $g_1'(B\bar{B}M)$, where $\bar{B}$, B, M are the usual antibaryon, baryon and meson octets.

We may now write down the SU_3 breaking part $\mathcal{L}_I'$ of the interaction Lagrangian $\mathcal{L}_I$ by introducing a spurion that behaves as the eighth component of an octet. We have 9 couplings (with one relation among the coefficient which is not of interest to us here):

$$\mathcal{L}_I' = a_1(\bar{B}SMB) + a_2(\bar{B}MBS) + a_3(\bar{B}BSM) +$$
$$+ a_4(\bar{B}SBM) + a_1^*(\bar{B}MSB) + a_3^*(\bar{B}BMS) +$$
$$+ a_7(\bar{B}M)(BS) + a_7^*(\bar{B}S)(BM) + a_9(\bar{B}B)(MS).$$

Hermicity requires the reality of a_2, a_4 and a_9, apart from the explicitly indicated relations above. From the above formula one deduces that all couplings involving pions and the η are C conserving, apart from the terms

$$\frac{1}{\sqrt{6}}(a_1+a_2+a_3+a_4+a_1^*+a_3^*+6a_7)(\bar{\Sigma}\,\Lambda)\,\pi$$
$$+\frac{1}{\sqrt{6}}(a_1+a_2+a_3+a_4+a_1^*+a_3^*+6\,a_7^*)(\bar{\Lambda}\,\Sigma)\,\pi$$

i.e., this coupling violates C through a_7 only. The couplings involving kaons do not have any special property in this respect.

Thus, from the above we do expect significant CP violating effects only if strange particles play a significant role (which of course may also be through virtual effects).

Inspecting now the existing evidence on CP or T violation in strong interactions we observe that T violation in usual experimental arrangements is not an easily measurable effect in view of final state interactions. The limit on T is of the order of 10%. But we must consider those strong interacting systems that are an eigenstate of C. These systems are the $^3S_1$ or $^1S_0$ proton-antiproton system, and the existing unstable particles. The p-$\bar{p}$ data are certainly not sufficiently accurate for this purpose. The most (and in fact the only - see the table for some examples) serious case is the decay of the η particle, supposedly going via an electromagnetic process [8]

$\eta \to \pi^+\pi^-\pi^0$	27 %
$\pi^+\pi^-\gamma$	5.5%
$3\pi^0$ } $\pi^0 2\gamma$	31 %
2γ	35 %

If I spin is conserved no $3\pi^0$ mode should be present and the Dalitz plot of the 3π system should be different from that which is observed. The above rates are of order e^4 in the electromagnetic interactions (why the e^2 process $\pi^+\pi^-\gamma$ is so small is not clear), as follows from the large $\eta \to 2\gamma$ rate, while a C violating process through the SU_3 breaking interactions could be of order f^2, i.e., a factor 10^3-10^4 larger. However, as pointed out above such effects may be suppressed in interactions not involving strange particles. Moreover, the isospin conservation implies a large angular momentum barrier; as is well known the simplest wave function involves 6 derivatives (i, j, k are isospin indices)

$$\eta\,\frac{\partial^3\pi^i}{\partial x_\mu\,\partial x_\nu\,\partial x_\kappa} - \frac{\partial^2\pi^j}{\partial x_\mu\,\partial x_\nu}\,\frac{\partial\pi^k}{\partial x_\kappa}\,\epsilon_{ijk}\,.$$

This, in connection with the small phase space for the 3π mode could give a very important suppression for this transition, in contrast to the C conserving I spin violating transition that involves only S waves in the final state. For this reason we do not consider the η decay modes as strong evidence for C conservation by the SU_3 breaking interactions.

Effects of the C and T violation proposed above would show up in non-leptonic weak decays, in strange particle leptonic decays, in proton-antiproton reactions, etc.

In leptonic decays with $\Delta S = 0$ the strong interaction currents are supposedly belonging to a $I = 1$ multiplet, and in our model this remains like that. As is well known the C and T violation can proceed then only via currents of the second class. For $\Delta S = 0$ currents therefore our theory is giving the same results as Cabibbo's model [4]. For $\Delta S = 1$ leptonic decays the non-renormalization theorem of Fubini and Furlan [9] is valid for the vector part, thus we expect there small effects (1%) only, in particular for $K_{\mu 3}$ decay.

In conclusion, let us state that apart from the difficulty connected with the η which can however be solved by an adequate model where the effects of barrier potential will be important, it is conceivable that the PC violation in weak interactions is due to a PC violation in medium strong interaction. It is then very important to re-examine again the problem of C and T conservation in strong interactions where effects possibly up to 10%, especially in strange particle physics, could be expected. Experiments of this kind should, in our opinion, be performed, and in particular the η decay should be very thoroughly re-examined. In weak interaction one expects then also effects of PC violation up to 10% in $\Delta S = 1$ transition, smaller in $\Delta S = 0$ for leptonic and non-leptonic decays. A very strong prediction of this model is the validity of the usual isospin selection rules also for PC violating effects in weak interactions.

Table
Consequences of C violation on some observed reactions

$\pi_0 \to \gamma$	No, parity.
$\pi_0 \to 3\gamma$	Presumably small, at most $1/1000 \times \pi_0 \to 2\gamma$. It would probably be interesting to look for it.
$\pi_0 \to e^+e^-$	Only in second order in electromagnetic interactions.
$K_{01} \to e^+e^-$, $\mu^+\mu^-$; $K_{02} \to e^+e^-$, $\mu^+\mu^-$	Only in second order in electromagnetic interactions in addition to first order in weak interactions.
$\omega_0 \to \pi^+\pi^-$	No, because of I spin and Bose statistics. No P wave with $I = 0$ possible.
$\omega_0 \to 3\pi$	Unchanged.
$\rho_0^{\pm} \to \pi^+\pi^-$	Unchanged.
$\rho_0^{\pm} \to 3\pi$	Negligible, experimentally difficult to establish.
$\varphi_0 \to K_{01} K_{02}$; $\varphi_0 \to K_{01} K_{01}$	Unchanged, due to statistics.
$\varphi_0 \to \pi\rho$	Unchanged.
$\varphi_0 \to \pi\omega$	No, I spin.

The authors are indebted to Drs. D. Amati, J. S. Bell and N. Cabibbo for numerous illuminating discussions, and to Professors T. D. Lee, M. Schwartz and L. Van Hove for fruitful comments. A clarifying discussion on the proton-antiproton experiment with Dr. L. Montanet has been most helpful.

References

1) J. H. Christenson, J. W. Cronin, V. L. Fitch and R. Turlay, Phys. Rev. Letters 13 (1964) 138.
2) J. S. Bell and J. K. Perring, Phys. Rev. Letters 13 (1964) 348.
J. Bernstein, N. Cabibbo and T. D. Lee, Physics Letters 12 (1964) 146.
S. Weinberg, Phys. Rev. Letters 13 (1964) 495.
3) L. Wolfenstein, Phys. Rev. Letters 13 (1964) 562.
4) N. Cabibbo, Physics Letters 12 (1964) 495.
5) R. G. Sachs, Phys. Rev. Letters 13 (1964) 286.
6) V. Soloviev, Zh. Eksperim i. Teor. Fiz. 33 (1957) 537, 796, Soviet Physics JETP 6 (1958) 419, 613.
7) A. Pais, Phys. Rev. Letters 12 (1964) 432.
8) A. H. Rosenfeld, A. Barbaro Galtieri, W. H. Barkas, P. L. Bastien, J. Kirz and M. Roos, Rev. of Mod. Phys. 36 (1964) 977.
9) S. Fubini and G. Furlan, CERN preprint 9832/TH. 487. The Ademollo-Gatto proof does not hold as they use CP invariance. M. Ademollo and R. Gatto, Phys. Rev. Letters 13 (1964) 264.

* * * * *

VIOLATION OF *CP* INVARIANCE AND THE POSSIBILITY OF VERY WEAK INTERACTIONS*

L. Wolfenstein
Carnegie Institute of Technology, Pittsburgh, Pennsylvania
(Received 31 August 1964)

The observation[1] of the decay $K_2{}^0 \to 2\pi$ provides evidence that the weak interactions are not invariant with respect to the operation CP. The required magnitude of this CP-violating term in the weak-interaction Hamiltonian and the implications for other possible observations depend upon the theoretical model for explaining the $K_2{}^0 \to 2\pi$ decay. Sachs[2] has provided such a model in which (1) the CP-violating term is maximal in the sense that the $\Delta Q = -\Delta S$ decay interaction is 90° out of phase with $\Delta Q = +\Delta S$ interaction, (2) CP violation would not occur in nonleptonic decays or in leptonic decays other than that of K^0, and (3) the strength of the CP-violating term is comparable to the CP-conserving term. In this note we wish to raise some possible objections to the model of Sachs and to suggest an alternative which essentially satisfies conditions (1) and (2) above but requires that CP-violating weak interactions are weaker than the CP-conserving ones by a factor of 10^7 to 10^8.

The Sachs model is equivalent to a weak interaction Hamiltonian of the following form[3]:

$$H_W = H_L + H_{L0} + H_{L1} + H_{N1} + H_{L,-1}', \tag{1}$$

where the first four terms are the usual weak Hamiltonian even under CP: H_L for purely leptonic processes, H_{L0} for $\Delta S = 0$ leptonic, H_{L1} for $\Delta Q = \Delta S$ leptonic, and H_{N1} for $|\Delta S| = 1$ nonleptonic. The additional term $H_{L,-1}'$ is odd under CP and yields $\Delta Q = -\Delta S$ leptonic decays. The K^0 eigenstates are written

$$|K_1^0\rangle = N(|K^0\rangle + r|\bar{K}^0\rangle),$$

$$|K_2^0\rangle = N(|K^0\rangle - r|\bar{K}^0\rangle),$$

where $|\bar{K}^0\rangle = CP|K^0\rangle$, and r is a complex admixture coefficient which would equal unity if CP invariance were valid. Since the CP invariance of H_{N1} assures that the amplitude for $K^0 \to 2\pi$ equals that for $\bar{K}^0 \to 2\pi$, the nonvanishing experimental ratio[1] of 2.3×10^{-3} for the amplitude of $K_2^0 \to 2\pi$ with respect to $K_1^0 \to 2\pi$ requires[2] $r = 1 + \epsilon$ with $|\epsilon| = 4.6 \times 10^{-3}$. With Eq. (1) a nonzero value of ϵ is associated with the self-energy term of the form

$$\Delta_L' = \sum_a \frac{\langle K^0|H_{L,-1}'|a\rangle\langle a|H_{L1}|\bar{K}^0\rangle}{E_K - E_a},$$

where the intermediate states $|a\rangle$ contain a lepton pair plus strong-interaction particles. In order that $|\epsilon| \ll 1$, it is required that Δ_L' be much smaller than the usual CP-invariant self-energy term given by

$$\Delta_N = \sum_b \frac{\langle K^0|H_{N1}|b\rangle\langle b|H_{N1}|\bar{K}^0\rangle}{E_K - E_b}.$$

Our possible objections to the Sachs model are these: (1) It is impossible to write the Hamiltonian in the form of a current interacting with itself.[4] (2) $\Delta Q = -\Delta S$ decays are assumed to be of comparable probability with $\Delta Q = \Delta S$ decays although contrary evidence exists, at least for the axial-vector interaction.[5] (3) The argument that $\Delta_L' \ll \Delta_N$ is based on the fact that the width associated with $K \to 2\pi$ which occurs via H_{N1} is known to be much larger than the widths associated with the leptonic decays $K \to \pi + e + \gamma$ which occur via H_{L1} or $H_{L,-1}'$. Arguments with respect to the real parts of Δ_L' and Δ_N are much more difficult to make since both involve divergent integrals. In this connection Ioffe has pointed out[6] that, while in the case of Δ_N it is reasonable to use a cutoff around the nucleon mass M due to the strong interactions among all the particles in states $|b\rangle$, in the case of Δ_L' the states $|a\rangle$ contain a lepton pair so that one of the integrations can only be cut off by a "weak interaction cutoff," Λ, presumably much larger than M. The ratio (Δ_L'/Δ_N) thus contains a factor $(\Lambda/M)^2$ and cannot be said to be small.[7]

As an alternative we suggest that H_W be written in the standard current-current form

$$H_W = (G/\sqrt{2})J^\mu J_\mu^\dagger,$$

$$J^\mu = L^\mu + \mathcal{J}^\mu + \mathcal{S}^\mu - i\alpha T^\mu, \tag{2}$$

where L^μ, $\mathcal{J}^\mu$, and $\mathcal{S}^\mu$ are the usual leptonic, strangeness-conserving, and $\Delta Q = \Delta S$ strangeness-changing currents, respectively, so that H_W contains H_L, H_{L0}, H_{L1}, and H_{N1} in the usual way. The additional current T^μ is a $\Delta Q = -\Delta S$ strangeness-changing current so defined that the factor i assures that the following terms in H_W are odd under CP:

$$H_{L,-1}' = (i\alpha G/\sqrt{2})(L^\mu T_\mu^\dagger - T^\mu L_\mu^\dagger), \tag{3a}$$

$$H_{N1}' = (i\alpha G/\sqrt{2})(\mathcal{J}^\mu T_\mu^\dagger - T^\mu \mathcal{J}_\mu^\dagger), \tag{3b}$$

$$H_{N2}' = (i\alpha G/\sqrt{2})(\mathcal{S}^\mu T_\mu^\dagger - T^\mu \mathcal{S}_\mu^\dagger). \tag{3c}$$

In addition to the CP-violating $\Delta Q = -\Delta S$ leptonic term $H_{L,-1}'$ present in the Sachs model, we have a $|\Delta S| = 1$ nonleptonic term H_{N1}' and a $|\Delta S| = 2$ nonleptonic term H_{N2}'. The last of these contributes a CP-violating contribution of order αG to the self-energy,

$$\Delta_N' = \langle\bar{K}^0|H_{N2}'|K^0\rangle.$$

In order that the CP-violating factor $\epsilon = r - 1$ have a magnitude of 5×10^{-3} it is necessary[8] that $|\Delta_N'/\Delta_N| \simeq 5 \times 10^{-3}$. Since both Δ_N and Δ_N' involve only strongly interacting particles, it is reasonable that mass factors and cutoffs be of the order of M so that as an order-of-magnitude result we have

$$|\Delta_N'/\Delta_N| \simeq \alpha G/G^2M^2 = 10^5\alpha,$$

or α is between 10^{-7} and 10^{-8}. While such order-of-magnitude arguments are notoriously unreliable, our results really depend only on the fact that α is extremely small.

The model given by Eq. (2) with this extremely small value of α has the following consequences:

(1) The probability of $\Delta Q = -\Delta S$ leptonic decays

is less than those of $\Delta Q=+\Delta S$ by a factor of the order of 10^{15}.

(2) Violations of CP invariance or of T invariance may occur in nonleptonic strangeness-changing decays (due to interference between H_{N1} and H_{N1}') but the relative amplitude of the violating term is only 10^{-7} to 10^{-8}.

(3) No violations of CP invariance or T invariance occur in leptonic decays except for the special case of the K^0. This follows as in reference 2, since H_{L1} and H_{L-1}' must interfere to give a violation, but this is not generally possible since they correspond to $\Delta Q=\Delta S$ and $\Delta Q=-\Delta S$, respectively. The conclusion would not be absolutely true if higher order weak interactions are considered.

(4) $|\Delta S|=2$ nonleptonic decays are allowed in lowest order via H_{N2}', but this first-order probability is most likely less than the usual second-order probability.

(5) The only observable violations of CP invariance that are not of the order 10^{-7} are consequences of $r\neq 1$. These are effects of order 10^{-2} to 10^{-3}. In addition to $K_2^0\to 2\pi$, for example, the ratio of the two charge states in the leptonic decays of K_2^0 would be about 1.006 instead of 1. However, even in these decays, the time-reversal violation would only be of order 10^{-7}. This is in striking contrast to reference 2, where the large effective coupling of $H_{L,-1}'$ allows for large time-reversal violations in K^0 decay.

If $\Delta Q=\Delta S$ is not to be violated at all, the most natural way to introduce CP violation is through a term H_{N1}' of order of magnitude $|\epsilon|$. In this case the CP-violating interaction would be intermediate in strength between that of the model discussed here and the "full strength" of the weak interaction as in reference 2.[9]

The most interesting point of the model discussed here lies in the possibility that the experiment of reference 1 may measure an interaction as much as 10^7 or 10^8 times weaker than the standard weak interactions. If this is the case it may prove extremely difficult to observe CP violation (or T violation) in independent ways.

*This work supported in part by U. S. Atomic Energy Commission.

[1]J. H. Christenson, J. W. Cronin, V. L. Fitch, and R. Turlay, Phys. Rev. Letters 13, 138 (1964).

[2]R. G. Sachs, Phys. Rev. Letters 13, 286 (1964).

[3]We use a prime superscript throughout to indicate CP-violating terms.

[4]The existence of a significant $\Delta Q=-\Delta S$ amplitude, which Sachs desires, rules out the simple current-current interaction by the Okun-Pontecorvo argument.

[5]R. W. Birge et al., Phys. Rev. Letters 11, 35 (1963); W. Willis et al., Phys. Rev. Letters 13, 291 (1964).

[6]B. L. Ioffe, Zh. Eksperim. i Teor. Fiz. 42, 1411 (1962) [translation: Soviet Phys.- JETP 15, 978 (1962)].

[7]Of course, in the Sachs model the ratio Δ_L'/Δ_N can be kept low by reducing the strength of $H_{L,-1}'$. A reduction by a large factor, however, would seriously change the qualitative picture given by this model.

[8]We assume here that the only important CP-violating contribution to the self-energy matrix comes from the first-order term Δ_N'. In particular, the second-order term Δ_L', the relative smallness of which in the Sachs model was questioned above, now contains the small factor α so that in our model we believe $\Delta_L'/\Delta_N \ll 10^{-3}$. It is also assumed that the factor $|\zeta-1| \ll 10^{-3}$, where ζ defined by Sachs[2] is $A(\bar{K}^0\to 2\pi)/A(K^0\to 2\pi)$, so that we obtain the same value for $|r-1|$ as Sachs. This assumption is based on the fact that $\zeta-1$ is proportional to the ratio of the contribution of H_{N1}' to $K^0\to 2\pi$ to that of H_{N1} and so is expected to be of order α. Since we find α to be between 10^{-7} and 10^{-8} our assumptions are justified even if our order-of-magnitude result for α is too low by three orders of magnitude.

[9]Models which explain the $K_2\to 2\pi$ process by terms of the character of H_{N1}' have been given recently by N. Cabibbo, Phys. Letters 12, 137 (1964); T. N. Truong, Phys. Rev. Letters 13, 358 (1964). These terms are of order of magnitude $|\epsilon|$ in the sense that the self-energy Δ_N', now given by the expression for Δ_N with H_{N1} replaced by H_{N1}', must be of order $|\epsilon|$ with respect to Δ_N. However, in each of the above models, the CP-violating term has a special transformation property such that in special cases it may produce effects of magnitude greater than $|\epsilon|$. Thus Truong predicts a ratio $K_2\to\pi^0+\pi^0$ to $K_2\to\pi^++\pi^-$ much larger than 0.5, whereas the present model predicts 0.5. Cabibbo's model allows for significant violations of T invariance in leptonic decays in contrast to the present model.

Note Concerning *CP* Parity

L. B. Okun'

Institute of Theoretical and Experimental Physics
State Atomic Energy Commission

Submitted to JNP editor December 22, 1964
J. Nucl Phys. (U.S.S.R.) **1**, 938–939 (May, 1965)

The reported observation [1] of the $K_2^0 \to 2\pi$ decay has given rise to several theoretical hypotheses.[2] The purpose of this letter is to note that the result of [1], if confirmed, can be explained by assuming the existence of an x-interaction approximately 10^3 times stronger than the weak w-interaction and approximately 10^3 times weaker than the strong s-interaction, for which $\Delta Y = 0$ and which is P-even, but C- and CP-odd.

The $K_2^0 \to 2\pi$ decay would occur in this case with participation of all three types of interactions (w, x, and s):

$$K_2^0 \xrightarrow{w} i_1 \xrightarrow{x} i_2 \xrightarrow{s} 2\pi$$

K_2^0	i_1	i_2
$CP = -1$	$CP = -1$	$CP = +1$
$C = +1$	$C = -1$	$C = +1$
$P = -1$	$P = +1$	$P = +1$
$Y = \pm 1$	$Y = 0$	$Y = 0$

(i_1 and i_2 are here intermediate virtual states).

The existence of x-interactions should lead to violation of C and CP invariance in the amplitudes of all processes with a participation of strongly interacting particles on the order of 0.1%. The experiments performed so far have established at best an upper limit for these amplitudes, on the order of 2–3%. The quantities measured in these experiments were: 1) the asymmetry and polarization of the scattered protons[3]; 2) the cross sections for the direct and inverse reactions[4]; 3) $\gamma\gamma$ and $\beta\gamma\gamma$ correlations[5]; 4) the correlations in β decay and in the decays of strange particles.[6]

An increase of approximately one order of magnitude in the accuracy of the most sensitive among these experiments would make it possible to verify the proposed hypothesis.

Searches for x-interaction in the decays of known mesons (of the type $\pi^0 \to 3\gamma$) are exceedingly difficult. The absence of decays of the type $\omega \to 2\pi^0$ or $\rho \to 2\pi^0$, if and when proved, will not contradict the discussed hypothesis, since these decays are forbidden not only by C-parity conservation but also by the Bose statistics of π^0 mesons.

In conclusion we note that the combination of s, x, and non-lepton w-interactions with $\Delta Y = 0$ should cause the neutron to have an electric dipole moment eD with $D \sim 10^{-23}$–10^{-24} cm, which is approximately 4–5 orders smaller than the experimental limit [7] 5×10^{-20} cm.

The author is sincerely grateful to B. L. Ioffe, I. Yu. Kobzarev, I. Ya. Pomeranchuk, M. V. Terent'ev, and I. S. Shapiro for discussions.

[1] Christenson, Cronin, Fitch, and Turlay, *Phys. Rev. Lett.* **13**, 138 (1964). Abashian, Abrams, Carpenter, Fisher, Nefkens, and Smith, *ibid.* **13**, 243 (1964).

[2] R. G. Sachs, *Phys. Rev. Lett.* **13**, 286 (1964). T. N. Truong, *ibid.* **13**, 358 (1964). Bernstein, Cabibbo, and Lee, *Phys. Lett.* **12**, 146 (1964). T. T. Wu and C. N. Yang, *Phys. Rev. Lett.* **13**, 380 (1964). J. L. Uretsky, *Phys. Lett.* **14**, 154 (1964). A. Abashian and H. Lipkin, ANL Preprint, 1964. L. Wolfenstein, *Phys. Rev. Lett.* **13**, 562 (1964). N. Cabibbo, *Phys. Lett.* **12**, 137 (1964). J. Bell and J. Perring, *Phys. Rev. Lett.* **13**, 348 (1964). S. Weinberg, *ibid.* **13**, 495 (1964). T. D. Lee, Preprint, Columbia University 1964. É. M. Lipmanov, *JETP* **48**, 750 (1965). *Soviet Phys. JETP* **21**, 496 (1965).

[3] A. Abashian and E. Hafner, *Phys. Rev. Lett.* **1**, 255 (1958). Hillman, Johansson, and Tiebell, *Phys. Rev.* **110**, 121 (1958). Hwang, Ophel, Throndike, and Willson, *Phys. Rev.* **119**, 352 (1960). R. Philips, *Nuovo cimento* **8**, 265 (1958). J. S. Bell and F. Mandi, *Proc. Phys. Soc.* **71**, 272 (1958).

[4] E. Henley and B. Jacobsohn, *Phys. Rev.* **108**, 502 (1957) and **113**, 225 (1959). B. A. Jacobsohn and E. M. Henley, *Phys. Rev.* **113**, 234 (1959). L. Rosen and J. Brolley, *Phys. Rev. Lett.* **2**, 98 (1959). Bodansky, Eccles, Farwell, Rickey, and Robinson, *ibid.* **2**, 101 (1959).

[5] S. P. Lloyd, *Phys. Rev.* **81**, 161 (1951). P. A. Krupchitskiĭ, ITEF Preprint No. 41, 1962. Fuschini, Gadjokov, Maroni, and Veronesi, *Nuovo cimento* **33**, 709 and 1309 (1964).

[6] Burgy, Krohn, Novey, Ringo, and Telegdi, *Phys. Rev. Lett.* **1**, 324 (1958). J. Cronin and O. Overseth, *Phys. Rev.* **129**, 1795 (1963).

[7] Smith, Purcell, and Ramsey, *Phys. Rev.* **108**, 120 (1957).

Translated by J. G. Adashko

3) F.R.Huson and W.B.Fretter, Nuovo Cimento 33 (1964) 1.
4) G.Bellini, M.Di Corato, E.Fiorini, P.Negri, S. Ratti, J.Hennessy, A.Lloret, J.J.Veillet, J. Crussard and J.Ginestet, Sienna Conference 1963, p. 621.
5) G.Goldhaber, J.Brown, S.Goldhaber, J.Kadyk, B. Shen and G.Trilling, Phys. Rev. Letters 12 (1964) 336.
6) R.Lander, M.Abolins, D.D.Carmony, T.Hendricks, Nguyen-huu-Xuong and P.Yager (University of California, San Diego, La Jolla, California, preprint).
7) Aachen, Birmingham, Bonn, Hamburg, London, München collaboration, Physics Letters 10 (1964) 226.
8) Suh Urk Chung, O.J.Dahl, L.M.Hardy, R.I.Hess, G.R.Kalbfleisch, J.Kirz, D.H.Miller and G.A. Smith, Phys. Rev. Letters 12 (1964) 621.
9) Nauenberg and Pais, Phys. Rev. Letters 8 (1962) 82 M.Nauenberg, private communication.
10) R.J.Oakes, Phys. Rev. Letters 12 (1964) 134.
11) S.Berman and S.Drell, Phys. Rev. Letters 11 (1963) 220.
12) C.Zemach, Phys. Rev. 133 (1964) B1201.
13) P.Dennery and A.Krzywicki (Orsay and Ecole Polytechnique, Paris, preprint); S.Berman, private communication.

* * * * *

CP INVARIANCE AND THE 2π DECAY MODE OF THE K_2^0 *

J. BERNSTEIN **, N. CABIBBO and T. D. LEE ***
CERN, Geneva, Switzerland

Received 20 August 1964

In a recent experiment Christenson et al. [1] have found definite indications that the long lived component of the K^0, the $K^0{}_2$, decays into two π mesons

$$K_2^0 \rightarrow 2\pi \,. \qquad (1)$$

The most direct interpretation of their experiment is to assume that *CP* invariance is violated [2] in this decay and hence in the weak interactions in general. However, it is important to realize that this experiment involves an entirely new domain of phenomena, those involving extremely small energies. The experiment of Christenson et al. is capable of detecting energy differences of the order of 10^{-8} eV. No other experiment in particle physics, up to now, has exhibited such sensitivity to small energies. On the other hand, none of the other experiments [3] on weak interactions that might have shown *CP* violation has done so. Therefore it is not unnatural to try to explain the new experiments in some other way. In this note we shall attempt to reconcile the two assumptions:

i) *CP* invariance holds for *all* interactions:
ii) the long lived component of the K^0, in (local) vacuum, or at least in what is usually called "vacuum", decays into 2π's.

Up to now vacuum is particle physics has been taken to mean the absence of neighbouring material objects, although not necessarily that of long range fields such as gravitational or electromagnetic fields. These conventional fields, which are produced by distant bodies such as Milky Way and the other galaxies, are sufficiently well understood so that one can assume, with some confidence, that their effects on the K^0, $\overline{K}^0$ system can be neglected, either because they are too small (we have in mind the possible electromagnetic conversion between $K^0{}_1$ and $K^0{}_2$ due to the difference in the charge radius of K^0 and $\overline{K}^0$) or because they have the same effect on particles and antiparticles (e.g., gravitational forces).

In this note we postulate that there is a new long range, extremely weak, field which can be neglected so long as one does not measure energies as small as those measured in the experiment of Christensen et al. This field, we assume, produces a potential energy which is equal in magnitude and opposite in sign for K^0 and $\overline{K}^0$, say $\frac{1}{2}V$ for K^0 but $-\frac{1}{2}V$ for $\overline{K}^0$. This new interaction is *CP* conserving and if we could maintain a given K^0 and $\overline{K}^0$ system, but transform the environment into its *CP* conjugate state, then this potential energy would reverse sign †.

* Work supported in part by the U.S.Atomic Energy Commission. the Army Research Office (Durham) and the National Science Foundation.

** Permanent address: New York University, New York, N.Y., U.S.A.

*** Permanent address: Columbia University, New York, N.Y., U.S.A.

† During the completion of this paper, we learnt of a preprint by J.S.Bell and J.K.Perring, who have in-

As we shall see, this possibility leads to unique predictions concerning other "apparent" *CP* violating results which are in general quite *different* from the alternative possibility [2] of assuming *CP* non-invariance of weak interactions. Our analysis does not depend on the precise nature of the new interaction except that it must produce a potential energy difference V between K^o and $\bar{K}^o$. [To give an example, we may envisage that this potential energy difference is transmitted by a four-vector field similar to the electromagnetic interaction. It can be generated by any strongly interacting particle and has an amplitude proportional to, say, I_z, or $\frac{1}{2}(N+S)$, of the particle where I_z = z component of isotopic spin and N = baryon number, S = strangeness. (We leave open the question of whether it can have leptonic couplings)].

Whatever the origin of this interaction with macroscopic bodies, the fact that it gives rise to a static potential energy implies that so far as energies of the order of 10^{-8} eV can be neglected in the *decays of all particles other than neutral* K *mesons, there are no observable "apparent" CP non-invariant effects.* For the neutral K meson, all other 'apparent" *CP* non-invariant effects are completely determined by the parameter V.

We first discuss the problem of a K meson in local isolation (i.e., in the absence of any neighbouring matter but taking into account the effects of the new long range interaction with distant objects). Let $\psi(t)$ be the wave function of a K meson in the laboratory system. In what follows, we regard t as the independent variable and the co-ordinate $\boldsymbol{r}$ of the K meson as a dependent variable given by $(\boldsymbol{r}_o + \boldsymbol{v}t)$ where $\boldsymbol{v}$ is the K meson velocity and $\boldsymbol{r}_o$ is its initial co-ordinate. The relative velocity of the laboratory and the distant macroscopic bodies will be neglected compared to the velocity of the K in the laboratory. We may write:

$$\psi(t) = a(t)|K^o\rangle + b(t)|\bar{K}^o\rangle \, , \tag{2}$$

or simply

$$\psi(t) = \begin{pmatrix} a(t) \\ b(t) \end{pmatrix} \tag{3}$$

where

$$|\bar{K}^o\rangle \equiv CP|K^o\rangle \, . \tag{4}$$

The time development of $\psi(t)$ can be determined from

$$i\frac{d\psi}{dt} = (\eta + \xi\,\sigma_x + \zeta\,\sigma_z)\psi \tag{5}$$

where

$$\sigma_x = \begin{pmatrix} 0 & 1 \\ 1 & 0 \end{pmatrix} , \quad \sigma_z = \begin{pmatrix} 1 & 0 \\ 0 & -1 \end{pmatrix} , \tag{6}$$

and, for a locally isolated system, η, ξ, ζ, are given by ($\hbar = c = 1$)

$$\eta = \eta_o = \frac{1}{2\gamma}[(m_1 + m_2) - i\tfrac{1}{2}(\Gamma_1 + \Gamma_2)] \, , \tag{7}$$

$$\xi = \xi_o = \frac{1}{2\gamma}[(m_1 - m_2) - i\tfrac{1}{2}(\Gamma_1 - \Gamma_2)] \tag{8}$$

and

$$\zeta = \zeta_o = \tfrac{1}{2}V \, . \tag{9}$$

The subscript o indicates that these are values for the K meson in a locally isolated system, and,

$$\gamma = (1 - v^2)^{-\frac{1}{2}} \tag{10}$$

where v = velocity of the K meson in the laboratory. In the above equations, $\zeta_o \sigma_z$ is the effect of the new interaction, m_1, Γ_1 and m_2, Γ_2 are the mass and width of K^o_1 and K^o_2 if $V = 0$. The solution of eq. (5) is well known. There are two eigensolutions $\psi_1(t)$ and $\psi_2(t)$ which are given by

$$\psi_1 = (1 + |\epsilon|^2)^{-\frac{1}{2}}[\psi_+ + \epsilon\psi_-]\, e^{i\lambda_1 t}$$

and

$$\psi_2 = (1 + |\epsilon|^2)^{-\frac{1}{2}}[\psi_- - \epsilon\psi_+]\, e^{i\lambda_2 t} \tag{11}$$

where

$$\psi_\pm = \frac{1}{\sqrt{2}}\begin{pmatrix} 1 \\ \pm 1 \end{pmatrix} , \tag{12}$$

$$\epsilon = \zeta^{-1}[-\xi + (\xi^2 + \zeta^2)^{\frac{1}{2}}] \, , \tag{13}$$

$$\lambda_1 = \eta + (\xi^2 + \zeta^2)^{\frac{1}{2}} \tag{14}$$

and

$$\lambda_2 = \eta - (\xi^2 + \zeta^2)^{\frac{1}{2}} \, . \tag{15}$$

Using eqs. (7)-(9) and assuming small V, we find

$$\epsilon = \epsilon_o \cong \tfrac{1}{2}\gamma V[(m_1 - m_2) - i\tfrac{1}{2}(\Gamma_1 - \Gamma_2)]^{-1} .$$

The value ϵ_o for the experiment of Christenson et al. is found to be

$$|\epsilon_o| \cong 2 \times 10^{-3} \, . \tag{16}$$

It is important to note that if V is due to the average of the fourth component of a vector field, then V is independent of the energy of the K meson. Thus *the parameter* ϵ_o *depends linearly on* γ. (On the other hand, V is proportional to γ^{-1} if it is due to a spin 0 field, and is porportional to γ if it is due to a spin 2 field. However, in the frame-

dependently considered the possibility of a long range vector interaction in connection with the experiment of Christenson et al.

work of a simple local field theory, both spin 0 and spin 2 fields give equal static potential for a particle and its antiparticle. Therefore it seems more attractive to assume that this new interaction is due to a vector field with an extremely small mass.)

The solution, eq. (11), also determines other "apparent" *CP* non-invariant effects in the K decay. For example, the ratio

$$r_i \equiv \frac{\text{rate } (K_i^0 \to \pi^- + l^+ + \nu_l)}{\text{rate } (K_i^0 \to \pi^+ + l^- + \bar{\nu}_l)}$$

can be different from 1, where i = 1 or 2, and l = e or μ. Assuming that the $\Delta Q = \Delta S$ rule holds, we find

$$r_1 = \left|\frac{1+\epsilon_0}{1-\epsilon_0}\right|^2$$

and

$$r_2 = \left|\frac{1-\epsilon_0}{1+\epsilon_0}\right|^2 \qquad (17)$$

which gives $|r_i - 1| \simeq 6\times10^{-3}$ for the ϵ_0 given by eq. (16).

Next, we discuss the problem of a K meson in a medium. Let n and n' denote, respectively, the complex indices of refraction of K^0 and $\bar{K}^0$ in the medium [4]. Eq. (5) remains valid where, instead of eqs. (7)-(9), the parameters η, ξ, ζ are now given by

$$\eta = \frac{1}{2\gamma}\,[(m_1+m_2) - i\tfrac{1}{2}(\Gamma_1+\Gamma_2)] - \tfrac{1}{2}(n+n'-2)\,kv\,, \qquad (18)$$

$$\xi = \frac{1}{2\gamma}[(m_1-m_2) - i\tfrac{1}{2}(\Gamma_1-\Gamma_2)]\,. \qquad (19)$$

$$\zeta = \tfrac{1}{2}[V - (n-n')\,kv]\,, \qquad (20)$$

and k is the momentum of the K meson.

The solution, eq. (11), also remains valid. For $|\zeta| \ll |\xi|$, the parameter ϵ is given approximately by

$$\epsilon \cong \tfrac{1}{2}\gamma[V - (n-n')\,kv][(m_1-m_2) - i\tfrac{1}{2}(\Gamma_1-\Gamma_2)]^{-1}\,. \qquad (21)$$

The absolute magnitude of ϵ in a medium can be obtained by measuring the corresponding rate of reaction (1). The interference term between V and $(n-n')\,kv$ in the expression for $|\epsilon|$ can also determine the sign of V.

It is interesting to speculate on the possible nature of such a long range interaction. As mentioned in the preceding discussion, a simple possibility is to make the ad hoc assumption of the existence of a four-vector field which interacts with all particles. For definiteness, let us assume that the effective "charge" for all strongly interacting particles is

$$f \times I_z \qquad (22)$$

or $f \times \frac{1}{2}(N+S)$, where f is a dimensionless coupling constant. The potential energy difference V between K^0 and $\bar{K}^0$ due to the Milky Way is simply given by *

$$V \approx \tfrac{1}{2}f^2(M_g/M_p)\,R^{-1} \qquad (23)$$

where M_g = mass of the galaxy, m_p = mass of proton, and R = effective radius of the galaxy, defined so that $G M_g R^{-1}$ = gravitational potential on earth due to the galaxy, G is the gravitation constant. The factor $\frac{1}{2}$ is due to the isospin value of p. It seems reasonable to assume that effects from very distant galaxies are small, and the main contribution is due only to the Milky Way. By taking $M_g \approx 10^{11}\times$ solar mass, $R \sim 10$ kpc and $V \sim 5\times10^{-4}$ cm^{-1}, we find

$$f^2 \sim 2.5\times10^{-49}\,. \qquad (24)$$

Such a long range force would also give rise to a slight difference between the observed "gravitational" mass and inertial mass **. The current experimental upper limit [6] on f^2 is

$$f^2 <\sim 2\times10^{-45} \qquad (25)$$

which is certainly compatible with eq. (24).

Two of us (J. B. and T. D. L.) wish to thank Dr. S. F. Tuan for discussions, and Professors V. F. Weisskopf and L. Van Hove for the hospitality extended to us at CERN.

References

1) J.H. Christenson, J.W. Cronin, V. L. Fitch and R. Turlay, Phys. Rev. Letters 13 (1964) 138.
2) N. Cabibbo, Physics Letters, 12 (1964) 137. R.G. Sachs, to be published.
3) M.T. Burgy, V.E. Krohn, T.B. Novey, G.R. Ringo and V.L. Telegdi, Phys. Rev. Letters 1 (1958) 324; see also: J.W. Cronin and O.E. Overseth, Proc. 1962 Int. Conf. on high energy physics, CERN, p. 453.
4) M. L. Good, Phys. Rev. 106 (1957) 591.
5) T.D. Lee and C.N. Yang, Phys. Rev. 98 (1955) 1501.
6) R.H. Dicke, Phys. Rev. 126 (1962) 1580.

* We implicitly assume that the mass of this vector field is smaller than, or comparable to, R^{-1}.

** The reasoning here is very similar to that given by Lee and Yang [5].

PHYSICS REPORTS (Section C of Physics Letters) 9, no. 2 (1974) 143–177. NORTH-HOLLAND PUBLISHING COMPANY

CP NONCONSERVATION AND SPONTANEOUS SYMMETRY BREAKING*

T.D. LEE

Columbia University, New York, N.Y. 10027, USA

Received 4 October 1973

Contents:

Abstract:

The observed *CP* violation is assumed to be due to the spontaneous symmetry-breaking mechanism; the Lagrangian is *CP* invariant but its particular solution is not. The general classification of such theories when coupled with different unified gauge models of the weak and electromagnetic interactions is given. All such theories lead naturally to a basically milliweak *CP* noninvariant solution. The possibility that for most weak transitions the result may resemble a superweak theory is analysed, and possible experiments to distinguish these two different types of theories are discussed. Detailed calculations for various *CP* violating amplitudes are carried out for a generalized Georgi–Glashow model.

Single orders for this issue

PHYSICS REPORTS (Section C of PHYSICS LETTERS) 9, No. 2 (1974) 143–177.

Copies of this issue may be obtained at the price given below. All orders should be sent directly to the Publisher. Orders must be accompanied by check.

Single issue price Dfl. 12.50, postage included.

* This research was supported in part by the U.S. Atomic Energy Commission.

1. *CP* nonconservation

1.1. *Present status*

Although the theoretical possibility of *CP* nonconservation and time reversal asymmetry was raised [1] as early as 1956, at about the same time parity violation was suggested, the actual experimental discovery came much later. In 1964 Christenson, Cronin, Fitch and Turlay [2] observed that the long-lived neutral kaon K_L^o can also decay into 2π:

$$K_L^o \to \pi^+\pi^- . \tag{1.1}$$

This is of great significance since K_L^o is known to decay into 3π with the final pions predominantly in the s states. The *CP* value of the $\pi^+\pi^-$ final state is clearly +1, but that of the $\pi^+\pi^-\pi^o$ is -1 (or, at least predominantly -1). The fact that a single non-degenerate state K_L^o which has a definite lifetime can decay into states with different *CP* values establishes the violation of *CP* conservation.

At present, *CP* violations have been observed *only* in the various decay modes of the $K^o - \overline{K}^o$ system; besides the $\pi^+\pi^-$ decay, there are [3–7] the decay $K_L^o \to 2\pi^o$ and the charge asymmetry measurements [8, 9] in the $K_{\pi l3}$ decay. Furthermore, from the detailed analysis [10] of various K decay amplitudes and especially the recent upper bound [11]

$$\mathrm{rate}(K_S^o \to 3\pi^o)/\mathrm{rate}(K_L^o \to 3\pi^o) < 1.2, \tag{1.2}$$

one can conclude that the time reversal symmetry *T* is also violated, while the product *CPT* symmetry remains valid.

From these experimental results, it follows that the state vectors of K_L^o and K_S^o are *not* eigenstates of *CP*. If one assumes *CPT* invariance, then these state vectors can depend only on one complex parameter [1]:

$$K_S^o = [2(1+|\epsilon|^2)]^{-1/2}\ [(1+\epsilon)K^o + (1-\epsilon)\overline{K}^o], \tag{1.3}$$

and

$$K_L^o = [2(1+|\epsilon|^2)]^{-1/2}\ [(1+\epsilon)K^o - (1-\epsilon)\overline{K}^o], \tag{1.4}$$

where $\overline{K}^o$ is defined to be the *CPT* conjugate of K^o. If *CP* invariance were true, then K_L^o should be orthogonal to K_S^o, and therefore ϵ would be zero. Thus, the parameter ϵ characterizes the extent of *CP* nonconservation in the description of the K_L^o and K_S^o state vectors.

The phase angle $\arg((1-\epsilon)/(1+\epsilon))$ depends on the relative phase between K^o and $\overline{K}^o$ which can be arbitrarily chosen. If we adopt the phase convention that the transition amplitude $K^o \to 2\pi$ in the isospin $I = 0$ final state is real, then the phase angle ϵ becomes approximately determined [12–14]

$$\arg\epsilon \approx \tan^{-1}\{2(m_L - m_S)/(\gamma_S - \gamma_L)\}, \tag{1.5}$$

where m_L, γ_L and m_S, γ_S refer to the mass and the width of K_L^o and K_S^o respectively. The magnitude of ϵ can then be measured from the charge asymmetry results in the $K_{\pi l3}$ decay. By using [1, 8–10]

$$\frac{\text{rate}\,(K_S^o \to \pi^- l^+ \nu_l)}{\text{rate}\,(K_S^o \to \pi^+ l^- \bar{\nu}_l)} \approx \frac{\text{rate}\,(K_L^o \to \pi^- l^+ \nu_l)}{\text{rate}\,(K_L^o \to \pi^+ l^- \bar{\nu}_l)} \approx 1 + 4\,\text{Re}\,\epsilon, \tag{1.6}$$

one finds

$$|\epsilon| \approx 2 \times 10^{-3}. \tag{1.7}$$

The present experimental situation can best be summarized in terms of the simple phenomenological description that all observed T and CP violations are completely determined by the single small parameter $\epsilon \neq 0$; i.e., to a good approximation, one may regard the transition matrix elements between K^o, $\bar{K}^o$ and 2π, 3π, $\pi l \nu$, etc., as T and CP invariant. For example, in this approximation one has

$$\eta_{+-} \approx \eta_{oo} \approx \epsilon, \tag{1.8}$$

where

$$\eta_{ab} \equiv \text{amp}(K_L^o \to \pi^a \pi^b)/\text{amp}(K_S^o \to \pi^a \pi^b). \tag{1.9}$$

Equation (1.8) is in good agreement with the present measurements of η_{+-} and η_{oo} [15, 16].

Since the K^o and $\bar{K}^o$ system is determined not only by the weak interaction but also by the strong and the electromagnetic interactions, there are several alternative theoretical possibilities concerning the origin of the CP violation [17]:

(i) superweak theory [18]: the CP violating interaction is assumed to violate the hypercharge conservation by two units and its strength is $\sim 10^{-9}\, G_F$ where G_F is the usual Fermi coupling constant for the weak interaction. Because of the extreme smallness of the superweak coupling constant, one may safely neglect any higher order effect. Equations (1.5) and (1.8) are then exact [17];

(ii) milliweak possibility: the CP violating interaction is assumed to violate the hypercharge conservation by one unit, and its strength is $\sim 10^{-3}\, G_F$. We note that the deviation from (1.8) is given by [19]

$$|\eta_{+-} - \epsilon| = |2^{-1/2}\, \text{Im}\, A_2/A_0| \equiv |\epsilon'|, \tag{1.10}$$

where A_0 and A_2 refer respectively to the transition amplitudes of $K^o \to 2\pi$ in the $I = 0$ and 2 states. At present, the origin of the usual $|\Delta I| = \frac{1}{2}$ rule is quite unclear. If one assumes that the same rule applies to the CP violating amplitude as well, then any such milliweak theory would imply the magnitude of $|\text{Im}\, A_2/A_2| \sim 10^{-3}$. Consequently, one has $|\text{Im}\, A_2/A_0| \sim 5 \times 10^{-5}$, which is $\ll |\epsilon|$ and is, therefore, consistent with (1.8). On the other hand, if the milliweak CP violating interaction does not satisfy the $|\Delta I| = \frac{1}{2}$ rule, then an additional explanation is needed for (1.8);

(iii) millistrong possibility [17, 20]: the *CP* violating interaction strength is assumed to conserve the hypercharge and its strength is $\sim 10^{-3}$ times the strong interaction strength. It is also possible that such a millistrong force may be electromagnetic in origin [21]. In any case, if such a millistrong force conserves isospin also, then one expects (1.8) to hold at least to the level where the $|\Delta I| = \frac{1}{2}$ rule holds for the usual (T conserving) weak interaction.

1.2. *Two fundamental questions*

At present, the full implications of *CP* nonconservation and *T* violation are far from being understood. There are at least two fundamental questions:

1. Why should *T* symmetry and *CP* symmetry be violated?
2. Why should the violation be so small?

In order to provide some possible answers to these questions, we shall examine in this paper a general class of theories [22] in which the total Lagrangian (which includes all interactions, strong, electromagnetic and weak) is assumed to satisfy:

(a) invariance under *CP* and *T*,
(b) renormalizability, and
(c) invariance under certain weak and electromagnetic gauge transformations.

We note that if (a) holds, then by identifying time reversal with *T* and space inversion with *CP* (not *P*), the full Poincaré group of space-time transformations, including these discrete ones, may be regarded as *exact* symmetries of all interactions. The dynamical equation is completely symmetrical with respect to *T* and *CP*, though its particular solution can be non-invariant. Such a mechanism will be referred to as the *spontaneous T violation*, or spontaneous *CP* violation. In physics, it is a familiar practice to assume that the dynamical equation is symmetric under a relatively large group of transformations, while its particular solution does not have to be an invariant under the same transformations. For example, rotational invariance of the physical law by no means implies that all physical solutions should be spherically symmetrical. From an esthetical point of view, it appears pleasing to assume a *CP*-invariant and *T*-invariant interaction, but to attribute all observed *CP* violations merely to the asymmetrical nature of the particular solution. Such a possibility gives an answer to our first question.

As we shall see, the requirement (b) of renormalizability makes all such *CP* and *T* noninvariant effects of the solution finite and therefore, at least in principle, computable. In addition, as we shall also show, because of the condition (c) the magnitude of *CP* violation must be quite small at low energy. This then provides a possible answer to our second question.

We emphasize that the above three conditions (a), (b), and (c) are logically independent. For example, as discussed by Pais [23], *CP* violation can be introduced into a gauge theory [24, 25] without assuming the spontaneous *CP* violation mechanism. However, since the entire approach of the unified gauge theory of the electromagnetic and weak interactions [26–30] is based on the spontaneous symmetry-breaking mechanism, it seems natural, and also esthetically more appealing, to include *CP* violation as an integral part of the same mechanism.

2. Spontaneous T violation

2.1. *A simple example*

To illustrate the phenomenon of spontaneous T violation, let us first examine a simple example which satisfies our conditions (a) and (b), but without the condition (c). In this way, we may also demonstrate the logical independence of the spontaneous T violation from the well-known mechanism of spontaneous gauge-symmetry violation [24–26].

This simple example consists of a spin $\frac{1}{2}$ Dirac field ψ and a spin 0 Hermitian field ϕ. The Lagrangian density is assumed to be invariant under T, C and P

$$L = -\tfrac{1}{2}\left(\frac{\partial\phi}{\partial x_\mu}\right)^2 - V(\phi) - \psi^\dagger\gamma_4\left(\gamma_\mu\frac{\partial}{\partial x_\mu} + m\right)\psi - \mathrm{i}g\,\psi^\dagger\gamma_4\gamma_5\psi\phi, \tag{2.1}$$

where on account of renormalizability the most general form of the potential $V(\phi)$ is a quartic function, given by

$$V(\phi) = -\tfrac{1}{2}\lambda\phi^2 + \tfrac{1}{4}\kappa\phi^4.$$

From Hermiticity, the parameters m, g, λ and κ must be real. It can be readily verified that under T, C and P, ϕ transforms according to

$$T\phi(\boldsymbol{r},t)T^{-1} = -\phi(\boldsymbol{r},-t), \tag{2.2a}$$

$$C\phi(\boldsymbol{r},t)C^{-1} = \phi(\boldsymbol{r},t), \tag{2.2b}$$

and

$$P\phi(\boldsymbol{r},t)P^{-1} = -\phi(-\boldsymbol{r},t). \tag{2.2c}$$

To generate a spontaneous T violation, we assume that the (renormalized) coupling constants λ and κ satisfy

$$\lambda > 0, \qquad \text{and} \qquad \kappa > 0.$$

Thus, as shown in fig. 1, the vacuum expectation value of ϕ is not zero;

$$\langle\phi\rangle_{\text{vac}} = \rho \neq 0.$$

Since according to (2.2), ϕ is of $T = -1$, $CP = -1$ and $P = -1$, a non-zero vacuum expectation value of ϕ implies a spontaneous violation of T (and also of CP and P). The T symmetry of the Lagrangian requires that if $\langle\phi\rangle_{\text{vac}} = \rho$ is a solution, then $\langle\phi\rangle_{\text{vac}} = -\rho$ must also be a solution. These two solutions transform into each other under T, but by itself either solution is not invariant under T. (It is also not invariant under CP and P, though it is invariant under C and CPT.)

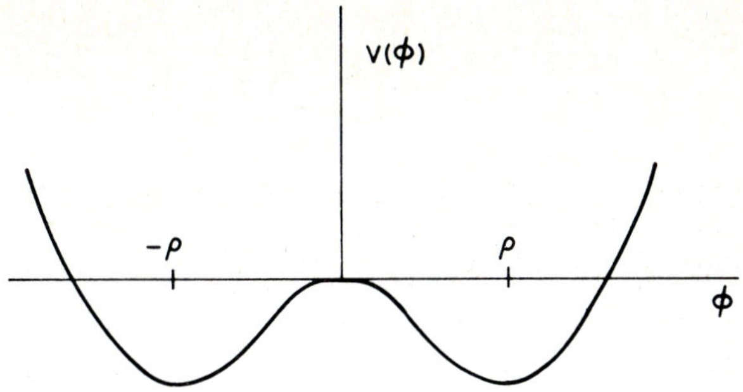

Fig. 1. The potential $V(\phi) = -\frac{1}{2}\lambda\phi^2 + \frac{1}{4}\kappa\phi^2$ for $\lambda > 0$ and $\kappa > 0$.

In the tree approximation, ρ is determined by the minimum of the c. number function V. Therefore

$$\rho^2 = (\lambda/\kappa). \tag{2.3}$$

Because of quantum effects, the field ϕ fluctuates around its vacuum expectation value. We may write

$$\phi = \rho + \delta\phi. \tag{2.4}$$

In terms of $\delta\phi$, the potential V becomes

$$V = V_0 + \tfrac{1}{2}\mu^2(\delta\phi)^2 + \kappa\rho(\delta\phi)^3 + \tfrac{1}{4}\kappa(\delta\phi)^4,$$

where $V_0 = -\frac{1}{4}\kappa^{-1}\lambda^2$ and $\mu^2 = 2\lambda$. In this case, since T is a discrete symmetry and there is no spontaneous gauge-symmetry violation, one does not encounter any Goldstone boson [24]. The mass μ of the field fluctuation $\delta\phi$ is not zero.

In order to exhibit more clearly the T-violating character of the solution, we may perform a unitary transformation under which ϕ is unchanged but

$$\psi \to [\exp(-\tfrac{1}{2}\mathrm{i}\gamma_5\alpha)]\,\psi. \tag{2.5}$$

Therefore, the quadratic expression

$$\psi^\dagger\gamma_4(m + \mathrm{i}g\rho\gamma_5)\psi \to \psi^\dagger\gamma_4 M\psi, \tag{2.6}$$

where

$$M = (m^2 + g^2\rho^2)^{1/2},$$

provided that the angle α is given by $\tan\alpha = m^{-1}g\rho$. Correspondingly, the Lagrangian density L becomes

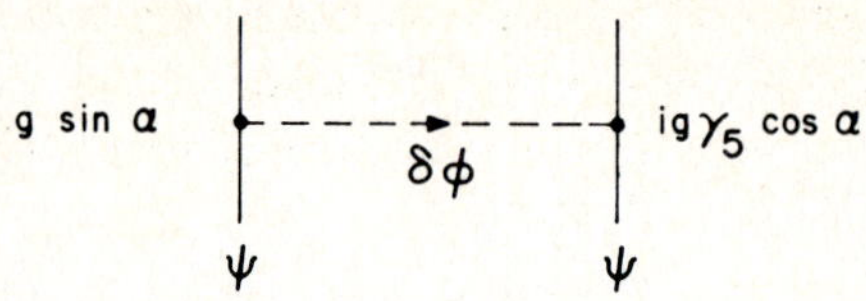

Fig. 2. A T violating scattering diagram due to the exchange of the quantum of $\delta\phi$, where $\delta\phi \equiv \phi - \langle\phi\rangle_{\mathrm{vac}}$.

$$-\tfrac{1}{2}\left[\frac{\partial}{\partial x_\mu}(\delta\phi)\right]^2 - V - \psi^\dagger\gamma_4\left(\gamma_\mu\frac{\partial}{\partial x_\mu} + M\right)\psi - g\psi^\dagger\gamma_4(\sin\alpha + \mathrm{i}\gamma_5\cos\alpha)\psi\cdot(\delta\phi). \tag{2.7}$$

Since the operator $\psi^\dagger\gamma_4\psi$ is of $P = 1$, $C = 1$ and $T = 1$ while the operator $\mathrm{i}\psi^\dagger\gamma_4\gamma_5\psi$ is of $P = -1$, $C = 1$ and $T = -1$, any exchange of the $\delta\phi$-quantum would give an interference term between these two operators that violates T, P and CP; but the product CPT remains invariant. For example, the T and CP violating amplitude A_- of the diagram given in fig. 2 is

$$A_- = g^2\sin\alpha\cos\alpha\,(k^2 + \mu^2)^{-1}, \tag{2.8}$$

where k denotes the 4-momentum transfer.

This simple example illustrates the basic mechanism of a spontaneous T violation. One assumes that the ground state of the system (called the vacuum state) has a nonzero expectation value $\langle\phi\rangle_{\mathrm{vac}}$, where ϕ is a $T = -1$ spin 0 field; thus, the vacuum state is noninvariant under T, even though the interaction satisfies T invariance. The fact that $\langle\phi\rangle_{\mathrm{vac}} \neq 0$ also makes the vacuum state resemble more closely a "medium", and this is why the fermion field ψ in moving through this "medium" acquires a (mass)^2-shift Δ^2, given by

$$\Delta^2 \equiv M^2 - m^2 = g^2\langle\phi\rangle^2_{\mathrm{vac}}. \tag{2.9}$$

For small Δ, and at zero momentum transfer ($k = 0$), the T violating amplitude (2.8) becomes simply

$$A_- \approx g^2\,\Delta/\mu^2 M. \tag{2.10}$$

The T invariance of the interaction implies that the vacuum state must have a double degeneracy. It is interesting to examine the barrier penetration between these two degenerate ground state solutions: $\langle\phi\rangle_{\mathrm{vac}} = \rho$ and $-\rho$. By assuming the entire system to be in a finite volume Ω and by following the standard procedure of quantization, one can easily show that as $\Omega \to \infty$, this barrier penetration probability becomes zero exponentially. On the other hand, in any realistic cosmological model the volume Ω should be finite. In particular, if one assumes an oscillating cosmological model in which Ω changes periodically, then when Ω is of a microscopic dimension, the system should undergo transitions between these two solutions $\langle\phi\rangle_{\mathrm{vac}} = \rho$ and $-\rho$. At different cycles of such Ω oscillations, the system may acquire different solutions, sometimes with $\langle\phi\rangle_{\mathrm{vac}} = \rho$ and sometimes $-\rho$. An examination of such cosmological consequences is exremely interesting, but lies outside the scope of the present paper.

In this simple example, the magnitude of the T violating amplitude A_- is, of course, entirely arbitrary. As we shall see, this arbitrariness may be limited if the spontaneous T violation is coupled with the spontaneous gauge-symmetry violation of the weak and electromagnetic interactions.

2.2. *A second example*

In order to illustrate how a discrete symmetry, such as T or CP, may be combined with a continuous gauge-symmetry so that both symmetries can be broken spontaneously and simultaneously, let us consider another simple example. While this second example is also *not* realistic, it nevertheless does contain most of the essential features. (All the so-called "realistic" gauge models [26–30] of weak and electromagnetic interactions are unfortunately rather complicated; these will be discussed in the next section.)

We assume that the Lagrangian density L of all interactions satisfies the three conditions (a), (b) and (c), stated in section 1.2. In this simple model, the gauge-symmetry group is assumed to be the SO_3 group, like that in the Georgi-Glashow model [27]. The generators of the group are

$$Q^{\rm wk}, \quad Q^{\gamma} \quad \text{and} \quad (Q^{\rm wk})^{\dagger}, \tag{2.11}$$

which are related to the weak interaction current $J_\mu^{\rm wk}$ and the electromagnetic current J_μ^{γ} by

$$Q^{\rm wk} = \int J_0^{\rm wk}\, d^3r, \qquad Q^{\gamma} = \int J_0^{\gamma}\, d^3r, \tag{2.12}$$

and $\dagger$ denotes the Hermitian conjugate. Corresponding to these three generators there is a triplet of spin 1 gauge fields $\boldsymbol{W}_\mu$:

$$W_\mu^+, \qquad W_\mu^{\rm o}, \qquad W_\mu^-, \tag{2.13}$$

where $W_\mu^{\rm o}$ is the usual electromagnetic field A_μ. In addition, there is a triplet of spin 0 Hermitian fields $\boldsymbol{\phi}$:

$$\phi^+, \qquad \phi^{\rm o}, \qquad \phi^-,$$

and a triplet of spin $\frac{1}{2}$ hadron fields

$$\psi = \begin{pmatrix} p^+ \\ n^{\rm o} \\ q^- \end{pmatrix}.$$

(The realistic Georgi-Glashow model [27] would require many additional fields.)

The Lagrangian density is

$$L = L_{\phi W} + L_{W\psi} + L_{\psi\phi}, \tag{2.14}$$

where

$$L_{\phi W} = -\tfrac{1}{4} W_{\mu\nu}^2 - \tfrac{1}{2}(\partial_\mu \boldsymbol{\phi})^2 - V(\phi), \tag{2.15}$$

$$L_{W\psi} = -\psi^\dagger \gamma_4 (\gamma_\mu \partial_\mu + m)\psi, \tag{2.16}$$

and

$$L_{\psi\phi} = -\mathrm{i}g\,\psi^\dagger \gamma_4 \gamma_5 \boldsymbol{I}\psi \cdot \boldsymbol{\phi}, \tag{2.17}$$

in which the components of $\boldsymbol{I}$ form the (3×3) Hermitian matrix representation of the generators of the SO_3 group,

$$\boldsymbol{W}_{\mu\nu} = \frac{\partial}{\partial x_\mu} \boldsymbol{W}_\nu - \frac{\partial}{\partial x_\nu} \boldsymbol{W}_\mu + e(\boldsymbol{W}_\mu \times \boldsymbol{W}_\nu), \tag{2.18}$$

$$\partial_\mu \boldsymbol{\phi} = \left(\frac{\partial}{\partial x_\mu} + e\boldsymbol{W}_\mu \times\right)\boldsymbol{\phi}, \tag{2.19}$$

$$\partial_\mu \psi = \left(\frac{\partial}{\partial x_\mu} - \mathrm{i}e\boldsymbol{I}\cdot \boldsymbol{W}_\mu\right)\psi, \tag{2.20}$$

and

$$V(\phi) = -\tfrac{1}{2}\lambda \boldsymbol{\phi}^2 + \tfrac{1}{4}\kappa(\boldsymbol{\phi}^2)^2\,. \tag{2.21}$$

In this Lagrangian, all constants e, g, λ and κ are real because of Hermiticity. One can easily verify that the above Lagrangian is invariant under T, C, P and the SO_3 transformation, and is also renormalizable [31, 32].

Just as in the previous example, we require the renormalized constants λ and κ to be both positive definite, so that $\langle\boldsymbol{\phi}\rangle_{\text{vac}}$ is not zero. Defining the direction of $\langle\boldsymbol{\phi}\rangle_{\text{vac}}$ to be the neutral direction, we may write

$$\langle\phi^{\text{o}}\rangle_{\text{vac}} \neq 0,$$

and

$$\langle\phi^{\pm}\rangle_{\text{vac}} = 0. \tag{2.22}$$

Since ϕ^{o} is the third component of a vector, any solution with $\langle\phi^{\text{o}}\rangle_{\text{vac}} \neq 0$ clearly is not invariant under a general SO_3 transformation. In addition, because of the pseudoscalar interaction in $L_{\psi\phi}$, ϕ^{o} is of $T = -1$, $P = -1$ and $C = +1$; such a solution is also not invariant under T, P and CP.

So far as the Lagrangian $L_{\phi W}$ is concerned, the structure is identical to that of the usual Higgs mechanism [25]. Since $L_{\phi W}$ is an even function of $\boldsymbol{\phi}$, the usual Goldstone-Higgs treatment makes no distinction between ϕ^{o} of $T = -1$ or $+1$. To understand the Higgs mechanism, let us consider the specific term $-\frac{1}{2}(\partial_\mu \boldsymbol{\phi})^2$ in the Lagrangian (2.15). By using (2.19) and the fact that $\langle\boldsymbol{\phi}\rangle_{\text{vac}} \neq 0$, we may write

$$-\tfrac{1}{2}(\partial_\mu \boldsymbol{\phi})^2 = -\tfrac{1}{2} e^2(\boldsymbol{W}_\mu \times \langle\boldsymbol{\phi}\rangle_{\rm vac})^2 - e\langle\boldsymbol{\phi}\rangle_{\rm vac} \cdot (\boldsymbol{W}_\mu \times \partial\boldsymbol{\phi}/\partial x_\mu) + \dots . \tag{2.23}$$

Let us examine the first term in the above expression: because of the vector product, only the components of $\boldsymbol{W}_\mu$ that are perpendicular to the direction of $\langle\boldsymbol{\phi}\rangle_{\rm vac}$ are involved. One sees that these components acquire a mass $m_{\rm W}$ given by

$$m_{\rm W}^2 = e^2\langle\boldsymbol{\phi}\rangle_{\rm vac}^2, \tag{2.24}$$

while the component of $\boldsymbol{W}_\mu$ that is parallel to $\langle\boldsymbol{\phi}\rangle_{\rm vac}$ remains massless and is identified to be the photon field A_μ; therefore, the direction of $\langle\boldsymbol{\phi}\rangle_{\rm vac}$ is indeed the neutral direction, as given by (2.22). The second term $\langle\boldsymbol{\phi}\rangle_{\rm vac} \cdot (\boldsymbol{W}_\mu \times \partial\boldsymbol{\phi}/\partial x_\mu)$ couples only the charged components $W_\mu^\pm$ with $\partial\phi^\pm/\partial x_\mu$. It is straightforward to verify that $\partial\phi^\pm/\partial x_\mu$ now combines with $W_\mu^\pm$ (which has only transverse components) to form a single massive spin 1 charged intermediate boson field; $\partial\phi^\pm/\partial x_\mu$ now becomes the longitudinal mode of the spin 1 field. Let $G_{\rm F}$ be the Fermi weak interaction constant in this simple example; one has

$$e^2/4m_{\rm W}^2 = G_{\rm F}/\sqrt{2}. \tag{2.25}$$

(The realistic Georgi-Glashow model [27] would have an additional factor $\sin^2\beta$ on the left-hand side.)

In order to exhibit more clearly the T violating character of the solution, let us concentrate on the $\phi^{\rm o}$-interaction part of $L_{\psi\phi}$. By using (2.17), one may write

$$L_{\psi\phi} = -{\rm i}g(p^\dagger\gamma_4\gamma_5 p - q^\dagger\gamma_4\gamma_5 q)\phi^{\rm o} + \dots .$$

Just as in the first example, similarly to (2.4) we may expand

$$\phi^{\rm o} = \langle\phi^{\rm o}\rangle_{\rm vac} + \delta\phi^{\rm o}. \tag{2.26}$$

The transformation (2.5) is slightly modified because of the SO_3 property

$$\psi \to [\exp(-\tfrac{1}{2}{\rm i}\gamma_5\alpha I_3)]\,\psi; \tag{2.27}$$

i.e., $p \to [\exp(-\frac{1}{2}{\rm i}\gamma_5\alpha)]p$, $n \to n$ and $q \to [\exp(\frac{1}{2}{\rm i}\gamma_5\alpha)]q$ where $\tan\alpha = m^{-1}g\rho$. The masses of p, n and q become

$$m_{\rm p}^2 = m_{\rm q}^2 = m_{\rm n}^2 + (g\rho)^2, \tag{2.28}$$

and $m_{\rm n} = m$. Correspondingly, after the transformation, the $\delta\phi^{\rm o}$-part of the interaction is given by

$$L_{\psi\phi} = -g[p^\dagger \gamma_4(\sin\alpha + i\gamma_5 \cos\alpha)p] \cdot \delta\phi^o$$
$$-g[q^\dagger \gamma_4(\sin\alpha - i\gamma_5 \cos\alpha)q] \cdot \delta\phi^o + \dots \tag{2.29}$$

Thus, just as in the previous example, the exchange of the $\delta\phi^o$ quantum leads explicitly to *T, P* and *CP* violations. In the single $\delta\phi^o$-exchange diagram the *T*-violating amplitude A_- remains as given by (2.8). However, because the spontaneous *T* violation is coupled to the spontaneous gauge-symmetry violation of the weak interaction, one finds that by using (2.24), (2.25) and (2.28), at $k = 0$

$$A_- \approx 2^{3/2} G_F \cdot \Delta^3/\mu^2 m_p, \tag{2.30}$$

where μ is the mass of the neutral spin 0 boson and $\Delta^2 = m_p^2 - m_n^2$. By definition, the ratio Δ/m_p is < 1.

Although the magnitude of A_- still depends on the unknown mass ratio $(\Delta^3/\mu^2 m_p)$, certain limits can already be estimated. The mass shift Δ must reflect in some way certain observed hadron mass differences; therefore Δ cannot be too large. In addition, independently of the question of *T* invariance, in order that the usual weak interaction be of the V–A character, it must be dominated by the exchanges of the intermediate spin 1 boson, *not* the spin 0 boson; consequently, μ must be fairly large. Thus, it seems reasonable to assume $(\Delta^3/\mu^2 m_p) \ll 1$, and therefore the *T* violating amplitude A_- is much smaller than the Fermi constant G_F.

This simple example illustrates the main feature of combining the spontaneous *T* violation with the spontaneous gauge symmetry violation of the weak and electromagnetic interactions. Because $\langle\phi\rangle_{vac} \neq 0$, the vacuum acts as a "medium"; it gives the hadron field a mass shift $\Delta = g\langle\phi^o\rangle_{vac}$ and the intermediate boson field a mass $m_W = e\langle\phi^o\rangle_{vac}$, which then leads the *T* violating amplitude A_- from eq. (2.10) of our first example to the above expression (2.30). Furthermore, since $g\langle\phi^o\rangle_{vac}$ contributes to the mass matrix of the hadrons, any exchange of $\delta\phi^o$ quantum satisfies the same selection rule as that particular piece of the mass matrix. As we shall see in subsequent sections, in a more realistic model this implies the selection rule $\Delta Y = 0$ where Y is the hypercharge. If one assumes

$$\Delta^3/\mu^2 m_p \sim 10^{-3},$$

then it gives a milliweak $\Delta Y = 0$ *T*-violating interaction.

In order to make a "realistic" model, with the inclusion of the Cabibbo angle, the usual V–A current, etc., one must at least increase the number of hadron fields to eight [27, 28], and the number of lepton fields also to eight. In addition, one must at least double the neutral spin 0 Hermitian fields; otherwise, there will be difficulties. For example, among the lepton fields, there is a triplet representation

$$\psi_l = \begin{pmatrix} X^+ \\ X^o \\ e^- \end{pmatrix}.$$

If there is only a single spin 0 triplet $\boldsymbol{\phi}$, the interaction between ψ_l and $\boldsymbol{\phi}$ would have the same form as $L_{\psi\phi}$ given by (2.17). By going through the same argument that leads to (2.28), one would have obtained the unacceptable conclusion that the heavy electron X^+ should have the same mass as that of the electron e^-. To avoid this difficulty, we must have at least *two* neutral spin 0 Hermitian fields ϕ_I^o and ϕ_R^o; one of them, say ϕ_I^o, has a pseudoscalar interaction, and the other, ϕ_R^o, a scalar interaction. Therefore, ϕ_I^o is of $T = -1$ and ϕ_R^o is of $T = +1$. We assume

$$\langle \phi_I^o \rangle_{vac} = \rho_I \neq 0$$

and

$$\langle \phi_R^o \rangle_{vac} = \rho_R \neq 0. \tag{2.31}$$

Under T, $\langle \phi_I^o \rangle_{vac}$ becomes $-\rho_I$, but $\langle \phi_R^o \rangle_{vac}$ remains the same. The T invariance of the interaction requires both solutions $+\rho_I$ and $-\rho_I$ to exist; but, as before, either solution is not T invariant, and it leads to a spontaneous T violation.

Under the SO_3 transformation, ϕ_I^o and ϕ_R^o must belong to two different representation spaces, which may or may not be of the same dimension. For simplicity we may consider the special case where ϕ_I^o belongs to a triplet $\boldsymbol{\phi}_I$, and ϕ_R^o to another triplet $\boldsymbol{\phi}_R$, different from $\boldsymbol{\phi}_I$. The field $\boldsymbol{\phi}_I$ may interact with the fermions through a pseudoscalar interaction like that in (2.17), and $\boldsymbol{\phi}_R$ through a scalar interaction like that in the original Georgi-Glashow model [27, 28]. The discussion of such a "realistic" Georgi-Glashow model with spontaneous T violation is straightforward, but somewhat complicated because of the relatively large number of fields involved. The details will be given in section 3.

2.3. *General classification*

Because of the large variety of different gauge models of weak and electromagnetic interactions available, in this section we shall try to study only the general question: how can each of these models be enlarged to include a spontaneous T violation, and what would be the general pattern of such a T violation? We assume the total Lagrangian to satisfy the three conditions (a), (b) and (c), stated in section 1.2. Let G be the relevant gauge-symmetry group. The neutral spin 0 field ϕ^o must belong to a representation of G.

We distinguish two different cases:

(i) ϕ^o belongs to a real representation, and

(ii) ϕ^o belongs to a complex representation.

In case (i), as illustrated in our second example, a single Hermitian ϕ^o with a non-zero vacuum expectation value is sufficient to generate a spontaneous T violation (although in the "realistic" Georgi-Glashow model, as mentioned before, because of other requirements there are actually two such Hermitian ϕ^o fields).

In case (ii), the situation is quite different. The complex representation means that ϕ^o must be a complex field. The gauge invariance of the interaction implies that the time reversal operator T is not defined up to a gauge transformation factor. Therefore, if there is only a single ϕ^o field, one has in general

$$T\phi^o T^{-1} = e^{i\alpha}\phi^o, \tag{2.32}$$

where a is arbitrary. Let the vacuum expectation value of ϕ^o be

$$\langle\phi^o\rangle_{vac} = \rho e^{ib},$$

where ρ is real. By choosing $a = -2b$, one can always arrange to have

$$T\langle\phi^o\rangle_{vac} T^{-1} = \langle\phi^o\rangle^*_{vac} = \langle T\phi^o T^{-1}\rangle_{vac}. \tag{2.33}$$

Therefore, with only a single complex ϕ^o field, it is not possible to generate a spontaneous T violation. The simplest way to produce a spontaneous T violation is to double the number of neutral spin 0 fields, so that there are two such complex fields, say ϕ^o_1 and ϕ^o_2. These two neutral fields must belong to two separate representation spaces, which may or may not be equivalent. In general, the relative phase between ϕ^o_1 and ϕ^o_2 is no longer arbitrary. For definiteness let us assume

$$T\phi^o_1 T^{-1} = e^{ia}\phi^o_1,$$

and

$$T\phi^o_2 T^{-1} = e^{ia}\phi^o_2$$

with the *same* phase angle a. The vacuum expectation values of ϕ^o_1 and ϕ^o_2 may be written as

$$\langle\phi^o_1\rangle_{vac} = \rho_1 \exp\{i(\theta + b)\},$$

and

$$\langle\phi^o_2\rangle_{vac} = \rho_2 e^{ib},$$

where ρ_1 and ρ_2 are both real and > 0. Just as in (2.32), the angle a is arbitrary since the time reversal operator T is not defined up to a gauge transformation factor. Furthermore, the absolute phase of ϕ^o_1 and ϕ^o_2 is not determined; therefore, the angle b is also arbitrary. Thus, we can always set

$$a = b = 0,$$

and consequently

$$\langle\phi^o_1\rangle_{vac} = \rho_1 e^{i\theta}, \qquad \text{and} \qquad \langle\phi^o_2\rangle_{vac} = \rho_2. \tag{2.34}$$

These vacuum expectation values may be represented geometrically by the triangle given in fig. 3. Under T, $\theta \to -\theta$. If $\theta \neq 0$ or π, then the T invariance of the interaction implies that both solutions θ and $-\theta$ must exist, but either solution is not T invariant.

The quantum fluctuations of ϕ^o_1 and ϕ^o_2 about their vacuum expectation values correspond pre-

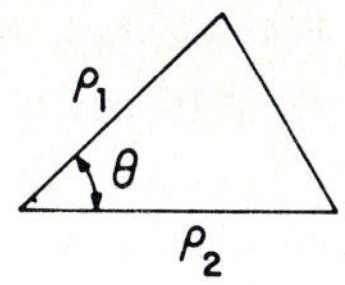

Fig. 3. Representation of $\langle\phi_2^0\rangle_{vac} = \rho_2$ and $\langle\phi_1^0\rangle_{vac} = \rho_1 e^{i\theta}$ by a triangle.

cisely to the vibrations of this triangle. Since there are three degrees of freedom: $\delta\rho_1$, $\delta\rho_2$ and $\delta\theta$, there must be three normal modes, just like the vibrations of a triangular molecule. The virtual exchange of these vibrational quanta can lead to explicit T and CP violations, similar to the exchange of the $\delta\phi^0$-quantum discussed before in the simple example.

As mentioned before, in the Georgi-Glashow model, although the spin 0 field belongs to a real representation (i.e., case (i)), because of other dynamical reasons, in order to construct a "realistic" theory, we also need to double the neutral spin 0 field; their vacuum values are given by (2.31), which can again be represented by a triangle, but with

$$\theta = 90°,$$

and the corresponding fluctuation $\delta\theta = 0$.

If one wishes, one may also regard our first simple example of only one Hermitian neutral field ϕ^0 (discussed in section 2.1) as a further degeneration of the triangle, with the length of one of its sides approaching zero. In this sense, the triangular representation offers the general characterization of the spontaneous T violation, applicable to all such cases. The detailed description of such a triangle is given in ref. [22].

In general, if the two sides of the triangle ρ_1 and ρ_2 are of comparable magnitude, one has

$$e\rho_1 \sim e\rho_2 \sim m_W,$$

$$(e/m_W)^2 \sim G_F,$$

and therefore, just as in (2.30), at low energy the T and CP violating amplitude is of the order of

$$A_- \sim G_F \cdot \Delta^3/\mu^2 m, \qquad (2.35)$$

where Δ^2 is some appropriate (mass)^2-difference between the hadrons (or leptons), m is the appropriate hadron (or lepton) mass, and μ is the mass of the relevant spin 0 quantum. As noted before, in any reasonable model, Δ cannot be too large because of the observed hadron spectrum, and μ cannot be too small since the observed weak interaction is dominated by the exchange of an intermediate boson of spin 1, not spin 0. In addition, the ratio (Δ/m) is by definition < 1. Thus, A_- must be quite small. If we assume $\Delta \lesssim 1$ GeV and $\mu \gtrsim 10$ GeV, then it seems reasonable to expect $(\Delta^3/\mu^2 m) \sim 10^{-3}$ and therefore that the T violating amplitude A_- is milliweak.

A typical pattern of such a theory is that any exchange of spin 0 bosons leads to T and P violation while any exchange of spin 1 bosons between the usual V–A currents gives only C and P violation. This is because a transformation of the form $\psi \to \exp(-\frac{1}{2}i\gamma_5\alpha)\psi$ leaves the vector

$i\psi^\dagger\gamma_4\gamma_\mu\psi$ and the axial-vector $i\psi^\dagger\gamma_4\gamma_\mu\gamma_5\psi$ invariant, but not the scalar $\psi^\dagger\gamma_4\psi$ and the pseudoscalar $i\psi^\dagger\gamma_4\gamma_5\psi$. To lowest order, T violation can be produced via the exchange of the neutral massive spin 0 boson; as already mentioned in section 2.2, such a T violation usually satisfies the selection rule $\Delta Y = 0$. In all the realistic models that we have examined, there is always the existence of some genuine (i.e., non-Goldstone) charged massive spin 0 bosons. Through the exchange of these charged bosons, T violation can also be generated. As we shall see in the next section, such exchange typically satisfies the $\Delta Y = 0$ and ± 1 rule. In both cases, the order of magnitude of the T violating amplitude is given by (2.35). If we assume the ratio $(\Delta^3/\mu^2 m)$ to be $\sim 10^{-3}$, then one has a *milliweak P, T* violating theory which satisfies the rule:

$$\Delta Y = 0, \qquad \text{and} \qquad \pm 1. \tag{2.36}$$

In principle, since all T violating effects in the theory are finite, one can calculate, e.g., the ϵ parameters in K_L^o and K_S^o states, and the electric dipole moment D_E of the neutron. However, in practice, there are many serious difficulties: there is no confidence that any of the presently available gauge theories is the correct one (especially when it is applied to hadrons). In any of these theories, one does not know how to properly include the strong interaction; reliable calculations can only be made for free quark-like hadron fields. The connection between a calculation for these free fields and that for the physical hadrons is extremely uncertain. Furthermore, the theoretical expressions for ϵ and D_E are in terms of some appropriate mass difference Δ of these quark-like hadrons and the mass μ of the relevant spin 0 boson, which are all unknown. Nevertheless, it seems possible that the order of magnitude of, say, $(\Delta^3/\mu^2 m)$ in (2.35) may be obtained by using the observed ϵ parameter in the $K_L^o - K_S^o$ system, and from that one may estimate the order of magnitude of D_E for the neutron. As remarked before, without any detailed calculation, one may well expect the resulting theory to be of the milliweak type. Just as in any milliweak theory, the order of magnitude of D_E may be estimated to be

$$D_E \sim |(4\pi)^{-1} G_F e m_N \epsilon| \sim e \times 10^{-23}\ \text{cm}, \tag{2.37}$$

which is consistent with the present experimental value [33]. To see how such a crude estimation may survive a more detailed calculation, we shall in the next section consider a specific model: the "realistic" Georgi-Glashow model, but generalized to include the spontaneous T violation mechanism. Both ϵ and D_E will be calculated explicitly.

Before discussing the details of such calculations, we note that in any "realistic" model, it is necessary to postulate the existence of some "charmed" hadrons [34]. If one wishes, one may arrange to have T violation *only* when there are either real or virtual "charmed" hadrons. (See, e.g., eq. (3.51) below.) In all observed lowest order weak transitions between known particles, it is a good approximation to neglect the "charmed" hadrons. Thus, even though the theory is basically of a milliweak character, so far as all known weak decays are concerned, the predictions can be almost identical to those of a superweak theory; i.e., in all lowest order $\Delta Y = 0$ and ± 1 weak reactions the T violating transition amplitudes can be much less than 10^{-3} times the corresponding T conserving amplitudes. However, as we shall see in section 3.5, the order of magnitude of the electric dipole moment of the neutron, or proton, should retain its milliweak character, and this may well offer the best experimental possibility for differentiating such a theory from a genuine superweak theory.

References

[1] T.D. Lee, R. Oehme and C.N. Yang, Phys. Rev. 106 (1957) 340.
[2] J.H. Christenson, J.W. Cronin, V.L. Fitch and R. Turlay, Phys. Rev. Letters 13 (1964) 138.
[3] D.F. Bartlett et al., Phys. Rev. Letters 21 (1968) 558.
[4] J.M. Gaillard et al., paper 139 contributed to the 14th Intern. Conf. on High Energy Physics, Vienna, 1968 (CERN, Geneva, 1968).
[5] M. Banner et al., Phys. Rev. Letters 21 (1968) 1107.
[6] R.J. Cence et al., paper 476 contributed to the 14th Intern. Conf. on High Energy Physics, Vienna, 1968 (CERN, Geneva, 1968).
[7] J. Chollet et al., Topical Conference on Weak Interactions (CERN, 1969) p. 309.
[8] D. Dorfan et al., Phys. Rev. Letters 19 (1967) 987.
[9] S. Bennett et al., Phys. Rev. Letters 19 (1967) 993.
[10] J. Steinberger, Topical Conference on Weak Interactions (CERN, 1969) p. 291.
[11] V.V. Barmin et al., paper 828b contributed to the XVI Intern. Conf. on High Energy Physics, National Accelerator Laboratory, Batavia, 1972.
[12] L. Wolfenstein, Nuovo Cimento 62A (1966) 17.
[13] J.S. Bell and J. Steinberger, Proc. Oxford Intern. Conf. on Elementary Particles (Rutherford Laboratory, 1965) p. 195.
[14] T.D. Lee and C.S. Wu, Annual Review of Nuclear Sciences 16 (1966) 511.
[15] G. Geweniger et al., Proc. Aix-en-Provence Conf. on Elementary Particles, to be published.
[16] T.A. Lasinski et al., Review of Particle Properties, Revs. Mod. Phys. 45 (1973).
[17] T.D. Lee and L. Wolfenstein, Phys. Rev. 138 (1965) B1490.
[18] L. Wolfenstein, Phys. Rev. Letters 13 (1964) 180.
[19] T.T. Wu and C.N. Yang, Phys. Rev. Letters 13 (1964) 180.
[20] L.B. Okun, Soviet J. Nucl. Phys. 1 (1965) 670;
J. Prentki and M. Veltman, Phys. Letters 15 (1965) 88.
[21] J. Bernstein, G. Feinberg and T.D. Lee, Phys. Rev. 138 (1965) B1650.
[22] T.D. Lee, Phys. Rev. D8 (1973) 1226; Topical Conference on Weak Interactions, Trieste, 1973 (Columbia University preprint CO-2271-18, unpublished).
[23] A. Pais, Phys. Rev. Letters 29 (1972) 1712; 30 (1973) 114(E).
[24] J. Goldstone, Nuovo Cimento 19 (1961) 154;
J. Goldstone, A. Salam and S. Weinberg, Phys. Rev. 127 (1962) 965.
[25] P.W. Higgs. Phys. Letters 12 (1964) 132; Phys. Rev. Letters 13 (1964) 508; and Phys. Rev. 145 (1966) 1156;
F. Englert and R. Brout, Phys. Rev. Letters 13 (1964) 321;
G.S. Guralnik, C.R. Hagen and T.W.B. Kibble, Phys. Rev. Letters 13 (1964) 585;
T.W.B. Kibble, Phys. Rev. 155 (1967) 1554.
[26] S. Weinberg, Phys. Rev. Letters 19 (1967) 1264; 27 (1971) 1688; and Phys. Rev. D5 (1972) 1412; D5 (1972) 1962;
A. Salam and J.C. Ward, Phys. Letters 13 (1964) 168;
A. Salam, in Elementary Particle Theory, ed. N. Svartholm (Almquist and Forlog A.B., Stockholm, 1968).
[27] H. Georgi and S.L. Glashow, Phys. Rev. Letters 28 (1972) 1494.
[28] B.W. Lee, J.R. Primack and S.B. Treiman, Phys. Rev. D7 (1973) 510.

[29] J. Prentki and B. Zumino, Nucl. Phys. B47 (1972) 99.
[30] J.D. Bjorken and C.H. Llewellyn Smith, Phys. Rev. D7 (1973) 887.
[31] G. 't Hooft, Nucl. Phys. B33 (1971) 173; B35 (1971) 167; and Phys. Letters 37B (1971) 195;
G. 't Hooft and M. Veltman, Nucl. Phys. B44 (1972) 189; B50 (1972) 318.
[32] B.W. Lee, Phys. Rev. B5 (1972) 823;
B.W. Lee and J. Zinn-Justin, Phys. Rev. D5 (1972) 3121, 3127, 3155.
[33] W.B. Dress, P.D. Miller and N.F. Ramsey, Phys. Rev. D7 (1973) 3147.
[34] S. Glashow, J. Iliopoulos and L. Maiani, Phys. Rev. D2 (1970) 1285.
[35] N. Cabibbo, Phys. Letters 10 (1963) 513.
[36] M. Gell-Mann and M. Levy, Nuovo Cimento 16 (1960) 705.
[37] T.D. Lee and C.N. Yang, Phys. Rev. 128 (1962) 885.
[38] L.D. Faddeev and V.N. Popov, Phys. Letters 25B (1967) 29; Kiev Report No. ITP-67-36, unpublished.
[39] K. Fugiwara, B.W. Lee and A.I. Sanda, Phys. Rev. D6 (1972) 2923.

Gauge Theory of *CP* Nonconservation*

Steven Weinberg
Lyman Laboratory of Physics, Harvard University, Cambridge, Massachusetts 02138
(Received 2 July 1976)

It is proposed that *CP* nonconservation arises purely from the exchange of Higgs bosons.

Ever since the discovery[1] that *CP* conservation is not exact, the mystery has been why it is so feebly violated.[2] In many proposed theories,[3] one must arrange to have *CP* to be approximately conserved, by making the appropriate constants in the Lagrangian to have sufficiently small values. However, one would prefer a more natural explanation.

Renormalizable gauge theories[4] of the weak and electromagnetic interactions provide a mechanism which could violate *CP* conservation with about the right strength: the Higgs boson. The coupling strength of a Higgs boson to a quark or lepton of mass m is of order $mG_F^{1/2}$ (where G_F is the Fermi coupling constant), so that the exchange of a Higgs boson of mass m_H produces an effective Fermi interaction with coupling of order $G_F m^2/m_H^2$. For reasonable mass values,[5] this is "milliweak." However, in order for the Higgs exchange to appear as a natural explanation for a feeble *CP* nonconservation, one must understand why *CP* conservation is strongly violated in the Higgs exchange, *and nowhere else*. In this paper, I wish to present a realistic gauge theory, in which *CP* nonconservation automatically arises in just this way.[6]

We assume an SU(2)$\otimes$U(1) gauge theory,[4] with the usual four quarks[7]: $\mathcal{P}_1$ and $\mathcal{P}_2$ have charge $+\frac{2}{3}$

and $\mathfrak{N}_1$ and $\mathfrak{N}_2$ have charge $-\frac{1}{3}$. The left-handed quarks form two doublets $(\mathcal{P}_1, \mathfrak{N}_2)$, while all right-handed quarks are SU(2) singlets. The only scalar bosons which can couple to the quarks then are the doublets $(\varphi_r^+, \varphi_r^0)$, and their charge conjugates. In order to insure that strangeness and charm conservation are not violated by φ^0 exchange, we must assume that there are *at most* two scalar doublets coupled to the quarks, with a Yukawa interaction of the form[8]

$$\mathcal{L}_Y = \sum_{i,j=1}^{2} \Gamma_{ij}^{(1)}\overline{\mathfrak{N}}_{iR}(\varphi_1^{+*}\mathcal{P}_{jL} + \varphi_1^{0*}\mathfrak{N}_{jL}) + \sum_{i,j=1}^{2} \Gamma_{ij}^{(2)}\overline{\mathcal{P}}_{iR}(\varphi_2^0\mathcal{P}_{jL} - \varphi_2^+\mathfrak{N}_{jL}) + \text{H.c.} \tag{1}$$

[A label L or R denotes multiplication with $(1+\gamma_5)/2$ or $(1-\gamma_5)/2$.] In general, φ_2 and φ_1 could be charge conjugates, but I assume here that they are not.[9] The interaction will automatically take this general form, provided we assume that there is some sort of discrete symmetry which only allows φ_1 (φ_2) to couple to $\mathfrak{N}_R$ $(\mathcal{P}_R)$ quarks.

I am not yet imposing either exact or approximate CP conservation on the Lagrangian. In particular, the matrices $\Gamma_{ij}^{(r)}$ are still perfectly arbitrary. The φ_r^0 will have vacuum expectation values λ_r, yielding mass matrices which in general are neither diagonal nor even Hermitian. However, we can always define new quark fields

$$\begin{pmatrix} u \\ c \end{pmatrix}_{L,R} = U_{L,R}\mathcal{P}_{L,R}, \quad \begin{pmatrix} d \\ s \end{pmatrix}_{L,R} = V_{L,R}\mathfrak{N}_{L,R} \tag{2}$$

(with unitary U and V matrices), so that the corresponding mass matrices are real and diagonal:

$$V_R\Gamma^{(1)}\lambda_1 V_L^{-1} = \begin{pmatrix} m_d & 0 \\ 0 & m_s \end{pmatrix}, \quad U_R\Gamma^{(2)}\lambda_2^* U_L^{-1} = \begin{pmatrix} m_u & 0 \\ 0 & m_c \end{pmatrix}. \tag{3}$$

Further, we can define the phases of the new quark fields so that the 2×2 matrix $U_L V_L^{-1}$ is real and orthogonal,

$$U_L V_L^{-1} = \begin{pmatrix} \cos\theta_c & \sin\theta_c \\ -\sin\theta_c & \cos\theta_c \end{pmatrix}. \tag{4}$$

The gauge interactions can therefore be rewritten in terms of the usual Cabibbo-mixed doublets, so that they automatically conserve CP whether or not we assume that CP is a good symmetry of the Lagrangian or the vacuum.[10]

The Yukawa interaction (1) can now be rewritten in terms of the new quarks. Using Eqs. (2)–(4), I find

$$\begin{aligned} \mathcal{L}_Y = {} & (\lambda_1^*)^{-1}\varphi_1^{0*}(m_d\bar{d}_R d_L + m_s\bar{s}_R s_L) + \lambda_2^{-1}\varphi_2^0(m_u\bar{u}_R u_L + m_c\bar{c}_R c_L) \\ & + (\lambda_1^*)^{-1}\varphi_1^{+*}[m_d\cos\theta_c\,\bar{d}_R u_L - m_d\sin\theta_c\,\bar{d}_R c_L + m_s\sin\theta_c\,\bar{s}_R u_L + m_s\cos\theta_c\,\bar{s}_R c_L) \\ & - \lambda_2^{-1}\varphi_2^+(m_u\cos\theta_c\,\bar{u}_R d_L + m_u\sin\theta_c\,\bar{u}_R s_L - m_c\sin\theta_c\,\bar{c}_R d_L + m_c\cos\theta_c\,\bar{c}_R s_L) + \text{H.c.} \end{aligned} \tag{5}$$

This interaction will also conserve CP, provided that one assigns to the scalar fields a suitable CP-transformation phase. [The CP properties of the quark fields are fixed by Eqs. (2)–(4).] However, for a wide class of theories, *CP will not be conserved by the scalar propagators.* In particular, the exchange of a single Higgs boson leads to an effective milliweak Fermi interaction

$$\begin{aligned} & -A(m_d\cos\theta_c\,\bar{d}_R u_L - m_d\sin\theta_c\,\bar{d}_R c_L + m_s\sin\theta_c\,\bar{s}_R u_L + m_s\cos\theta_c\,\bar{s}_R c_L) \\ & \quad \times(m_u\cos\theta_c\,\bar{u}_R d_L + m_u\sin\theta_c\,\bar{u}_R s_L - m_c\sin\theta_c\,\bar{c}_R d_L + m_c\cos\theta_c\,\bar{c}_R s_L) + \text{H.c.} \end{aligned} \tag{6}$$

with a single unknown parameter

$$A \equiv \langle T\{\varphi_1^{+*}\varphi_2^+\}\rangle_{0,q=0}/\lambda_1^*\lambda_2. \tag{7}$$

This interaction will violate CP conservation, because A is in general complex.

The nonconservation of CP can be easily checked for a fairly broad class of models. In order to preserve the rather successful SU(2)⊗U(1) predictions[4] for the strengths of neutral currents, I now assume that *all* the scalar fields form SU(2) doublets $(\varphi_r^+, \varphi_r^0)$. For simplicity, I also suppose that the Lagrangian is invariant under separate reflections under which any one of the φ_r (and perhaps some fermions) changes sign. Discrete symmetries of this kind are needed anyway to insure that it is only φ_1 and φ_2 that couple to $\mathfrak{N}_R$ and $\mathcal{P}_R$. The most general quartic polynomial that is invariant under these reflections and SU(2)⊗U(1) is of the form

$$P(\varphi)=\sum_r M_r^2(\varphi_r^\dagger\varphi_r)+\sum_{rs}a_{rs}(\varphi_r^\dagger\varphi_r)(\varphi_s^\dagger\varphi_s)+\sum_{rs}b_{rs}(\varphi_r^\dagger\vec{\tau}\varphi_r)(\varphi_s^\dagger\vec{\tau}\varphi_s)+\sum_{rs}c_{rs}(\varphi_r^\dagger\varphi_s)(\varphi_r^\dagger\varphi_s). \quad (8)$$

Hermiticity requires that M_r^2 be real and a_{rs} and b_{rs} be real and symmetric, while c_{rs} need only be Hermitian.

For three or more φ doublets, this polynomial need not conserve CP. It is invariant under a CP transformation $\varphi_r \to \exp(i\theta_r)\varphi_r^*$ if and only if c_{rs} has phase $\pm\exp(i\theta_r - i\theta_s)$. For this to be possible for some set of θ_r, all the quantities $c_{rs}c_{st}c_{tr}$ would have to be real. A direct calculation for the case of three doublets shows that if $c_{12}c_{23}c_{31}$ is not real, then CP is actually not conserved, and the quantity (7) is complex. Further, even if $P(\varphi)$ *is* required to conserve CP, it can be shown that there will always (for three or more doublets) be at least a finite range of the parameters a, b, c, and M for which CP conservation is spontaneously broken.

Let us now consider the experimental implications of this sort of theory:

(1) The Higgs exchange produces a CP-nonconserving milliweak interaction with $\Delta S=1$, $\Delta C=0$. Inspection of (6) shows that $\Delta I=\frac{1}{2}$ terms dominate by large factors $m_c/m_{u,d}$, as required[11] if a milliweak interaction is to imitate the successful predictions of the superweak theory. The effect of terms involving c quarks may be dynamically suppressed, but if we ignore such effects, the dominant interaction here is $Am_s m_c \sin\theta_c \cos\theta_c \bar{s}_R c_L \bar{c}_R \times d_L$. In addition, there is a superweak interaction produced by exchange of a Higgs plus a W boson; this contribution to $\bar{s}d \to \bar{d}s$ (at energies and momentum transfers of order m_s) is less than the usual $2W$ contribution by a factor of order m_s^2A/G_F. It is not clear how to use these remarks to calculate observable parameters, but if we take (6) at face value, we would expect $m_s m_c \times \mathrm{Im}A/G_F$ to be of the order of the phase of $K^0 \to 2\pi$ (relative to $\langle \bar{K}^0|M|K^0\rangle^{1/2}$), or 3.25×10^{-3}. If we also assume a maximal CP violation then $\mathrm{Im}A \approx |A| \approx G_F/m_H^2$; with $m_c=3m_s=1500$ MeV, this gives $m_H \approx 15$ GeV, not an unreasonable value.[5]

(2) There is also a CP- and P-nonconserving milliweak interaction with $\Delta S=\Delta C=0$, so that one expects the neutron to have a detectable electric dipole moment. To get a rough idea of its magnitude, we can neglect the strong interactions, and calculate the electric dipole moments of the d and u quarks produced by the virtual Higgs exchange. The results[12] for $m_H^2 \gg m_{\rm quark}$ are

$$D_d=\frac{em_d\mathrm{Im}A}{24\pi^2}\left(m_u^2\cos^2\theta_c \ln\frac{m_H^2}{m_u^2}+m_c^2\sin^2\theta_c\ln\frac{m_H^2}{m_c^2}\right),$$

$$D_u=-\frac{em_u\mathrm{Im}A}{48\pi^2}\left(m_d^2\cos^2\theta_c \ln\frac{m_H^2}{m_d^2}+m_s^2\sin^2\theta_c\ln\frac{m_H^2}{m_s^2}\right).$$

For illustration I take $m_u \approx m_d \approx 300$ MeV, $m_s \approx 500$ MeV, and $m_c \approx 1500$ MeV. Using previous estimates $\mathrm{Im}A \approx 3.2\times10^{-3}G_F/m_s m_c$ and $m_H \approx 15$ GeV, I find $D_d \simeq 1.6\times10^{-24}$ $e\cdot$cm and $D_u \simeq -0.5\times10^{-24}$ $e\cdot$cm. A simple quark-model calculation then gives the neutron electric dipole moment as $(4D_d - D_u)/3$, or about 2.3×10^{-24} $e\cdot$cm. The most recent experimental result[13] is $(0.4\pm1.1)\times10^{-24}$ $e\cdot$cm. In view of the great uncertainties surrounding the calculation, it would be premature to call this a real contradiction, but any appreciable further improvement in the precision of these measurements will be likely to confirm or refute the existence of a milliweak interaction of the type predicted here.

(3) By applying to the leptons the same considerations as to the quarks, we can infer that the leptons will have CP-conserving gauge interactions, and a CP-nonconserving Higgs interaction of the form

$$\mathcal{L}_Y'=(\lambda_3^*)^{-1}m_\mu(\varphi_3^{0*}\bar{\mu}_R\mu_L+\varphi_3^{+*}\bar{\mu}_R\nu_{\mu L})+\text{H.c.}$$

(Higgs couplings are negligible for electron-type leptons.) Here φ_3 may or may not be independent of φ_1 and φ_2. It is tempting to suppose that φ_3 is the third independent doublet needed to allow CP violation, but this is not necessary. In general the quantities $\langle T\{\varphi_{1,2}^{+*},\varphi_3^+\}\rangle/\lambda_{1,2}^*\lambda_3$ will be complex, so that there will be a semileptonic CP-nonconserving milliweak interaction. This will produce small triple-scalar-product correlations in muonic decay modes. In particular, the Higgs exchange would produce an effective f_- form factor in $K_{\mu3}$ decay, with $\mathrm{Im}(f_-/f_+)$ being of order 10^{-3}. The observation of such a CP violation in $K_{\mu3}$ decay, together with the nonobservation of similar CP-violating effects in electronic processes such as the β decay, would be a clear indication that it is the Higgs bosons that are responsible for CP violation.

(4) The acceptability of this theory of CP violation could be affected by the experimental evidence on the number of quark flavors. At present, the situation is unclear. Even if it were found that new quarks must be introduced in order to account for bumps in e^+e^- or $\mu^+\mu^-$ channels or to cancel Bell-Jackiw anomalies associated with new heavy leptons, the essential features of the present theory could still easily be preserved, provided that there is some unbroken conservation law which prevents mixing of the additional quarks with u, d, s, and c. The present theory would, however, be ruled out if the reported anomalies in high-energy antineutrino-nucleon scattering were found definitely to be associated with new quarks or right-handed currents.

Suppose that Higgs exchange really does turn out to be the source of CP violation. Gauge theories would then provide a very attractive interpretation of all the inversions. It may be that none of them, C, P, T, or CP, is an *a priori* symmetry of the Lagrangian. The fact that the strong and electromagnetic interactions conserve C, P, and T would be explained by the circumstance that the right- and left-handed fermions happen to furnish isomorphic representations of the unbroken gauge subgroups of color and charge.[14] The fact that the weak interactions conserve CP approximately would be explained by the circumstances that there are four quarks, and that the quarks are lighter than the Higgs bosons. The polynomial part of the Lagrangian would violate CP conservation strongly, because there is no reason for it to behave otherwise.

I am grateful for valuable conversations with many colleagues, especially Alvaro De Rújula, Howard Georgi, Sheldon Glashow, Roman Jackiw, Ben Lee, T. D. Lee, and Gerard 't Hooft.

---

*Work supported in part by the National Science Foundation under Grant No. MPS75-20427.

[1]J. H. Christensen, J. W. Cronin, V. L. Fitch, and R. Turlay, Phys. Rev. Lett. 13, 138 (1964).

[2]The various possible interpretations of the small $(K_L \to 2\pi)/(K_S \to 2\pi)$ ratio are discussed by T. D. Lee and L. Wolfenstein, Phys. Rev. 138, B1490 (1965).

[3]Gauge theories of CP nonconservation are reviewed by R. N. Mohapatra, in *Particles and Fields—1974*, edited by C. E. Carlson, AIP Conference Proceedings No. 23 (American Institute of Physics, New York, 1974), p. 127.

[4]S. Weinberg, Phys. Rev. Lett. 19, 1264 (1967); A. Salam, in *Elementary Particle Physics*, edited by N. Svartholm (Almquist and Wiksells, Stockholm, 1968).

[5]The masses of the Higgs bosons are unknown, but they are expected to be comparable to those of the intermediate vector bosons if the scalar self-coupling is comparable with the gauge couplings. In one class of theories, there is one "light" Higgs boson with mass 7–10 GeV, but this boson's couplings necessarily conserve CP; see E. Gildener and S. Weinberg, Phys. Rev. D 13, 3333 (1976). There is in any case a lower bound of order 5 GeV on the Higgs boson masses; see S. Weinberg, Phys. Rev. Lett. 36, 294 (1976); A. D. Linde, to be published.

[6]The idea that the Higgs bosons are responsible for a milliweak CP violation was suggested a few years ago by T. D. Lee, Phys. Rev. D 8, 1226 (1973), and Phys. Rep. 9C, 143 (1974). There are several significant differences between Lee's work and the present paper: (i) Lee developed his ideas most fully in the context of a "modified Georgi-Glashow" model, which was subsequently ruled out by the discovery of neutral currents. He also considered a general class of $SU(2)\otimes U(1)$ models, but not specifically with four quarks, so that the problem of strangeness-changing neutral currents was left unresolved. In particular, Lee attributed CP violation to the exchange of the *neutral* Higgs bosons, which as shown here is only possible if the neutral Higgs exchange is allowed to violate strangeness and/or charm conservation. (ii) In some of the models considered by Lee, the CP violation does *not* occur only in the Higgs exchange; it also occurs in the gauge couplings of light to heavy quarks. Such couplings can produce large CP-violation effects involving only the light quarks when the strong interactions are taken into account; see, e.g., H. Fritzsch and P. Minkowski, California Institute of Technology Report No. CALT-68-537 (to be published). (iii) Lee assumed that CP is a good symmetry of the Lagrangian, while the present paper leaves open the possibility that CP is not an *a priori* symmetry at all.

[7]S. L. Glashow, J. Iliopoulos, and L. Maiani, Phys. Rev. D 2, 1285 (1970). The incorporation of a fourth quark in renormalizable gauge theories was suggested by S. Weinberg, Phys. Rev. D 5, 1412 (1972).

[8]S. L. Glashow and S. Weinberg, to be published. This paper points out that we do not yet know (but soon will!) whether neutral-current processes conserve charm, so it is possible that there is more than one scalar doublet coupled to the $\mathcal{P}_R$ quarks. This would complicate the theory developed here, but would not change its general features.

[9]If φ_2 is the charge conjugate of φ_1, then CP nonconservation can still occur in Higgs self-couplings, but not in the charged Higgs propagators. It is difficult to see in this case how the $K_L \to 2\pi$ decay could have its observed rate.

[10]This observation is originally due to M. Kobayashi and K. Maskawa, Prog. Theor. Phys. 49, 652 (1973); see also L. Maiani, unpublished; and S. Pakvasa and

H. Sugawara, to be published. It has led some authors to suppose that there must be more than four quarks (or right-handed currents), to allow for *CP*-nonconserving gauge couplings. From the viewpoint of the present paper, it is hoped that there are *not* more than four quarks, to insure that *CP* violation arises only from Higgs exchange.

[11]T. T. Wu and C. N. Yang, Phys. Rev. Lett. 13, 180 (1964).

[12]These formulas are obtained by making suitable adjustments in the results of T. D. Lee, Ref. 6. Terms of order unity are neglected in comparison with the logarithms; in particular, m_H is an average Higgs mass, and I neglect logarithms of Higgs boson mass ratios.

[13]N. F. Ramsey, in *Proceedings of the 1975 Neutrino Conference* (Hungarian Physical Society, Budapest, 1975), Vol. I, p. 307.

[14]See, e.g., S. Weinberg, Phys. Rev. Lett. 31, 494 (1973). One possible objection to this viewpoint arises from the fact that it is possible to add a term to the strong interaction Lagrangian (the curl of the gluon field times its dual) that violates *CP* and *P* conservation. This term is formally a total divergence, but triangle anomalies may give it a finite effect; see G. 't Hooft, to be published; R. Jackiw and C. Rebbi, to be published; C. Callen, R. Dashen, and D. Gross, to be published.

PHYSICAL REVIEW D VOLUME 22, NUMBER 11 1 DECEMBER 1980

Spontaneous *CP* nonconservation and natural flavor conservation: A minimal model

Gustavo C. Branco
Carnegie-Mellon University, Pittsburgh, Pennsylvania 15213
(Received 4 June 1980)

A minimal model based on the standard SU(2)×U(1) gauge theory, with spontaneous *CP* violation and natural flavor conservation, is presented in some detail. In particular it is shown that one on the minima of the classical Higgs potential leads to spontaneous *CP* breaking. The generalized Cabibbo-type matrix is real and there is no *CP* violation in the manner of Kobayashi-Maskawa. The *CP* nonconservation is entirely due to Higgs-boson exchange, and can be characterized by a single phase for an arbitrary number of quark generations.

I. INTRODUCTION

It has been recently pointed out[1] that the constraints of spontaneous *CP* breaking and natural flavor conservation[2] (NFC) lead to a class of theories where *CP* nonconservation[3] arises *exclusively* through Higgs boson exchange[4] (for an arbitrary number of fermion generations). A key problem in incorporating *CP* violation in a unified gauge theory is to understand the smallness of the violation.[5] The class of theories considered here provides an attractive scenario for understanding the strength of *CP* breaking: Since Higgs-boson exchange leads to a Fermi interaction of strength $G_F m_F^2/m_H^2$, the smallness of *CP* violation simply reflects the fact that Higgs bosons are much heavier[6] than light quarks.

In this paper, it will be shown how a minimal model with spontaneous *CP* breaking and NFC can be realized in the context of the standard SU(2) ×U(1) gauge theory. The present paper is organized as follows: In Sec. II we discuss the requirements which should be satisfied in order to have spontaneous *CP* breaking. In particular, it is demonstrated that if one imposes NFC, then at least three Higgs doublets are needed in order to violate *CP*. We then classify the various minima of the corresponding Higgs potential according to their *CP* transformation properties. It is shown that for an appropriate range of the free parameters of the Lagrangian the absolute minimum is not *CP* invariant. In Sec. III we exhibit explicitly the fermion interactions with the physical Higgs boson in the context of a minimal model. In particular, we compute the *CP*-violating component of the Fermi interaction originated through Higgs-boson exchange. Finally, in Sec. IV, we draw our conclusions.

II. A MINIMAL MODEL

Let us consider the standard $SU(2)_L\times U(1)$ gauge theory,[7] including an arbitrary number of quarks, with the left-handed components forming SU(2) doublets $\psi_{iL}=(\mathcal{N}_i,\mathcal{P}_i)_L$, while the right-handed components are singlets. In this case, quarks acquire mass through their Yukawa couplings with Higgs doublets $\Phi_i=(\Phi_i^+,\Phi_i^0)$. It turns out that if one adheres to the principal of NFC in the Higgs sector, then a minimum of three Higgs doublets are necessary in order to have spontaneous *CP* violation.[8] This can be easily seen by considering a model with only two Higgs doublets. It is well known that in this case the constraints of NFC necessitate the introduction of some extra symmetry, the simplest choice being

$$R:\ \Phi_1\to-\Phi_1,\ \mathcal{N}_{iR}\to-\mathcal{N}_{iR}\,, \tag{1}$$

with all the other fields unchanged. The most general SU(2)×U(1)×R-invariant Yukawa interactions can then be written

$$\mathcal{L}=\sum_{i,j}(\bar\psi_{iL}\Gamma^1_{ij}\Phi_1\mathcal{N}_{jR}+\bar\psi_{iL}\Gamma^2_{ij}\tilde\Phi_2\,\mathcal{P}_{jR}+\text{H.c.}, \tag{2}$$

where $\tilde\Phi_2=i\sigma_2\Phi_2^*$. The coupling constants Γ_{ij} are chosen to be real so that the Lagrangian is *CP* invariant. The most general Higgs potential consistent with (1) is given by

$$\begin{aligned}V(\Phi)=&m_1^2\Phi_1^\dagger\Phi_1+m_2^2\Phi_2^\dagger\Phi_2+a_{11}(\Phi_1^\dagger\Phi_1)^2+a_{22}(\Phi_2^\dagger\Phi_2)^2\\&+a_{12}(\Phi_1^\dagger\Phi_1)(\Phi_2^\dagger\Phi_2)+b_{12}(\Phi_1^\dagger\Phi_2)(\Phi_2^\dagger\Phi_1)\\&+[c_{12}(\Phi_1^\dagger\Phi_2)(\Phi_1^\dagger\Phi_2)+\text{H.c.}].\end{aligned} \tag{3}$$

Using the SU(2) gauge invariance of the theory, one can assume, without loss of generality, that the minimum of the potential is at

$$\langle\Phi_1\rangle=\frac{1}{\sqrt2}\begin{pmatrix}0\\v_1e^{i\theta_1}\end{pmatrix},\quad \langle\Phi_2\rangle=\frac{1}{\sqrt2}\begin{pmatrix}w\\v_2\end{pmatrix}, \tag{4}$$

with v_i, w real. It can be easily shown that for an appropriate range of the parameters of the Higgs potential, the minimum occurs at $w=0$, as needed in order to have an unbroken U(1) for electromagnetism. Furthermore, it is clear from (3) that

$V(\Phi)$ depends on θ only through the last term. It follows then that

$$(\theta_1)_{\min} = n\pi \quad \text{for } c_{12} < 0, \tag{5}$$

$$(\theta_1)_{\min} = \left(\frac{2n+1}{2}\right)\pi \quad \text{for } c_{12} > 0 \,. \tag{6}$$

In the case of (5), there is clearly no spontaneous CP breaking. It turns out that solution (6) is also CP invariant. To show this, assume that $c_{12} > 0$ and let the tree-approximation minimum of $V(\Phi)$ be at

$$\langle 0|\phi_1^0|0\rangle = \frac{v_1}{\sqrt{2}}\, e^{i\pi/2}, \quad \langle 0|\phi_2^0|0\rangle = \frac{v_2}{\sqrt{2}} \,. \tag{7}$$

Under time reversal T, the Higgs fields transform as

$$T\phi_i^0 T^{-1} = e^{i\alpha_i}\phi_i^0 \,. \tag{8}$$

A T-invariant minimum (i.e., $T|0\rangle = |0\rangle$) implies

$$\langle 0|\phi_i^0|0\rangle = \langle 0|\,T\phi_i T^{-1}|0\rangle^* = e^{-i\alpha_i}\langle 0|\phi_i^0|0\rangle^* \,, \tag{9}$$

where we have used the fact that T is antiunitary. It follows then from (7) and (9) that ϕ_i^0 should transform as

$$T\phi_1^0 T^{-1} = -\phi_1^0 \,, \; T\phi_2^0 T^{-1} = \phi_2^0 \,. \tag{10}$$

Since the Higgs potential is invariant under (10), it follows that (7) corresponds to a CP-invariant minimum. We have thus shown that in a model with NFC and two Higgs doublets, CP is conserved by the tree-approximation minima. Furthermore, one can invoke the Georgi-Pais theorem[9] to conclude that CP will still be respected by higher orders in the perturbation expansion.

At this point, it is worth emphasizing the key role played by NFC in our analysis. Had we not insisted on NFC, $V(\Phi)$ would contain terms such as $(\Phi_i^\dagger\Phi_i)(\Phi_1^\dagger\Phi_2)$, which in general would lead to spontaneous CP breaking.

In view of the previous result, we consider next the case of three Higgs doublets, which turns out to be the minimal structure required to achieve spontaneous CP violation. The simplest way to conform[10] to NFC is by preventing the third Higgs doublet Φ_3 from coupling to quarks (this is easily achieved by introducing a reflection symmetry R' under which $\Phi_3 \to -\Phi_3$ while all other fields remain unchanged). The most general gauge-invariant Higgs potential consistent with $R\times R'$ can be written

$$V(\Phi) = \sum_{i=1}^{3}[m_i^2\Phi_i^\dagger\Phi_i + a_{ii}(\Phi_i^\dagger\Phi_i)^2] + \sum_{i<j}\{a_{ij}(\Phi_i^\dagger\Phi_i)(\Phi_j^\dagger\Phi_j) + b_{ij}(\Phi_i^\dagger\Phi_j)(\Phi_j^\dagger\Phi_i) + [c_{ij}(\Phi_i^\dagger\Phi_j)(\Phi_i^\dagger\Phi_j) + \text{H.c.}]\} \,, \tag{11}$$

where all the coupling constants are real, so that CP invariance holds at the Lagrangian level. We assume the minimum to be at

$$\Phi_i = \frac{1}{\sqrt{2}}\begin{pmatrix} 0 \\ v_i e^{i\theta_i} \end{pmatrix} \tag{12}$$

as required by charge conservation. Since only relative phases are physically meaningful, we set $\theta_2 = 0$. From (11) and (12) one obtains the following stationarity conditions:[8]

$$\begin{aligned}
&m_1^2 + a_{11}v_1^2 + \tfrac{1}{2}a_{12}v_2^2 + \tfrac{1}{2}a_{13}v_3^2 + \tfrac{1}{2}b_{12}v_2^2 + \tfrac{1}{2}b_{13}v_3^2 + c_{12}v_2^2\cos 2\theta_1 + c_{13}v_3^2\cos^2(\theta_1 - \theta_3) = 0,\\
&m_2^2 + a_{22}v_2^2 + \tfrac{1}{2}a_{12}v_1^2 + \tfrac{1}{2}a_{23}v_3^2 + \tfrac{1}{2}b_{12}v_1^2 + \tfrac{1}{2}b_{23}v_3^2 + c_{12}v_1^2\cos(2\theta_1) + c_{23}v_3^2\cos(2\theta_3) = 0 \,,\\
&m_3^2 + a_{33}v_3^2 + \tfrac{1}{2}a_{13}v_1^2 + \tfrac{1}{2}a_{23}v_2^2 + \tfrac{1}{2}b_{13}v_1^2 + \tfrac{1}{2}b_{23}v_2^2 + c_{13}v_1^2\cos 2(\theta_1 - \theta_3) + c_{23}v_2^2\cos 2\theta_3 = 0 \,,\\
&c_{12}v_2^2\sin 2\theta_1 + c_{13}v_3^2\sin 2(\theta_1 - \theta_3) = 0,\\
&c_{13}v_1^2\sin 2(\theta_1 - \theta_3) - c_{23}v_2^2\sin 2\theta_3 = 0 \,,
\end{aligned} \tag{13}$$

We now classify the various solutions of (13), according to their CP transformation properties.

(a) CP-conserving solutions. These correspond to

$$\theta_1 = \tfrac{1}{2}n\pi, \quad \theta_3 = \tfrac{1}{2}m\pi \,, \tag{14}$$

with n, m integers. It can be easily verified, following arguments entirely analogous to those leading to Eqs. (9) and (10), that these solutions are CP conserving.

(b) CP-violating solution. Fortunately, there is another solution of (13), namely,

$$\begin{aligned}
\cos 2\theta_1 &= \frac{1}{2}\left(\frac{D_{13}D_{23}}{D_{12}^2} - \frac{D_{23}}{D_{13}} - \frac{D_{13}}{D_{23}}\right),\\
\cos 2\theta_3 &= \frac{1}{2}\left(\frac{D_{13}D_{12}}{D_{23}^2} - \frac{D_{12}}{D_{13}} - \frac{D_{13}}{D_{12}}\right),
\end{aligned} \tag{15}$$

where $D_{ij} = c_{ij}v_i^2 v_j^2$. Since in general $2\theta_i \neq n\pi$, this solution leads to spontaneous CP breaking. In order that this solution exists, the following inequalities have to be satisfied:

$$|D_{13}D_{23}| < |D_{12}D_{23}| + |D_{13}D_{12}| ,$$
$$|D_{12}D_{23}| < |D_{13}D_{23}| + |D_{13}D_{12}| , \quad (16)$$
$$|D_{13}D_{12}| < |D_{13}D_{23}| + |D_{12}D_{23}| .$$

These relations have a simple geometrical interpretation: The three quantities $|D_{12}D_{23}|$, $|D_{12}D_{13}|$, and $|D_{13}D_{23}|$ should form a closed triangle. In the Appendix we show that if $V(\Phi)$ is bounded below and if we assume all D_{ij} to be positive, then the CP-violating solution is the absolute minimum of $V(\Phi)$.

III. INTERACTION OF HIGGS BOSONS AND FERMIONS

The quark Yukawa interactions are still given by Eq. (2), since the third Higgs doublet does not couple to quarks. From (2) and (12) one obtains, after spontaneous symmetry breaking, the following mass terms:

$$\sum_{i,j} \frac{1}{\sqrt{2}} [\bar{\mathfrak{N}}_{iL}(v_1\Gamma^1_{ij})e^{i\theta_1}\,\mathfrak{N}_{jR} + \bar{\mathcal{P}}_{iL}v_2\Gamma^2_{ij})\mathcal{P}_{jR}] . \quad (17)$$

As noticed before,[1] by making a phase redefinition

$$n'_{jR} = e^{i\theta_1}\mathfrak{N}_{jR} \quad (18)$$

one obtains real mass matrices, which are diagonalized by the usual biorthogonal transformations:

$$\mathfrak{N}_{iL} = (O^d_L)_{ij}\, d_{jL},\ \mathfrak{N}'_{iR} = (O^d_R)_{ij}\, d_{jR} ,$$
$$\mathcal{P}_{iL} = (O^u_L)_{ij}\, u_{jL},\ \mathcal{P}_{iR} = (O^u_R)_{ij} u_{jR} . \quad (19)$$

The weak charged current is then given by

$$J_{\mu L} = \bar{u}_{iL}\gamma_\mu (O_C)_{ij}\, d_{jL} , \quad (20)$$

with the generalized Cabibbo matrix given by $O_C = (O^u_L)^T O^d_L$. The reality of O_C guarantees that the vector gauge interactions conserve CP, for an arbitrary number of quark generations. It has been previously shown[1] that this result is much more general and applies to *any model* based on SU(2)×U(1), provided one conforms to NFC and spontaneous symmetry breaking. We now exhibit the Yukawa interactions with the charged Higgs boson. Using (2) and (19), one obtains

$$\mathcal{L}(\phi^\pm) = \sqrt{2}\left(\bar{u}_L O_C M_d\, d_R \frac{\phi_1'^+}{v_1} - \bar{u}_R M_u O_C\, d_L \frac{\phi_2'^+}{v_2} + \text{H.c.}\right) , \quad (21)$$

where $\phi_i'^+ = e^{-i\theta_i}\phi_i^+$ and M_u, M_d are diagonal matrices:

$$M_d = \begin{bmatrix} m_d & & & \\ & m_s & & \\ & & m_b & \\ & & & \ddots \end{bmatrix} , \quad M_u = \begin{bmatrix} m_u & & & \\ & m_c & & \\ & & m_t & \\ & & & \ddots \end{bmatrix} . \quad (22)$$

There are three charged[11] Higgs states. One component $G^\pm$ remains massless and is just the Goldstone boson which is absorbed as the longitudinal component of $W^\pm_\mu$. The other two components $H^\pm_i$ acquire mass and remain as physical charged scalars. In order to find their Yukawa interactions, we analyze next the Higgs-boson mass matrix. From (11), (12), and (13) one obtains the following mass matrix in the basis of Φ'_j:

$$M_H = \begin{bmatrix} \frac{1}{v_1^2}(X_{12}+X_{12}) & -\frac{1}{v_1v_2}(X_{12}+iY) & -\frac{1}{v_1v_3}(X_{13}-iY) \\ -\frac{1}{v_1v_2}(X_{12}-iY) & \frac{1}{v_2^2}(X_{12}+X_{23}) & -\frac{1}{v_2v_3}(X_{23}+iY) \\ -\frac{1}{v_1v_3}(X_{13}+iY) & -\frac{1}{v_2v_3}(X_{23}-iY) & \frac{1}{v_3^2}(X_{13}+X_{23}) \end{bmatrix} , \quad (23)$$

where

$$X_{ij} = [\tfrac{1}{2}b_{ij} + c_{ij}\cos 2(\theta_i - \theta_j)v_i^2 v_j^2] \quad (24)$$

and

$$Y = c_{12}v_1^2v_2^2 \sin 2\theta_1$$
$$= -c_{13}v_1^2v_3^2 \sin 2(\theta_1 - \theta_3)$$
$$= -c_{23}v_2^2v_3^2 \sin 2\theta_3 . \quad (25)$$

The Higgs-boson mass matrix is diagonalized by performing a unitary transformation

$$\begin{pmatrix} G^+ \\ H_1^+ \\ H_2^+ \end{pmatrix} = U_H \begin{pmatrix} \phi_1' \\ \phi_2' \\ \phi_3' \end{pmatrix} . \quad (26)$$

From (23), one obtains by inspection

$$G^{\pm}=\frac{1}{v}\left(v_1\phi_1'+v_2\phi_2'+v_3\phi_3'\right), \tag{27}$$

where $v^2=\sum v_i^2$. By appropriate choice of the H_i^+ phases, the matrix U_H assumes the conventional Kobayashi-Maskawa[12] (KM) form, with the first row and first column real. Two of the KM angles[12] θ_i are determined by (26):

$$\sin\theta_1=\frac{(v_2^2+v_3^2)^{1/2}}{v}, \qquad \sin\theta_3=\frac{v_3}{(v_2^2+v_3^2)^{1/2}}. \tag{28}$$

Similarly, $\sin\theta_2$ and the CP-violating phase δ can be determined in terms of the parameters of M_H. Using (21) and (26), one obtains the quark interactions with the physical Higgs boson H_i. Note that since the generalized Cabibbo matrix O_C is real, the only CP-violating phase comes from U_H. Exchange of charged Higgs bosons leads to an effective Fermi interaction, whose CP-violating component is given by

$$\mathcal{H}^{CP}=-\operatorname{Im}\left(\frac{m^2}{(m^4-\Delta^4)}A_+ +\frac{2}{(m^4-\Delta^4)}A_-\right)\frac{2}{v_1v_2}H, \tag{29}$$

with

$$H=(\bar{d}_R M_D O_C U_L)(\bar{u}_R M_u O_C d_L) \tag{30}$$

and

$$A_{\pm}=u_{21}^*u_{22}\pm u_{31}^*u_{32}, \tag{31}$$

$$m^2=\tfrac{1}{2}(m_{H_1}^2+m_{H_2}^2), \tag{32}$$

$$\Delta^2=\tfrac{1}{2}(m_{H_2}^2-m_{H_1}^2), \tag{33}$$

where m_{H_i} are the charged-Higgs boson masses and u_{ij} denote the elements of u_H. Using the fact that u_H is unitary, it follows that $A_+=-v_1v_2/v_2$, and A_+ is therefore real. The CP-violating part of the amplitude comes only from the A_- term, and vanishes in the limit of degenerate Higgs-boson masses.[13] Computing the imaginary part of A_- from (23), (26), and (31), one finally obtains for the CP-violating component of the Fermi interaction

$$\mathcal{H}^{CP}=2\sqrt{2}\,G_F\frac{vv_3}{v_1v_2}\left(\frac{\Delta^2}{m^4-\Delta^4}\right)\cos\xi H, \tag{34}$$

where

$$\sin^2\xi=\frac{\Sigma^2}{\Sigma^2+(4v^2/v_1^2v_2^2v_3^2)Y^2},$$

and Σ^2 is a quadratic form of the X_{ij}. Note that in expression (34) $|\cos\xi|$ can vary from zero to one, independently of the v_i. It is clear that $\mathcal{H}^{CP}$ can alternatively by expressed in terms of the KM-type mixing angles θ_i and the CP-violating phase δ. If one assumes that only one of the Higgs bosons is relatively light (i.e., $m_{H_1}^2 \ll m_{H_2}^2$), expression (34) becomes

$$\mathcal{H}^{CP}\cong\frac{\sqrt{2}\,G_F}{m_{H_1}^2}\left(\frac{vv_3}{v_1v_2}\right)(\cos\xi)H. \tag{35}$$

IV. CONCLUSION

We have shown how a minimal model with spontaneous CP breaking and NFC can be realized in the context of a SU(2)×U(1) gauge theory. In particular, we have verified that there is a solution for the minimum of the Higgs potential which leads to spontaneous CP violation. The generalized Cabibbo matrix is real, and there is no CP violation in the manner of Kobayashi-Maskawa.[12] The breaking of CP invariance is solely generated by Higgs-boson exchange and can be characterized by a *unique phase* for an *aribtrary* number of fermion generations.

ACKNOWLEDGMENTS

I wish to thank Professor Lincoln Wolfenstein for his valuable advice and for a careful reading of the manuscript. I have also benefited from discussions with Professor Ling-Fong Li. This work was supported in part by the U. S. Department of Energy.

APPENDIX

We show here that there is a range of the free parameters of the Higgs potential for which the CP-violating solution (15) is indeed an absolute minimum of the potential. We do this by proving the following theorem:

Theorem. If the function $V(\Phi)$ is bounded below and if all the c_{ij} are positive, then the CP-violating solution given by Eq. (15) is the absolute minimum of $V(\Phi)$.

Proof. From (11) and (12), it follows that $V(v_i,\theta_i)$ can be written

$$V=U(v_i)+W(v_i,\theta_i), \tag{A1}$$

where U depends only on v_i and W is given by

$$W=\tfrac{1}{2}\left[c_{12}v_1^2v_2^2\cos2\theta_1+c_{13}v_1^2v_3^2\cos2(\theta_1-\theta_3)+c_{23}v_2^2v_3^2\cos2\theta_3\right]. \tag{A2}$$

In the case of the CP-breaking solution, one obtains, using Eq. (15),

$$W=-\frac{1}{4}\left(\frac{c_{13}c_{12}}{c_{23}}v_1^4+\frac{c_{12}c_{23}}{c_{13}}v_2^4+\frac{c_{13}c_{23}}{c_{12}}v_3^4\right). \tag{A3}$$

If the function $V(\Phi)$ is bounded below, one of its stationary points has to be an absolute minimum.

It can be easily shown that if $c_{ij} > 0$, none of the solutions corresponding to (13) can be an absolute minimum. Consider, for example, the stationary point

$$\begin{aligned} V^{(0)} &= V(v_i^{(0)},\ \theta_1 = \theta_3 = 0) \\ &= U(v_i^0) + \tfrac{1}{2}c_{12}v_1^{(0)2}v_2^{(0)2} \\ &\quad + \tfrac{1}{2}c_{13}v_1^{(0)2}v_3^{(0)2} + \tfrac{1}{2}c_{23}v_2^{(0)2}v_3^{(0)2}\ . \end{aligned} \tag{A4}$$

It is clear that for $c_{ij} > 0$, $V^{(0)}$ cannot be the absolute minimum, since V assumes a lower value at any point with $v = v_i^{(0)}$, $\theta_i \neq 0$. Consider another stationary point, corresponding to (13):

$$\begin{aligned} V^{(1)} &= V(v_i^{(1)},\ \theta_1 = 0,\ \theta_3 = \tfrac{1}{2}\pi) \\ &= U(v_i^{(1)}) + \tfrac{1}{2}c_{12}v_1^{(1)2}v_2^{(1)2} \\ &\quad - \tfrac{1}{2}c_{13}v_1^{(1)2}v_3^{(1)2} - \tfrac{1}{2}c_{23}v_2^{(1)2}v_3^{(1)2}\ . \end{aligned} \tag{A5}$$

It turns out that V assumes a lower value when $v_i = v_i^{(1)}$ and θ_i assume the values of Eq. (15). This is seen by observing that

$$\begin{aligned} &(\tfrac{1}{2}c_{12}v_1^2v_2^2 - \tfrac{1}{2}c_{13}v_1^2v_3^2 - \tfrac{1}{2}c_{23}v_2^2v_3^2) + \frac{1}{4}\left(\frac{c_{13}c_{12}}{c_{23}}v_1^4 + \frac{c_{12}c_{23}}{c_{13}}v_2^4 + \frac{c_{13}c_{23}}{c_{12}}v_3^4\right) \\ &\qquad = \frac{1}{4c_{12}c_{13}c_{23}}\left[c_{12}(c_{13}v_1^2 + c_{23}v_2^2) - c_{23}c_{13}v_3^2\right]^2 \geq 0\ . \end{aligned} \tag{A6}$$

The same argument can be applied, *mutatis mutandis*, to the other *CP*-conserving solutions given by (13). Therefore, we conclude that Eq. (15) corresponds to the absolute minimum.

[1] Gustavo C. Branco, Phys. Rev. Lett. 44, 504 (1980).

[2] S. L. Glashow and S. Weinberg, Phys. Rev. D 15, 1958 (1977). See also E. Paschos, *ibid.* 15, 1966 (1977); L. T. Trueman, F. Paige, and E. Paschos, *ibid.* 15, 3416 (1977); K. Kang and J. E. Kim, Phys. Lett. 64B, 93 (1976).

[3] For a review of gauge theories of *CP* nonconservation, see R. N. Mohapatra, in *Particles and Fields—1974*, edited by C. E. Carlson (American Institute of Physics, New York, 1974).

[4] T. D. Lee, Phys. Rep. 9C, 143 (1974); S. Weinberg, Phys. Rev. Lett. 37, 657 (1976).

[5] For a discussion of the possible origin of the small *CP* nonconservation in *K* decays, see T. D. Lee and L. Wolfenstein, Phys. Rev. B 138, 1490 (1965).

[6] There is a lower bound of about 5 GeV on the Higgs-boson masses; see S. Weinberg, Phys. Rev. Lett. 36, 294 (1976).

[7] S. Weinberg, Phys. Rev. Lett. 19, 1264 (1967); A. Salam, in *Elementary Particle Theory. Relativistic Groups and Analyticity (Nobel Symposium No. 8*, edited by N. Svartholm (Wiley, New York, 1969).

[8] See N. G. Deshpande and E. Ma, Phys. Rev. D 16, 1583 (1977).

[9] H. Georgi and A. Pais, Phys. Rev. D 16, 3520 (1977); the theorem states that if a Lagrangian is invariant under a (discrete or continuous) symmetry and if the vacuum expectation values respect this symmetry in the tree approximation, then the symmetry will still hold in higher order, provided that at the tree level there are no massless scalar meson fields which are not Goldstone bosons.

[10] For a classification of the generation symmetries that lead to NFC, see R. Gatto, G. Morchio, and F. Strocchi, Phys. Lett. 80B, 265 (1979); G. Segrè and H. Weldon, Ann. Phys. (N.Y.) 124, 37 (1980).

[11] The properties of charged Higgs particles in the context of a SU(2) × U(1) gauge theory have been recently studied by various authors. See, for example, J. F. Donoghue and L. F. Li, Phys. Rev. D 19, 945 (1979); L. F. Abbott, P. Sikivie, and M. B. Wise, *ibid.* 21, 1393 (1980); C. H. Albright, J. Smith, and S.-H. H. Tye, *ibid.* 21, 1711 (1980).

[12] M. Kobayashi and K. Maskawa, Prog. Theor. Phys. 49, 652 (1973).

[13] This was pointed out to me by Professor L. Wolfenstein.

PHYSICAL REVIEW D VOLUME 34, NUMBER 5 1 SEPTEMBER 1986

Weinberg *CP*-violation model revisited

Hai-Yang Cheng*
Physics Department, Brandeis University, Waltham, Massachusetts 02254
(Received 28 January 1986)

Chiral properties of the weak amplitudes in the Weinberg model of *CP* violation are carefully treated. The *CP*-odd $K\to 2\pi$ amplitude is chiral-symmetry nonlinear-realization dependent, and so is the soft-pion relation. This realization dependence is canceled by the pole contribution arising from the combination of a strong $K\pi$ vertex and a K-vacuum tadpole, so that to lowest order in chiral symmetry the physical $K\to 2\pi$ amplitude is chiral-model independent. The vacuum-insertion method is valid only in a particular realization scheme. Various *CP*-violating effects are then reexamined in this model. When the η-η' mixing is taken into account, $\epsilon'/\epsilon \approx -0.007$ is obtained. The electric dipole moment of a neutron due to neutral-Higgs-boson exchange and the current experimental limit are used to set an upper bound for neutral-Higgs-boson mixing. Finally, *CP* violation in $K\to 3\pi$ decays is discussed.

I. INTRODUCTION

Recently there has been renewed interest in the Weinberg model of *CP* violation.[1] On the one hand, Donoghue and Holstein[2] showed how to correctly treat the chiral properties of the weak amplitudes within the framework of chiral perturbation theory. As a result, the *CP*-violating parameter ϵ'/ϵ is estimated to be -0.006 (see also Ref. 3), in contrast with the conventional result $\epsilon'/\epsilon=-\frac{1}{20}$ (Ref. 4). Since the new estimate is within the experimental limit, this model seems to be viable. On the other hand, Anselm *et al.*[5] showed that the dominant contribution to the neutron electric dipole moment (EDM) arises from neutral-Higgs-boson exchange rather than from the charged Higgs boson. This dominant contribution exceeds the present experimental upper bound by 2–3 orders of magnitude, which appears to be a serious argument against the Weinberg model of *CP* nonconservation unless the neutral-Higgs-boson mixing is smaller than the charged-Higgs-boson mixing.

Since a K-vacuum tadpole can be induced by the *CP*-odd Lagrangian in the Weinberg model, a pole contribution has to be included in the calculation of the $K\to 2\pi$ amplitude. Because the strong $K\pi$ scattering amplitude is chiral-symmetry nonlinear-realization dependent, the inclusion of an additional pole term seems to give a puzzle; the on-shell physical amplitude to lowest order in chiral symmetry ought to be chiral-model independent. After briefly reviewing the Weinberg model in Sec. II we study the chiral properties of the weak amplitudes and solve the aforementioned puzzle. Some constraints are obtained for the charged-Higgs-boson mixing angles and the vacuum expectation values.

Section III is devoted to the discussions of *CP*-odd effects which can be manifested in this model. Section IV gives the conclusions.

II. MODEL

In the Weinberg model of *CP* violation there are at least three Higgs doublets in order to implement *CP* violation and conservation of natural flavor at the tree level. The breakdown of *CP* symmetry can arise from the complex vacuum expectation values of Higgs fields and the complex Yukawa interactions. If *CP* nonconservation is realized spontaneously, the Kobayashi-Maskawa (KM) matrix of the quark mixing at the tree level is real[6] and *CP* violation comes solely from the Higgs sector. Higher-order radiative correction would yield a small *CP*-violating phase δ in the KM matrix in the gauge interaction. Since δ is calculable and naturally small, this has been regarded for some time as one of the attractive features of this type of *CP*-odd model. Nowadays we know that $\sin\delta$ is of order unity, inferred from the measurement of the b-quark lifetime and its semileptonic decay.[7]

The Yukawa interaction of the charged Higgs boson in the unitary gauge is given by[8,9]

$$\mathscr{L}_Y^{\pm}=(2\sqrt{2}G_F)^{1/2}\sum_{i=1}^{2}(\alpha_i\bar{U}_LKM_DD_R+\beta_i\bar{U}_RM_UKD_L+\gamma_i\bar{N}_LM_EE_R)H_i^{\dagger}+\text{H.c.}\,, \tag{1}$$

where K is a real KM matrix, M_U, M_D, and the M_E are diagonal mass matrices for the up quarks, down quarks, and leptons, respectively, and

$$\begin{aligned}
&\alpha_1=-\frac{\tilde{s}_1\tilde{c}_3}{\tilde{c}_1},\quad \beta_1=\frac{-\tilde{c}_1\tilde{c}_2\tilde{c}_3+\tilde{s}_2\tilde{s}_3e^{i\delta_H}}{\tilde{s}_1\tilde{c}_2}\,,\\
&\gamma_1=\frac{\tilde{c}_1\tilde{s}_2\tilde{c}_3+\tilde{c}_2\tilde{s}_3e^{i\delta_H}}{\tilde{s}_1\tilde{s}_2}\,,\\
&\alpha_2=-\frac{\tilde{s}_1\tilde{s}_3}{\tilde{c}_1},\quad \beta_2=\frac{-\tilde{c}_1\tilde{c}_2\tilde{s}_3-\tilde{s}_2\tilde{c}_3e^{i\delta_H}}{\tilde{s}_1\tilde{c}_2}\,,\\
&\gamma_2=\frac{\tilde{c}_1\tilde{s}_2\tilde{s}_3-\tilde{c}_2\tilde{c}_3e^{i\delta_H}}{\tilde{s}_1\tilde{s}_2}\,,
\end{aligned} \tag{2}$$

with $\tilde{s}_i$, $\tilde{c}_i$, and δ_H being the Higgs-boson mixing angles and phase defined in complete analogy to the KM matrix of quark mixing. From (2) this gives

$$\mathrm{Im}(\alpha_2\beta_2^*)=-\mathrm{Im}(\alpha_1\beta_1^*)\,,$$
$$\mathrm{Im}(\alpha_2\gamma_2^*)=-\mathrm{Im}(\alpha_1\gamma_1^*)\,, \qquad (3)$$
$$\mathrm{Im}(\beta_2\gamma_2^*)=-\mathrm{Im}(\beta_1\gamma_1^*)\,.$$

The Yukawa interaction of the neutral Higgs boson in the diagonalized basis reads[9]

$$\mathscr{L}_Y^0=(\sqrt{2}\,G_F)^{1/2}\sum_{i=1}^{5}(\zeta_{1i}\bar{D}M_D D+i\zeta_{2i}\bar{D}M_D\gamma_5 D +\zeta_{3i}\bar{U}M_U U+i\zeta_{4i}\bar{U}M_U\gamma_5 U +\zeta_{5i}\bar{E}M_E E+i\zeta_{6i}\bar{E}M_E\gamma_5 E)H_i\,, \qquad (4)$$

where the coupling constants ζ_{ij} are real. Since $\bar{\psi}\psi$ and $i\bar{\psi}\gamma_5\psi$ have opposite P, T, and CP transformation properties, P and CP can be violated through the exchange of the neutral Higgs boson. In this model CP violation is usually characterized by the parameter

$$\mathrm{Im}\,A\equiv\mathrm{Im}(\langle\phi_1^{\dagger *}\phi_2^{\dagger}\rangle/v_1^* v_2)$$
$$\equiv\frac{G_F}{m_0^{\,2}}=2\sqrt{2}\,G_F\sum_{i=1}^{2}\frac{\mathrm{Im}\alpha_i\beta_i^*}{m_{H_i}^{\,2}}\,. \qquad (5)$$

For later purposes we rewrite the equation for $m_0^{\,2}$:

$$\frac{1}{m_0^{\,2}}=\frac{\sqrt{2}\,vv_3}{v_1 v_2}\sin 2\tilde{\theta}_3\sin\delta_H\left[\frac{1}{m_{H_1}^{\,2}}-\frac{1}{m_{H_2}^{\,2}}\right], \qquad (6)$$

where v_1,v_2,v_3 are the vacuum expectation values of the Higgs fields ϕ_1, ϕ_2, and ϕ_3, respectively, $v=(v_1^{\,2}+v_2^{\,2}+v_3^{\,2})^{1/2}=(2\sqrt{2}\,G_F)^{-1/2}$. In deriving (6), Eqs. (2) and (3.12) of Ref. 9 have been used.

A. Chiral properties of the weak amplitude and ϵ'/ϵ

In $K\to 2\pi$ decays the CP-violating parameters ϵ and ϵ' are given by Ref. 10:

$$\epsilon=\frac{1}{2\sqrt{2}}(\epsilon_m+2\xi_0)e^{i\pi/4}\,,$$
$$\epsilon'=\frac{i}{\sqrt{2}}\frac{\mathrm{Re}a_2}{\mathrm{Re}a_0}(\xi_2-\xi_0)e^{i(\delta_2-\delta_0)}\,, \qquad (7)$$
$$\epsilon_m\equiv\mathrm{Im}M_{12}/\mathrm{Re}M_{12}\,,$$

where $\xi_i=\mathrm{Im}a_i/\mathrm{Re}a_i$, $a_0(a_2)$ is the isospin-zero (-two) amplitude of $K\to 2\pi$. As in the KM model, $\mathrm{Im}a_2$ also vanishes in the Weinberg-model phase convention; hence

$$\frac{\epsilon'}{\epsilon}\cong-\frac{1}{20}\left[\frac{2\xi}{\epsilon_m+2\xi}\right], \qquad (8)$$

where the subscript 0 of ξ has been dropped for convenience. The dominant contribution to the imaginary part of the $K\to 2\pi$ amplitude comes from the Higgs penguin diagram. The effective $\Delta S=1$ CP-odd bilinear operator[11,12] from $s\to d+$gluon is

$$\mathscr{L}_-^{\Delta S=1}=i\tilde{f}\,\bar{d}\sigma^{\mu\nu}(1+\gamma_5)\lambda^a s G_{\mu\nu}^a \qquad (9)$$

with[13]

$$\tilde{f}=\frac{G_F}{\sqrt{2}}\frac{g_s}{32\pi^2}m_s m_c^{\,2}\cos\theta_C\sin\theta_C \times\sum_{i=1}^{2}\frac{\mathrm{Im}\alpha_i\beta_i^*}{m_{H_i}^{\,2}}\left[\ln\frac{m_{H_i}^{\,2}}{m_c^{\,2}}-\frac{3}{2}\right]. \qquad (10)$$

The Higgs-boson penguin diagram does not correspond to a local four-quark operator since the loop integral does not give a factor of k^2 canceling the pole in the gluon propagator. However, using the vacuum-saturation approximation to evaluate the matrix element $\langle 2\pi|\mathscr{L}_-^{\Delta S=1}|K\rangle$ one encounters an effective four-quark operator of the form[4]

$$\tilde{\mathscr{L}}^{\Delta S=1}=\sum_q^{u,s,d}\bar{q}_R d_L\bar{s}_R q_L\,, \qquad (11)$$

where $q_L=(1-\gamma_5)q/2$.

It was noted by Donoghue and Holstein[2] that the amplitude $K\to 2\pi$ induced by the CP-odd Lagrangian $\Theta=\mathscr{L}_-^{\Delta S=1}$ or $\tilde{\mathscr{L}}^{\Delta S=1}$ involves an additional pole contribution which arises from the combination of a four-point strong K-K-π-π vertex and a weak K-vacuum vertex (i.e., a tadpole). More precisely,

$$P=A(K\pi\to K\pi)\langle 0|\Theta|K\rangle/m_K^{\,2}\,, \qquad (12)$$

where $A(K\pi\to K\pi)$ is the strong-interaction $K\pi$ scattering amplitude. The total CP-violating $K\to 2\pi$ amplitude is then given by

$$A(K^0\to\pi^+\pi^-)=\langle\pi^+\pi^-|\Theta|K^0\rangle+P\,. \qquad (13)$$

The kaon-pion scattering amplitude can be easily computed using chiral perturbation theory. The strong-interaction Lagrangian to lowest order in momenta including chiral-SU(3)×SU(3)-symmetry breaking by the meson masses is

$$\mathscr{L}_s^{(2)}=\frac{f^2}{8}[\mathrm{Tr}(\partial_\mu U\partial^\mu U^\dagger)+\mathrm{Tr}M(U+U^\dagger)]\,, \qquad (14)$$

where f is the common decay constant $\sim$130 MeV, $M_{ij}=0$ $(i\neq j)$, $m_\pi^{\,2}=M_{11}=M_{22}$, $m_K^{\,2}=(M_{11}+M_{33})/2$, and the general form of the meson matrix U is[14]

$$U=1+2i\phi/f-2\phi^2/f^2-ia_3\phi^3/f^3 +2(a_3-1)\phi^4/f^4+\cdots\,,$$
$$\phi\equiv\phi^a\lambda^a\,, \qquad (15)$$

with λ^a the Gell-Mann matrices. Up to order four in ϕ there is only one arbitrary constant a_3 in the expansion of U. For $a_3=\frac{4}{3}$ the meson matrix becomes the familiar form $U=\exp(2i\phi/f)$. From Eqs. (12)–(15), it follows that the pole term on shell is

$$P=\langle 0|\Theta|K\rangle a_3/(2f^2)\,. \qquad (16)$$

The matrix element $\langle 2\pi|\Theta|K\rangle$ can be evaluated either by current algebra or by the vacuum-insertion method. If we use the soft-pion relations

$$\langle\pi^+\pi^-|\Theta|K^0\rangle=-i\sqrt{2}\langle\pi^0|\Theta|K^0\rangle/f\,,$$
$$\langle\pi^0|\Theta|K^0\rangle=-i\langle 0|\Theta|K^0\rangle/(\sqrt{2}f)\,, \tag{17}$$

we will encounter two related puzzles. From Eqs. (13), (16), and (17) we obtain

$$A(K^0\to 2\pi)=-(1-a_3/2)\langle 0|\Theta|K^0\rangle/f^2\,. \tag{18}$$

However, this is in contradiction with the theorem of Chisholm:[15] S-matrix elements (i.e., on-shell amplitudes) are independent of the value of a_3. Furthermore, as we will see later the Feinberg-Kabir-Weinberg (FKW) theorem[16] requires that no physical transition amplitudes be induced to lowest order in chiral perturbation theory. For $a_3=\frac{1}{2}$, as originally chosen by Ref. 2, or $a_3=\frac{4}{3}$, as often used in the literature, the CP-odd $K\to 2\pi$ amplitude does not vanish on shell. These puzzles are solved based on the observation that in the chiral limit the operator Θ transforms as $(\bar{3},3)$ under $SU_L(3)\times SU_R(3)$, and hence the corresponding effective Lagrangian is of the form

$$\Theta\sim \mathrm{Tr}(\lambda_6 U)+\mathrm{H.c.} \tag{19}$$

From Eq. (15) it becomes clear that the weak amplitude $\langle 2\pi|\Theta|K\rangle$ is a_3 dependent. Consequently, the soft-pion reduction of the $K\to 2\pi$ amplitude is nonlinear-realization dependent and should read

$$\langle\pi^+\pi^-|\Theta|K^0\rangle=-i\frac{\sqrt{2}}{f}\left[\frac{a_3}{2}\right]\langle\pi^0|\Theta|K^0\rangle\,. \tag{20}$$

So, if we use the usual soft-pion relations (17), the values of a_3 for the strong $K\pi$ scattering amplitude has to be 2 rather than $\frac{1}{2}$ or $\frac{4}{3}$. Thus,

$$\begin{aligned}A(K^0\to\pi^+\pi^-)&=\left[-i\frac{\sqrt{2}}{f}\langle\pi^0|\Theta|K^0\rangle+\frac{1}{f}\langle 0|\Theta|K^0\rangle\right]\frac{a_3}{2}\\&=0\,,\end{aligned} \tag{21}$$

in accordance with both Chisholm and FKW theorems. The fundamental reason for the physical CP-odd $K\to 2\pi$ amplitude vanishing to lowest order in chiral symmetry is that the effective chiral representation for the $\Delta S=1$ operator $\mathcal{L}_-^{\Delta S=1}$ [Eq. (19)] is similar to the mass terms in the strong-interaction Lagrangian (14). As a result, its effect can be rotated away by an appropriate chiral transformation of the on-shell amplitudes.[17] We also evaluate $A(K^0\to\pi^+\pi^-)$ using the vacuum-insertion approximation:

$$\begin{aligned}&\langle\pi^+\pi^-|\tilde{\mathcal{L}}^{\Delta S=1}|K^0\rangle\\&\quad=\langle\pi^+\pi^-|\bar{d}_R d_L|0\rangle\langle 0|\bar{s}_R d_L|K^0\rangle\\&\qquad+\langle\pi^+|\bar{u}_R d_L|0\rangle\langle\pi^-|\bar{s}_R u_L|K^0\rangle\\&\qquad+\langle 0|\bar{d}_R d_L+\bar{s}_R s_L|0\rangle\langle\pi^+\pi^-|\bar{s}_R d_L|K^0\rangle\end{aligned} \tag{22}$$

and

$$\begin{aligned}&\langle 0|\tilde{\mathcal{L}}^{\Delta S=1}|K^0\rangle\\&\quad=\langle 0|\bar{d}_R d_L+\bar{s}_R s_L|0\rangle\langle 0|\bar{s}_R d_L|K^0\rangle\,,\end{aligned} \tag{23}$$

where the third term in (22) and the tadpole contribution (23) are missing in the original computation in Ref. 4. Keeping the vertices independent of the momentum

$$\begin{aligned}&\langle 0|\bar{q}q|0\rangle=-f^2v/2\,,\\&\langle 0|\bar{s}_L d_R|K^0\rangle=-\langle\pi^+|\bar{u}_R d_L|0\rangle\\&\qquad=ifv/2\,,\\&\langle\pi^+\pi^-|\bar{d}_R d_L|0\rangle=\langle\pi^-|\bar{s}_L u_R|K^0\rangle=v/2\,,\end{aligned} \tag{24}$$

with

$$v=\frac{m_{\pi^+}^2}{m_u+m_d}=\frac{m_{K^+}^2}{m_u+m_s}=\frac{3m_\eta^2}{m_u+m_d+4m_s}\,,$$

and noting that the matrix element $\langle\pi^+\pi^-|\bar{s}(1-\gamma_5)d|K\rangle$ can be evaluated either by current algebra or by the effective Lagrangian $\mathrm{Tr}[(\lambda_6-i\lambda_7)U]$ for the bilinear operator $\bar{s}(1-\gamma_5)d$ (Ref. 18), so that

$$\langle\pi^+\pi^-|\bar{s}(1-\gamma_5)d|K^0\rangle=i\frac{v}{f}\left[\frac{a_3}{2}\right], \tag{25}$$

the reader can then check that the vacuum-insertion approximation yields the same results as the current algebra only for $a_3=2$. Indeed, Eq. (20) is respected by the vacuum-insertion method only if $a_3=2$.

We have thus demonstrated that in the Weinberg model of CP violation the physical CP-violating $K\to 2\pi$ amplitude vanishes to lowest order in chiral symmetry in both current-algebra and vacuum-insertion calculations. It is striking that the KM model also shares this important chiral property though for a different reason: the matrix element of the penguin operator, and hence ϵ'/ϵ, vanishes to lowest order in the low-energy expansion of chiral perturbation theory.[19]

Beyond the lowest-order chiral limit, $\mathrm{Im}a_0$ and ϵ'/ϵ may receive contributions from the following two sources.

(i) Next-order tree-graph and one-loop corrections to the weak amplitudes $\langle 2\pi|\Theta|K\rangle$, $\langle 0|\Theta|K\rangle$, and the strong amplitude of $K\pi$ scattering. For example, the next order strong-interaction Lagrangian is of the form

$$\mathcal{L}_s^{(4)}=\frac{f^2}{\Lambda^2}(3\text{ four-derivative terms}+\text{mass terms})\,,$$

where $\Lambda\approx 1$ GeV is the scale of chiral symmetry.[23] Since the counterterms necessary to renormalize one-loop graphs are of the same structures as $\mathcal{L}_s^{(2)}$ (Ref. 24), both four-derivative and one-loop corrections are suppressed by a factor q^2/Λ^2 at low energies.

(ii) Current-quark masses can induce an (8,1) term to the CP-odd Lagrangian, which cannot be transformed away by any chiral rotation. This SU(3)-braking contribution to ϵ'/ϵ was first discussed by Dupont and Pham.[3] Lacking reliable techniques for handling the higher-order chiral terms and loop corrections to the weak amplitudes, we follow Donoghue and Holstein[2] in giving a crude estimate:

$$A(K^0 \to 2\pi^0) = \langle 2\pi^0 | \mathcal{L}_-^{\Delta S=1} | K^0 \rangle m_K^2/\Lambda^2$$

$$= -i\frac{\sqrt{2}}{f}\langle \pi^0 | \mathcal{L}_-^{\Delta S=1} | K^0 \rangle \frac{m_K^2}{\Lambda^2} . \quad (26)$$

The dispersive effect on the kaon mass matrix has been discussed in Refs. 3 and 25. The long-distance contribution to the matrix element $\mathrm{Im}M_{12}$ due to the π,η,η' poles are approximately given by

$$2m_K(\mathrm{Im}M_{12})_{\mathrm{LD}} = \sum_i^{\pi,\eta,\eta'} \frac{\mathrm{Im}(\langle K^0 | H_W | i \rangle \langle i | H_W | \bar{K}^0 \rangle)}{m_K^2 - m_i^2} . \quad (27)$$

Following Refs. 2 and 3, and taking into account the $\eta - \eta'$ mixing, we obtain

$$2m_K(\mathrm{Im}M_{12})_{\mathrm{LD}} \approx -3\langle K^0 | H_- | \pi \rangle \langle \pi | H_+ | \bar{K}^0 \rangle , \quad (28)$$

where use of the relation

$$\langle \eta_1 | H_+ | K \rangle = -2\sqrt{2/3}\rho \langle \pi | H_+ | K \rangle$$

[η_1 being the SU(3)-singlet state] has been made. The parameter ρ can be determined from $K_L \to 2\gamma$ decays and is found to be 0.96 (with 30% error) (Ref. 26)], which is close to unity as expected from the quark model. Without the $\eta - \eta'$ mixing, it has been noted that the η and π poles cancel each other due to the FKW theorem.[2] However, the inclusion of the $\eta - \eta'$ mixing alters the situation. An almost complete cancellation occurs between π and η' poles, so the η' dispersive pole is not the dominant contribution of long-distance effects. Since the long-distance contribution to $\mathrm{Im}M_{12}$ is quite large compared to the short-distance one,[2] it is clear that $2\xi/\epsilon_m$, and ϵ'/ϵ are Higgs-boson-mixing and *CP*-violating angle independent. Using the experimental value $\langle 2\pi^0 | H_+ | K \rangle = 2.63 \times 10^{-7}$ GeV, we find[27]

$$2\xi/\epsilon_m \approx 0.16, \quad \epsilon'/\epsilon \approx -0.007 . \quad (29)$$

Because of the crude estimate made in (26), the result (29) should be regarded as order-of-magnitude estimates. At any rate, ϵ'/ϵ obtained here is consistent with the recent measurements[29]

$$\epsilon'/\epsilon = -0.0046 \pm 0.0053 \pm 0.0024 \text{ (Chicago-Saclay)}$$

$$= 0.0017 \pm 0.0082 \text{ (Yale-BNL)} .$$

Finally, we would like to remark that the relation $(\mathrm{Im}M_{12}/\mathrm{Re}M_{12})_{\mathrm{LD}} = -2\xi$ does not hold in the Weinberg model of *CP* violation even for the π,η intermediate states because of the pole contribution in Eq. (13). Accordingly, for ϵ the long-distance contribution to $\mathrm{Im}M_{12}$ is not canceled by the one occurring in $2\xi\,\mathrm{Re}M_{12}$. In fact, if the aforementioned relation were true, then from Eq. (8) together with $(\mathrm{Im}M_{12})_{\mathrm{LD}} > (\mathrm{Im}M_{12})_{\mathrm{box}}$, one would have[30]

$$\frac{\epsilon'}{\epsilon} = -\frac{1}{20}\frac{\Delta m}{(2\,\mathrm{Re}M_{12})_{\mathrm{box}}} .$$

However, ϵ'/ϵ is now Higgs-boson-mixing-angle dependent through the $(\mathrm{Re}M_{12})_{\mathrm{box}}$ term, as it should not be.

B. Constraints on Im *A* and the vacuum expectation values

From the experimental limit for the mass of the charged Higgs boson and the value of ϵ, we can obtain some constraints on $\mathrm{Im}A$ [Eq. (5)] and the vacuum expectation values. The matrix element

$$\langle \pi^0 | \mathcal{L}_-^{\Delta S=1} | K^0 \rangle \equiv \tilde{f} A_{K\pi}$$

has been computed in the MIT bag model and was found to be $A_{K\pi} = 0.4$ GeV^3 (for $\alpha_s = 1$) (Ref. 11). From Eqs. (3), (7), (26), (27), and the experimental value for ϵ, we obtain

$$\mathrm{Im}(\alpha_1\beta_1^*)\left[\frac{1}{m_{H_1}^2}\left[\ln\frac{m_{H_1}^2}{m_c^2} - \frac{3}{2}\right] - \frac{1}{m_{H_2}^2}\left[\ln\frac{m_{H_2}^2}{m_c^2} - \frac{3}{2}\right]\right] = \frac{1}{2\sqrt{2}\tilde{m}_0^2} \quad (30)$$

with $\tilde{m}_0 = 3.3$ GeV. The present experimental lower bound for the mass of the charged Higgs boson is[32]

$$m_H \geq 16 \text{ GeV} . \quad (31)$$

With our convention that $m_{H_2} > m_{H_1}$, the term in large square brackets in (30) is $\leq 1.3 \times 10^{-2}$; thus we have a lower limit on $\mathrm{Im}(\alpha_1\beta_1^*)$:

$$\mathrm{Im}(\alpha_1\beta_1^*) > 2.6 . \quad (32)$$

From Eq. (6) we also have a lower bound on $(1/m_{H_1}^2 - 1/m_{H_2}^2)$:

$$\left[\frac{1}{m_{H_1}^2} - \frac{1}{m_{H_2}^2}\right] \geq \frac{v_1 v_2}{\sqrt{2}\, v v_3 m_0^2} . \quad (33)$$

This together with Eq. (5) yields

$$\frac{v v_3}{v_1 v_2} \geq 2\,\mathrm{Im}(\alpha_1\beta_1^*) = 5.2 \quad (34)$$

and

$$m_0 > 5.9 \text{ GeV} . \quad (35)$$

For a given m_0 (typically 10 GeV), the existence of a light-charged Higgs boson is in general required in the Weinberg model unless v_3 is unexpectedly high. We remark that without the Higgs-boson penguin contribution (hence $\xi = 0$) a small value $m_0 = 2$ GeV was obtained in Ref. 33. It is of interest to note that the constraint (34) on the vacuum expectation values (VEV's) is satisfied only if $v_3 > v_1, v_2$. In other words not all three VEV's are of the same order of magnitude. Indeed, the hierarchy of the VEV's $v_3 > v_2 \gg v_1$ is assumed in Ref. 31 in order to obtain an ϵ'/ϵ consistent with the experimental value. While in our case, ϵ'/ϵ is small because of the small *CP* violation in the $K \to 2\pi$ decay amplitude due to the chiral properties of the *CP*-odd Lagrangian, and because of the large long-distance contribution to the imaginary part of M_{12}. For later purposes of calculations, we choose $m_{H_1} = 20$ GeV, $m_{H_2} = 100$ GeV, which corresponds to

$$m_0 = 6.2 \text{ GeV}, \quad \mathrm{Im}(\alpha_1\beta_1^*) = 3.8 . \quad (36)$$

III. APPLICATIONS

In this section we reexamine and update the discussions of other *CP*-violating effects in the Weinberg model of *CP* violation. The reason for doing this is attributed to the fact that some parameters in this model are changed because of the chiral properties of the *CP*-odd Lagrangian. Hence it is worthwhile to update this model's predictions by using the new determined constraints on the fundamental parameters in this theory, Eqs. (32)—(36).

A. Electric dipole moment of the neutron

The complete expression of the EDM of a quark due to the charged-Higgs-boson exchange is given in Ref. 34. Using $m_d=7.6$ MeV, $m_c=1.5$ GeV, $m_t=45$ GeV, $K_{cd}=0.227$, $K_{td}=0.0174$, and Eq. (36), we find

$$D_n \sim 1\times 10^{-25}\ e\,\text{cm}\ , \tag{37}$$

which is to be compared with the current experimental limit $D_n < 5\times 10^{-25}$ e cm (Ref. 35).

The contributions from the neutral Higgs boson are proportional to the cube of the small light-quark masses and thus are negligible.[36] Nevertheless, as pointed out by Anselm *et al.*,[5] the correct estimate of the nucleon coupling with the neutral Higgs boson at low momenta leads to a value which is proportional to the nucleon mass and does not vanish in the chiral limit m_u, $m_d \to 0$. The number $D_n=10^{-22}$ e cm is obtained by Anselm *et al.*, with $m_0=2$ GeV under the assumption that the neutral- and charged-Higgs-boson mixing are of the same order of magnitude (see also Ref. 37):

$$\frac{\langle Hx\rangle}{v^2} \sim \text{Im}\frac{\langle \phi_1^{\dagger *}\phi_2^{\dagger}\rangle}{v_1^* v_2} \tag{38}$$

[i.e., $g^{(1)}g^{(2)}$ is of the same order as $\text{Im}(\alpha\beta^*)$ in our notation]. With Eq. (36) their result is modified to

$$D_n \le 1.1\times 10^{-23}\ e\,\text{cm}\ . \tag{39}$$

However, no definite conclusion can be drawn about whether the Weinberg model is ruled or not, since nothing is known about the neutral-Higgs-boson mixing,[38] this can be interpreted as an equation for the upper bound on $g^{(1)}g^{(2)}$. The current experimental limit sets

$$g^{(1)}g^{(2)} \le 4.5\times 10^{-2}\ . \tag{40}$$

B. Strong *CP*

The theoretical constraint for the effective strong *CP*-violating parameter $\bar\theta=\theta_{\text{QCD}}+\theta_{\text{QFD}}$ (QFD denotes quantum flavor dynamics) inferred from the upper bound on D_n is $\bar\theta < 10^{-9}$ (Ref. 39). In models in which *CP* symmetry is spontaneously broken, $\theta_{\text{QCD}}=0$ appears naturally since the initial Lagrangian is *CP* invariant. Also

$$\theta_{\text{QFD}}(=\arg\det M_R = \arg\det(M_U)_R + \arg\det(M_D)_R)$$

is finite and calculable to all orders in the perturbation theory. At the tree level

$$\bar\theta_{\text{tree}} = N(\theta_1-\theta_2)\ , \tag{41}$$

where θ_1 and θ_2 are the phases of the VEV of ϕ_1 and ϕ_2, respectively, and N is the number of generations. To have weak *CP* violation, $(\theta_1-\theta_2)$ cannot be set to $m\pi/2$, with m integer.[6] Thus even at the zeroth order, $\bar\theta$ in the Weinberg model is of order unity. Even if one imposes a discrete symmetry on the Higgs potential so that $N(\theta_1-\theta_2)=n\pi (n\neq 0)$ and thus strong *CP* is conserved, one still has trouble with the radiative corrections to $\bar\theta$. At the one-loop level, the exact expression of the radiative correction from the charged Higgs bosons is given in Ref. 42. From Eq. (36), it follows that numerically

$$\bar\theta_{1\,\text{loop}} \sim 3\times 10^{-3} \tag{42}$$

which is also too large at the one-loop order. In fact, this is a general feature for spontaneous *CP*-violating models, though $\theta_{\text{QCD}}=0$ and θ_{QFD} is finite and calculable, $\bar\theta$ usually turns out too large unless one imposes an additional *ad hoc* discrete symmetry (for instance, the Peccei-Quinn symmetry, as done in Ref. 43) or enlarges the gauge group to protect a small $\bar\theta$. Before proceeding, we comment that the contribution of the neutral Higgs bosons to $\bar\theta_{1\,\text{loop}}$ is roughly of the same order as that of the charged one.

C. *T*-odd correlation in $K_{\mu 3}$ decays

A test of the *T*-odd correction is the transverse polarization of the muon in $K_{\mu 3}$ decays. The degree of the muon transverse polarization is usually expressed in terms of the parameter $\xi=f_-/f_+$, where f_+ and f_- are the form factors of the vector and scalar parts, respectively, of the $K_{\mu 3}$ amplitude. The imaginary part of ξ governs the size of the *T*-odd polarization. In the Weinberg model it is given by[9,43]

$$\text{Im}\xi = -\frac{m_K{}^2}{2\sqrt{2}\,m_0{}^2}\frac{v_2{}^2}{v_3{}^2}\ , \tag{43}$$

where v_2 and v_3 are the vacuum expectation values of the Higgs fields ϕ_2 and ϕ_3, whose neutral components couple only to down quarks and leptons, respectively. From Eqs. (31), (33), and with $v_3 > v_1 \sim v_2$ we obtain an upper bound

$$\text{Im}\xi \le 5\times 10^{-4} \tag{44}$$

while experimentally[44]

$$K_{\mu 3}^{+}:\ \text{Im}\xi = -0.017\pm 0.025\ ,$$

$$K_{\mu 3}^{0}:\ \text{Im}\xi = -0.020\pm 0.022\ .$$

This *T*-violating polarization is a unique prediction of the Weinberg model; all other known viable models of *CP* nonconservation predict a null result at the tree level.[36]

D. $K\to 3\pi$

Effects of *CP* nonconservation in $K\to 3\pi$ decays can be studied using the standard current-algebra technique to relate to the better known *CP*-violating $K\to 2\pi$ decays. It has been shown that[45,46] as long as the commutation relation

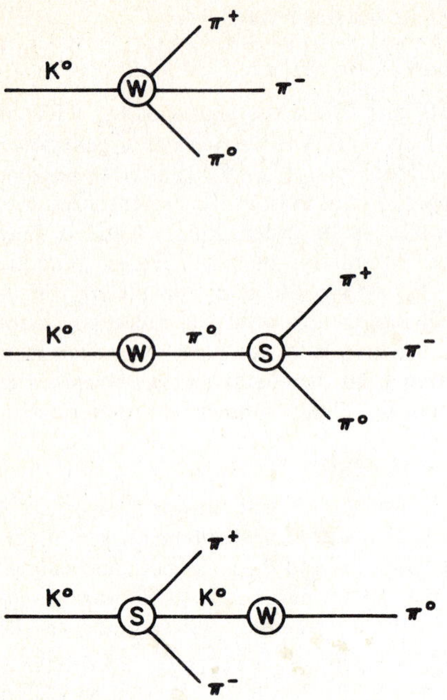

FIG. 1. $K^0 \to \pi^+\pi^-\pi^0$ decay due to the Lagrangian (9). The $S(W)$ denotes a strong (weak) vertex.

$$[Q_5^3, H^{\Delta S=1}] = [Q^3, H^{\Delta S=1}] \quad (45)$$

holds for both *CP*-conserving and -violating interactions, then at the center of the Dalitz plot (i.e., $s_1 = s_2 = s_3$)

$$\eta_{+-0}(0) - \epsilon = -2\epsilon' e^{i(\delta_2 - \delta_0)} . \quad (46)$$

The above result holds in the KM model since only the left-handed quarks are involved in weak interactions. In the Weinberg model *CP* violation in the decay amplitude $K \to 3\pi$ also vanishes to lowest order in chiral symmetry (i.e., for vertices independent of the momentum) as in the case of $K \to 2\pi$ decays due to the FKW theorem. A direct calculation of Fig. 1 using Eqs. (14), (15), and (19) shows

$$A(K^0(k) \to \pi^+(k_1)\pi^-(k_2)\pi^0(k_3)) \propto k \cdot (k_3 - k_2)/(m_K{}^2 - m_\pi{}^2) . \quad (47)$$

The momentum-independent terms are thus canceled out. Again, higher-order chiral terms and SU(3) breaking can generate *CP* violation in the decay amplitude at the center of the Dalitz plot. Since the *CP*-violating Lagrangian (9) does respect the commutation relation (45) to lowest order in chiral perturbation theory, the result (46) is valid in the Weinberg model of *CP* nonconservation. Of course, ϵ' and hence η_{+-0} in Eq. (46) is model dependent.

IV. CONCLUSIONS

Motivated by the recent work on the reanalyses of ϵ'/ϵ in the Weinberg model of *CP* violation, we have reexamined the chiral properties of the *CP*-violating weak amplitudes. In $K \to 2\pi$ decays there is a new pole contribution which arises from the combination of a strong $K\pi$ vertex and a K-vacuum tadpole. The strong $K\pi$ scattering amplitude depends on the chiral-symmetry nonlinear realization. However, the weak $K \to 2\pi$ transition is also realization dependent (so is the soft-pion relation), which renders the physical $K \to 2\pi$ amplitude nonlinear-realization independent, consistent with the Chisholm theorem. Consequently, the *CP*-odd amplitude vanishes to lowest order in the low-energy expansion of chiral perturbation theory. Although this is well known to experts in the field, the realistic calculation of the lowest-order amplitude has not been treated correctly in the literature. This is the primary purpose of this paper to clarify this problem and to show how to obtain the correct results. It is also pointed out that the vacuum-insertion method works only in the realization scheme in which $a_3 = 2$.

From the experimental value of ϵ, some constraints can be set on the Higgs-boson mixing and *CP*-violating angles and on the relative magnitudes of the vacuum expectation values. It turns out not all three vacuum expectation values are of the same order of magnitude. In fact, this is the starting assumption of Ref. 31 to obtain a small ϵ'/ϵ.

We also have reexamined various *CP*-violating effects which can be manifested in this model, these are the electric dipole moment of a neutron, the strong *CP*-violating parameter θ, the transverse polarization of the muon in $K_{\mu 3}$ decays, *CP* violation in the decay amplitude of $K \to 3\pi$ decays, and the charge asymmetry in $K_L \to 3\pi$ decays. The result $\epsilon'/\epsilon \approx -0.007$ is barely compatible with the present experimental measurements. The electric dipole moment of the neutron due to the neutral-Higgs-boson exchange and the current experimental limit are used to set an upper bound for the neutral-Higgs-boson mixing. The strong *CP*-violating parameter θ is too large (of order unity) at the tree level and does not get improved at the one-loop correction. As in the KM model, this difficulty may be avoided by imposing the Peccei-Quinn symmetry (so there would be four Higgs doublets), as already done in Ref. 43. A direct calculation also shows that no on-shell $K \to 3\pi$ transition amplitude can be induced by the *CP*-odd Lagrangian to lowest order in chiral symmetry, in accordance with the Feinberg-Kabir-Weinberg theorem.

ACKNOWLEDGMENTS

I wish to thank Dr. Ling-Lie Chau for valuable discussions. This research was supported in part by DOE contract DE-AC03-ER03230 and the Chester Davis fellowship.

---

*Present address: Physics Department, Indiana University, Bloomington, IN 47405.

[1]S. Weinberg, Phys. Rev. Lett. 37, 657 (1976).

[2]J. F. Donoghue and B. R. Holstein, Phys. Rev. D 32, 1152 (1985); J. F. Donoghue, in *Proceedings of the Santa Fe Meeting of the Division of Particles and Fields of the American Physical Society, 1984,* edited by T. Goldman and M. M. Nieto (World Scientific, Singapore, 1985).

[3]Y. Dupont and T. N. Pham, Phys. Rev. D **28**, 2169 (1983).

[4]A. I. Sanda, Phys. Rev. D **23**, 2647 (1981); N. G. Deshpande, *ibid.* **23**, 2654 (1981).

[5]A. A. Anselm, V. E. Bunakov, V. P. Gudkov, and N. G. Uraltsev, Pis'ma Zh. Eksp. Teor. Fiz. **40**, 231 (1984) [JETP Lett. **40**, 1102 (1984)]; Phys. Lett. **152B**, 116 (1985).

[6]G. C. Branco, Phys. Rev. Lett. **44**, 504 (1980); Phys. Rev. D **22**, 2901 (1980).

[7]See, e.g., L. L. Chau and W. Y. Keung, Phys. Rev. D **29**, 592 (1984).

[8]C. Albright, J. Smith, and S.-H. H. Tye, Phys. Rev. D **21**, 711 (1980); K. Shizuya and S.-H. H. Tye, *ibid.* **23**, 1613 (1981).

[9]H. Y. Cheng, Phys. Rev. D **26**, 143 (1982). In Eqs. (3.25) and (3.27) of this reference $2\sqrt{2}/m_0^2$ should be replaced by $1/(2\sqrt{2}m_0^2)$.

[10]L. L. Chau, Phys. Rep. **95**, 1 (1983).

[11]J. F. Donoghue, J. S. Hagelin, and B. R. Holstein, Phys. Rev. D **25**, 195 (1982).

[12]Y. Dupont, Phys. Lett. **125B**, 211 (1983).

[13]This expression of $\tilde{f}$ is derived in the four-quark model; the generalization of the six-quark case is given in Ref. 11. For $m_{H_2}=100$ GeV and $m_{H_1}=20$ GeV, the t quark only contributes about 4% to $\tilde{f}$. Hence, Eq. (10) is sufficient for our purposes.

[14]J. A. Cronin, Phys. Rev. **161**, 1483 (1967).

[15]J. S. R. Chisholm, Nucl. Phys. **26**, 469 (1961).

[16]G. Feinberg, P. K. Kabir, and S. Weinberg, Phys. Rev. Lett. **3**, 527 (1959).

[17]At the quark level the FKW theorem states that the Lagrangian $\bar{q}[A_1+\gamma_5A_2+(B_1+i\gamma_5B_2)\not{p}]q$ gives the same S-matrix elements to all orders as the Lagrangian in which the mass matrix is diagonalized. In chiral perturbation theory, the effect of a $(\bar{3},3)$ or $(3,\bar{3})$ interaction is guaranteed to be transformed away only to lowest-order expansion. As an application of the FKW theorem, no $K\to 2\pi$ transition can be induced by the interaction $\Theta=\bar{s}(1-\gamma_5)d$; the weak amplitude due to Θ is exactly canceled by the pole contribution. In the literature it is often misclaimed that $\langle 2\pi|\Theta|K\rangle$ vanishes because of the equation of motion for $\bar{s}d$ and conservation of momentum.

[18]C. Bernard, T. Draper, A. Soni, H. D. Politzer, and M. B. Wise, Phys. Rev. D **32**, 2343 (1985); J. Bijnens, Phys. Lett. **152B**, 226 (1985).

[19]This was first observed by Shifman, Vainshtein, and Zakharov (Ref. 20) that the penguin amplitude vanishes in the chiral limit if it is evaluated in the vacuum insertion with no momentum dependence (i.e., only vertices independent of the momentum are used, see also Ref. 21). This feature is easily seen in the effective chiral Lagrangian approach (Ref. 22).

[20]A. I. Vainshtein, V. I. Zakharov, and M. A. Shifman, Zh. Eksp. Teor. Fiz. **72**, 1275 (1977) [Sov. Phys. JETP **45**, 670 (1977)].

[21]J. F. Donoghue, Phys. Rev. D **30**, 1499 (1984); B. Gevela, A. Le Yaouanc, L. Oliver, O. Pène, and J. C. Raynal, Phys. Lett. **148B**, 225 (1984).

[22]T. N. Pham, Phys. Lett. **145B**, 113 (1984).

[23]A. Manohar and H. Georgi, Nucl. Phys. **B234**, 189 (1984); J. F. Donoghue, E. Golowich, and B. R. Holstein, Phys. Rev. D **30**, 587 (1984).

[24]J. Gasser and H. Leutwyler, Ann. Phys. (N.Y.) **158**, 142 (1984).

[25]D. Chang, Phys. Rev. D **25**, 1318 (1982); J. Hagelin, Phys. Lett. **117B**, 441 (1982).

[26]L. L. Chau and H. Y. Cheng, Phys. Rev. Lett. **54**, 1768 (1985).

[27]Our results for ϵ'/ϵ are numerically the same as Ref. 3, but disagrees with the first reference of Ref. 28. This may be due to the fact that the parameter ρ used in Ref. 28 is negative.

[28]A. I. Sanda, Report No. RU85/B/112, 1985 (unpublished); J. M. Frere, J. Hagelin, and A. I. Sanda, Phys. Lett. **151B**, 161 (1985).

[29]J. K. Black *et al.*, Phys. Rev. Lett. **54**, 1628 (1985); R. H. Bernstein *et al.*, *ibid.* **54**, 1631 (1985).

[30]This is the equation used by Ref. 31 for evaluating ϵ'/ϵ.

[31]G. C. Branco, A. J. Buras, and J.-M. Gerard, Phys. Lett. **155B**, 192 (1985); Nucl. Phys. **B259**, 306 (1985); J.-M. Gerard, Report No. MPI-PAE/PTh 5/85, 1985 (unpublished).

[32]C. A. Blocker *et al.*, Phys. Rev. Lett. **49**, 271 (1982); W. Bartel *et al.*, Phys. Lett. **114B**, 211 (1982); B. Adeva *et al.*, *ibid.* **115B**, 345 (1982).

[33]A. A. Anselm and D. I. Dyakonov, Nucl. Phys. **145**, 271 (1978).

[34]G. Beall and N. G. Deshpande, Phys. Lett. **132B**, 427 (1984).

[35]I. S. Altarev *et al.*, Phys. Lett. **102B**, 13 (1981); J. M. Pendlebury *et al.*, *ibid.* **136B**, 327 (1984).

[36]H. Y. Cheng, Phys. Rev. D **28**, 150 (1983).

[37]N. G. Deshpande and E. Ma, Phys. Rev. D **16**, 1583 (1977).

[38]A neutral Higgs boson can also contribute to the *CP*-violating $\eta\to 2\pi$ decays. However, no information on the neutral-Higgs-boson mixing can be contracted from the available experimental results.

[39]V. Baluni, Phys. Rev. D **19**, 2070 (1979); R. J. Crewther, P. di Vecchia, G. Veneziano, and E. Witten, Phys. Lett. **88B**, 123 (1979).

[40]A. A. Anselm and N. G. Uraltsev, Yad. Fiz. **30**, 465 (1979) [Sov. J. Nucl. Phys. **30**, 240 (1979)].

[41]H. Y. Cheng, Z. Phys. C **24**, 93 (1984).

[42]Akhoury and I. I. Bigi, Nucl. Phys. **B234**, 459 (1984).

[43]A. R. Zhitnitsky, Yad. Fiz. **31**, 1024 (1980) [Sov. J. Nucl. Phys. **31**, 529 (1980)].

[44]Particle Data Group, Rev. Mod. Phys. **56**, S1 (1984).

[45]L. F. Li and L. Wolfenstein, Phys. Rev. D **21**, 178 (1980).

[46]H. Y. Cheng, Phys. Lett. **129B**, 357 (1983).

Left-right gauge symmetry and an "isoconjugate" model of *CP* violation*

Rabindra N. Mohapatra and Jogesh C. Pati
Department of Physics and Astronomy, University of Maryland, College Park, Maryland 20742
(Received 16 April 1974)

Left-right symmetry in the starting gauge interactions provides the basis for an "isoconjugate" model of *CP* violation with the consequence that $\eta_{+-}=\eta_{00}$. The magnitude of *CP* violation is naturally suppressed at least to the extent that $V+A$ interactions are suppressed in nature compared to the $V-A$ interactions. This allows the possibility that intrinsic *CP*-violating phase (arising through spontaneous symmetry breaking, for example) may have a maximal character, which may reveal itself at intermediate high energies to possibly disappear at still higher energies. *CP* violation in leptonic, semileptonic, and $\Delta Y=0$ parity-violating nonleptonic interactions (which could contribute to the electric dipole moment of the neutron) arise on the one hand in fourth order of the weak gauge interactions and on the other via Yukawa interactions between fermions and leftover Higgs mesons as well as via W_L-W_R mixing. The magnitude of the latter contributions may be limited to be less than or of order $G_F\times 10^{-4}$ in order that the experimentally observed relation $\eta_{+-}=\eta_{00}$ may hold at least to a few percent. Thus the electric dipole moment of the neutron d_n is expected to be less than or of order 10^{-24} e cm. *CP* violation in all $|\Delta S|=1$ nonleptonic decays ($Y\to N+\pi$, etc.) should, in general, have the same order of magnitude as $|\eta_{+-}|$.

The desirability of an *"isoconjugate"* relation of the form $[I_3, P^{(-)}]=\frac{1}{2}i\tan\xi\, P^{(+)}$, where $P^{(+)}$ and $P^{(-)}$ are the CP-even and CP-odd interactions contributing to $K_{L,S}\to 2\pi$ decays and I_3 is the third component of the $\vec{I}$-spin generator, has been remarked upon.[1] It was shown that the class of theories satisfying the above equation automatically satisfy[2] the relation $\eta_{+-}=\eta_{00}$, with the phase ϕ_{+-} given to a good approximation[3] by $\tan^{-1}(2\Delta m/\Gamma_S)$, where $\Delta m = m_{K_L}-m_{K_S}$ and Γ_S is the width of the short-lived kaon. Both these results are compatible with experiments. The purpose of this note is to point out that a particularly simple scheme of CP violation distinct from previous attempts[4-6] and satisfying the said relation may be realized within a gauge theory of the weak, electromagnetic, *and* strong interactions by assuming that the left- and right-handed gauges[7] enter into the theory in a *symmetrical* fashion,[8] the corresponding currents being coupled to *distinct* gauge mesons W_L and W_R. The observed left-right asymmetry with predominance of left-handed $V-A$ interactions at low energies and breakdown of CP symmetry will arise in such a scheme due to heavier masses of the W_R gauge mesons compared to those of W_L's and the complex character of the fermion mass matrix, respectively, both of which in turn will be attributed to spontaneous symmetry breaking. The main results of such a scheme are the following:

(1) *Regardless* of the detailed structure of the mass matrix, its complex nature transcribed into the gauge interactions straightforwardly leads to the isoconjugate scheme of CP violation mentioned above and hence to the relation $\eta_{+-}=\eta_{00}$.

(2) The magnitude of CP violation is naturally suppressed at least to the extent that the right-handed $V+A$ interactions are known to be suppressed compared to the left-handed ones at present energies. Thus it is possible to allow "intrinsic" CP violation to bear a maximal character.

(3) CP violation in $|\Delta S|=1$ processes (i.e., $K_L\to 2\pi$, $Y\to N+\pi$, etc.) should in general exhibit milliweak strength while that in $\Delta S=0$ processes may be considerably weaker than milliweak. Thus the electric dipole moment of the neutron is predicted to be less than or of order 10^{-24} e cm.

We exhibit such a scheme of CP violation within a gauge theory of the weak, electromagnetic, and strong interactions based on the anomaly-free gauge group[8]

$$\mathcal{G}_0 \equiv SU(2)_L\times SU(2)_R\times SU(3)'_{L+R}\times U(1)'_{L+R}\ ,$$

whose one distinctive feature is its left-right symmetry. Its physical realization may be provided by a sixteenplet of left and sixteenplet of right fermions $\Psi_{L,R}$ comprising quarks and leptons:

$$\Psi_{L,R}=\begin{bmatrix}\mathcal{P}_a^0 & \mathcal{P}_b^0 & \mathcal{P}_c^0\\ \mathfrak{N}_a^0 & \mathfrak{N}_b^0 & \mathfrak{N}_c^0\\ \lambda_a^0 & \lambda_b^0 & \lambda_c^0\\ \chi_a^0 & \chi_b^0 & \chi_c^0\end{bmatrix}_{L,R}+\begin{bmatrix}\nu_e=\mathcal{P}_d^0\\ e^-=\mathfrak{N}_d^0\\ \mu^-=\lambda_d^0\\ \nu_\mu=\chi_d^0\end{bmatrix}_{L,R}, \qquad (1)$$

$$\Psi_L=(2+2,1,3,1)+(2+2,1,1,1)\,,$$
$$\Psi_R=(1,2+2,3,1)+(1,2+2,1,1)\,, \qquad (2)$$

where the representations are characterized by their dimensions. The superscript zeros designate that the corresponding fields are linear combinations of the physical fields to be defined by the mass matrix. The gauge groups $SU(2)_{L,R}$ generate chiral weak interactions, while $SU(3)'_{L+R}$ generates vector strong[9] interactions acting on the 3-component color index (a, b, c) of the quarks (with either integer or fractional charges). The Abelian $U(1)'$ generates[10] a vector gauge interaction, which needs to be introduced to generate electromagnetism.

Our considerations here would be equally applicable to the extended gauge group[8]

$$\mathcal{G}=SU(2)_L\times SU(2)_R\times SU(4)'_{L+R}\,,$$

comprising left-right symmetry as well as baryon-lepton unification. [SU(4)′ operates on the four-color index (a, b, c, d).] We conjecture that a qualitative link between breakdown of CP invariance and and tiny but nonzero neutrino[11] masses may emerge in such an extended scheme, which may be considered elsewhere.

For considerations of CP violation, we will be concerned only with the fermion–gauge-boson weak interaction generated by the local group $SU(2)_L \times SU(2)_R$ and the structure of the fermion mass matrix. The former is given by

$$\mathcal{L}_I=\frac{g_L}{2}\sum_{\alpha=a,b,c,d}\sum_{i=1,2}(\bar{\psi}_i^\alpha)_L\vec{\tau}\gamma_\mu(\psi_i^\alpha)_L(\vec{W}_\mu)_L +(L\to R)\,, \qquad (3)$$

where

$$\psi_1^\alpha=\begin{pmatrix}\mathcal{P}^0_\alpha\\ \mathfrak{N}^0_\alpha\end{pmatrix},\quad \psi_2^\alpha=\begin{pmatrix}\chi^0_\alpha\\ \lambda^0_\alpha\end{pmatrix}. \qquad (4)$$

Note the left-right *symmetric* nature of the gauge interactions.

For simplicity of writing we consider below the mass matrix of the baryonic quartets only; similar considerations[12] apply to the lepton quartet. Furthermore, we suppress the color index (a, b, c); it is to be understood that identical Yukawa coupling constants and mass parameters apply for all three colors if the relevant Higgs multiplet is a color singlet. A complex mass matrix for the fermions may arise in two alternative ways:

(1) The simplest possibility is that the Higgs-meson–fermion Yukawa coupling constants are complex. Introduce a suitable multiplet $\Phi=(2,2,1,1)$ whose vacuum expectation value may be taken to be

$$\langle\Phi\rangle=\begin{pmatrix}\kappa & 0\\ 0 & \kappa'\end{pmatrix},$$

κ and κ' being real. The general Yukawa coupling permitted by gauge invariance is given by

$$\sum_{i,j=1,2}\bar{\psi}_{iL}(h_{ij}\phi+f_{ij}\tilde{\phi})\psi_{jR}+\text{H.c.}\,,$$

where $\tilde{\phi}=\tau_2\phi^*\tau_2$. This leads to the mass matrix

$$\left[(\bar{\mathcal{P}}^0\bar{\chi}^0)_L\begin{pmatrix}m_{11} & m_{12}\\ m_{21} & m_{22}\end{pmatrix}\begin{pmatrix}\mathcal{P}^0\\ \chi_0\end{pmatrix}_R +(\bar{\mathfrak{N}}^0\bar{\lambda}^0)_L\begin{pmatrix}M_{11} & M_{12}\\ M_{21} & M_{22}\end{pmatrix}\begin{pmatrix}\mathfrak{N}^0\\ \lambda^0\end{pmatrix}_R\right]+\text{H.c.}\,, \qquad (5)$$

where $m_{ij}=\kappa h_{ij}+\kappa' f_{ij}$ and $M_{ij}=\kappa f_{ij}+\kappa' h_{ij}$. In general all h_{ij}'s and f_{ij}'s are nonzero and complex.

(2) Alternatively, and perhaps more elegantly, a complex mass matrix may also arise through complex vacuum expectation values of scalar fields. In this case, it is necessary to introduce at least two[6] distinct Higgs multiplets, ϕ_1 and ϕ_2, each transforming in the same manner[13] [i.e., (2, 2, 1, 1)]. One may choose (with appropriate choice of potential)

$$\langle\phi_{1,2}\rangle=\begin{pmatrix}\kappa_{1,2}e^{i\xi_{1,2}} & 0\\ 0 & \kappa'_{1,2}e^{i\xi'_{1,2}}\end{pmatrix}. \qquad (6)$$

Allowing for the general gauge-invariant Yukawa interaction,

$$\sum_{i,j}\bar{\psi}_{iL}(h_{ij}\phi_1+h'_{ij}\phi_2+f_{ij}\tilde{\phi}_1+f'_{ij}\tilde{\phi}_2)\psi_{jR}+\text{H.c.}\,,$$

one may again obtain a complex fermion mass matrix of the form (5) with *real* Yukawa coupling constants. The major difference between alternatives (1) and (2) is that CP violation will have a genuine spontaneous origin in the case (2). (The two cases may be distinguishable at higher energies; see remarks later.)

Turning to the mass matrix of the form (5), physical $\mathcal{P}$ and χ fields are obtained by transforming the corresponding (2×2) complex matrix into real diagonal form[14]:

$$U_L\begin{pmatrix}m_{11} & m_{12}\\ m_{21} & m_{22}\end{pmatrix}V_R^{-1}=\begin{pmatrix}m_{\mathcal{P}} & 0\\ 0 & m_\chi\end{pmatrix}, \qquad (7)$$

where U_L and V_R define U(2) transformations on $(\mathcal{P}^0_L, \chi^0_L)$ and $(\mathcal{P}^0_R, \chi^0_R)$ bases, respectively. Even though U_L and V_R define U(2) transformations, the physical consequences of such transformations may be seen to be equivalent to SU(2) transformations of the form

$$\begin{pmatrix} \mathcal{P}^0 \\ \chi^0 \end{pmatrix}_{L,R} = \begin{pmatrix} \cos\theta_{L,R} & -\sin\theta_{L,R}\, e^{i\delta_{L,R}} \\ \sin\theta_{L,R}\, e^{-i\delta_{L,R}} & \cos\theta_{L,R} \end{pmatrix} \begin{pmatrix} \mathcal{P} \\ \chi \end{pmatrix}_{L,R}, \tag{8}$$

where $\theta_{L,R}$ and $\delta_{L,R}$ are real parameters. Similar transformations will also apply to the $(\mathfrak{N}^0, \lambda^0)_{L,R}$ quarks and observable consequences will be given by the combined transformations in $(\mathcal{P}^0, \chi^0)$ and $(\mathfrak{N}^0, \lambda^0)$ spaces. For ease of writing and without loss of generality, assume that $(\mathfrak{N}^0, \lambda^0)$ are physical fields, i.e., $M_{12} = M_{21} = 0$ and M_{11} and M_{22} are real, so that $(\mathfrak{N}^0, \lambda^0)_{L,R} = (\mathfrak{N}, \lambda)_{L,R}$. Substituting Eq. (7) into Eq. (3), and suppressing the color index (a, b, c), the fermion-weak-gauge-boson interactions take the form

$$\mathcal{L}_I = \frac{g_L}{\sqrt{2}}[\bar{\mathcal{P}}_L\gamma_\mu(\mathfrak{N}_L\cos\theta_L + e^{i\delta_L}\lambda_L\sin\theta_L)\,W^+_{\mu L} + \bar{\chi}_L\gamma_\mu(-e^{-i\delta_L}\mathfrak{N}_L\sin\theta_L + \lambda_L\cos\theta_L)\,W^+_{\mu L}] + \text{H.c.} + (L \to R) + \mathcal{L}_{\text{neutral}} + \mathcal{L}_{\text{lepton}}, \tag{9}$$

where $\mathcal{L}_{\text{neutral}}$ is composed of neutral-current interactions, is strangeness conserving, and is CP-even. Note that due to $\kappa\kappa'$ interference, which gives rise to W_L-W_R mixing, the $W^{\pm}_{L,R}$ fields written above are not really the eigenstates of the mass matrix. The mixing term is given by $g_L g_R \kappa\kappa'$ for case (1), and a similar term is given for case (2). For our considerations, we will consistently *assume* that the net effect of such mixing is small and neglect such a mixing. This amounts to choosing κ' to be at least an order of magnitude smaller than κ.[15]

The $|\Delta S| = 1$ charm-conserving effective nonleptonic Hamiltonian following from Eq. (9) is then given by

$$\mathcal{H}^{|\Delta S|=1}_{\text{wk}} = (f_L e^{-i\delta_L} + f_R e^{-i\delta_R})S + (f_L e^{-i\delta_L} - f_R e^{-i\delta_R})P + \text{H.c.}, \tag{10}$$

where

$$\begin{aligned} S &= (V_\pi - V_{K^+} + A_\pi - A_{K^+})_{\text{symm}} - (V'_\pi - V'_{K^+} + A'_\pi - A'_{K^+})_{\text{symm}}, \\ P &= (V_\pi - A_{K^+} + A_\pi - V_{K^+})_{\text{symm}} - (V'_\pi A'_{K^+} + A'_\pi - V'_{K^+})_{\text{symm}}; \end{aligned} \tag{11}$$

here $V_{\pi^-} = \bar{\mathcal{P}}\gamma_\mu\mathfrak{N}$, $A_{\pi^-} = \bar{\mathcal{P}}\gamma_\mu\gamma_5\mathfrak{N}$, $V'_{\pi^-} = \bar{\chi}\gamma_\mu\mathfrak{N}$, $V_{K^+} = \bar{\lambda}\gamma_\mu\mathcal{P}$, etc. We define $(V_\pi - V_{K^+})_{\text{symm}} \equiv (V_\pi - V_{K^+} + V_{K^+}V_{\pi^-})$, and $f_{L,R} \equiv g_{L,R}{}^2 \sin 2\theta_{L,R}/4 m_{W_{L,R}}{}^2$. Redefining the phase of the λ field[16] (which has no observable consequence), we may express (10) as

$$H^{|\Delta S|=1}_{\text{wk}} = S^{(+)} + [P^{(+)} + P^{(-)}], \tag{12}$$

where

$$\begin{aligned} S^{(+)} &= G_S(S + S^\dagger), \\ P^{(+)} &= G_P\cos\xi(P + P^\dagger), \quad P^{(-)} = iG_P\sin\xi(P - P^\dagger), \\ G_S &= f_L(1 + \epsilon^2 + 2\epsilon\cos\delta)^{1/2}, \quad \delta = \delta_R - \delta_L \\ G_P &= f_L(1 + \epsilon^2 - 2\epsilon\cos\delta)^{1/2}, \\ \tan\xi &= \frac{2\epsilon\sin\delta}{1-\epsilon^2}, \quad \epsilon = (f_R/f_L). \end{aligned} \tag{13}$$

Clearly the interaction given by (12) is in general CP-violating. Note that it would be CP-conserving if either f_L or f_R were zero and/or $\delta = \delta_L - \delta_R = 0$. The scheme possesses CP violation not only through the gauge interactions (12), but also through the Yukawa interaction between fermions and leftover Higgs mesons. We will *assume* that the effects of the latter are suppressed compared to those of the former by the masses of the Higgs mesons (see estimates later). We are led to make such an assumption in view of the simple results (i.e., $\eta_{+-} = \eta_{00}$) that follow from Eq. (12) in agreement with experiments, which will be affected in general in the presence of a Higgs-boson contribution.

We now remark that the parity-violating CP-even $(P^{(+)})$ and CP-odd $(P^{(-)})$ operators satisfy the desired relation

$$[I_3, P^{(-)}] = \tfrac{1}{2}i\tan\xi\, P^{(+)} \tag{14}$$

and hence[2] $\eta_{+-} = \eta_{00}$. Using Eq. (14) we may also *calculate* η_{+-}, which for small $\tan\xi$ is given[17] by

$$\eta_{+-} \simeq i\tan\xi(1-y), \tag{15}$$

where $y = A_2/(A_1 + A_2)$, with $A_i \equiv \int dt\,\langle K^0|T(\Omega_i(t)\Omega_i(0))|\bar{K}^0\rangle$ and $\Omega_1 = S^{(+)}$, $\Omega_2 = P^{(+)}$. Thus y is a *complex* number with magnitude of order 1 and, barring accidental cancellation, $|\eta_{+-}|$ is of order $\tan\xi$. Thus [see Eq. (13)], the suppression of CP violation may arise entirely due to the smallness of (f_R/f_L), leaving the possibility that "intrinsic" CP violation in weak processes may be maximal (i.e., $\delta = \frac{1}{2}\pi$, say), which ought to manifest itself at intermediate high energies (i.e., energies comparable to M_{W_R}). At still higher energies and momentum transfers ($\gg M_{W_R}$), one may expect CP violation [if it is of spontaneous origin, as in case (2)] as well as parity violation (if the bare coupling constants $g_L^{(0)}$ and $g_R^{(0)}$ are equal) to disappear[18] altogether.

We should remark that we have chosen the phase of the λ field[16] such that CP violation in the $|\Delta S| = 1$ nonleptonic Hamiltonian is entirely in the parity-violating part [see Eq. (12)]. With this choice of phase, CP violation in $K_{L,S} \to 3\pi$ decays arises only through the contributions of $P^{(-)}$ to the $(K_1$-$K_2)$ mass matrix, and one thus obtains the relation $\eta_{+-0} = \eta_{000} = \rho$ as in a superweak theory, where ρ is the CP-even mixing parameter in K_L, i.e.,

$$K_L = (K_2 + \rho K_1)/(1 + |\rho|^2)$$

and

$$\eta_{ijk} = M(K_S^0 \to \pi^i \pi^j \pi^k)/M^0(K_L^0 \to \pi^i \pi^j \pi^k).$$

Alternatively, we may have chosen the phase of the λ field such that CP violation would be shifted entirely into the parity-conserving sector with a relation analogous to Eq. (14) between $S^{(-)}$ and $S^{(+)}$. In this case CP violation in $K_{L,S} \to 2\pi$ decays would arise only through the mass matrix, while that in $K_{L,S} \to 3\pi$ decays would arise both through the mass matrix and the decay amplitudes. All observable consequences are, of course, *independent* of the choice of phase. It is worth noting that *unlike* the superweak theory, in the present scheme $\eta_{+-0} \neq \eta_{+-}$. In fact (see Ref. 2), one obtains in the present scheme[19] $(\eta_{+-} - \eta_{+-0}) = [(R + \rho)/(1 - \rho R)] - \rho = R = i \tan\xi$ (neglecting ρR compared to unity).

It is easy to see from Eq. (9) that the CP-violating part of the gauge interactions is purely[20] $\Delta S = 1$ and therefore its contribution to the electric dipole moment of the neutron d_n is small ($\sim e G_F^2 \sin^2\theta \sin\xi\, m_p^3 \approx 10^{-29}$ e cm). The Higgs-boson interactions contain $\Delta Y = 0$ P- and T-violating parts. However, as their contributions affect the relation $\eta_{+-} = \eta_{00}$, which is experimentally valid up to 5–10%, one may expect the effective strength of such interactions (h^2/m_σ^2) to be[21] $\lesssim G_F \times 10^{-4}$. Thus their contributions to d_n is of order $e h^2 m_p / m_\sigma^2 \lesssim 10^{-24}$ e cm.

Analogous to $\Delta Y = 0$ nonleptonic transitions, CP violation in leptonic and semileptonic processes (including $\Delta Q = 0$ and $|\Delta S| = 1$ transitions) may arise in the scheme only via exchange of Higgs mesons and W_L-W_R mixing, whose effective strengths are, however, limited to be $\lesssim G_F \times 10^{-4}$ (for reasons mentioned above).[20,21]

In summary, from a theoretical point of view, the unique feature of the scheme is the isoconjugate relation (14), and we stress that the fact that the left- and right-handed currents enter into the theory in a *symmetrical* fashion, being coupled to *distinct* gauge mesons W_L and W_R, has played an *essential* role[22,23] in its realization. Having assumed that the leftover Higgs mesons are sufficiently massive, the only parameter of the scheme relevant for CP violation at present energies is $\tan\xi$.

We should remark that parameters of the fermion mass matrix, which characterize the scales of SU(2), SU(3), SU(4), and chiral $SU(4)_L \times SU(4)_R$ breaking and Cabibbo rotation on the one hand, as well as CP violation on the other, are not at present calculable deviations from "natural" symmetries. This is a pressing problem and to our knowledge is not resolved in a satisfactory manner in any full-fledged realistic scheme. Perhaps the solution may lie in a better understanding of the mechanism of spontaneous symmetry breaking, in particular if it is dynamical. It is to be hoped that the eventual nature of such symmetry breaking, while it will constrain and thereby interrelate the parameters of the theory, does not alter the character of CP violation in the scheme as presented here. We should stress once again that the nature of CP violation in the scheme as reflected by Eqs. (9) and (14) follows straightforwardly from the gauge structure and does *not* depend upon precise values of the parameters (once the relevant parameters are nonzero). Furthermore, it should be noted that despite the arbitrariness in $\sin\delta$, the CP-violating parameter $\tan\xi \simeq (f_R/f_L)\sin\delta$ is *guaranteed to be small by the experimental fact that* (f_R/f_L) *is known to be small.*

As regards the experimental consequences, the present scheme is distinct from a superweak theory; the two schemes coincide only in the predictions $\eta_{+-} = \eta_{00}$ and $\eta_{+-0} = \eta_{000}$. They may be distinguished by measurements of (i) the phase ϕ_{+-}, (ii) the strength of CP violation in processes such as $Y \to N + \pi$ decays, and (iii) most important, the electric dipole moment of the neutron, if it turns out to lie between 10^{-24} and 10^{-28} e cm. The present scheme may also be distinguished from the usual milliweak theories (see Ref. 1 for example, and other references cited therein), since the latter generally predict a neutron electric dipole moment d_n of order 10^{-22} to 10^{-23} e cm. It is also distinct from other gauge-theory models[4-6] of CP violation, since it combines the predictions $\eta_{+-} = \eta_{00}$ with a *small* electric dipole moment of the neutron ($d_n \lesssim 10^{-24}$ e cm).[24]

Added note: Considerably after the submission of our paper it was brought to our attention by Professor L. Wolfenstein that a somewhat similar model of CP violation has also been proposed recently by Frenkel and Ebel.[25] There is one major difference: Our lepton sector is left-right symmetric as much as the hadron sector is; i.e., the right-handed leptons as well as the hadrons couple to W_R's; owing to this, if any U(1) is introduced into the gauge group to generate the photon, it turns out to be pure vector gauge and the complete gauge group is anomaly-free. In the case of Frenkel and Ebel the known leptons are coupled only to $SU(2)_L$ but not to $SU(2)_R$; hence their charges *need* contribution from an Abelian U(1) gauge, which is not pure vector. This leads to anomalies, unless one introduces heavy leptons to cancel the anomalies. Some of the consequences of the model in the hadronic sector, i.e., the isoconjugate relation and the exactness of $\eta_{+-}=\eta_{00}$, $\eta_{+-}=\eta_{000}$ as being properties of the gauge interactions of the model, are not noted in Ref. 25.

We thank Professor Abdus Salam for illuminating discussions on CP nonconservation and Professor L. Wolfenstein and Professor J. Prentki for many helpful comments.

---

*Work supported in part by the National Science Foundation under Grant No. NSF GP 20709.

[1]R. N. Mohapatra and J. C. Pati, Phys. Rev. D 8, 2317 (1973).

[2]To repeat the arguments of Ref. 1, take matrix elements of both sides of $[I_3, P^{(-)}]=\frac{1}{2}i\tan\xi P^{(+)}$ between $\langle\pi^i\pi^j|$ and $|K_1\rangle$ states, where $(i,j)=(+-)$ or $(0,0)$ and $CP|K_1\rangle=|K_1\rangle$ and $CP|K_2\rangle=-|K_2\rangle$. Since $I_3|\pi^i\pi^j\rangle=0$ and $I_3|K_1\rangle=-\frac{1}{2}|K_2\rangle$, the ratio $R\equiv\langle\pi^i\pi^j|P^{(-)}|K_2\rangle/\langle\pi^i\pi^j|P^{(+)}|K_1\rangle=i\tan\xi$ *independent* of (i,j), from which it follows that $\eta_{+-}=\eta_{00}=(R+\rho)/(1-\rho R)$, where ρ is the CP-even mixing parameter in K_L, i.e., $K_L=(K_2+\rho K_1)/(1+|\rho|^2)$.

[3]It holds to the extent that one may neglect $|3\pi\rangle$ compared to the $|2\pi\rangle$ real intermediate state in the ratio of off-diagonal to diagonal elements of the K_1-K_2 width matrix; on the other hand, $\tan\phi_{+-}=2\Delta m/\Gamma_s$ is an exact prediction of the superweak theory [L. Wolfenstein, Phys. Rev. Lett. 13, 562 (1964)].

[4]R. N. Mohapatra, Phys. Rev. D 6, 2023 (1972).

[5]A. Pais, Phys. Rev. D 8, 625 (1973).

[6]T. D. Lee, Phys. Rep. 9C, 143 (1974).

[7]Right-handed currents (without, however, left-right symmetry) were suggested in Ref. 4 for introducing CP violation into gauge theories.

[8]Such a left-right symmetric gauge structure has been proposed by J. C. Pati and Abdus Salam, Phys. Rev. Lett. 31, 661 (1973); Phys. Rev. D 10, 275 (1974). The details of the Higgs mechanism giving adequate masses (i.e., $m_{W_R}\gg m_{W_L}$, etc.) are given in this reference.

[9]J. C. Pati and Abdus Salam, Phys. Rev. D 8, 1240 (1973).

[10]The corresponding fermion current may be taken to be

$$(\tfrac{1}{24})^{1/2}\Big[\sum_{\alpha=a,b,c}(\bar{\mathcal{P}}^0_\alpha\mathcal{P}^0_\alpha+\bar{\mathfrak{N}}^0_\alpha\mathfrak{N}^0_\alpha+\bar{\lambda}^0_\alpha\lambda^0_\alpha+\bar{\chi}^0_\alpha\chi^0_\alpha)_{L+R}-3(\bar{\nu}_e\nu_e+\bar{e}e+\bar{\mu}\mu+\bar{\nu}_\mu\nu_\mu)_{L+R}\Big],$$

generated by the 15th generator of SU(4)′ (see Ref. 8).

[11]Note that CP violation in the q_α sector vanishes as $(m_{q_\alpha})_{\alpha=a,b,c}\to 0$ for any $q=(\mathcal{P},\mathfrak{N},\lambda,\text{ or }\chi)$, since one can independently redefine the phases of $q_{\alpha L}$ and $q_{\alpha R}$. On the other hand, quarks and lepton masses are related (see Ref. 8) within the extended gauge group, expecially if the Higgs multiplet is a color singlet. This leads us to conjecture that CP violation in this extended scheme is linked on the one hand to nonvanishing quark *and* lepton masses (including neutrinos) and to Cabibbo angles $\theta_{L,R}$ on the other. (See later that CP violation in gauge interactions would disappear if θ_L or $\theta_R\to 0$.)

[12]Note that the Yukawa coupling parameters and therefore the mass matrices for the baryonic (a,b,c) and leptonic quartets are unrelated (related) if the gauge group is $\mathcal{G}_0$ ($\mathcal{G}$).

[13]As emphasized in Ref. 6, one needs invariant couplings of the form $a(\phi_1^\dagger\phi_2)+b(\phi_1^\dagger\phi_2)^2+\text{H.c.}$ to retain observable consequences of complex vacuum expectation values. This necessitates ϕ_1 and ϕ_2 having the same representation. [We should remark that duplication of Higgs fields is by no means unnatural if they are to be regarded as $(\bar{\psi}_L\psi_R)$ composites, which give rise to four $\phi_i=(2,2,1,1)$.] Defining the fields $\phi_3\equiv\tau_2\phi_1^*\tau_2$ and $\phi_4\equiv\tau_2\phi_2^*\tau_2$, it is easy to verify that the full potential, which gives the desired pattern of vacuum expectation values $\langle\phi_{1,2}\rangle$ (chosen in the text) for appropriate choice of signs of the coefficients, is given by

$$\sum_{i,j,k,l=1,2,3,4}\Big[a_{ij}\operatorname{Tr}(\phi_i^\dagger\phi_j)+b_{ijkl}\operatorname{Tr}(\phi_i^\dagger\phi_j\phi_k^\dagger\phi_l)+c_{ijkl}\operatorname{Tr}(\phi_i^\dagger\phi_j)\operatorname{Tr}(\phi_k^\dagger\phi_l)+d_{ij}\operatorname{Tr}(\phi_i^\dagger\phi_j)(T_L^\dagger T_L+T_R^\dagger T_R)\Big]$$
$$+\sum_{\alpha=L,R}[\mu_\alpha^2(T_\alpha^\dagger T_\alpha)+f'(T_\alpha^\dagger T_\alpha)^2]+e'(T_L^\dagger T_L)(T_R^\dagger T_R)+\text{H.c.},$$

where we have introduced two additional fields $T_L=(2,1,1,1)$ and $T_R=(1,2,1,1)$ to provide desired mass differences between $W_L^\pm$ and $W_R^\pm$ (solely) due to differences between μ_L^2 and μ_R^2, all other terms being $L\leftrightarrow R$ symmetric. Note that such a pattern is not affected in the presence of SU(3)′ or SU(4)′ color-gauge interactions. We thank Professor A. Pais for discussions on the choice of the potential.

[14]The reader may convince himself that such U(2) transformations can always be found so as to transform an arbitrary 2×2 complex matrix to real diagonal form.

[15]This follows from the observed validity of the relation $\eta_{+-}=\eta_{00}$ to a few percent. Note that the CP-nonconserving amplitude due to the *mixing term* introduces corrections of order (κ'/κ) to the above relation, which holds exactly in the absence of such a mixing. The choice $\kappa'\ll\kappa$ (with κ' and $\kappa\neq 0$) is compatible with the general Higgs potential (Ref. 13) and the requirement of renormalizability of the theory.

[16]I.e.,

$$\lambda_{L,R}\to e^{i\phi}\lambda_{L,R};$$
$$\tan\phi=(f_L\sin\delta_L+f_R\sin\delta_R)/(f_L\cos\delta_L+f_R\cos\delta_R).$$

[17]This follows by noting that $(p^2/q^2)=\langle\bar{K}^0|T|K^0\rangle/\langle K^0|T|\bar{K}^0\rangle=M^{(-)}/M^{(+)}$, where $M^{(\pm)}\equiv A_1+(1\pm 2i\tan\xi-\tan^2\xi)A_2$ and $(p/q)=(1+\rho)/(1-\rho)$. See Ref. 2.

[18]Note that the Higgs sector cannot be completely left-right symmetric in order that M_{W_R} be different from M_{W_L} after spontaneous symmetry breaking. However, if the asymmetry is only through mass terms of the scalar fields (see footnote 13), it will be unimportant at high energies and in such a case (g_L-g_R) would be

computable. This will be discussed elsewhere. Note that if the Higgs mechanism is eventually replaced by dynamical symmetry breaking, then parity violation would be entirely of spontaneous origin (assuming $g_L^{(0)} = g_R^{(0)}$. Spontaneous parity violation has been considered by P. Fayet [Orsay report, 1974 (unpublished)] in a different context.

[19]We thank Professor L. Wolfenstein for pointing out that the difference between η_{+-} and η_{+-0} provides, at least in principle, an avenue for measuring the most relevant parameter of our scheme ($\tan\xi$) in a model-independent manner.

[20]There would be a $\Delta Y = 0$ P- and T-violating term in the gauge interactions due to W_L- W_R mixing which is proportional to $g_L g_R \kappa\kappa'$; this contribution is limited to be $\lesssim 10^{-24}$ e cm if (κ'/κ) is less than (1/10), which (as explained in footnote 15) is required to preserve the relation $\eta_{+-} = \eta_{00}$.

[21]Since the Yukawa coupling constants (h's and f's) are of order $e(m_q/m_{W_L})$, if one requires Higgs-boson exchange amplitudes to be $\lesssim G_F \times 10^{-4}$, it would follow that $m_\sigma \gtrsim (m_q/m_{W_L}) \times 3 \times 10^3 \approx 30$ GeV if $(m_q/m_W) \simeq 10^{-2}$ (say). Such scales of masses for the Higgs mesons are at least not unreasonable, considering that m_W's are of order 100 GeV.

[22]Note that an alternative scheme (based on phase angles between vector and axial-vector currents *outside* the gauge-theory approach) also leads to the relation (14) and $\eta_{+-} = \eta_{00}$ only for a special relation between $\Delta S = 0$ and $\Delta S = 1$ phase angles as pointed out in Ref. 1. Such a scheme still differs from the present note in the structure of $S^{(\pm)}$ and does not predict $\eta_{+-0} = \eta_{000}$. Furthermore, it should lead to a neutron electric dipole moment d_n of order 10^{-23} e cm.

[23]We note that if we chose

$$\begin{pmatrix} \mathcal{P}^0 \\ \chi^0 \end{pmatrix}_R$$

and

$$\begin{pmatrix} \mathfrak{N}^0 \\ \lambda^0 \end{pmatrix}_R$$

as $SU(2)_R$ doublets with

$$\begin{pmatrix} \mathcal{P}^0 \\ \mathfrak{N}^0 \end{pmatrix}_L$$

and

$$\begin{pmatrix} \chi^0 \\ \lambda^0 \end{pmatrix}_L$$

as the $SU(2)_L$ doublets, we may again introduce CP violation into the gauge interactions (after spontaneous symmetry breaking) in a manner similar to that discussed in the text; the resulting scheme, however, does not satisfy the isoconjugate relation (14). Furthermore, assuming that $\mathfrak{N}^0_{R,L}$ and $\lambda^0_{R,L}$ are related to the physical fields by Cabibbo rotations, one obtains $\Delta S = 2$ transitions in order f_R. One must thus choose

$$f_R = (g_R^2/4m_{W_R}^2)(\sin 2\theta_R) < G_F^2 m_p^2$$

to avoid conflict with $(m_{K_L} - m_{K_S})$. In this case the gauge interactions provide a genuine superweak scheme of CP violation. Note that in this case, one *must* introduce an Abelian U(1) contribution to electric charge [in spite of the full SU(4)'-color gauging] since the $SU(2)_R$ mentioned above can not contribute to electric charge. The resulting scheme is still anomaly-free.

[24]See L. Wolfenstein, Nucl. Phys. B77, 375 (1974) for a review on estimates of the electric dipole moment of the neutron in various gauge-theory and other models.

[25]J. Frenkel and M. Ebel, Univ. of Wisconsin report (unpublished).

Nuclear Physics B258 (1985) 328–360

CP VIOLATION AND LEFT-RIGHT SYMMETRY*

G. ECKER and W. GRIMUS

Institut für Theoretische Physik, Universität Wien, Wien, Austria

Received 22 March 1985

We investigate minimal L-R symmetric gauge models with spontaneous *CP* violation. The effective $|\Delta S| = 1, 2$ hamiltonians are calculated giving special attention to a careful evaluation of QCD corrections. Two classes of models are distinguished by the sign of the L-R contribution $\Delta m^{LR}_{K_L K_S}$ to the $K_L K_S$ mass difference. Requiring $|\Delta m^{LR}_{K_L K_S}| \leqslant \Delta m^{exp}_{K_L K_S}$ leads to the lower bounds $M_2 \gtrsim 2.5$ TeV, $M_H \gtrsim 10$ TeV for the heavy charged vector boson and neutral Higgs mass, respectively. ε, ε' and the neutron electric dipole moment d_n are computed in the small-phase approximation. The experimental value of ε can be reproduced with M_2 in the several TeV range. ε'/ε may have either sign and is strongly correlated with d_n. Depending on the sign of $\Delta m^{LR}_{K_L K_S}$, we find $|\varepsilon'/\varepsilon| \lesssim (4 \text{ or } 6) \times 10^{-3}$ and $|d_n| \lesssim (4 \times 10^{-25} \text{ or } 3 \times 10^{-26})$ e · cm.

1. Introduction

CP violation has recently become the focus of increasing experimental and theoretical activity (e.g. [1]). Measurements of the bottom lifetime [1,2] and of the semileptonic branching ratio $\overline{R} = \Gamma(b \to ue\nu)/\Gamma(b \to ce\nu)$ [3] have pinned down the weak mixing angles to such an extent that the old problem of explaining the smallness of *CP* violation in the context of the three-generation standard model has been completely inverted to a problem of maximal *CP* violation [4]. According to the majority view [1] the standard six-quark model is still compatible with experiment. It takes little imagination, however, to anticipate two possible future problems for the standard model in the neutral kaon sector:

(i) ε problem: the only experimentally established *CP*-violating parameter ε comes out too small for $m_{top} < 50$ GeV [5] and $\overline{R} < 0.01$;

(ii) ε'/ε problem: the standard model predicts [1] a positive lower bound $\varepsilon'/\varepsilon \gtrsim 2 \times 10^{-3}$ to be compared with the present experimental values:

$$\varepsilon'/\varepsilon = \begin{cases} (-4.6 \pm 5.3 \pm 2.4) \times 10^{-3} & \text{Chicago-Saclay [6]} \\ (1.7 \pm 8.2) \times 10^{-3} & \text{BNL-Yale [7]}. \end{cases} \tag{1.1}$$

* Partially supported by "Fonds zur Förderung der wissenschaftlichen Forschung", project number 5444.

The experimental situation supplies ample motivation to investigate alternative or at least additional mechanisms for *CP* violation. There are in addition rather convincing theoretical arguments. The standard model mechanism of breaking *CP* explicitly in terms of complex Yukawa couplings provides a possible parametrization, but not a fundamental understanding of *CP* violation. This shortcoming is removed in models with spontaneous *CP* violation where both the gauge symmetry breaking and *CP* violation have a common origin. Although the mechanism of gauge symmetry breaking needs to be clarified such a unified treatment is almost unavoidable if the Higgs fields are composite fields in an underlying purely fermionic gauge theory [8].

To implement spontaneous *CP* violation the scalar sector must be enlarged beyond the single Higgs doublet of the standard model. Once we opt for spontaneous *CP* violation it seems unnatural to leave parity aside. Spontaneous *P* and *CP* violation together with the minimal number of two scalar doublets leads directly to the minimal L-R symmetric models [9] with gauge group $SU(2)_L \times SU(2)_R \times U(1)$.

The minimal L-R models with spontaneous *P* and *CP* violation fall into two classes [10] distinguished by the structure of their Yukawa coupling matrices. The first class (manifest *CP*) has real and symmetric coupling matrices and will be the main subject of the present paper. The second class (non-manifest *CP*) consists of a single model where the coupling matrices are hermitian, but not real and where quark masses and weak mixing angles are constrained by *P* and *CP* invariance of the lagrangian. The *CP* phenomenology of this model was discussed in [11]. The model exhibits manifest *CP* in a certain limit and it will therefore also serve as a test for the general treatment of this paper.

The purpose of the present paper is twofold. First, we collect all the necessary ingredients for an investigation of *CP* violation in L-R models such as the dominant Feynman diagrams, a rather complete discussion of QCD corrections to the effective $|\Delta S| = 1,2$ hamiltonians and estimates of all relevant hadronic matrix elements. We stress the importance of a correct evaluation of short-distance corrections which allow for the first reliable determination of lower bounds for the heavy vector boson mass M_2 and for the neutral flavour-changing Higgs mass M_H.

The second aim of our work is a detailed discussion of the *CP* phenomenology of models with manifest *CP*. The minimal L-R models with manifest *CP* possess the useful property that all *CP*-violating observables can be expressed in terms of a single phase, a ratio of scalar vacuum expectation values, quark masses and weak mixing angles, but not on unconstrained quantities such as Yukawa couplings or additional phases. Quantitative estimates will be given in the small-phase approximation extending the work of Chang [12]. The models with manifest *CP* can be divided into two subclasses characterized by the sign of the genuine L-R contributions to the $K_L K_S$ mass difference. These two subclasses have different *CP* properties especially concerning the neutron electric dipole moment d_n.

In sect. 2 the dominant Feynman diagrams for the effective $|\Delta S| = 1,2$ hamiltonians are calculated in the minimal L-R model. We comment on the problem of non-gauge invariance of box diagrams recently raised by Chang et al. [13]. The short-distance QCD corrections to the weak hamiltonians of sect. 2 are performed in sect. 3. We complete and partly correct existing calculations. The dominant effects due to the large masses M_2, M_H are evaluated exactly whereas we use approximate methods to integrate out the heavy quark fields. Unlike in the standard model, the QCD corrections enhance both the $|\Delta S| = 1$ and 2 hamiltonians. In sect. 4 we calculate all relevant hadronic matrix elements using the vacuum insertion technique and we set up the necessary equations to determine the CP violating parameters ε, ε'. A comment is added on the $\Delta I = \frac{1}{2}$ rule. While sects. 3 and 4 apply to L-R models in general we return to the minimal model in sect. 5 to discuss the CP phenomenology of models with manifest CP. After recalling the general classification of spontaneous CP violation we show for an arbitrary number of generations that in addition to in principle known quantities such as quark masses and mixing angles all CP-violating parameters depend on a single phase and on a ratio of scalar vacuum expectation values. The relevant equations are solved in the small-phase approximation. We distinguish two different cases depending on the sign of the L-R contributions to $\Delta m_{K_L K_S}$. Lower bounds on M_2 and M_H are derived from the requirement that the absolute value of L-R contributions should not exceed $\Delta m^{\exp}_{K_L K_S}$. After making sure that the models can account for the observed ε regardless of R we calculate ε'/ε and d_n in the small-phase approximation supplemented by exact numerical calculations of d_n in the non-manifest CP model [11]. We comment on the sensitivity of our predictions for ε'/ε to the $\Delta I = \frac{1}{2}$ problem. Our conclusions are summarized in sect. 6.

2. $|\Delta S| = 1,2$ effective hamiltonians

Throughout this paper we consider the minimal $SU(2)_L \times SU(2)_R \times U(1)$ gauge model [9] with three fermion generations and with a single scalar multiplet:

$$\Phi = \underbrace{\begin{pmatrix} \Phi_{11}^0 & \Phi_{12}^+ \\ \Phi_{21}^- & \Phi_{22}^0 \end{pmatrix}}_{\overleftarrow{SU(2)_R}} \downarrow SU(2)_L , \tag{2.1}$$

coupling to quarks. Although the relevant diagrams would be the same with an enlarged Higgs sector, only in the minimal case are the couplings of scalar fields to quarks determined by masses and mixing angles only. In this case, there is a single charged scalar field $\Phi^{\pm}$ and two neutral scalar fields* Φ_2, Φ_3 with flavour-changing

* There is also a flavour-conserving neutral scalar field Φ_1, the analogue of the Higgs field in the standard model.

couplings to quarks [14]. The Yukawa interaction is

$$-\mathcal{L}_{\mathrm{Y}} = \bar{u}(G_{\mathrm{R}}P_{\mathrm{R}} + G_{\mathrm{L}}P_{\mathrm{L}})\,\mathrm{d}\Phi^{+} + \sum_{i=2}^{3}\left(\bar{u}G_i^{\mathrm{u}}P_{\mathrm{R}}u + \bar{d}G_i^{\mathrm{d}}P_{\mathrm{R}}d\right)\Phi_i + \mathrm{h.c.}$$

$$P_{\substack{\mathrm{L}\\ \mathrm{R}}} = \tfrac{1}{2}(1 \mp \gamma_5)\,, \tag{2.2}$$

where u, d are quark mass eigenfields (flavour and colour indices are suppressed) and the three-dimensional coupling matrices $G_{\mathrm{R}}, G_{\mathrm{L}}, G_i^{\mathrm{u}}, G_i^{\mathrm{d}}$ can be found in the appendix of [15]. As suggested by a study of the most general Higgs potential we assume a common mass M_{H} for the scalar fields $\Phi^{\pm}, \Phi_2, \Phi_3$.

The charged current lagrangian is given by

$$\begin{aligned}-\mathcal{L}_{\mathrm{CC}} &= \sqrt{\tfrac{1}{2}}\, g\bar{u}\left(\cos(\xi_{\mathrm{LR}})K_{\mathrm{L}}\gamma^{\mu}P_{\mathrm{L}} - \mathrm{e}^{-i\lambda}\sin(\xi_{\mathrm{LR}})K_{\mathrm{R}}\gamma^{\mu}P_{\mathrm{R}}\right) dW_{1\mu}^{+} \\ &\quad + \sqrt{\tfrac{1}{2}}\, g\bar{u}\left(\mathrm{e}^{i\lambda}\sin(\xi_{\mathrm{LR}})K_{\mathrm{L}}\gamma^{\mu}P_{\mathrm{L}} + \cos(\xi_{\mathrm{LR}})K_{\mathrm{R}}\gamma^{\mu}P_{\mathrm{R}}\right) dW_{2\mu}^{+} + \mathrm{h.c.}\,,\end{aligned} \tag{2.3}$$

with charged vector boson fields W_1, W_2 (masses M_1, M_2) and left (right)-handed mixing matrices K_{L} (K_{R}). The L-R mixing parameters $\xi_{\mathrm{LR}}, \lambda$ are related to the scalar vacuum expectation values

$$\langle\Phi\rangle_0 = \sqrt{\tfrac{1}{2}}\begin{pmatrix} v & 0 \\ 0 & w\end{pmatrix} \tag{2.4}$$

as

$$\begin{aligned}\xi_{\mathrm{LR}} &\simeq \frac{2|vw|}{|v|^2 + |w|^2}\frac{M_1^2}{M_2^2}\,, \\ \mathrm{e}^{i\lambda} &= -vw^*/|vw|\,.\end{aligned} \tag{2.5}$$

It can be argued and explicitly demonstrated in specific examples [16] that a natural explanation of the smallness of weak mixing angles requires $|w/v| = \mathrm{O}(m_{\mathrm{b}}/m_{\mathrm{t}})$. The natural magnitude of ξ_{LR} is therefore [14]

$$\xi_{\mathrm{LR}} \simeq \frac{2m_{\mathrm{b}}}{m_{\mathrm{t}}}\frac{M_1^2}{M_2^2} \sim 2\times 10^{-3}\,(M_2/\mathrm{TeV})^{-2}\,, \tag{2.6}$$

to be compared with the phenomenological upper bound [17] $\xi_{\mathrm{LR}} \lesssim 4\times 10^{-3}$. It will thus be sufficient to calculate all amplitudes to leading order in ξ_{LR}.

There exists already an extensive literature on the $|\Delta S| = 1, 2$ effective interactions [18] in L-R symmetric gauge models. In this section we present the dominant electroweak contributions deferring a discussion of QCD corrections to sect. 3.

The $|\Delta S| = 1$ effective hamiltonian is to a very good approximation determined by W_1, W_2 exchange:

$$\mathcal{H}_{\text{eff}}^{|\Delta S|=1} = 2\sqrt{2}\, G_F \{ \lambda_u^{LL}\, \bar{d}\gamma^\mu P_L u \bar{u}\gamma_\mu P_L s + \beta\lambda_u^{RR}\, \bar{d}\gamma^\mu P_R u \bar{u}\gamma_\mu P_R s - \xi_{LR} (e^{-i\lambda}\lambda_u^{LR} \bar{d}\gamma^\mu P_L u \bar{u}\gamma_\mu P_R s + e^{i\lambda}\lambda_u^{RL} \bar{d}\gamma^\mu P_R u \bar{u}\gamma_\mu P_L s) \} + \text{h.c.}, \tag{2.7}$$

$$\beta = M_1^2/M_2^2,$$

$$\lambda_i^{AB} = K_{A,id}^* K_{B,is} \qquad (i = u, c, t;\ A, B = L, R).$$

The Yukawa couplings (2.2) imply a strong suppression of scalar exchange compared to (2.7) as long as $M_H \gtrsim M_2$. Neutral scalar exchange is suppressed by approximately $10^{-4}\, (M_2/M_H)^2$ compared to the L-R mixing terms in (2.7) which will turn out to dominate *CP* violation. Charged Higgs exchange is even smaller.

The flavour-changing neutral Higgs couplings in (2.2) induce a tree-level contribution to the $|\Delta S| = 2$ effective hamiltonian [14, 19]:

$$\mathcal{H}_{\text{Higgs}}^{|\Delta S|=2} = -\frac{2(|v|^2 + |w|^2)}{M_H^2 (|v|^2 - |w|^2)^2} \sum_{i,j=u,c,t} m_i m_j \lambda_i^{LR} \lambda_j^{RL} \bar{d} P_L s \bar{d} P_R s + \text{h.c.} \tag{2.8}$$

Assuming $M_H \gtrsim M_2$ again, the leading contributions at the one-loop level are given by the box diagrams in fig. 1. All diagrams were computed in the 't Hooft-Feynman gauge including, of course, the box graphs with unphysical scalars S_1, S_2 which are omitted in fig. 1.

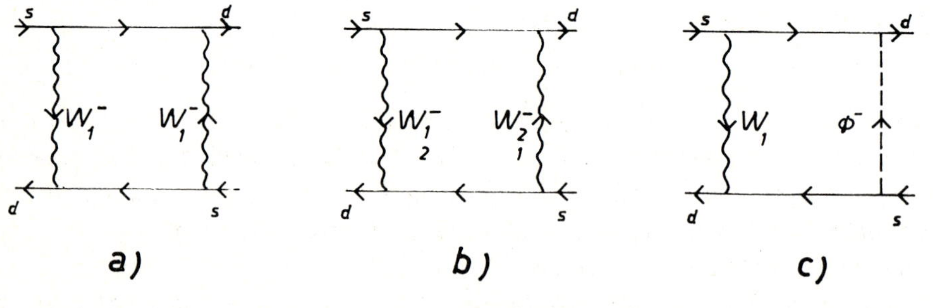

Fig. 1. Dominant box diagrams for the $|\Delta S| = 2$ amplitude. Φ^- is the charged Higgs boson of the minimal model.

The L-R box diagrams (fig. 1b) yield

$$\mathcal{H}^{|\Delta S|=2}_{(W_1+S_1)(W_2+S_2)} = \frac{G_F^2 M_1^2 \beta}{2\pi^2} \bar{d} P_L s \bar{d} P_R s \sum_{i,j=u,c,t} \lambda_i^{LR} \lambda_j^{RL} \sqrt{x_i x_j}$$

$$\times \left\{ (4 + x_i x_j \beta) I_1(x_i, x_j, \beta) - (1+\beta) I_2(x_i, x_j, \beta) \right\} + \text{h.c.}, \tag{2.9}$$

$$I_1(x_i, x_j, \beta) = \frac{x_i \ln x_i}{(1-x_i)(1-x_i\beta)(x_i - x_j)} + (i \leftrightarrow j) - \frac{\beta \ln \beta}{(1-\beta)(1-x_i\beta)(1-x_j\beta)},$$

$$I_2(x_i, x_j, \beta) = \frac{x_i^2 \ln x_i}{(1-x_i)(1-x_i\beta)(x_i - x_j)} + (i \leftrightarrow j) - \frac{\ln \beta}{(1-\beta)(1-x_i\beta)(1-x_j\beta)},$$

$$x_i = m_i^2 / M_1^2,$$

whereas the $(W_1 + S_1)\Phi^{\pm}$ diagram (fig. 1c) contributes

$$\mathcal{H}^{|\Delta S|=2}_{(W_1+S_1)\Phi^{\pm}} = \frac{G_F^2 M_1^2 \beta_H (1-\beta)}{2\pi^2} \bar{d} P_L s \bar{d} P_R s \sum_{i,j=u,c,t} \lambda_i^{LR} \lambda_j^{RL} \sqrt{x_i x_j}$$

$$\times \left\{ -I_2(x_i, x_j, \beta_H) + x_i x_j I_1(x_i, x_j, \beta_H) \right\} + \text{h.c.}, \tag{2.10}$$

$$\beta_H = M_1^2 / M_H^2.$$

With the same notation, the standard left-left box diagram (fig. 1a) gives [20]

$$\mathcal{H}^{|\Delta S|=2}_{(W_1+S_1)(W_1+S_1)} = \frac{G_F^2 M_1^2}{4\pi^2} \bar{d} \gamma^\mu P_L s \bar{d} \gamma_\mu P_L s \sum_{i,j=u,c,t} \lambda_i^{LL} \lambda_j^{LL}$$

$$\times \left\{ \left(1 + \tfrac{1}{4} x_i x_j\right) I_2(x_i, x_j, 1) - 2 x_i x_j I_1(x_i, x_j, 1) \right\} + \text{h.c.} \tag{2.11}$$

It is straightforward to convince oneself that the $W_2 W_2$, $W_2 \Phi^{\pm}$ and $\Phi^{\pm}\Phi^{\pm}$ box amplitudes may safely be neglected. There are additional box diagrams where neutral gauge bosons and the flavor-changing neutral Higgs bosons Φ_2, Φ_3 are exchanged. Such diagrams are genuine radiative corrections to the tree-level hamiltonian (2.8) and may also be omitted.

Chang et al. have recently observed [13] that the sum of box diagrams is in general not gauge invariant in L-R models. Although there is a priori no reason for any

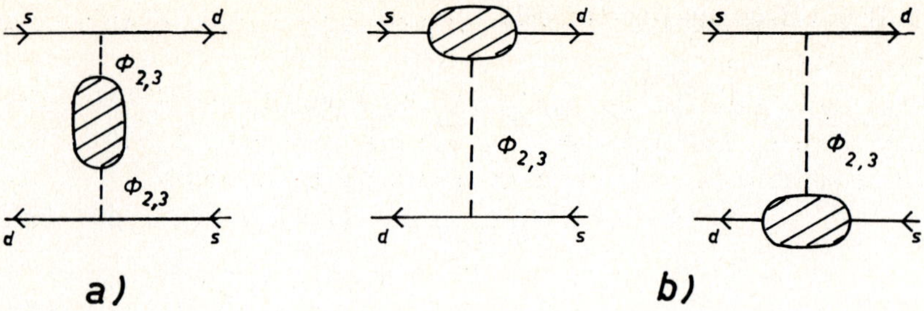

Fig. 2. Self-energy (a) and vertex corrections (b) to neutral Higgs exchange contributing to the $|\Delta S| = 2$ amplitude. $\Phi_{2,3}$ are the two flavour-changing Higgs bosons of the minimal model and the blobs denote all possible one-loop insertions.

subset of one-loop diagrams to be gauge invariant the standard model box graph of fig. 1a is actually gauge invariant by itself. The diagrams restoring gauge invariance are the one-loop self-energy correction (fig. 2a) and the vertex corrections (fig. 2b) to neutral Higgs exchange. The self-energy correction can be lumped into an effective Higgs mass M_H at zero-momentum transfer. The parameter M_H in (2.8) is therefore, strictly speaking, not the physical Higgs mass but differs from it by a one-loop radiative correction. The vertex corrections of fig. 2b require more work but it turns out that in the 't Hooft-Feynman gauge all contributions are suppressed by at least a factor $g^2\beta \ln \beta$ compared to (2.8). Essentially the same conclusion is reached in two very recent papers [21] where, moreover, the restoration of gauge invariance is explicitly demonstrated in a general R_ξ gauge.

For $M_H \gtrsim M_2$ the total effective $|\Delta S| = 2$ hamiltonian without QCD corrections is therefore well approximated by

$$\mathcal{H}_{\text{eff}}^{|\Delta S|=2} = \mathcal{H}_{\text{Higgs}}^{|\Delta S|=2} + \mathcal{H}_{(W_1+S_1)(W_2+S_2)}^{|\Delta S|=2} + \mathcal{H}_{(W_1+S_1)\Phi^\pm}^{|\Delta S|=2} + \mathcal{H}_{(W_1+S_1)(W_1+S_1)}^{|\Delta S|=2} . \tag{2.12}$$

3. Short-distance QCD corrections

The method of obtaining the dominant short-distance QCD corrections to weak amplitudes is in principle well known (for a review see, e.g. [22]). By means of an operator product expansion one successively integrates out the heavy fields to arrive at low-energy effective hamiltonians in terms of only light fields. The appropriate Wilson coefficients are usually computed in the leading logarithm approximation.

It will turn out that QCD corrections in the L-R model enhance the effective hamiltonians for both $|\Delta S| = 1$ and 2. Moreover, those corrections are bigger than in the standard model for two reasons: four-quark operators with greater anomalous dimensions are involved and there are larger masses (M_2, M_H) in the theory. Since the dominant effects are due to these large masses we treat the vector and Higgs bosons exactly and content ourselves with approximate methods to integrate out the heavy quark fields t, b, c.

The dominant corrections to the $|\Delta S| = 1$ hamiltonian have all been calculated before [23] so that we can be very brief. The operators appearing in (2.7) can be expressed in terms of the multiplicatively renormalizable operators

$$O_{\pm}^{\substack{\mathrm{LL}\\ \mathrm{RR}}} = \bar{d}\gamma^{\mu}P_{\substack{\mathrm{L}\\ \mathrm{R}}}u\bar{u}\gamma_{\mu}P_{\substack{\mathrm{L}\\ \mathrm{R}}}s \pm \bar{d}\gamma^{\mu}P_{\substack{\mathrm{L}\\ \mathrm{R}}}s\bar{u}\gamma_{\mu}P_{\substack{\mathrm{L}\\ \mathrm{R}}}u$$

$$\gamma\left(O_{+}^{\substack{\mathrm{LL}\\ \mathrm{RR}}}\right) = g_{\mathrm{s}}^{2}/4\pi^{2}, \qquad \gamma\left(O_{-}^{\substack{\mathrm{LL}\\ \mathrm{RR}}}\right) = -g_{\mathrm{s}}^{2}/2\pi^{2},$$

$$O_{+}^{\substack{\mathrm{LR}\\ \mathrm{RL}}} = \bar{d}\gamma^{\mu}P_{\substack{\mathrm{L}\\ \mathrm{R}}}u\bar{u}\gamma_{\mu}P_{\substack{\mathrm{R}\\ \mathrm{L}}}s + \tfrac{2}{3}\bar{d}P_{\substack{\mathrm{R}\\ \mathrm{L}}}s\bar{u}P_{\substack{\mathrm{L}\\ \mathrm{R}}}u,$$

$$\gamma\left(O_{+}^{\substack{\mathrm{LR}\\ \mathrm{RL}}}\right) = g_{\mathrm{s}}^{2}/8\pi^{2},$$

$$O_{-}^{\substack{\mathrm{LR}\\ \mathrm{RL}}} = \tfrac{2}{3}\bar{d}P_{\substack{\mathrm{R}\\ \mathrm{L}}}s\bar{u}P_{\substack{\mathrm{L}\\ \mathrm{R}}}u,$$

$$\gamma\left(O_{-}^{\substack{\mathrm{LR}\\ \mathrm{RL}}}\right) = -g_{\mathrm{s}}^{2}/\pi^{2}, \tag{3.1}$$

where g_{s} is the strong coupling constant and the γ's denote the appropriate anomalous dimensions. Integrating out the vector boson fields yields the QCD corrected weak hamiltonian

$$\begin{aligned}\mathcal{H}^{|\Delta S|=1}_{\text{W-exchange}} = \sqrt{2}\,G_{\mathrm{F}}\Big\{&\lambda_{\mathrm{u}}^{\mathrm{LL}}\Big(\big[\alpha_{\mathrm{s}}(\mu^{2})/\alpha_{\mathrm{s}}(M_{1}^{2})\big]^{-2/b}O_{+}^{\mathrm{LL}} + \big[\alpha_{\mathrm{s}}(\mu^{2})/\alpha_{\mathrm{s}}(M_{1}^{2})\big]^{4/b}O_{-}^{\mathrm{LL}}\Big)\\
&+\beta\lambda_{\mathrm{u}}^{\mathrm{RR}}\Big(\big[\alpha_{\mathrm{s}}(\mu^{2})/\alpha_{\mathrm{s}}(M_{2}^{2})\big]^{-2/b}O_{+}^{\mathrm{RR}}\\
&+\big[\alpha_{\mathrm{s}}(\mu^{2})/\alpha_{\mathrm{s}}(M_{2}^{2})\big]^{4/b}O_{-}^{\mathrm{RR}}\Big) + 2\xi_{\mathrm{LR}}\lambda_{\mathrm{u}}^{\mathrm{LR}}\mathrm{e}^{-i\lambda}\\
&\times\Big(\big[\alpha_{\mathrm{s}}(\mu^{2})/\alpha_{\mathrm{s}}(M_{1}^{2})\big]^{8/b}O_{-}^{\mathrm{LR}}\\
&-\big[\alpha_{\mathrm{s}}(\mu^{2})/\alpha_{\mathrm{s}}(M_{1}^{2})\big]^{-1/b}O_{+}^{\mathrm{LR}}\Big) + 2\xi_{\mathrm{LR}}\lambda_{\mathrm{u}}^{\mathrm{RL}}\mathrm{e}^{i\lambda}\\
&\times\Big(\big[\alpha_{\mathrm{s}}(\mu^{2})/\alpha_{\mathrm{s}}(M_{1}^{2})\big]^{8/b}O_{-}^{\mathrm{RL}} - \big[\alpha_{\mathrm{s}}(\mu^{2})/\alpha_{\mathrm{s}}(M_{1}^{2})\big]^{-1/b}O_{+}^{\mathrm{RL}}\Big)\Big\} + \text{h.c.},\end{aligned} \tag{3.2}$$

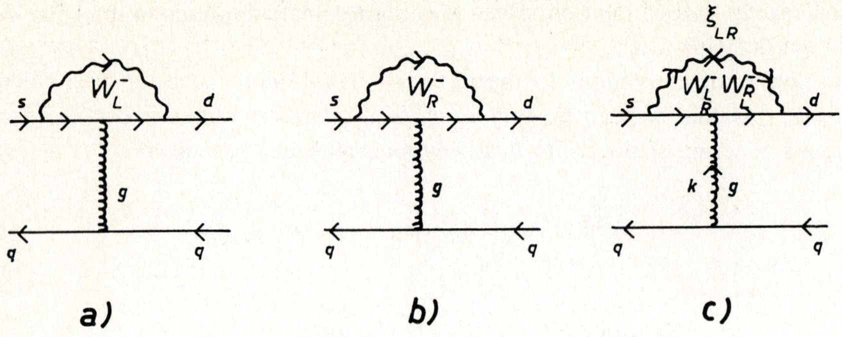

Fig. 3. One-loop penguin diagrams for the $|\Delta S|=1$ amplitude. The direction of the gluon momentum k entering eq. (3.7) is indicated in diagram c.

where

$$\alpha_s(m^2)=\frac{4\pi}{b\ln(m^2/\Lambda_f^2)}, \qquad b=11-\tfrac{2}{3}N_f. \tag{3.3}$$

Λ_f is the QCD parameter for N_f quark flavours in the $\overline{\text{MS}}$ scheme and μ^2 is a renormalization point below M_1^2.

As the scale μ^2 is lowered to a typical hadronic scale new operators appear upon integrating out the heavy quark fields which are not included in (3.2). Those operators will in general mix with the ones contained in $\hat{\mathcal{H}}^{|\Delta S|=1}_{\text{W-exchange}}$. Instead of performing a complete operator product expansion for the L-R model we propose the following approximate method: we compute all one-loop penguin diagrams and normalize the standard model penguin (fig. 3a) to the exact operator product expansion result [24]. This procedure yields an effective $\alpha_s \simeq 0.6$ which we then insert into the genuine L-R penguins (fig. 3b, c).

The L-L and R-R penguin diagrams lead to effective local hamiltonians

$$\mathcal{H}_{\text{P}}^{\text{LL}}=\frac{\sqrt{2}\,G_F\alpha_s}{12\pi}\,\bar{d}\gamma^\mu P_L\lambda^a s\left(\bar{u}\gamma_\mu\lambda^a u+\bar{d}\gamma_\mu\lambda^a d+\bar{s}\gamma_\mu\lambda^a s\right)\sum_{i=u,c,t}\lambda_i^{\text{LL}}f_{\text{P}}(x_i)+\text{h.c.}, \tag{3.4}$$

$$\mathcal{H}_{\text{P}}^{\text{RR}}=\frac{\sqrt{2}\,G_F\alpha_s\beta}{12\pi}\,\bar{d}\gamma^\mu P_R\lambda^a s\left(\bar{u}\gamma_\mu\lambda^a u+\bar{d}\gamma_\mu\lambda^a d+\bar{s}\gamma_\mu\lambda^a s\right)\sum_{i=u,c,t}\lambda_i^{\text{RR}}f_{\text{P}}(x_i\beta)+\text{h.c.}, \tag{3.5}$$

with [25]

$$f_{\text{P}}(x)=(1-x)^{-4}\left[-\tfrac{9}{4}x+\tfrac{29}{8}x^2-\tfrac{5}{4}x^3-\tfrac{1}{8}x^4+\left(1-4x+\tfrac{9}{4}x^2\right)\ln x\right]. \tag{3.6}$$

The analogous gauge-invariant calculation of the L-R penguin graph in fig. 3c gives a non-local operator in coordinate space. In momentum space, one obtains

$$\mathcal{H}_{\mathrm{P}}^{\mathrm{LR}} = \frac{\sqrt{2}\, G_{\mathrm{F}} \alpha_{\mathrm{s}} \xi_{\mathrm{LR}}}{4\pi k^2} \sum_{i=\mathrm{u,c,t}} m_i f_{\mathrm{LR}}(x_i)\, \bar{d} i \sigma_{\mu\nu} k^{\nu}$$
$$\times \left(\mathrm{e}^{i\lambda} \lambda_i^{\mathrm{RL}} P_{\mathrm{L}} + \mathrm{e}^{-i\lambda} \lambda_i^{\mathrm{LR}} P_{\mathrm{R}}\right) \lambda^a s \left(\bar{u}\gamma^{\mu}\lambda^a u + \bar{d}\gamma^{\mu}\lambda^a d + \bar{s}\gamma^{\mu}\lambda^a s\right) + \mathrm{h.c.} \quad (3.7)$$

to leading order in ξ_{LR} and β. k is the gluon momentum as defined in fig. 3c and

$$f_{\mathrm{LR}}(x) = (1-x)^{-3}\left(1 - \tfrac{3}{4}x - \tfrac{1}{4}x^3 + \tfrac{3}{2}x \ln x\right). \quad (3.8)$$

The one-loop calculation of penguin operators is sufficient for our purposes for two reasons. In the standard model the penguin-type operators mix only little with the operators in (3.2) and it is therefore a rather good approximation to simply add $\mathcal{H}_{\mathrm{P}}^{\mathrm{LL}}$ to (3.2) with a properly chosen α_{s}. Secondly, in contrast to the standard model the matrix elements of

$$O_{-}^{\substack{\mathrm{LR} \\ \mathrm{RL}}}$$

are not helicity suppressed as will be discussed in sect. 4. As a consequence, the contributions of (3.5) and (3.7) to the CP-violating parameter ε' will turn out to be rather small compared to the dominant contributions from (3.2).

The short-distance corrections to the $|\Delta S| = 2$ Higgs exchange hamiltonian (2.8) are relatively simple because the operator $O_{\mathrm{S}} = \bar{d} P_{\mathrm{L}} s \bar{d} P_{\mathrm{R}} s$ is multiplicatively renormalizable at the one-loop level. Its anomalous dimension can be inferred from (3.1) to be

$$\gamma(O_{\mathrm{S}}) = -g_{\mathrm{s}}^2/\pi^2 = 2\gamma_{\mathrm{m}}, \quad (3.9)$$

where γ_{m} is the anomalous dimension of the mass operator. Thus, the operator $m_i m_j O_{\mathrm{S}}$ is renormalization group invariant to one-loop order and therefore its matrix elements do not depend on the heavy mass M_{H}. If we insert in (2.8) running quark masses $m_i(m_i^2)$ which can be related to constituent quark masses [26], compensating Wilson coefficients appear because the matrix elements of O_{S} are to be evaluated at the hadronic mass scale $\mu \simeq 1$ GeV. The QCD corrected neutral Higgs exchange hamiltonian can therefore be written

$$\hat{\mathcal{H}}_{\mathrm{Higgs}}^{|\Delta S|=2} = -\frac{2\left(|v|^2 + |w|^2\right)}{M_{\mathrm{H}}^2\left(|v|^2 - |w|^2\right)^2} \sum_{i,j=\mathrm{c,t}} \lambda_i^{\mathrm{LR}} \lambda_j^{\mathrm{RL}} m_i(m_i^2)\, m_j(m_j^2)$$
$$\times \left[\alpha_{\mathrm{s}}(\mu^2)^2 / \left(\alpha_{\mathrm{s}}(m_i^2)\, \alpha_{\mathrm{s}}(m_j^2)\right)\right]^{4/b} O_{\mathrm{S}} + \mathrm{h.c.} \quad (3.10)$$

leaving out the negligibly small u-quark contribution.

Calculating short-distance corrections to (2.9)–(2.11) is more complicated because these hamiltonians originate from one-loop diagrams instead of the tree graphs considered so far. A systematic operator product expansion was performed in [27] to compute QCD corrections for the standard model hamiltonian (2.11). For the genuine L-R hamiltonians (2.9) and (2.10) we follow instead a physically intuitive, albeit only approximate method of Vysotskii [28] which was also used in [29]. In accordance with our treatment of $|\Delta S| = 1$ the dominant effects due to vector and scalar fields are taken into account exactly whereas the heavy quark effects are computed only approximatively.

Since our results differ from those of Bigi and Frère [29] we discuss in some detail the calculations for the pure W_1W_2 box diagram of fig. 1b, i.e. without unphysical scalars S_1, S_2. Before QCD corrections this diagram corresponds to the first term in (2.9), i.e.

$$\frac{2G_F^2M_1^2\beta}{\pi^2} O_S \sum_{i,j=c,t} \lambda_i^{LR}\lambda_j^{RL}\sqrt{x_ix_j}\, I_1(x_i, x_j, \beta), \tag{3.11}$$

omitting once more the negligible u-quark contribution.

The procedure of Vysotskii consists in keeping first the loop momentum k fixed to determine the various Wilson coefficients as functions of k^2 and other mass scales in the problem by dressing the box diagram with gluons in all possible ways. Finally, one has to compute integrals of the form

$$\int d^4k f(k^2)\left[\alpha_s(k^2)/\alpha_s(\mu^2)\right]^{\gamma}, \tag{3.12}$$

where $\gamma = 0$ corresponds to the original loop integrals. Two cases are now distinguished: if the original loop integral is dominated by a single mass m, i.e.

$$\int d^4k f(k^2) \sim (m^2)^c, \tag{3.13}$$

with some exponent c, one may approximate (3.12) by

$$\int d^4k f(k^2)\left[\alpha_s(k^2)/\alpha_s(\mu^2)\right]^{\gamma} \sim (m^2)^c\left[\alpha_s(m^2)/\alpha_s(\mu^2)\right]^{\gamma}. \tag{3.14}$$

This is the situation for the standard W_1W_1 box diagram where two quarks of the same type (cc or tt) are exchanged. For ct exchange in the standard box diagram and for all genuine L-R box diagrams, however, the corresponding loop integrals exhibit a logarithmic behaviour:

$$\int d^4k f(k^2) \sim \ln(m_2/m_1), \tag{3.15}$$

with two different masses m_1, m_2. In this case, the Wilson coefficients

$[\alpha_s(k^2)/\alpha_s(\mu^2)]^\gamma$ are to be averaged with respect to the measure

$$\left(\ln\left(m_2^2/m_1^2\right)\right)^{-1}\int_{m_1^2}^{m_2^2} dk^2/k^2. \tag{3.16}$$

A convenient way to parametrize such integrals is to define a function $R(\gamma, m_1, m_2)$ by

$$\left(\ln\left(m_2^2/m_1^2\right)\right)^{-1}\int_{m_1^2}^{m_2^2}\frac{dk^2}{k^2}\left[\alpha_s(k^2)/\alpha_s(\mu^2)\right]^\gamma = R(\gamma, m_1, m_2)\left[\alpha_s\left(m_1^2\right)/\alpha_s\left(\mu^2\right)\right]^\gamma. \tag{3.17}$$

From (3.17) one obtains

$$R(\gamma, m_1, m_2) = \frac{\ln\left(m_1/\Lambda_f\right)}{\ln\left(m_2/m_1\right)}\left\{\left[\alpha_s\left(m_2^2\right)/\alpha_s\left(m_1^2\right)\right]^{\gamma-1} - 1\right\}/(1-\gamma), \qquad \gamma \neq 1,$$

$$R(1, m_1, m_2) = \frac{\ln\left(m_1/\Lambda_f\right)}{\ln\left(m_2/m_1\right)}\ln\left[\alpha_s\left(m_1^2\right)/\alpha_s\left(m_2^2\right)\right], \tag{3.18}$$

with Λ_f the QCD parameter for N_f active flavours in the range m_1 to m_2. The Vysotskii procedure for logarithmically behaved loop integrals is therefore to compute the Wilson coefficients at the lower scale m_1^2 and to multiply by the appropriate factors $R(\gamma, m_1, m_2)$.

The first step in calculating QCD corrections to (3.11) is analogous to the $|\Delta S| = 1$ case. The four-quark operators appearing in the T-product

$$T(x, y; v, w) = T\left\{\bar{d}(x)\gamma^\mu P_R q(x)\bar{q}(y)\gamma_\mu P_R s(y),\right.$$
$$\left.\bar{d}(v)\gamma^\lambda P_L q(v)\bar{q}(w)\gamma_\lambda P_L s(w)\right\}, \tag{3.19}$$

where $q = c$ or t, are not multiplicatively renormalizable for $x \to y$, $v \to w$. In terms of the operators defined in (3.1) one may write

$$T(x, y; v, w) = \tfrac{1}{4}\left\{T_{++}(x, y; v, w) + T_{+-}(x, y; v, w)\right.$$
$$\left. + T_{-+}(x, y; v, w) + T_{--}(x, y; v, w)\right\}, \tag{3.20}$$

with ($\varepsilon_{\substack{R\\L}} = \pm 1$)

$$T_{\varepsilon_R\varepsilon_L}(x, y; v, w) = T\left\{\bar{d}(x)\gamma^\mu P_R q(x)\bar{q}(y)\gamma_\mu P_R s(y)\right.$$
$$+\varepsilon_R \bar{d}(x)\gamma^\mu P_R s(y)\bar{q}(y)\gamma_\mu P_R q(x), \bar{d}(v)\gamma^\lambda P_L q(v)\bar{q}(w)\gamma_\lambda P_L s(w)$$
$$\left.+\varepsilon_L \bar{d}(v)\gamma^\lambda P_L s(w)\bar{q}(w)\gamma_\lambda P_L q(v)\right\}. \tag{3.21}$$

Integrating out the vector boson fields W_1, W_2 then produces corresponding Wilson coefficients:

$$c_{\varepsilon_R\varepsilon_L}(k^2) = c_{\varepsilon_R}(k^2)c_{\varepsilon_L}(k^2) = [\alpha_s(k^2)/\alpha_s(M_2^2)]^{\gamma_{\varepsilon_R}}[\alpha_s(k^2)/\alpha_s(m_1^2)]^{\gamma_{\varepsilon_L}} \tag{3.22}$$

in momentum space with $\gamma_+ = -2/b$, $\gamma_- = 4/b$. Contracting the quark fields q in (3.21) gives rise to the same loop integral for all combinations $\varepsilon_R, \varepsilon_L$ multiplied by the four-quark operators

$$O_S - \tfrac{1}{2}(\varepsilon_R + \varepsilon_L + 3\varepsilon_R\varepsilon_L)O_V^{LR}, \tag{3.23}$$

where $O_V^{LR} = \bar{d}\gamma^\mu P_L s \bar{d}\gamma_\mu P_R s$. While O_S is multiplicatively renormalizable with anomalous dimension (3.9), O_V^{LR} is not. In analogy to (3.1) one defines

$$O_{\tilde{V}} = O_V^{LR} + \tfrac{2}{3}O_S, \tag{3.24}$$

with

$$\gamma(O_{\tilde{V}}) = g_s^2/8\pi^2. \tag{3.25}$$

The original T-product (3.19) thus gives rise to the short-distance corrected operator structure

$$\begin{aligned}&\tfrac{1}{3}[2c_{++}(k^2) + c_{--}(k^2)][\alpha_s(k^2)/\alpha_s(\mu^2)]^{-8/b}O_S \\ &\quad + \tfrac{1}{8}[-5c_{++}(k^2) + 3c_{+-}(k^2) + 3c_{-+}(k^2) - c_{--}(k^2)][\alpha_s(k^2)/\alpha_s(\mu^2)]^{1/b}O_{\tilde{V}}.\end{aligned} \tag{3.26}$$

Finally, there is a k^2-dependence due to gluonic corrections to the quark propagators producing a factor $m_i(k^2)m_j(k^2)$ instead of the bare quark masses in (3.11). Since the loop integral $I_1(x_i, x_j, \beta)$ has a dominant logarithmic behaviour

$$I_1(x_i, x_j, \beta) \sim \ln x_i \qquad (\text{for } x_i \geqslant x_j), \tag{3.27}$$

the appropriate factors $R(\gamma, m_i, M_1)$ must still be applied.

Putting everything together, we obtain instead of (3.11) the short-distance corrected effective hamiltonian

$$
\begin{aligned}
\mathcal{H}^{|\Delta S|=2}_{W_1W_2} = \frac{2G_F^2M_1^2\beta}{\pi^2} \\
&\times\Big\{\tfrac{1}{3}O_S\Big[\lambda_c^{LR}\lambda_c^{RL}x_cI_1(x_c,x_c,\beta)c_S(m_c^2) \\
&\times(2c_{++}(m_c^2)R(-4/b,m_c,M_1)+c_{--}(m_c^2)R(8/b,m_c,M_1)) \\
&+\big\{(\lambda_c^{LR}\lambda_t^{RL}+\lambda_t^{LR}\lambda_c^{RL})\sqrt{x_cx_t}\,I_1(x_c,x_t,\beta)\sqrt{c_S(m_c^2)c_S(m_t^2)} \\
&+\lambda_t^{LR}\lambda_t^{RL}x_tI_1(x_t,x_t,\beta)c_S(m_t^2)\big\} \\
&\times(2c_{++}(m_t^2)R(-4/b,m_t,M_1) \\
&+c_{--}(m_t^2)R(8/b,m_t,M_1))\Big] \\
&+\tfrac{1}{8}O_{\tilde{V}}\Big[\lambda_c^{LR}\lambda_c^{RL}x_cI_1(x_c,x_c,\beta)c_{\tilde{V}}(m_c^2) \\
&\times\{-5c_{++}(m_c^2)R(5/b,m_c,M_1) \\
&+3(c_{+-}(m_c^2)+c_{-+}(m_c^2))R(11/b,m_c,M_1) \\
&-c_{--}(m_c^2)R(17/b,m_c,M_1)\} \\
&+\big\{(\lambda_c^{LR}\lambda_t^{RL}+\lambda_t^{LR}\lambda_c^{RL})\sqrt{x_cx_t}\,I_1(x_c,x_t,\beta) \\
&\times\sqrt{c_S(m_c^2)/c_S(m_t^2)}+\lambda_t^{LR}\lambda_t^{RL}x_tI_1(x_t,x_t,\beta)\big\}c_{\tilde{V}}(m_t^2) \\
&\times\{-5c_{++}(m_t^2)R(5/b,m_t,M_1)+3(c_{+-}(m_t^2)+c_{-+}(m_t^2)) \\
&\times R(11/b,m_t,M_1)-c_{--}(m_t^2)R(17/b,m_t,M_1)\}\Big]\Big\}, \qquad (3.28)
\end{aligned}
$$

where

$$c_S(m^2) = [\alpha_s(m^2)/\alpha_s(\mu^2)]^{-8/b},$$

$$c_{\tilde{V}}(m^2) = [\alpha_s(m^2)/\alpha_s(\mu^2)]^{1/b}. \tag{3.29}$$

All quark masses in (3.28) are running quark masses $m_i(m_i^2)$.

In a similar way QCD corrections can be calculated for the box diagrams with one or two vector bosons replaced by scalars. We limit ourselves to the dominant diagrams in the 't Hooft-Feynman gauge, i.e. the S_1W_2 and $W_1\Phi^{\pm}$ box graphs. Both diagrams are proportional to $I_2(x_i, x_j, \beta$ or $\beta_H)$ and their QCD corrections are very similar. The final results are

$$\begin{aligned}\mathcal{H}^{|\Delta S|=2}_{S_1W_2} = -\frac{G_F^2M_1^2\beta}{2\pi^2}\Big\{&\lambda_c^{LR}\lambda_c^{RL}x_cI_2(x_c, x_c, \beta)c_S(m_c^2) + (\lambda_c^{LR}\lambda_t^{RL} + \lambda_t^{LR}\lambda_c^{RL})\\ &\times\sqrt{x_cx_t}\,I_2(x_c, x_t, \beta)\sqrt{c_S(m_c^2)c_S(m_t^2)}\\ &+\lambda_t^{LR}\lambda_t^{RL}x_tI_2(x_t, x_t, \beta)c_S(m_t^2)\Big\}\\ \times\Big\{&\tfrac{1}{3}O_S[2c_+(M_1^2)R(-2/b, M_1, M_2) + c_-(M_1^2)R(4/b, M_1, M_2)]\\ &+\frac{O_{\tilde{V}}c_{\tilde{V}}(M_1^2)}{4c_S(M_1^2)}[c_-(M_1^2)R(13/b, M_1, M_2) - c_+(M_1^2)R(7/b, M_1, M_2)]\Big\},\end{aligned} \tag{3.30}$$

where the functions $c_{\varepsilon_R}(M_1^2)$ are defined in (3.22) and

$$\begin{aligned}\mathcal{H}^{|\Delta S|=2}_{W_1\Phi^{\pm}} = -\frac{G_F^2M_1^2\beta_H}{2\pi^2}\Big\{&\lambda_c^{LR}\lambda_c^{RL}x_cI_2(x_c, x_c, \beta_H)c_S(m_c^2) + (\lambda_c^{LR}\lambda_t^{RL} + \lambda_t^{LR}\lambda_c^{RL})\\ &\times\sqrt{x_cx_t}\,I_2(x_c, x_t, \beta_H)\sqrt{c_S(m_c^2)c_S(m_t^2)}\\ &+\lambda_t^{LR}\lambda_t^{RL}x_tI_2(x_t, x_t, \beta_H)c_S(m_t^2)\Big\}\\ \times\Big\{&\tfrac{1}{3}O_S[2R(-2/b, M_1, M_H) + R(4/b, M_1, M_H)]\\ &+\frac{O_{\tilde{V}}c_{\tilde{V}}(M_1^2)}{4c_S(M_1^2)}[R(13/b, M_1, M_H) - R(7/b, M_1, M_H)]\Big\}.\end{aligned} \tag{3.31}$$

The hamiltonians (3.28), (3.30) and (3.31) together with (3.10) represent the short-distance corrected genuine L-R $|\Delta S| = 2$ effective interactions. Adding (2.11) including QCD corrections [27] finally yields the total effective $|\Delta S| = 2$ hamiltonian.

The matrix elements of all four-quark operators appearing in the $|\Delta S| = 1, 2$ hamiltonians are to be calculated in an effective three-flavour theory at a typical hadronic mass scale μ. Since the various hamiltonians in this section must be independent of the arbitrary renormalization point μ^2 we assume that whatever approximate method is used to calculate hadronic matrix elements applies to the renormalization group-invariant operators like $[\alpha_s(\mu^2)]^{4/b} O_-^{LL}$, $[\alpha_s(\mu^2)]^{8/b} O_S$, etc. In this way, our ignorance about the correct value of μ^2 is hidden in the fundamental ignorance how to calculate hadronic matrix elements of quark operators. In practice, such a procedure amounts to setting $\alpha_s(\mu^2) = 1$ in all Wilson coefficients.

4. Hadronic matrix elements

In this section we calculate the appropriate hadronic matrix elements of the four-quark operators contained in the effective hamiltonians of sect. 3 using the vacuum insertion technique [30]. Much work has been done on the hadronic matrix elements in the standard model employing different methods with rather different results [31]. In particular, corrections to the vacuum insertion coming from one-particle intermediate states have been computed for the $K^0\bar{K}^0$ transition in the standard model [32]. It turns out that those additional contributions cancel more or less so that the vacuum insertion result remains roughly valid. Although such a cancellation need not necessarily hold for other matrix elements, the vacuum insertion method is at least expected to give the right order of magnitude. It also has the additional merits of simplicity and general applicability.

4.1. $|\Delta S| = 2$

Three operators determine the effective $|\Delta S| = 2$ hamiltonian of sect. 3:

$$O_V^{LL} = \bar{d}\gamma^\mu P_L s \bar{d}\gamma_\mu P_L s\,,$$

$$O_S = \bar{d}P_L s \bar{d} P_R s\,,$$

$$O_{\tilde{V}} = \bar{d}\gamma^\mu P_L s \bar{d}\gamma_\mu P_R s + \tfrac{2}{3} O_S\,. \tag{4.1}$$

We define $|\bar{K}^0(-\boldsymbol{p})\rangle = -CP|K^0(\boldsymbol{p})\rangle$ and use

$$\langle 0|\bar{d}\gamma^\mu\gamma_5 s|\bar{K}^0(p_K)\rangle = -\langle K^0(p_K)|\bar{d}\gamma^\mu\gamma_5 s|0\rangle = i\sqrt{2} f_K p_K^\mu\,, \tag{4.2}$$

with $f_K \simeq 1.22\ f_\pi$ and $f_\pi = 93$ MeV. Vacuum insertion then gives the matrix elements

$$\langle K^0|O_V^{LL}|\bar{K}^0\rangle = \tfrac{4}{3}f_K^2 m_K^2,$$

$$\langle K^0|O_S|\bar{K}^0\rangle = \left(\frac{m_K^2}{(m_s+m_d)^2}+\tfrac{1}{6}\right)f_K^2 m_K^2,$$

$$\langle K^0|O_{\tilde{V}}|\bar{K}^0\rangle = -\tfrac{8}{9}f_K^2 m_K^2, \tag{4.3}$$

with $m_K^2/(m_s+m_d)^2 \simeq 11$. Note that in contrast to O_S both O_V^{LL} and $O_{\tilde{V}}$ have helicity-suppressed matrix elements. It is therefore a reasonable approximation to drop the terms containing $O_{\tilde{V}}$ in the hamiltonians (3.28), (3.30) and (3.31).

4.2. $|\Delta S| = 1$

In this case we have to consider matrix elements of the type $\langle \pi\pi(I=0,2)|O|\bar{K}^0\rangle$ where $I=0,2$ denotes the isospin of the two pions in the final state. In addition to (4.2) we also need

$$\langle \pi^-(p_-)|\bar{d}\gamma^\mu\gamma_5 u|0\rangle = -i\sqrt{2}f_\pi p_-^\mu,$$

$$\langle \pi^+(p_+)|\bar{u}\gamma^\mu s|\bar{K}^0(p_K)\rangle = f_+(t)(p_K+p_+)^\mu + f_-(t)(p_K-p_+)^\mu, \tag{4.4}$$

$$p_K = p_+ + p_-,$$

$$t = (p_K-p_+)^2 = m_\pi^2,$$

with $f_+(m_\pi^2)\simeq 1$. $f_-(m_\pi^2)$ can be neglected because the relevant ratio is

$$|f_-(m_\pi^2)/f_+(m_\pi^2)|m_\pi^2/m_K^2 \ll 1. \tag{4.5}$$

We also make use of vector current conservation in the form

$$\langle \pi^+\pi^-|\bar{u}\gamma_\mu u|0\rangle p_K^\mu = \langle \pi^+\pi^-|\bar{d}\gamma_\mu d|0\rangle p_K^\mu = 0. \tag{4.6}$$

For the W_1W_2-mixing amplitudes and the penguins the following scalar amplitudes will be needed:

$$\langle \pi^+\pi^-|\bar{d}d|0\rangle = \langle \pi^+\pi^-|\bar{u}u|0\rangle = A\left(1-m_K^2/m_\sigma^2\right)^{-1},$$

$$\langle \pi^-|\bar{d}\gamma_5 u|0\rangle = -i\sqrt{2}f_\pi A\left(1-m_\pi^2/m_\sigma^2\right)^{-1},$$

$$\langle 0|\bar{d}\gamma_5 s|\bar{K}^0\rangle = -i\sqrt{2}f_K A\left(1-m_K^2/m_\sigma^2\right)^{-1},$$

$$\langle \pi^+|\bar{u}s|\bar{K}^0\rangle = f_+(m_\pi^2)A\left(1-m_\pi^2/m_\sigma^2\right)^{-1}, \tag{4.7}$$

with

$$A = \frac{m_\pi^2}{m_u + m_d} = \frac{m_K^2}{m_s + m_d} = \frac{m_K^2 - m_\pi^2}{m_s - m_u} . \tag{4.8}$$

The scalar matrix elements are corrected [33] for momentum dependence and we take $m_\sigma = 700$ MeV.

It is convenient to parametrize the matrix elements of $|\Delta S| = 1$ operators in terms of the three products of basic matrix elements which appear in the calculation:

$$\begin{aligned} X &= -\langle \pi^- | \bar{d}\gamma^\mu\gamma_5 u | 0 \rangle \langle \pi^+ | \bar{u}\gamma_\mu s | \bar{K}^0 \rangle = i\sqrt{2}\, f_\pi f_+ (m_\pi^2)(m_K^2 - m_\pi^2), \\ Y &= -\langle \pi^+\pi^- | \bar{u}u | 0 \rangle \langle 0 | \bar{d}\gamma_5 s | \bar{K}^0 \rangle = i\sqrt{2}\, f_K A^2 (1 - m_K^2/m_\sigma^2)^{-2}, \\ Z &= -\langle \pi^- | \bar{d}\gamma_5 u | 0 \rangle \langle \pi^+ | \bar{u}s | \bar{K}^0 \rangle = i\sqrt{2}\, f_\pi f_+ (m_\pi^2) A^2 (1 - m_\pi^2/m_\sigma^2)^{-2}. \end{aligned} \tag{4.9}$$

The matrix element X is obviously helicity suppressed compared to Y and Z. In terms of (4.9) we find

$$\begin{aligned} \langle I=0 | O_+^{\substack{LL\\RR}} | \bar{K}^0 \rangle &= \pm \tfrac{1}{3}\sqrt{\tfrac{1}{6}}\, X, \\ \langle I=2 | O_+^{\substack{LL\\RR}} | \bar{K}^0 \rangle &= \pm \tfrac{2}{3}\sqrt{\tfrac{1}{3}}\, X, \\ \langle I=0 | O_-^{\substack{LL\\RR}} | \bar{K}^0 \rangle &= \pm \tfrac{1}{2}\sqrt{\tfrac{1}{6}}\, X, \\ \langle I=2 | O_-^{\substack{LL\\RR}} | \bar{K}^0 \rangle &= 0, \\ \langle I=0 | O_+^{\substack{LR\\RL}} | \bar{K}^0 \rangle &= \pm \tfrac{4}{9}\sqrt{\tfrac{1}{6}}\, X, \\ \langle I=2 | O_+^{\substack{LR\\RL}} | \bar{K}^0 \rangle &= \pm \tfrac{2}{9}\sqrt{\tfrac{1}{3}}\, X, \\ \langle I=0 | O_-^{\substack{LR\\RL}} | \bar{K}^0 \rangle &= \mp \sqrt{\tfrac{1}{6}}\left(\tfrac{1}{18}X + \tfrac{1}{2}Y + \tfrac{1}{6}Z\right), \\ \langle I=2 | O_-^{\substack{LR\\RL}} | \bar{K}^0 \rangle &= \mp \tfrac{1}{6}\sqrt{\tfrac{1}{3}}\left(\tfrac{1}{6}X - Z\right). \end{aligned} \tag{4.10}$$

For the L-L and R-R penguins the relevant operators are

$$O_{\mathrm{P}}^{\substack{\mathrm{L}\\\mathrm{R}}} = \tfrac{1}{4}\bar{d}\gamma^{\mu}P_{\substack{\mathrm{L}\\\mathrm{R}}}\lambda^{a}s\left(\bar{u}\gamma_{\mu}\lambda^{a}u + \bar{d}\gamma_{\mu}\lambda^{a}d + \bar{s}\gamma_{\mu}\lambda^{a}s\right). \tag{4.11}$$

Since all penguin operators have $I = \frac{1}{2}$ we only need

$$\langle I=0|O_{\mathrm{P}}^{\substack{\mathrm{L}\\\mathrm{R}}}|\bar{K}^{0}\rangle = \pm\tfrac{2}{3}\sqrt{\tfrac{1}{6}}\left(\tfrac{1}{2}X - Y + Z\right). \tag{4.12}$$

From (3.7) the relevant operators for the L-R penguin are

$$O_{\mathrm{LR-P}}^{\substack{\mathrm{L}\\\mathrm{R}}} = \bar{d}\frac{i\sigma_{\mu\nu}k^{\nu}}{4k^{2}}P_{\substack{\mathrm{L}\\\mathrm{R}}}\lambda^{a}s\left(\bar{u}\gamma^{\mu}\lambda^{a}u + \bar{d}\gamma^{\mu}\lambda^{a}d + \bar{s}\gamma^{\mu}\lambda^{a}s\right). \tag{4.13}$$

Replacing $-k^{2}$ by the square of a "typical" hadronic mass μ_{h} involved in the problem (we take $\mu_{\mathrm{h}} = m_{\mathrm{K}}$) we obtain [34]

$$\langle I=0|O_{\mathrm{LR-P}}^{\substack{\mathrm{L}\\\mathrm{R}}}|\bar{K}^{0}\rangle = \pm\frac{\sqrt{2}\,m_{\mathrm{const}}}{\sqrt{3}\,\mu_{\mathrm{h}}^{2}}(Y+Z), \tag{4.14}$$

where $m_{\mathrm{const}} \simeq 300$ MeV is a constituent quark mass.

We are now in a position to calculate the amplitudes

$$M_{12} = \langle K^{0}|\mathcal{H}_{\mathrm{eff}}^{|\Delta S|=2}|\bar{K}^{0}\rangle/2m_{\mathrm{K}},$$

$$A_{0,2}^{*} = \langle\pi\pi(I=0,2)| - i\mathcal{H}_{\mathrm{eff}}^{|\Delta S|=1}|\bar{K}^{0}\rangle, \tag{4.15}$$

which determine the *CP*-violating observables ε, ε'. To keep track of possible long-distance contributions [35] we define

$$M_{12}^{\mathrm{theor}} = M_{12}^{\mathrm{Higgs}} + M_{12}^{W_1W_1} + M_{12}^{W_1W_2} + M_{12}^{S_1W_2} + M_{12}^{W_1\Phi^{\pm}}, \tag{4.16}$$

corresponding to the QCD corrected hamiltonians of sect. 3 and

$$\tilde{M}_{12} = \frac{2M_{12}^{\mathrm{theor}}}{\Delta m_{\mathrm{K_LK_S}}}, \tag{4.17}$$

with $\Delta m_{\mathrm{K_LK_S}} = m_{\mathrm{K_L}} - m_{\mathrm{K_S}}$. Then

$$\mathrm{Re}\,\tilde{M}_{12} = 1 - \Delta m_{\mathrm{K_LK_S}}^{\mathrm{LD}}/\Delta m_{\mathrm{K_LK_S}}, \tag{4.18}$$

where the superscript LD denotes long-distance contributions. Long-distance effects could also be present [36] in

$$\varepsilon_{\mathrm{m}} = \frac{\operatorname{Im} M_{12}}{\operatorname{Re} M_{12}} = \operatorname{Im} \tilde{M}_{12} + \varepsilon_{\mathrm{m}}^{\mathrm{LD}} . \tag{4.19}$$

Normalizing $A_{0,2}$ by the experimental values of their real parts

$$\tilde{A}_0 = A_0/\operatorname{Re} A_0 , \qquad \tilde{A}_2 = A_2/\operatorname{Re} A_2 , \tag{4.20}$$

we get for the phases

$$\xi_{0,2} = -\operatorname{Im} \tilde{A}^*_{0,2} . \tag{4.21}$$

The CP-violating parameters ε, ε' can then be written

$$\varepsilon = \tfrac{1}{2}\sqrt{\tfrac{1}{2}}\, \mathrm{e}^{i\pi/4}(\varepsilon_{\mathrm{m}} + 2\xi_0) ,$$

$$\varepsilon' = \sqrt{\tfrac{1}{2}} \exp\left(i\left(\tfrac{1}{2}\pi + \delta_2 - \delta_0\right)\right) \frac{\operatorname{Re} A_2}{\operatorname{Re} A_0} (\xi_2 - \xi_0) , \tag{4.22}$$

where $\delta_{0,2}$ are the strong interaction $\pi\pi$ phase shifts and $\operatorname{Re} A_2/\operatorname{Re} A_0 \simeq \frac{1}{22}$. Experimentally, $\delta_0 - \delta_2 \simeq \frac{1}{4}\pi$ and ε'/ε is therefore approximately real.

We close this section with a comment on the $\Delta I = \frac{1}{2}$ rule. Since we have normalized $A_{0,2}$ by the experimental values of $\operatorname{Re} A_{0,2}$ we should get $\operatorname{Re} \tilde{A}_{0,2} = 1$ if the theory was correct. Putting everything together we find indeed $\operatorname{Re} \tilde{A}_2 \simeq 1$, but $\operatorname{Re} \tilde{A}_0 < 1$. This is just the well-known problem of the standard model to reproduce the $\Delta I = \frac{1}{2}$ rule because L-R symmetry does not help in this case: all terms involving the right-handed sector are proportional to β or ξ_{LR} and can therefore not contribute appreciably to the real parts. It is beyond the scope of the present paper to discuss the origin of the $\Delta I = \frac{1}{2}$ problem but we will come back to this point in sect. 5 to investigate how it influences our predictions for ε'/ε.

5. Spontaneous *CP* violation in the small-phase approximation

L-R symmetric models have the attractive feature that both P and CP can be broken spontaneously already with a minimal Higgs sector [9]. If the Yukawa interaction is written

$$-\mathcal{L}_{\mathrm{Y}} = \bar{q}_{\mathrm{L}} \Gamma \Phi q_{\mathrm{R}} + \bar{q}_{\mathrm{L}} \Delta \tilde{\Phi} q_{\mathrm{R}} + \mathrm{h.c.} ,$$

$$\tilde{\Phi} = \tau_2 \Phi^* \tau_2 , \tag{5.1}$$

where $q_{\mathrm{L,R}}$ are weak eigenfields of quarks and Φ is defined in (2.1), P invariance restricts the coupling matrices Γ, Δ to be hermitian in a certain basis. In such a basis the additional requirement of CP invariance leads to two different cases [10]:

(i) *Manifest CP*: Γ, Δ are real and therefore symmetric implying $K_R = K_L^*$ with a certain phase convention, but no further constraints on quark masses and mixing angles.

(ii) *Non-manifest CP*: P and CP transformations are non-aligned in flavour space in this case allowing for a single model given by

$$\Gamma = \mathrm{diag}(g_1, g_2, g_3), \qquad \Delta = i\begin{pmatrix} 0 & h_1 & h_2 \\ -h_1 & 0 & h_3 \\ -h_2 & -h_3 & 0 \end{pmatrix}, \tag{5.2}$$

$$h_i, g_i \text{ real} \qquad |g_1| \leqslant |g_2| \leqslant |g_3| .$$

If all $g_i \neq 0$, $h_i \neq 0$ there is no choice of quark phases which would make Γ, Δ real and symmetric but the lagrangian nevertheless conserves CP prior to spontaneous symmetry breaking. The CP phenomenology of this model was discussed in [11]. For the special case $h_2 = 0$ the model is equivalent to the unique model from horizontal symmetries [16]. Moreover, for $h_2 = 0$ phase transformations can be performed to make Γ and Δ real and symmetric. In other words, we recover manifest CP in this special case. We also want to emphasize that the mass matrices of model (5.2) lead to similar physical consequences as the ones proposed and investigated by Stech [37].

The purpose of this section is to discuss in detail the CP phenomenology of the remaining case of manifest CP. At first sight, this case seems to comprise a tremendous number of different models. What we intend to show, however, is that once the quark masses and mixing angles (excluding the phases) are fixed there is very little freedom left: the Higgs phase $\alpha = \arg(vw)$ can still be chosen and in principle the ratio $|w/v|$. However, in view of the discussion of sect. 2 we always assume $|w/v| = \mathrm{O}(m_b/m_t)$.

In order to simplify the counting of parameters which we will perform for an arbitrary number n_G of generations we choose a gauge and an energy scale such that $v = 1$ and, furthermore, a quark basis where Γ is diagonal. Since the quark mass matrices are symmetric in the case of manifest CP they can be diagonalized in the following way:

$$\hat{M}_u = \sqrt{\tfrac{1}{2}}\, U_L^{p\dagger}(\Gamma + w^*\Delta) U_L^{p*},$$

$$\hat{M}_d = \sqrt{\tfrac{1}{2}}\, U_L^{n\dagger}(w\Gamma + \Delta) U_L^{n*}, \tag{5.3}$$

with diagonal, positive definite $\hat{M}_u$, $\hat{M}_d$. This corresponds to a choice of phases where the mixing matrices are related as

$$K_L = U_L^{p\dagger} U_L^n = K_R^* . \tag{5.4}$$

We now claim that at least as far as the number of free parameters is concerned Γ, Δ and all phases in K_L (and therefore in K_R) can be expressed in terms of $\hat{M}_u, \hat{M}_d, w$ and the $\frac{1}{2}n_G(n_G-1)$ generalized Euler (mixing) angles. Since the matrices in (5.3) are symmetric, (5.3) consists of $2n_G(n_G+1)$ real equations. Keeping $\hat{M}_u, \hat{M}_d, w$ and the mixing angles fixed there are n_G real parameters in Γ, $\frac{1}{2}n_G(n_G+1)$ in Δ, n_G^2 in U_L^p and $\frac{1}{2}n_G(n_G+1)$ phases in K_L. Since U_L^n can be expressed by K_L and U_L^p through (5.4), we have altogether $2n_G(n_G+1)$ unknowns matching exactly the number of equations. Of course, it is in general impossible to solve (5.3) exactly and we will have to be satisfied with approximate solutions.

CP violation is characterized by the parameter

$$y = \mathrm{Im}\, w = |w| \sin\alpha . \tag{5.5}$$

Following Chang [12] we are going to calculate the phases of K_L to first order in y. This should certainly be a good approximation for small α and could in principle be valid more generally keeping in mind the natural relation $|w| = \mathrm{O}(m_b/m_t)$.

For $y=0$ the quark mass matrices $M_{p,n}^0$ are real and symmetric. They can therefore be diagonalized with orthogonal matrices $O_{p,n}$:

$$O_p^T M_p^0 O_p = \hat{\mu}_u , \qquad O_n^T M_n^0 O_n = \hat{\mu}_d . \tag{5.6}$$

The diagonal matrices $\hat{\mu}_u, \hat{\mu}_d$ contain the quark masses* up to signs

$$\begin{aligned} &\mu_i = \eta_i m_i \qquad (i = u, c, t) \\ &\mu_j = \eta'_j m_j \qquad (j = d, s, b) \qquad \eta_i = \eta'^2_j = 1 . \end{aligned} \tag{5.7}$$

We want to emphasize the importance of keeping track of the sign factors η_i, η'_j which were disregarded in [12]. Their physical significance will become apparent shortly.

With an appropriate choice of signs

$$K = O_p^T O_n \tag{5.8}$$

is the KM matrix [38] with $e^{i\delta} = \pm 1$. A straightforward calculation to first order in y gives

$$\begin{aligned} K_{L,ij} &= \zeta_i \zeta'^*_j \left(K_{ij} + iy K_{ij}^{(1)} \right), \\ K_R &= K_L^* , \\ (1 - w_\alpha^2) K_{ij}^{(1)} &= \sum_{k=u,c,t} \left(K \hat{\mu}_d K^T \right)_{ik} K_{kj} / (\mu_i + \mu_k) \\ &\quad + \sum_{l=d,s,b} K_{il} \left(K^T \hat{\mu}_u K \right)_{lj} / (\mu_l + \mu_j) - w_\alpha K_{ij} , \\ \zeta_i^2 &= \eta_i , \qquad \zeta'^2_j = \eta'_j , \qquad w_\alpha = \mathrm{Re}\, w = |w| \cos\alpha . \end{aligned} \tag{5.9}$$

* Corrections to the relations (5.7) are of order y^2.

Instead of the basis defined by (5.9) it will be more convenient to use a basis where K_{L} has the standard form of the KM matrix:

$$K_{\mathrm{L}} = K_{\mathrm{KM}} = \mathrm{e}^{-i\beta}\left(K + iyK^{(1)}\right)\mathrm{e}^{i\gamma},$$
$$K_{\mathrm{R}} = \eta\,\mathrm{e}^{-i\beta}\left(K - iyK^{(1)}\right)\mathrm{e}^{i\gamma}\eta' = \eta\,\mathrm{e}^{-2i\beta}K^{*}_{\mathrm{KM}}\mathrm{e}^{2i\gamma}\eta'. \tag{5.10}$$

$\mathrm{e}^{i\beta}, \mathrm{e}^{i\gamma}$ denote diagonal phase matrices with

$$\beta_i = y\left(K^{(1)}_{i\mathrm{d}}/K_{i\mathrm{d}} - K^{(1)}_{\mathrm{ud}}/K_{\mathrm{ud}}\right),$$
$$\gamma_j = -yK^{(1)}_{\mathrm{u}j}/K_{\mathrm{u}j}, \tag{5.11}$$

and η, η' are diagonal sign matrices with elements defined in (5.7). Also to first order in y the KM phase δ can be derived from the relation

$$\mathrm{Im}\,K_{\mathrm{KM},ij} = y\left[K^{(1)}_{ij} - K_{ij}\left(K^{(1)}_{i\mathrm{d}}/K_{i\mathrm{d}} + K^{(1)}_{\mathrm{u}j}/K_{\mathrm{u}j} - K^{(1)}_{\mathrm{ud}}/K_{\mathrm{ud}}\right)\right]. \tag{5.12}$$

Within this approximation we have therefore obtained all the phases of K_{L} and K_{R}.

Before proceeding we take note of an intrinsic limitation of the small-phase approximation described above. Taking the limit $s_3 \to 0$ in K does not imply vanishing CP violation in $K_{\mathrm{L}} = K_{\mathrm{KM}}$ because the expansion in y generates an induced s_3 of the form

$$s_3^{\mathrm{ind}} = |yK^{(1)}_{\mathrm{ub}}|/s_1 \sim \frac{0.3|y|}{1 - w_\alpha^2}s_2 \ll s_2. \tag{5.13}$$

There is no problem in practice, however: for $s_3 \lesssim s_3^{\mathrm{ind}}$ the standard model share of CP violation is negligible and the right-handed sector is practically unaffected by such a small s_3.

We define the normalized $|\Delta S| = 2$ genuine L-R amplitude as (see (4.16))

$$\tilde{M}_{12}^{\mathrm{LR}} = 2\left(M_{12}^{\mathrm{Higgs}} + M_{12}^{W_1W_2} + M_{12}^{S_1W_2} + M_{12}^{W_1\Phi^\pm}\right)/\Delta m_{\mathrm{K_L K_S}}. \tag{5.14}$$

From sect. 3 $\tilde{M}_{12}^{\mathrm{LR}}$ is seen to have the general structure

$$\tilde{M}_{12}^{\mathrm{LR}} = \sum_{i,j=\mathrm{u,c,t}} \lambda_i^{\mathrm{LR}}\lambda_j^{\mathrm{RL}}g(x_i, x_j), \tag{5.15}$$

where $g(x_i, x_j)$ is a symmetric function of x_i, x_j. In the small-phase approximation one therefore obtains

$$\tilde{M}_{12}^{\mathrm{LR}} = \left[1 + 2i(\gamma_{\mathrm{s}} - \gamma_{\mathrm{d}})\right]\eta'_{\mathrm{d}}\eta'_{\mathrm{s}}\sum_{i,j} g(x_i, x_j)\eta_i\eta_j K_{i\mathrm{d}}K_{i\mathrm{s}}K_{j\mathrm{d}}K_{j\mathrm{s}}, \tag{5.16}$$

or

$$\operatorname{Im}\tilde{M}_{12}^{\mathrm{LR}} = 2(\gamma_s - \gamma_d)\operatorname{Re}\tilde{M}_{12}^{\mathrm{LR}}. \tag{5.17}$$

We can also express $\gamma_s - \gamma_d$ in terms of quark masses and mixing angles. The result depends crucially on sign($\eta'_d\eta'_s$) and we find approximately

$$\gamma_s - \gamma_d \simeq \frac{y}{1-w_\alpha^2} \times \begin{cases} \mu_c/\mu_s, & \eta'_s = \eta'_d \\ \dfrac{c_1^2\mu_u}{2\mu_d}(1+2m_d/m_s) - \mu_s/\mu_c + \dfrac{\mu_t s_3^2}{2\mu_s} & \\ \quad + \dfrac{(\mu_c + s_2^2\mu_t)(c_1^2 - s_1^2 m_s/m_d)(1+2m_d/m_s)}{2\mu_s}, & \eta'_s = -\eta'_d \end{cases} \tag{5.18}$$

From (5.5), (5.18) and with the estimate

$$|w| \sim m_b/m_t \sim m_s/m_c \tag{5.19}$$

we can establish a relation between $\gamma_s - \gamma_d$ and the Higgs phase α. For $\eta'_s = \eta'_d$ we get immediately $|\gamma_s - \gamma_d| \sim |\sin\alpha|$ and thus the approximation is expected to break down once α becomes big. For $\eta'_s = -\eta'_d$ the relation depends very much on the sign factors in (5.7) and on the precise value of $c_1^2 - s_1^2 m_s/m_d \sim 0$. In any case,

$$|\gamma_s - \gamma_d| \lesssim \tfrac{1}{10}|\sin\alpha|, \qquad \eta'_s = -\eta'_d, \tag{5.20}$$

so that $\gamma_s - \gamma_d$ is at least one order of magnitude smaller for given α. Since $\gamma_s - \gamma_d$ will essentially determine all *CP*-violating parameters for $\eta'_s = -\eta'_d$ we conclude from (5.20) that the small-y approximation may be expected to make sense regardless of the size of α.

With the definition

$$\varepsilon_m = \varepsilon_m^{\mathrm{LR}} + \varepsilon_m^{\mathrm{SM}}, \qquad \varepsilon_m^{\mathrm{LR}} = \operatorname{Im}\tilde{M}_{12}^{\mathrm{LR}}, \tag{5.21}$$

where the standard model contribution $\varepsilon_m^{\mathrm{SM}}$ includes $\varepsilon_m^{\mathrm{LD}}$ from (4.19), we can write (5.17) as

$$\gamma_s - \gamma_d = \frac{\varepsilon_m - \varepsilon_m^{\mathrm{SM}}}{2\operatorname{Re}\tilde{M}_{12}^{\mathrm{LR}}} \simeq \frac{\sqrt{2}\,e^{-i\pi/4}(\varepsilon - \varepsilon_{\mathrm{SM}})}{\operatorname{Re}\tilde{M}_{12}^{\mathrm{LR}}}. \tag{5.22}$$

For the second, approximate equality we have neglected the ξ_0 contribution to ε defined in (4.22). From the definition (5.14) $\operatorname{Re}\tilde{M}_{12}^{\mathrm{LR}}$ is seen to denote the genuine

L-R contribution to the normalized K_LK_S mass difference. Although $g(x_i, x_j)$ in (5.15) is a very complicated function we can write down an approximate relation

$$\mathrm{Re}\,\tilde{M}_{12}^{\mathrm{LR}} \simeq -\eta'_{\mathrm{d}}\eta'_{\mathrm{s}}\{3f_{\mathrm{QCD}}(M_2/\mathrm{TeV})^{-2}(1+0.05\ln(M_2/2.5\ \mathrm{TeV})) + 57g_{\mathrm{QCD}}(M_{\mathrm{H}}/\mathrm{TeV})^{-2}\} \quad (5.23)$$

using the fact that $\mathrm{Re}\,\tilde{M}_{12}^{\mathrm{LR}}$ is dominated by the c-quark contribution because of the small s_2. In (5.23) f_{QCD}, g_{QCD} exhibit the size of QCD corrections* as calculated in sect. 3 with $f_{\mathrm{QCD}} \simeq 2.3$, $g_{\mathrm{QCD}} \simeq 1.8$ for $\Lambda_4 = 200$ MeV and we have taken $m_{\mathrm{c}}(m_{\mathrm{c}}^2) = 1.27$ GeV [26].

The relation (5.23) is an especially clear example for the physical relevance of the sign factors (5.7). For $\eta'_{\mathrm{d}} = -\eta'_{\mathrm{s}}$ the L-R terms yield a positive contribution to the experimentally positive $\Delta m_{\mathrm{K_LK_S}}$ while they have the opposite sign for $\eta'_{\mathrm{d}} = \eta'_{\mathrm{s}}$. Only the latter case is usually encountered in the literature because it also arises in so-called manifestly L-R symmetric models characterized by $K_{\mathrm{R}} = K_{\mathrm{L}}$. Although the notorious long-distance contributions [35] prevent a definite conclusion the case $\eta'_{\mathrm{d}} = -\eta'_{\mathrm{s}}$ is an attractive possibility in view of the possibly too small standard model contribution. In either case, we may estimate lower bounds

$$M_2 \gtrsim 2.5\ \mathrm{TeV}, \qquad M_{\mathrm{H}} \gtrsim 10\ \mathrm{TeV} \quad (5.24)$$

from (5.23) by demanding $|\mathrm{Re}\,\tilde{M}_{12}^{\mathrm{LR}}| \leqslant 1$ which amounts to assuming that the absolute value of L-R contributions is at most equal to the experimental value of $\Delta m_{\mathrm{K_LK_S}}$.

(5.18), (5.22) and (5.23) can also be used to correlate $\sin\alpha$ to M_2 and M_{H} depending once more on $\mathrm{sign}(\eta'_{\mathrm{d}}\eta'_{\mathrm{s}})$. As long as M_2, M_{H} are close to the lower bounds (5.24), α must be quite small, especially for $\eta'_{\mathrm{s}} = \eta'_{\mathrm{d}}$.

This shows that unlike the standard model the minimal L-R model can easily account for ε regardless of the value of s_3 or $\bar{R}$. The small values of α also imply that the standard model contribution $\varepsilon_{\mathrm{SM}}$ which is governed by

$$\mathrm{Im}\,\lambda_{\mathrm{c}}^{\mathrm{LL}} = -\mathrm{Im}\,\lambda_{\mathrm{t}}^{\mathrm{LL}} \simeq -\frac{ys_1}{1-w_\alpha^2}\{s_2(c_1s_2 \pm s_3)(\mu_{\mathrm{t}}\mu_{\mathrm{s}}/\mu_{\mathrm{b}}^2 + \mu_{\mathrm{s}}/\mu_{\mathrm{c}}) + c_1s_3(c_1s_3 \pm s_2)\mu_{\mathrm{c}}/\mu_{\mathrm{s}}\} \quad (5.25)$$

is negligible in that regime. As M_2, M_{H} increase also α becomes bigger till one reaches a domain where the genuine L-R contributions to ε become negligible and where one recovers essentially the standard model. We refrain from more quantitative conclusions such as upper bounds on M_2, M_{H} and only note that the relevant range in M_2, M_{H} is expected to be smaller for $\eta'_{\mathrm{d}} = -\eta'_{\mathrm{s}}$.

* While g_{QCD} is completely independent of M_2 and M_{H}, f_{QCD} depends very weakly on M_2.

For the $|\Delta S| = 1$ amplitudes we need the following quantities in the small-phase approximation in addition to (5.25):

$$\lambda_i^{RR} = \eta_d' \eta_s' [1 + 2i(\gamma_s - \gamma_d)] \lambda_i^{LL*}, \qquad i = u, c, t,$$

$$\operatorname{Im} \Lambda_i^{LR} = (\eta_d' + \eta_s') \eta_i \{ \cos\alpha (2\beta_i K_{id} K_{is} + \operatorname{Im} \lambda_i^{LL}) - \sin\alpha K_{id} K_{is} \}$$
$$- 2\eta_i \cos\alpha K_{id} K_{is} (\eta_d' \gamma_d + \eta_s' \gamma_s), \tag{5.26}$$

where

$$\Lambda_i^{LR} = e^{-i\lambda} \lambda_i^{LR} - e^{i\lambda} \lambda_i^{RL} = e^{-i\alpha} \lambda_i^{RL} - e^{i\alpha} \lambda_i^{LR} \tag{5.27}$$

characterizes the size of the $W_L W_R$-mixing amplitudes. In terms of quark masses and mixing angles

$$\operatorname{Im} \lambda_u^{RR} \simeq 2\eta_d' \eta_s' s_1 c_1 (\gamma_s - \gamma_d),$$

$$\operatorname{Im} \lambda_c^{RR} \simeq -\operatorname{Im} \lambda_u^{RR} - \eta_d' \eta_s' \operatorname{Im} \lambda_c^{LL}, \tag{5.28}$$

and, for $\eta_s' = \eta_d'$,

$$\operatorname{Im} \Lambda_u^{LR} \simeq -2\eta_u \eta_s' s_1 c_1 [\sin\alpha - 2y(\mu_d/\mu_u)\cos\alpha],$$

$$\eta_c \operatorname{Im} \Lambda_c^{LR} \simeq -\eta_u \operatorname{Im} \Lambda_u^{LR} + 2\eta_s' [\operatorname{Im} \lambda_c^{LL} - s_1 c_1 y(\mu_c/\mu_s - \mu_d/\mu_u)], \tag{5.29}$$

while for $\eta_s' = -\eta_d'$

$$\operatorname{Im} \Lambda_u^{LR} \simeq -2\eta_u \eta_s' s_1 c_1 (\gamma_s - \gamma_d)\cos\alpha,$$

$$\eta_c \operatorname{Im} \Lambda_c^{LR} \simeq -\eta_u \operatorname{Im} \Lambda_u^{LR}. \tag{5.30}$$

$\gamma_s - \gamma_d$ is given in (5.18) and we have neglected w_α compared to one. We have not written down the corresponding equations for $\operatorname{Im} \lambda_t^{RR}, \operatorname{Im} \Lambda_t^{LR}$ because the t-quark contributions to the R-R and L-R penguins are negligible due to the small s_2.

Using the assumptions and results of this and previous sections we obtain the following approximate equations for the various contributions to ξ_0 and ξ_2:

$$\xi_0(W_R\text{-exchange}) \sim 0.9\beta \operatorname{Im} \lambda_u^{RR},$$

$$\xi_2(W_R\text{-exchange}) \sim 4.2\beta \operatorname{Im} \lambda_u^{RR},$$

$$\xi_0(W_L W_R\text{-mixing}) \sim 260 \xi_{LR} \operatorname{Im} \Lambda_u^{LR},$$

$$\xi_2(W_L W_R\text{-mixing}) \sim -460 \xi_{LR} \operatorname{Im} \Lambda_u^{LR},$$

$$\xi_0(\text{LL-penguin}) \sim -5.4(1 + 0.3 \ln(m_t/35\ \text{GeV})) \operatorname{Im} \lambda_c^{LL},$$

$$\xi_0(\text{RR-penguin}) \sim 1.5\beta \operatorname{Im} \lambda_u^{RR},$$

$$\xi_0(\text{LR-penguin}) \sim 16 \xi_{LR} \operatorname{Im} \Lambda_c^{LR}. \tag{5.31}$$

The QCD corrections were computed for $\Lambda_4 = 200$ MeV and $M_2 = 5$ TeV but the results are again quite insensitive to M_2.

From (5.28), (5.29) and (5.30) one finds that $\mathrm{Im}\,\lambda_u^{RR}$, $\mathrm{Im}\,\Lambda_u^{LR}$ and $\mathrm{Im}\,\Lambda_c^{LR}$ are of the same order of magnitude. Unless the natural relation (2.6) grossly overestimates ξ_{LR} we conclude from (5.31) that the W_LW_R-mixing terms dominate ε' by far. Keeping in mind the definition (4.22) this conclusion is corroborated by the observation that the W_R-exchange and R-R penguin contributions tend to cancel in ε' while the W_LW_R-mixing terms interfere constructively. Keeping track of the standard contribution ξ_0(LL-penguin), which can become relevant for increasing M_2, M_H and not too small s_3, by defining a corresponding quantity ε'_{SM}, one arrives at the approximate final result

$$\frac{\varepsilon' - \varepsilon'_{SM}}{\varepsilon} \simeq \frac{15\eta_u \cos(\alpha)\xi_{LR}(1-\varepsilon_{SM}/\varepsilon)}{|\mathrm{Re}\,\tilde{M}_{12}^{LR}|} \times \begin{cases} (2\mu_d/\mu_u - w_\alpha^{-1})m_s/\mu_c, & \eta'_s = \eta'_d \\ \eta'_s, & \eta'_s = -\eta'_d. \end{cases} \tag{5.32}$$

For the following discussion of (5.32) we neglect ε_{SM}, ε'_{SM} which is justified for M_2 in the several TeV range and in any case for sufficiently small s_3. The first conclusion from (5.32) is that ε'/ε can have either sign depending on the sign factors (5.7) of the particular model. This is still another manifestation of the physical relevance of those sign factors. Using (2.6), (5.19) and (5.23), one can derive the approximate upper bounds

$$|\varepsilon'/\varepsilon| \lesssim \begin{cases} 6\times 10^{-3}, & \eta'_s = \eta'_d \\ 4\times 10^{-3}, & \eta'_s = -\eta'_d, \end{cases} \tag{5.33}$$

to be compared with the experimental results (1.1). These upper bounds correspond to infinitely heavy Higgs bosons. As M_H decreases, $|\varepsilon'/\varepsilon|$ gets smaller in accordance with (5.23).

How does our lack of understanding the $\Delta I = \frac{1}{2}$ rule influence the predictions (5.32)? Eqs. (5.31) were derived under the implicit assumption that the difference $A_0^{exp} - A_0^{theor}$ was purely real. This is probably a too naive assumption and it is more likely [39] that either the matrix elements of O_-^{LL} and/or O_P^L or the QCD corrections were underestimated. In this case ξ_0(W_R exchange) and/or ξ_0(R-R penguin) in (5.31) would be underestimated as well and we expect

$$\Delta\xi_0 \sim f\beta\,\mathrm{Im}\,\lambda_u^{RR} \tag{5.34}$$

for the corresponding error $\Delta\xi_0$ with a numerical factor $f \lesssim 4$. Comparing (5.34) with (5.32) leads to the conclusion that the ensuing relative error in ε'/ε is at most $f \times 10^{-2}$. Thus, the $\Delta I = \frac{1}{2}$ problem has rather little impact on the predictions for ε'/ε.

Finally, we turn to a discussion of the neutron electric dipole moment d_n [40]. In the L-R model with minimal Higgs sector the neutral Higgs contribution to d_n is totally negligible and also the charged Higgs contribution can be neglected as long as $M_H \gtrsim M_2$ [15]. We can therefore use the approximate formula

$$d_n \simeq 3.7 \times 10^{-21}\,\mathrm{e \cdot cm}\, \xi_{LR} \Big\{ 5 \sum_{i=u,c,t} m_i f(x_i) \mathrm{Im}\big(e^{i\alpha} K^*_{L,id} K_{R,id}\big) - \sum_{j=d,s,b} m_j g(x_j) \mathrm{Im}\big(e^{i\alpha} K^*_{L,uj} K_{R,uj}\big) + 1.1\,\mathrm{Im}\big(e^{i\alpha} K^*_{L,ud} K_{R,ud}\big)\Big\}, \tag{5.35}$$

where the quark masses are measured in GeV and $f(x), g(x)$ are kinematic functions [15] with $f(0) = g(0) = 1$. The three terms in (5.35) are due to the one-loop contributions of d- and u-quarks and to the so-called exchange contributions, respectively.

The imaginary parts in (5.35) can be calculated in the small-phase approximation producing an overall factor y. The product $y \times \xi_{LR}$ can now be expressed in terms of $\varepsilon' - \varepsilon'_{SM}$ with the help of (5.17), (5.18) and (5.32). Omitting all details of the calculation and retaining only the dominant terms we obtain the final approximate result

$$d_n \simeq 2D \frac{\varepsilon' - \varepsilon'_{SM}}{\varepsilon} \times 10^{-25}\,\mathrm{e \cdot cm}, \tag{5.36}$$

with

$$D = (1 - 2w_\alpha \mu_d/\mu_u)^{-1} \{\eta_u \eta_c m_c(\mathrm{GeV})(1 - 3w_\alpha \mu_c/\mu_s) + 4[1 - w_\alpha(\mu_c/\mu_s + 2\mu_d/\mu_u)]\} \quad \text{for} \quad \eta'_s = \eta'_d, \tag{5.37}$$

$$D = -\{[\mu_u/\mu_d - 2\mu_s/\mu_c - s_3^2 \mu_t/\mu_s + (c_1^2 - s_1^2 m_s/m_d)\mu_c/\mu_s] w_\alpha/2\}^{-1} \times \{\eta_u \eta_c m_c(\mathrm{GeV})(1 - w_\alpha \mu_c/\mu_s) + 4(1 + w_\alpha \mu_c/\mu_s)\} \quad \text{for} \quad \eta'_s = -\eta'_d. \tag{5.38}$$

Although the denominator of (5.38) is somewhat difficult to estimate in general we choose $s_3 \ll s_2$ and $c_1^2 - s_1^2 m_s/m_d = 0$ to derive the rough upper bounds

$$|d_n| \lesssim \left|\frac{\varepsilon' - \varepsilon'_{SM}}{\varepsilon}\right| \times \begin{cases} 5 \times 10^{-24}\,\mathrm{e \cdot cm}, & \eta'_s = \eta'_d \\ 10^{-22}\,\mathrm{e \cdot cm}, & \eta'_s = -\eta'_d. \end{cases} \tag{5.39}$$

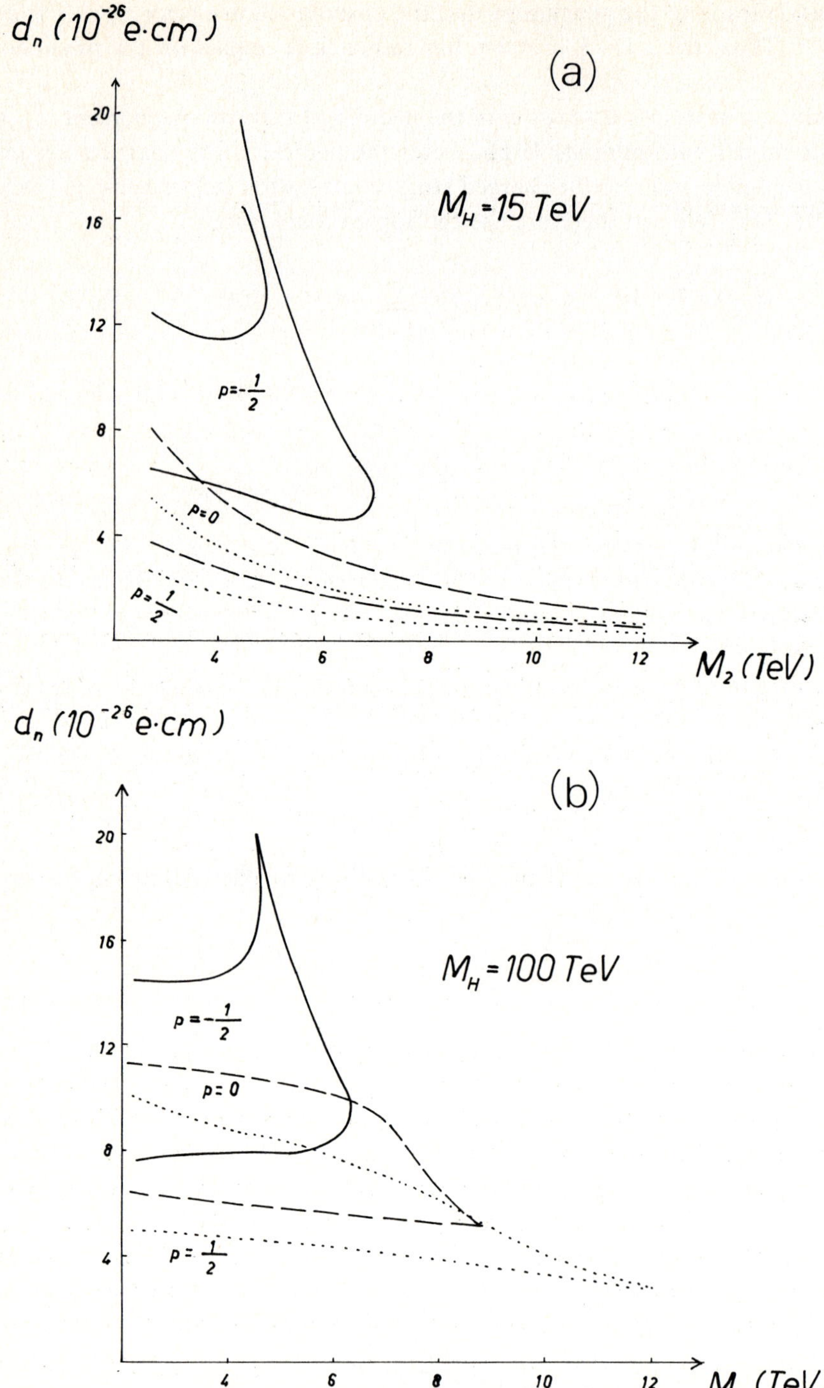

Fig. 4. Allowed domains for the neutron electric dipole moment d_n in model (5.2) with $M_H = 15$ TeV (a) and $M_H = 100$ TeV (b) for three values of the parameter $p = h_2/h_1$. $p = 0$ (dashed curves) corresponds to manifest CP with $\eta'_s = -\eta'_d$.

Together with (5.33) one obtains

$$|d_{\rm n}| \lesssim \begin{cases} 3\times 10^{-26}\ {\rm e\cdot cm}, & \eta'_{\rm s}=\eta'_{\rm d} \\ 4\times 10^{-25}\ {\rm e\cdot cm}, & \eta'_{\rm s}=-\eta'_{\rm d}, \end{cases} \tag{5.40}$$

where the bound for $\eta'_{\rm s}=\eta'_{\rm d}$ is rather insensitive to reasonable variations of quark masses and mixing angles. (5.40) should be compared to the present experimental limit [41]

$$|d_{\rm n}^{\rm exp}| < 4\times 10^{-25}\ {\rm e\cdot cm}\quad (90\%\ {\rm c.l.}). \tag{5.41}$$

In order to get some feeling for the validity of the various approximations made we have calculated $d_{\rm n}$ for the model (5.2) exactly, i.e. without using the small-phase approximation. In fig. 4 the results are displayed for three values of the free parameter $p=h_2/h_1$ and $M_{\rm H}=15$ and 100 TeV. The procedure is the same as in [11]: for given M_2, $M_{\rm H}$ we determine the allowed range in α which gives $\varepsilon_{\rm theor}=\varepsilon_{\rm exp}$ $\pm 30\%$ to account for possible long-distance contributions to ε [36]. The allowed range in α leads in turn to a predicted range for $d_{\rm n}$. As already mentioned in the beginning of this section, the case $p=h_2=0$ corresponds to a case of manifest CP and, moreover, has $\eta'_{\rm s}=-\eta'_{\rm d}$. The results in fig. 4 show that $|d_{\rm n}|$ is below the theoretical bound (5.40) derived in the small-phase approximation for all M_2. In addition, the curves exhibit clearly the effect of varying $M_{\rm H}$ in accordance with (5.39) and the remarks after (5.33). Of course, the same features can be made out in the numerical results for ε'/ε presented in [11]. Altogether, the numerical evidence supports the earlier conjecture that for $\eta'_{\rm s}=-\eta'_{\rm d}$ the so-called small-phase approximation can actually be trusted for practically all values of the Higgs phase α.

5. Conclusions

L-R symmetry provides a promising scenario for CP violation. Quite independent of phenomenological considerations, L-R symmetric gauge models possess the attractive feature that CP can be violated spontaneously already with the minimal Higgs sector. In the framework of such minimal models we have discussed in [11] the phenomenology of a special case (non-manifest CP) where the quark mass matrices are constrained by P and CP. In the present paper we have concentrated on the remaining class of models (manifest CP) with real and symmetric Yukawa coupling matrices.

Our calculations of QCD corrections and hadronic matrix elements are more generally applicable. QCD corrections substantially modify the effective $|\Delta S|=1,2$ hamiltonians, to a greater extent than in the standard model. A careful treatment of short-distance corrections as performed in this paper is therefore indispensable for reliable estimates of hadronic amplitudes. Hadronic matrix elements were computed with vacuum insertion techniques and are, of course, subject to some uncertainties.

In the minimal models with spontaneous *CP* violation all *CP*-violating parameters depend on a single phase of scalar vacuum expectation values. We have calculated ε, ε' and d_n in the small-phase approximation extending the work of Chang [12]. Two different groups of models must be distinguished depending on the sign of the genuine L-R contribution to $\Delta m_{K_L K_S}$. The requirement that the absolute value of $\Delta m^{LR}_{K_L K_S}$ does not exceed $\Delta m^{exp}_{K_L K_S}$ leads to the lower bounds $M_2 \gtrsim 2.5$ TeV and $M_H \gtrsim 10$ TeV.

In both classes of models the correct value for ε can be reproduced as long as M_2 or M_H are not much bigger than their lower bounds. The domain in M_2 and M_H where L-R contributions to ε are significant is in general smaller for $\Delta m^{LR}_{K_L K_S} > 0$. ε' turns out to be dominated by L-R mixing. Unlike in the standard model, ε'/ε may have either sign depending on certain sign factors characteristic of the specific model. The approximate upper bounds $|\varepsilon'/\varepsilon| \lesssim 4 \times 10^{-3}$ (6×10^{-3}) were derived for $\Delta m^{LR}_{K_L K_S} > 0$ (< 0), respectively. Thus, for M_2 in the several TeV range the models under consideration evade both problems of the standard model referred to in the introduction.

Since d_n is due to L-R mixing in lowest order, d_n and ε'/ε can be related to each other. Forthcoming experiments [42] may be sensitive enough to test this correlation. The predictions are again different for the two classes of models: for $\Delta m^{LR}_{K_L K_S} > 0$ one obtains $|d_n| \lesssim 4 \times 10^{-25}$ e·cm equal to the present experimental upper limit, but $|d_n| \lesssim 3 \times 10^{-26}$ e·cm in the other case. The upper bounds for both $|\varepsilon'/\varepsilon|$ and $|d_n|$ correspond to infinitely heavy Higgs bosons. Both $|\varepsilon'/\varepsilon|$ and $|d_n|$ decrease monotonically as M_H is lowered. The results in the small-phase approximation agree well with exact numerical calculations in the special model investigated in [11].

If the present indications of a possible crisis of the standard six-quark model should be confirmed by more precise future measurements, especially of ε'/ε and $\Gamma(b \to ue\nu)/\Gamma(b \to ce\nu)$, L-R symmetric models will be very serious candidates for an alternative understanding of *CP* violation.

We are grateful to I.I. Bigi and J.-M. Frère for discussions on the problem of QCD corrections in L-R symmetric models.

References

[1] Proc. 5th Moriond Workshop on Heavy quarks, flavour mixing and *CP* violation, La Plagne (Jan. 1985), ed. J. Tran Thanh Van, to be published
[2] R. Klanner, Proc. XXII. Int. Conf. on High energy physics, Leipzig (1984)
[3] C. Klopfenstein et al. (CUSB Coll.), Phys. Lett. 130B (1983) 444;
A. Chen et al. (CLEO Coll.), Phys. Rev. Lett. 52 (1984) 1084

[4] D. Hochberg and R.G. Sachs, Phys. Rev. D27 (1983) 606;
B. Stech, Phys. Lett. 130B (1983) 189;
L. Wolfenstein, Phys. Lett. 144B (184) 425;
M. Roos, Univ. Helsinki preprint HU-TFT-84-38 (Aug. 1984);
M. Gronau and J. Schechter, Phys. Rev. Lett. 54 (1985) 385
[5] G. Arnison et al. (UA1 Coll.), Phys. Lett. 147B (1984) 493
[6] R.H. Bernstein et al. (Chicago-Saclay Coll.), Phys. Rev. Lett. 54 (1985) 1631
[7] J.K. Black et al. (BNL-Yale Coll.), Phys. Rev. Lett. 54 (1985) 1628
[8] A. Masiero, R.N. Mohapatra and R.D. Peccei, Nucl. Phys. B192 (1981) 66
[9] R.N. Mohapatra and G. Senjanović, Phys. Rev. Lett. 40 (1980) 912;
Phys. Rev. D23 (1981) 165
[10] G. Ecker, W. Grimus and H. Neufeld, Nucl. Phys. B247 (1984) 70
[11] G. Ecker and W. Grimus, Phys. Lett. 153B (1985) 279
[12] D. Chang, Nucl. Phys. B214 (1983) 435
[13] D. Chang, J. Basecq, L.-F. Li and P.B. Pal, Phys. Rev. D30 (1984) 1601
[14] G. Ecker, W. Grimus and H. Neufeld, Phys. Lett. 127B (1983) 365; (E) 132B (1983) 467
[15] G. Ecker, W. Grimus and H. Neufeld, Nucl. Phys. B229 (1983) 421
[16] G. Ecker, W. Grimus and W. Konetschny, Phys. Lett. 94B (1980) 381; Nucl. Phys. B177 (1981) 489
[17] I.I. Bigi and J.-M. Frère, Phys. Lett. 110B (1982) 255;
J.F. Donoghue and B.R. Holstein, Phys. Lett. 113B (1982) 382;
L. Wolfenstein, Phys. Rev. D29 (1984) 2130
[18] G. Beall, M. Bander and A. Soni, Phys. Rev. Lett. 48 (1982) 848;
J. Trampetić, Phys. Rev. D27 (1983) 1565;
R.N. Mohapatra, G. Senjanović and M.D. Tran, Phys. Rev. D28 (1983) 546;
A. Datta and Amitava Raychaudhuri, Phys. Rev. D28 (1983) 1170;
P. Basak, A. Datta and Amitava Raychaudhuri, Z. Phys. C20 (1983) 305;
F.J. Gilman and M.H. Reno, Phys. Lett. 127B (1983) 426; Phys. Rev. D29 (1984) 937;
D. Chang, ref. [12];
G.C. Branco, J.-M. Frère and J.-M. Gérard, Nucl. Phys. B221 (1983) 317;
H. Harari and M. Leurer, Nucl. Phys. B233 (1984) 221;
M. Hwang and R.J. Oakes, Phys. Rev. D29 (1984) 127;
F.I. Olness and M.E. Ebel, Phys. Rev. D30 (1984) 1034;
R. Decker and U. Türke, Z. Phys. C26 (1984) 117;
R.N. Mohapatra, Univ. of Maryland preprint 85-124 (1985)
[19] B. Grzadkowski, Z. Phys. C22 (1984) 361
[20] M.K. Gaillard and B.W. Lee, Phys. Rev. D10 (1974) 897;
T. Inami and C.S. Lim, Prog. Theor. Phys. 65 (1981) 297;
A.J. Buras, Phys. Rev. Lett. 46 (1981) 1354
[21] J. Basecq, L.-F. Li and P.B. Pal, Carnegie-Mellon Univ. preprint CMU-HEP 85-1 (1985);
W.-S. Hou and A. Soni, Univ. of California preprint UCLA/85/TEP/1 (1985)
[22] R.D.C. Miller and B.H.J. McKellar, Phys. Reports 106 (1984) 171
[23] M.K. Gaillard and B.W. Lee, Phys. Rev. Lett. 33 (1974) 108;
G. Altarelli and L. Maiani, Phys. Lett. 52B (1974) 351
[24] M.A. Shifman, A.I. Vainshtein and V.I. Zakharov, Nucl. Phys. B120 (1977) 316;
F.J. Gilman and M.B. Wise, Phys. Rev. D20 (1979) 2392;
B. Guberina and R.D. Peccei, Nucl. Phys. B163 (1980) 289;
A.J. Buras and W. Słominski, Nucl. Phys. B253 (1985) 231
[25] Y. Meurice, Ann. Soc. Sci. Bruxelles 96 (1982) 28;
S.-P. Chia, Phys. Lett. 130B (1983) 315
[26] J. Gasser and H. Leutwyler, Phys. Reports 87C (1982) 77
[27] F.J. Gilman and M.B. Wise, Phys. Rev. D27 (1983) 1128
[28] M.I. Vysotskii, Yad. Fiz. 31 (1980) 1535 [Sov. J. Nucl. Phys. 31 (1980) 797]
[29] I.I. Bigi and J.-M. Frère, Phys. Lett. 129B (1983) 469

[30] M.K. Gaillard and B.W. Lee, ref. [20]
[31] K. Kang and J.E. Kim, Phys. Rev. D14 (1976) 1903;
R.E. Shrock and S.B. Treiman, Phys. Rev. D19 (1979) 2148;
J.F. Donoghue, E. Golowich and B.R. Holstein, Phys. Lett. 119B (1982) 412;
P. Colić, B. Guberina, D. Tadić and J. Trampetić, Nucl. Phys. B221 (1983) 141;
B. Guberina, B. Machet and E. de Rafael, Phys. Lett. 128B (1983) 269;
N. Cabibbo, G. Martinelli and R. Petronzio, Nucl. Phys. B244 (1984) 381;
R. Brower, G. Maturana, M.B. Gavela and R. Gupta, Phys. Rev. Lett. 53 (1984) 1318;
N. Bilić and B. Guberina, Phys. Lett. 136B (1984) 440; Phys. Lett. 150B (1985) 311; Univ. of Bielefeld preprint BI-TP84/15 (June 1984);
B. Machet, Z. Phys. C26 (1984) 449;
F. Wagner, MPI-Munich preprint MPI-PAE/PTh89/83 (1983)
[32] P. Cea, G. Nardulli and G. Preparata, Phys. Lett. 148B (1984) 477
[33] M.A. Shifman, A.I. Vainshtein and V.I. Zakharov, ZhETP 72 (1977) 1275 [Sov. Phys. JETP 45 (1977) 670];
J.F. Donoghue, Phys. Rev. D30 (1984) 1499
[34] A.I. Sanda, Phys. Rev. D23 (1981) 2647;
N.G. Deshpande, Phys. Rev. D23 (1981) 2654
[35] L. Wolfenstein, Nucl. Phys. B160 (1979) 501;
C.T. Hill, Phys. Lett. 97B (1980) 275;
I.I. Bigi and A.I. Sanda, Phys. Lett. 148B (1984) 205;
P. Cea and G. Nardulli, CNRS-Marseille preprint CPT-84/P. 1698 (Nov. 1984)
[36] J.F. Donoghue and B.R. Holstein, Phys. Rev. D29 (1984) 2088;
G. Ecker, Phys. Lett. 147B (1984) 369;
J.-M. Frère, J. Hagelin and A.I. Sanda, Phys. Lett. 151B (1985) 161
[37] B. Stech, Phys. Lett. 130B (1983) 189; Proc. Adv. Summer Institute Munich, 1983 (Plenum); Proc. Europhysics Conf. on Flavour mixing in weak interactions, Erice, 1984
[38] M. Kobayashi and T. Maskawa, Prog. Theor. Phys. 49 (1973) 652
[39] G. Nardulli and G. Preparata, Phys. Lett. 104B (1981) 399;
G. Nardulli, G. Preparata and D. Rotondi, Phys. Rev. D27 (1983) 557
[40] G. Beall and A. Soni, Phys. Rev. Lett. 47 (1981) 552;
G. Ecker, W. Grimus and H. Neufeld, ref. [15]
[41] V.M. Lobashev, Proc. Neutrino '82, Balatonfüred, vol. 2, eds. A. Frenkel and L. Jenik (Central Research Inst. for Physics, Budapest, 1982) 107
[42] P. Miranda and R. Turlay, ref. [1]

4. The Standard Model

Introduction

Within the standard SU(2) × U(1) electroweak model with a single Higgs boson doublet, the only way to introduce CP violation is through complex Yukawa couplings between the fermions and the Higgs bosons. When the neutral Higgs boson acquires its vacuum expectation value (vev) breaking the SU(2) symmetry, the Yukawa interactions turn into mass matrices for the quarks. When the mass matrices are diagonalized, the original Yukawa couplings determine the mass ratios among the quarks and the mixing matrices relating the flavor eigenstates to the mass eigenstates. The only place the CP violation in the original Yukawa interaction can show up, is in complex phases in the mixing matrices. Many of these phases can be removed by redefining the phase of various fields, such as $s \to s e^{i\alpha}$, where s is the strange quark field. The reason that this is possible, is that strangeness is conserved in the strong and electromagnetic interactions. It can be shown that with two generations of quarks all the phases can be removed so that no CP violation remains. Kobayashi and Maskawa [1] (KM) were the first to point out that if there were three generations there was one, and only one, phase that could not be removed. With the discovery of the Υ quarkonium and the B mesons, it became clear there was a third generation of quarks. Thus the KM model became the standard model of CP violation.

The original KM paper received relatively little attention until detailed discussions of its consequences were presented by Ellis et al. [2] and also by Pakvasa and Sugawara [3], and Maiani [3]. An important point of the Ellis et al. paper was to show that the major contribution to CP violation came from the mass matrix so that ϵ' was very small. The question of the quantitative evaluation of ϵ' was raised by Gilman and Wise [4], who argued that so-called penguin diagrams were the dominant contribution. They predicted a value of $|\epsilon'/\epsilon|$ of the order 0.02, a value large enough to inspire new rounds of experiments. A similar analysis carried out by Guberina and Peccei [5] gave an order-of-magnitude smaller value. At the present time, it is clear that a quantitative calculation is quite uncertain because of the difficulty of calculating hadronic matrix elements. Present methods of doing these calculations have been reviewed by Donoghue et al. [6]. The possibility exists that in the future numerical lattice gauge calculations will provide more definite results.

The 3 × 3 Cabibbo–Kobayashi–Maskawa (CKM) mixing matrix is defined by three mixing angles and one phase. One of the mixing angles, the Cabibbo angle θ_c, is well measured, but the other two must be determined from measurements on B mesons. From the lifetime and branching ratios one finds that if $\sin\theta_c$ is called λ, then one can write $U_{cb} = A\lambda^2$ and $U_{ub} = A\lambda^3 \rho e^{i\delta}$ with $A \approx 1$ and ρ less than one. This hierarchical form was first given by Wolfenstein [7] (in the original form $\rho - i\eta$ is used rather than $\rho e^{i\delta}$); the phase δ measures CP violation. It can be shown [8] that all CP-violating observables are proportional to a phase-convention-independent quantity

$$J = Im(U_{us} U_{cb} U_{ub}^* U_{cs}^*) \simeq A^2 \lambda^6 \rho \sin\delta.$$

Direct measurements provide only an upper limit for the value of ρ because the decays corresponding to $b \to u$ have not been definitively observed. The determination of δ depends on a quantitative evaluation of the box graph contribution to ϵ. Both ρ and δ are constrained by the evaluation of the box graph that contributes to the $B^0 - \bar{B}^0$ mixing. (B^0 is the state $\bar{b}d$ analogous to $K^0 = \bar{s}d$.) A large $B^0 - \bar{B}^0$ mixing corresponding to $\Delta M_B/\Gamma = 0.73 \pm 0.18$ was observed by the ARGUS group and subsequently confirmed by the CLEO collaboration [9]. Numerical analyses giving possible values of ρ and δ have been given in many papers [10]. The experimental determination of the mass m_t of the top quark would constrain the possibilities. From the figures of ref. [10] the quantity ρ sin δ appears to be within a factor two equal to 0.3. This leads to a value of ϵ'/ϵ between 5×10^{-4} and 6×10^{-3} which is quite consistent with the recent experiment presented in chapter 2.

The possibility of observing CP violation in the B^0 system analogous to that in K^0 system was first seriously analyzed by Bigi and Sanda [11]. Because the lifetimes of the eigenstates are expected to be approximately equal, it is necessary to depend on interference effects when comparing B^0 and $\bar{B}^0$ decays. The simplest case is that in which the final state is a CP eigenstate such as ψK_S. Once the hierarchical character of the CKM matrix was determined experimentally, it became clear that the small value of ϵ arose because ϵ was proportional to λ^4 and that in order to fit ϵ, the CP-violating phase δ could not be very small. It was then pointed out [12] that there were CP violating effects in the B^0 system that were not suppressed by any power of λ and would be sizeable. A recent review of possible experiments has been given by Bigi and Sanda [13]. The magnitude of the CP violating effect is shown in figs. 5 and 6 of Krawczyk et al. [10]. These relatively large effects are a unique feature of the KM model. Unfortunately, because of the low branching ratios of B^0 to a particular state, like ψK_s, it appears that an experiment with the order of 10^8 $B\bar{B}$ pairs would be needed to observe this effect.

References

[1] M. Kobayashi and T. Maskawa, Prog. Theor. Phys. (Japan) 49 (1973) 652. Paper 23, this volume.
[2] J. Ellis, M.K. Gaillard and D.V. Nanopoulos, Nucl. Phys. B 109 (1976) 213. Paper 24, this volume.
[3] S. Pakvasa and H. Sugawara, Phys. Rev. D 14 (1976) 305;
L. Maiani, Phys. Lett. B 62 (1976) 183.
[4] F.J. Gilman and M.B. Wise, Phys. Lett. B 83 (1979) 83. Paper 25, this volume; Phys. Rev. D 20 (1979) 2392.
[5] B. Guberina and R.D. Peccei, Nucl. Phys. B 163 (1980) 289. Paper 26, this volume.
[6] J.F. Donoghue, E. Golowich and B.R. Holstein, Phys. Rep. 131 (1986) 320. The relevant section is pp. 388–401.
[7] L. Wolfenstein, Phys. Rev. Lett. 51 (1983) 1945.
[8] C. Jarlskog, Phys. Rev. Lett. 55 (1985) 1039;
O.W. Greenberg, Phys. Rev. D 32 (1985) 1841;
D.D. Wu, Phys. Rev. D 33 (1986) 866.
[9] H. Albrecht et al., Phys. Lett. B 192 (1987) 245;
M. Artuso et al., Phys. Rev. Lett. 62 (1989) 2233.
[10] P. Krawczyk, D. London, R.D. Peccei and H. Steger, Nucl. Phys. B 307 (1988) 19. Paper 27, this volume.
See also J. Donoghue, T. Nakada, E.A. Paschos and D. Wyler, Phys. Lett. B 195 (1987) 285;
J. Ellis, J.S. Hagelin and S. Rudaz, Phys. Lett. B 192 (1987) 201;
V. Barger, T. Han, D.V. Nanopoulos and R.J.N. Phillips, Phys. Lett. B 194 (1987) 312;
A.J. Buras, W. Slominski and H. Steger, Nucl. Phys. B 238 (1984) 529; B. 245 (1984) 369.
[11] I.I. Bigi and A.I. Sanda, Nucl. Phys. B 193 (1981) 85. Paper 28, this volume.
[12] L. Wolfenstein, Nucl. Phys. B 246 (1983) 45. Paper 29, this volume.
[13] I.I. Bigi and A.I. Sanda, Nucl. Phys. B 281 (1987) 41.

Progress of Theoretical Physics, Vol. 49, No. 2, February 1973

CP-Violation in the Renormalizable Theory of Weak Interaction

Makoto KOBAYASHI and Toshihide MASKAWA

Department of Physics, Kyoto University, Kyoto

(Received September 1, 1972)

In a framework of the renormalizable theory of weak interaction, problems of *CP*-violation are studied. It is concluded that no realistic models of *CP*-violation exist in the quartet scheme without introducing any other new fields. Some possible models of *CP*-violation are also discussed.

When we apply the renormalizable theory of weak interaction[1] to the hadron system, we have some limitations on the hadron model. It is well known that there exists, in the case of the triplet model, a difficulty of the strangeness changing neutral current and that the quartet model is free from this difficulty. Furthermore, Maki and one of the present authors (T.M.) have shown[2] that, in the latter case, the strong interaction must be chiral $SU(4)\times SU(4)$ invariant as precisely as the conservation of the third component of the iso-spin I_3. In addition to these arguments, for the theory to be realistic, *CP*-violating interactions should be incorporated in a gauge invariant way. This requirement will impose further limitations on the hadron model and the *CP*-violating interaction itself. The purpose of the present paper is to investigate this problem. In the following, it will be shown that in the case of the above-mentioned quartet model, we cannot make a *CP*-violating interaction without introducing any other new fields when we require the following conditions: a) The mass of the fourth member of the quartet, which we will call ζ, is sufficiently large, b) the model should be consistent with our well-established knowledge of the semi-leptonic processes. After that some possible ways of bringing *CP*-violation into the theory will be discussed.

We consider the quartet model with a charge assignment of Q, $Q-1$, $Q-1$ and Q for p, n, λ and ζ, respectively, and we take the same underlying gauge group $SU_{\text{weak}}(2)\times SU(1)$ and the scalar doublet field φ as those of Weinberg's original model.[1] Then, hadronic parts of the Lagrangian can be devided in the following way:

$$\mathcal{L}_{\text{had}}=\mathcal{L}_{\text{kin}}+\mathcal{L}_{\text{mass}}+\mathcal{L}_{\text{strong}}+\mathcal{L}',$$

where $\mathcal{L}_{\text{kin}}$ is the gauge-invariant kinetic part of the quartet field, q, so that it contains interactions with the gauge fields. $\mathcal{L}_{\text{mass}}$ is a generalized mass term of q, which includes Yukawa couplings to φ since they contribute to the mass of q through the spontaneous breaking of gauge symmetry. $\mathcal{L}_{\text{strong}}$ is a strong-interaction part which conserves I_3 and therefore chiral $SU(4)\times SU(4)$ invariant.[2] We assume C- and P-invariance of $\mathcal{L}_{\text{strong}}$. The last term denotes residual interaction parts if they exist. Since $\mathcal{L}_{\text{mass}}$ includes couplings with φ, it has possibilities of violating CP-conservation. As is known as Higgs phenomena,[3] three massless components of φ can be absorbed into the massive gauge fields and eliminated from the Lagrangian. Even after this has been done, both scalar and pseudoscalar parts remain in $\mathcal{L}_{\text{mass}}$. For the mass term, however, we can eliminate such pseudoscalar parts by applying an appropriate constant gauge transformation on q, which does not affect on $\mathcal{L}_{\text{strong}}$ due to gauge invariance.

Now we consider possible ways of assigning the quartet field to representations of the $SU_{\text{weak}}(2)$. Since this group is commutative with the Lorentz transformation, the left and right components of the quartet field, which are respectively defined as $q_L=\frac{1}{2}(1+\gamma_5)q$ and $q_R=\frac{1}{2}(1-\gamma_5)q$, do not mix each other under the gauge transformation. Then, *each* component has three possibilities:

$$A)\quad \mathbf{4}=\mathbf{2}+\mathbf{2}\,,$$

$$B)\quad \mathbf{4}=\mathbf{2}+\mathbf{1}+\mathbf{1}\,,$$

$$C)\quad \mathbf{4}=\mathbf{1}+\mathbf{1}+\mathbf{1}+\mathbf{1}\,,$$

where on the r.h.s., $\mathbf{n}$ denotes an n-dimensional representation of $SU(2)$. The present scheme of charge assignment of the quartet does not permit representations of $n\geq 3$. As a result, we have nine possibilities which we will denote by (A, A), (A, B), $\cdots$, where the former (latter) in the parentheses indicates the transformation properties of the left (right) component. Since all members of the quartet should take part in the weak interaction, and size of the strangeness changing neutral current is bounded experimentally to a very small value, the cases of (B, C), (C, B) and (C, C) should be abandoned. The models of (B, A) and (C, A) are equivalent to those of (A, B) and (A, C), respectively, except relative signs between vector and axial vector parts of the weak current. Since g_A/g_V ratios are measured only for composite states, this difference of the relative signs would be reduced to a dynamical problem of the composite system. So, we investigate in detail the cases of (A, A), (A, B), (A, C) and (B, B).

i) *Case* (A, C)

This is the most natural choice in the quartet model. Let us denote two ($SU_{\text{weak}}(2)$) doublets and four singlets by L_{d1}, L_{d2}, $R_{s1}^{(p)}$, $R_{s2}^{(p)}$, $R_{s1}^{(n)}$ and $R_{s2}^{(n)}$, where superscript $p(n)$ indicates p-like (n-like) charge states. In this case, $\mathcal{L}_{\text{mass}}$ takes, in general, the following form:

$$\mathcal{L}_{\text{mass}}=\sum_{i,j=1,2}[M_{ij}^{(n)}\bar{L}_{di}\varphi R_{sj}^{(n)}+M_{ij}^{(p)}\bar{L}_{di}\varepsilon\varphi^* R_{sj}^{(p)}]+\text{h.c.}\,,$$

$$\varphi^*=\begin{pmatrix}\varphi^-\\ \bar{\varphi}^0\end{pmatrix},\qquad \varepsilon=\begin{pmatrix}0 & 1\\ -1 & 0\end{pmatrix},\tag{1}$$

where $M_{ij}^{(n)}$ and $M_{ij}^{(p)}$ are arbitrary complex numbers. We can eliminate three Goldstone modes ϕ_i by putting

$$\varphi = e^{i\phi_i\tau_i}\begin{pmatrix} 0 \\ \lambda+\sigma \end{pmatrix}, \tag{2}$$

where λ is a vacuum expectation value of φ^0 and σ is a massive scalar field. Thereafter, performing a diagonalization of the remaining mass term, we obtain

$$\mathcal{L}_{\text{mass}} = \bar{q}mq\left(1+\frac{\sigma}{\lambda}\right),$$

$$m = \begin{pmatrix} m_p & 0 & 0 & 0 \\ 0 & m_n & 0 & 0 \\ 0 & 0 & m_\zeta & 0 \\ 0 & 0 & 0 & m_\lambda \end{pmatrix}, \quad q = \begin{pmatrix} p \\ n \\ \zeta \\ \lambda \end{pmatrix}. \tag{3}$$

Then, the interaction with the gauge field in $\mathcal{L}_{\text{kin}}$ is expressed as

$$\sum_{j=1}^{3} A_\mu{}^j i\bar{q}\Lambda_j\gamma_\mu\frac{1+\gamma_5}{2}q\,. \tag{4}$$

Here, Λ_j is the representation matrix of $SU_{\text{weak}}(2)$ for this case and explicitly given by

$$\Lambda_+ = \frac{\Lambda_1+i\Lambda_2}{2} = K\begin{pmatrix} 0 & U \\ 0 & 0 \end{pmatrix}K^{-1}, \quad \Lambda_3 = \begin{pmatrix} 1 & 0 & 0 & 0 \\ 0 & 1 & 0 & 0 \\ 0 & 0 & -1 & 0 \\ 0 & 0 & 0 & -1 \end{pmatrix}, \quad K = \begin{pmatrix} 1 & 0 & 0 & 0 \\ 0 & 0 & 1 & 0 \\ 0 & 1 & 0 & 0 \\ 0 & 0 & 0 & 1 \end{pmatrix}, \tag{5}$$

where U is a 2×2 unitary matrix. Here and hereafter we neglect the gauge field corresponding to $U(1)$ which is irrelevant to our discussion. With an appropriate phase convention of the quartet field we can take U as

$$U = \begin{pmatrix} \cos\theta & \sin\theta \\ -\sin\theta & \cos\theta \end{pmatrix}. \tag{6}$$

Therefore, if $\mathcal{L}'=0$, no *CP*-violations occur in this case. It should be noted, however, that this argument does not hold when we introduce one more fermion doublet with the same charge assignment. This is because all phases of elements of a 3×3 unitary matrix cannot be absorbed into the phase convention of six fields. This possibility of *CP*-violation will be discussed later on.

ii) *Case* (A, B)

This is a rather delicate case. We denote two left doublets, one right doublet and two singlets by L_{d1}, L_{d2}, R_d, $R_s^{(p)}$ and $R_s^{(n)}$, respectively. The general form of $\mathcal{L}_{\text{mass}}$ is given by

$$\mathcal{L}_{\text{mass}}=\sum_{i=1,2}[m_i\bar{L}_{di}R_d+M_i^{(n)}\bar{L}_{di}\varphi R_s^{(n)}+M_i^{(p)}\bar{L}_{di}\varepsilon\varphi^* R_s^{(p)}]+\text{h.c.},$$

where m_i, $M_i^{(n)}$ and $M_i^{(p)}$ are arbitrary complex numbers. After diagonalization of mass terms (in this case, the *CP*-odd part of coupling with σ does not disappear in general) each multiplet can be expressed as follows:

$$L_{d1}=\frac{1+\gamma_5}{2}\begin{pmatrix} p \\ \cos\theta e^{i\alpha}n+\sin\theta e^{i\beta}\lambda\end{pmatrix},\qquad L_{d2}=\frac{1+\gamma_5}{2}\begin{pmatrix} e^{i\gamma}\zeta \\ -\sin\theta e^{i\alpha}n+\cos\theta e^{i\beta}\lambda\end{pmatrix},$$

$$R_d=\frac{1-\gamma_5}{2}\begin{pmatrix}\sin\xi\cdot p+\cos\xi\cdot\zeta \\ \sin\eta\cdot n+\cos\eta\cdot\lambda\end{pmatrix},\qquad R_s^{(p)}=\frac{1-\gamma_5}{2}(\cos\xi\cdot p-\sin\xi\cdot\zeta),$$

$$R_s^{(n)}=\frac{1-\gamma_5}{2}(\cos\eta\cdot n-\sin\eta\cdot\lambda), \qquad (7)$$

where phase factors α, β and γ satisfy two relations with the masses of the quartet:

$$e^{i\gamma}m_\zeta\sin\theta\cos\xi=m_p\cos\theta\sin\xi-e^{i\alpha}m_n\sin\eta\,,$$

$$e^{i\gamma}m_\zeta\cos\theta\cos\xi=-m_p\sin\theta\cos\xi+e^{i\beta}m_\lambda\cos\eta\,. \qquad (8)$$

Owing to the presence of phase factors, there exists a possibility of *CP*-violation also through the weak current. However, the strangeness changing neutral current is proportional to $\sin\eta\cos\eta$ and its experimental upper bound is roughly

$$\sin\eta\cos\eta<10^{-2\sim-3}. \qquad (9)$$

Thus, making an approximation of $\sin\eta\sim0$ (the other choice $\cos\eta\sim0$ is less critical) we obtain from Eq. (8)

$$m_\zeta/m_p\sim\cot\theta\cdot\tan\xi\,,$$

$$m_\lambda/m_n\sim\sin\xi/\sin\theta\,. \qquad (10)$$

We have no low-lying particle with a quantum number corresponding to ζ, so that m_ζ, which is a measure of chiral $SU(4)\times SU(4)$ breaking, should be sufficiently large compared to the masses of the other members. However, the present experimental results on the g_A/g_V ratios of the octet baryon β-decay would not permit $\sin\xi>\sin\theta$. Thus, it seems difficult to reconcile the hierarchy of chiral symmetry breaking with the experimental knowledge of the semileptonic processes.

iii) *Case* (B, B)

As a previous one, in this case also, occurrence of *CP*-violation is possible, but in order to suppress $|\Delta S|=1$ neutral currents, coefficients of the axial-vector part of $\Delta S=0$ and $|\Delta S|=1$ weak currents must take signs oppossite to each other. This contradicts again the experiments on the baryon β-decay.

iv) *Case* (A, A)

In a similar way, we can show that no CP-violation occurs in this case as far as $\mathcal{L}'=0$. Furthermore this model would reduce to an exactly $U(4)$ symmetric one.

Summarizing the above results, we have no realistic models in the quartet scheme as far as $\mathcal{L}'=0$. Now we consider some examples of CP-violation through $\mathcal{L}'$. Hereafter we will consider only the case of (A, C). The first one is to introduce another scalar doublet field ψ. Then, we may consider an interaction with this new field

$$\mathcal{L}'=\bar{q}\boldsymbol{\psi}\boldsymbol{C}\frac{1-\gamma_5}{2}q+\text{h.c.}, \tag{11}$$

$$\boldsymbol{\psi}\equiv\begin{pmatrix}\bar{\psi}^0 & \psi^+ & 0 & 0\\ -\psi^- & \psi^0 & 0 & 0\\ 0 & 0 & \bar{\psi}^0 & \psi^+\\ 0 & 0 & -\psi^- & \psi^0\end{pmatrix},\qquad \boldsymbol{C}\equiv\begin{pmatrix}c_{11} & 0 & c_{12} & 0\\ 0 & d_{11} & 0 & d_{12}\\ c_{21} & 0 & c_{22} & 0\\ 0 & d_{21} & 0 & d_{22}\end{pmatrix},$$

where c_{ij} and d_{ij} are arbitrary complex numbers. Since we have already made use of the gauge transformation to get rid of the CP-odd part from the quartet mass term, there remains no such arbitrariness. Furthermore, we note that an arbitrariness of the phase of ψ cannot absorb all the phases of c_{ij} and d_{ij}. So, this interaction can cause a CP-violation.

Another one is a possibility associated with the strong interaction. Let us consider a scalar (pseudoscalar) field S which mediates the strong interaction. For the interaction to be renormalizable and $SU_{\text{weak}}(2)$ invariant, it must belong to a $(4, 4^*)+(4^*, 4)$ representation of chiral $SU(4)\times SU(4)$ and interact with q through scalar and pseudoscalar couplings. It also interacts with φ and possible renormalizable forms are given as follows:

$$\begin{aligned}&\text{tr}\{G_0S^+\boldsymbol{\varphi}\}+\text{h.c.},\\ &\text{tr}\{G_1S^+\boldsymbol{\varphi}G_2\boldsymbol{\varphi}^+S\}+\text{h.c.},\\ &\text{tr}\{G_1'S^+\boldsymbol{\varphi}G_2'S^+\boldsymbol{\varphi}\}+\text{h.c.},\end{aligned} \tag{12}$$

with

$$\boldsymbol{\varphi}\equiv\begin{pmatrix}\bar{\varphi}^0 & \varphi^+ & 0 & 0\\ -\varphi^- & \varphi^0 & 0 & 0\\ 0 & 0 & \bar{\varphi}^0 & \varphi^+\\ 0 & 0 & -\varphi^- & \varphi^0\end{pmatrix},$$

where G_i is a 4×4 complex matrix and we have used a 4×4 matrix representation for S. It is easy to see that these interaction terms can violate CP-conservation.

Next we consider a 6-plet model, another interesting model of *CP*-violation. Suppose that 6-plet with charges $(Q, Q, Q, Q-1, Q-1, Q-1)$ is decomposed into $SU_{\text{weak}}(2)$ multiplets as $2+2+2$ and $1+1+1+1+1+1$ for left and right components, respectively. Just as the case of (A, C), we have a similar expression for the charged weak current with a 3×3 instead of 2×2 unitary matrix in Eq. (5). As was pointed out, in this case we cannot absorb all phases of matrix elements into the phase convention and can take, for example, the following expression:

$$\begin{pmatrix} \cos\theta_1 & -\sin\theta_1\cos\theta_3 & -\sin\theta_1\sin\theta_3 \\ \sin\theta_1\cos\theta_2 & \cos\theta_1\cos\theta_2\cos\theta_3-\sin\theta_2\sin\theta_3e^{i\delta} & \cos\theta_1\cos\theta_2\sin\theta_3+\sin\theta_2\cos\theta_3e^{i\delta} \\ \sin\theta_1\sin\theta_2 & \cos\theta_1\sin\theta_2\cos\theta_3+\cos\theta_2\sin\theta_3e^{i\delta} & \cos\theta_1\sin\theta_2\sin\theta_3-\cos\theta_2\sin\theta_3e^{i\delta} \end{pmatrix}. \tag{13}$$

Then, we have *CP*-violating effects through the interference among these different current components. An interesting feature of this model is that the *CP*-violating effects of lowest order appear only in $\Delta S\neq0$ non-leptonic processes and in the semi-leptonic decay of neutral strange mesons (we are not concerned with higher states with the new quantum number) and not in the other semi-leptonic, $\Delta S=0$ non-leptonic and pure-leptonic processes.

So far we have considered only the straightforward extensions of the original Weinberg's model. However, other schemes of underlying gauge groups and/or scalar fields are possible. Georgi and Glashow's model[4)] is one of them. We can easily see that *CP*-violation is incorporated into their model without introducing any other fields than (many) new fields which they have introduced already.

References

1) S. Weinberg, Phys. Rev. Letters **19** (1967), 1264; **27** (1971), 1688.
2) Z. Maki and T. Maskawa, RIFP-146 (preprint), April 1972.
3) P. W. Higgs, Phys. Letters **12** (1964), 132; **13** (1964), 508.
G. S. Guralnik, C. R. Hagen and T. W. Kibble, Phys. Rev. Letters **13** (1964), 585.
4) H. Georgi and S. L. Glashow, Phys. Rev. Letters **28** (1972), 1494.

Nuclear Physics B109 (1976) 213–243

LEFT-HANDED CURRENTS AND *CP* VIOLATION

John ELLIS, Mary K. GAILLARD * and D.V. NANOPOULOS **
CERN, Geneva

Received 9 February 1976
(Revised 26 April 1976)

An analysis is presented of a model for *CP* violation due to Kobayashi and Maskawa, which has six quarks and purely left-handed currents. The model can reproduce approximately results of the Glashow, Iliopoulos and Maiani four-quark model in *CP* conserving processes. It gives a close approximation to superweak predictions for *CP* violation in common K decays, whatever the masses of the fifth and sixth quarks. In principle there are substantial deviations from superweak predictions in rare K decays, but these seem difficult to observe in practice. There are also deviations from superweak results for *CP* violation in charmed particle decays. The neutron electric dipole moment is very sensitive to the masses of the b and t quarks, lying in a range containing the superweak prediction.

1. Introduction

Shortly after the discovery of the new particles, several authors [1, 2] discussed models with three charge $\frac{2}{3}$ quarks and three charge $-\frac{1}{3}$ quarks, and proposed that more than one of these new quarks might be jointly responsible for the happenings in e^+e^- annihilation and elsewhere. More recently, phenomenological analyses have tended to favour the excitation of just one new quark degree of freedom in present data [3]. However, interest in six-quark models has been maintained by the possibility that the $e^{\pm}\mu^{\mp}$ events discovered [4] at SPEAR may come from the decays of a new heavy lepton with its own new neutrino. If this were true, it would be natural to restore some lepton-hadron symmetry by increasing the number of (coloured) quarks also to six. This idea has a calculational basis, in that unless the new leptons and neutrinos (surprisingly) have pure V or pure A couplings, then the 4-quark model's delicate cancellation [5] of triangle anomalies will be destroyed. With it would go the renormalizability of the theory. The anomaly cancellation is restored in the usual Weinberg-Salam [6] SU(2) × U(1) model if there are two more quarks t and b with charges $\frac{2}{3}$ and $-\frac{1}{3}$ respectively. They and the three conventional quarks u, d and s

* And Laboratoire de Physique Théorique et Particules Elémentaires, laboratoire associé au CNRS.
** Present address: Service de Physique Théorique, Ecole Normale Supérieure, 24 rue Lhomond, 75231 Paris Cedex 05.

and the charmed quark c would together form three doublets under the weak SU(2) group. The anomaly cancellation is probably only a shallow reflection of a more profound reason for lepton-quark symmetry. But why so many quarks or leptons in the first place?

A possible reason to introduce six quarks was pointed out some time ago by Kobayashi and Maskawa [7]. They observed that in such a model the unitary matrix representing charged weak currents has one phase parameter in addition to the familiar real angles, such as the Cabibbo angle, that describe quark mixing. This extra phase parameter introduces *CP* violating effects, and links the problems of the origin of the Cabibbo angle and of *CP* violation.

The Kobayashi-Maskawa model has recently been revived by Pakvasa and Sugawara [8], and independently by Maiani [9]. Both papers point out that under suitable conditions the model reduces approximately to the superweak [10] form of *CP* violation for strange particle decays. Pakvasa and Sugawara [8] also propose a model for the new particles involving all three new quarks, whereas Maiani [9] assumes no degeneracy between quark masses.

In this paper we discuss systematically the possible patterns of *CP* violation in the Kobayashi-Maskawa [7] model. We analyse the validity of the superweak approximation in ordinary K decays, rare K decays and charmed particle decays. We also estimate the neutron electric dipole moment, finding it to be

$$\left|\frac{D}{e}\right|_{\mathrm{n}} \sim \begin{Bmatrix} (10^{-27} - 10^{-28})\ \mathrm{cm} \\ (10^{-30} - 10^{-31})\ \mathrm{cm} \end{Bmatrix} \qquad \text{for} \qquad m_{\mathrm{t,b}} \sim \begin{cases} m_{\mathrm{W}} \\ 5\ \mathrm{GeV} \end{cases} .$$

As suggested by Pakvasa and Sugawara, superweak results for common K decays can be approximately recovered if the t and c quarks are almost degenerate, but we disfavour this possibility on phenomenological grounds as mentioned below. We in fact find that approximate superweak results obtain for common K decays independently of the masses of the t and b quarks. The model predicts large deviations from superweak theory in rare K decays, and also has deviations from superweak predictions in charmed particles, but all these seem difficult to observe.

Our calculations are collected in sect. 2. The model is introduced in subsect. 2.1, and applied to common K decays in subsects. 2.2, 2.3 and 2.4. Subsect. 2.5 contains a discussion of rare K decays, charmed meson decays are studied in subsect. 2.6, and the neutron electric dipole moment estimated in subsect. 2.7. Our conclusions are summarized and discussed in sect. 3.

2. Calculations

2.1. The model

Let us consider models with N quarks of charge $\frac{2}{3}$ and N of charge $-\frac{1}{3}$ *, and sup-

* Each type of quark comes in three colours, but the weak and electromagnetic interactions are presumed to be colour-blind.

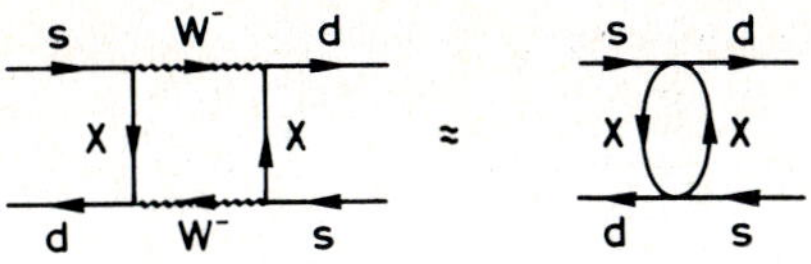

Fig. 1. Contributions to the $K^0-\bar{K}^0$ mass mixing matrix.

pose their weak and electromagnetic interactions are described by the SU(2) × U(1) Weinberg-Salam model, with the "left" components of the quark fields collected into N doublets under the weak SU(2) and the "right" components left as singlets. The (V – A) charged weak current is then described by an $N \times N$ unitary matrix U:

$$J_\mu^- = \bar{q}_P \gamma_\mu L U q_N \qquad \text{where} \qquad L \equiv \tfrac{1}{2}(1-\gamma_5) \,. \tag{2.1}$$

Glashow et al. [11] pointed out that in such a model the neutral weak current is purely diagonal in quark fields, being a mixture of the electromagnetic current and

$$J_\mu^0 = \bar{q}_P \gamma_\mu L U^\dagger U q_P + \bar{q}_N \gamma_\mu L U^\dagger U q_N \,. \tag{2.2}$$

In any such model $\Delta S = 1$ neutral currents and $\Delta S = 2$ transitions are very strongly suppressed. In particular, free quark model calculations of diagrams like those of fig. 1 in the case $N = 2$ (with just X = u, c quarks exchanged) led to

$$\Delta m \equiv m_{K_L} - m_{K_S} \simeq \frac{G_F}{\sqrt{2}} f_K^2 \frac{\alpha}{4\pi} \left(\frac{m_c}{38\ \text{GeV}}\right)^2 \sin^2\theta_C \cos^2\theta_C \,. \tag{2.3}$$

The experimental value of Δm implies [12] $m_c = 0(1.5 \text{ to } 2)$ GeV, an estimate not greatly altered by asymptotic freedom effects [13].

The unitary matrix U is generally characterized by N^2 parameters, 1 of which is an unobservable overall phase, and $2(N-1)$ of which can be removed by suitable choices of phases for the quark fields. The matrix U therefore has $(N-1)^2$ observable parameters. In the case $N = 2$, the SU(4) model of Glashow et al. [11], there is just one free parameter and U can be taken as an orthogonal matrix

$$J_\mu^- = (\bar{u}, \bar{c})\gamma_\mu L \begin{pmatrix} \cos\theta_C & \sin\theta_C \\ -\sin\theta_C & \cos\theta_C \end{pmatrix} \begin{pmatrix} d \\ s \end{pmatrix}, \tag{2.4}$$

with θ_C the Cabibbo angle. In the case $N = 3$, U has 4 observable parameters, only 3 of which can be absorbed into the 3 angles parametrizing 3-dimensional orthogonal matrices *. As pointed out originally by Kobayashi and Maskawa [7], there is another parameter which can be taken as a complex phase, and which introduces *CP* violation. Introducing new "top" t ($Q = \frac{2}{3}$) and "bottom" b ($Q = -\frac{1}{3}$) quarks, and

* In general an $N \times N$ orthogonal matrix is determined by $\frac{1}{2}[N(N-1)]$ real parameters leaving $\frac{1}{2}[(N-1)(N-2)]$ complex parameters in the general case. The 6-quark model is unique in that there is only one *CP* violating parameter and our subsequent analysis is based on this assumption. However, we expect it to be approximately valid even if there are more quarks beyond t and b, just as previous estimates [12] of m_c are not altered by going to 6 quarks.

parametrizing U as done by Kobayashi and Maskawa, we have:

$$J_\mu^- = (\bar{u}, \bar{c}, \bar{t})\gamma_\mu L \begin{pmatrix} c_1 & -s_1c_3 & -s_1s_3 \\ s_1c_2 & c_1c_2c_3 - s_2s_3e^{i\delta} & c_1c_2s_3 + s_2c_3e^{i\delta} \\ s_1s_2 & c_1s_2c_3 + c_2s_3e^{i\delta} & c_1s_2s_3 - c_2c_3e^{i\delta} \end{pmatrix} \begin{pmatrix} d \\ s \\ b \end{pmatrix}, \qquad (2.5)$$

where $c_i(s_i) \equiv \cos\theta_i$ ($\sin\theta_i$), $i = 1, 2, 3$.

The model (2.5) reduces to the Glashow-Iliopoulos-Maiani model (2.4) in the limit $\theta_1 \to \theta_C$, $\sin\theta_2$, $\sin\theta_3 \to 0$, $\delta \to 0$, and there is nothing to prevent the Kobayashi-Maskawa (KM) model from containing the Glashow Iliopoulos and Maiani (GIM) model as a good approximation to the weak interactions of the (u,d,s,c) sector.

Inspection of the weak current (2.5) suggests that the experimental validity of Cabibbo theory for u, d and s quarks would restrict possible values of the new angle parameters. The best estimate of the relative magnitude of $\Delta S = 0$ and $\Delta S = 1$ couplings seems to come [14] from semileptonic decays of baryons, and yields $\sin\theta_C = 0.230 \pm 0.003$. From a recent compilation of Jarlskog [15] we can estimate $\cos^2\theta_C$, by comparing nuclear β decay with μ decay. We find $\cos^2\theta_C = 0.9481 \pm 0.004$, where the dominant errors come from differing estimates of the nuclear quantity Ft, and from uncertainties in the (in principle calculable) higher order weak and electromagnetic corrections at the strong decay vertex. We then find $\cos^2\theta_C + \sin^2\theta_C = 1.001 \pm 0.004$ experimentally. Allowing one standard deviation, this implies that in the KM model (2.5) there must be small "leakage" of the weak coupling of the u quark to the b quark: $s_1^2 s_3^2 < 0.003$. Since $s_1^2 \approx \sin^2\theta_C \simeq 0.05$, this means

$$s_3^2 < 0.06\,, \qquad (2.6)$$

so that $|s_3|$ cannot be much bigger than $\sin\theta_C$.

Even in the continued absence of charmed particles, there are some more restrictions on s_2 and c_2. Consider first the proposal that just one new quark c has a mass O(1.5 to 2) GeV. The experimental absence of visible tracks of charmed particles suggest a lifetime not much more than 10^{-12} sec. As particles with masses ~2 GeV and unit weak couplings to light quarks probably have lifetimes $O(10^{-13})$ sec, and since the $\Delta C = 1$, $\Delta S = 0$ coupling in (2.5) is limited in magnitude to $O(\sin^2\theta_C) \simeq 0.05$, this means charm decays to states containing strange quarks must dominate, so that c_2^2 cannot be too small. Consider now the possibility that both c and t have masses O(1.5 to 2) GeV: then by the previous lifetime argument both c_2^2 and s_2^2 must be $\gtrsim 1/10$, and strange quarks must dominate the decays both of naked charm and of naked top states. The possibility that b has a low mass O(1.5 to 2) GeV seems untenable, because its decays to a u quark are suppressed by $s_1^2 s_3^2 < 0.003$, and decays to states containing c or t quarks would be strongly suppressed by phase space, though in principle the couplings could be of magnitude ~1. We cannot rule out a low top mass on the same grounds, but the observed spectroscopy of new states at SLAC and DESY [3] disfavours this possibility.

It is clear that *CP* violation can be introduced *a fortiori* into models [16] with right-handed currents also. We dislike such models purely for aesthetic reasons. First because there are no direct traces of right-handed couplings between u, d and s quarks, and there are theoretical problems [17] with a right-handed coupling between the c and d quarks. Secondly because e – μ universality is experimentally very good [18] ($g_e/g_\mu = 1.007 \pm 0.014$), and this is natural only if any new heavy lepton has a left-handed coupling to its own massless neutrino [19]. The most natural way to cancel triangle anomalies and restore renormalizability is to introduce a third left-handed quark doublet. The Higgs content is more complicated in theories with right-handed currents as well as left. In particular there are several charged and neutral physical Higgs bosons *, and one is quite bad enough. As we show in the appendix, *CP* violation can be incorporated in a left-handed current model using just the original complex doublet of Higgs fields, and hence just one physical Higgs particle.

The KM model has recently been revived by Pakvasa and Sugawara [8], and independently by Maiani [9], who presented interesting estimates of *CP* violation in K decays and of the neutron electric dipole moment. We now proceed to our calculations of these and other *CP* violating quantities.

2.2. CP violation in the neutral K mass matrix

It is apparent from fig. 1 that the effective higher order $\Delta S = 2$ transition between K^0 and $\bar{K}^0$ introduces *CP* violation when X = c or t. Indeed these diagrams give the superweak [10] state mixing which experimentally [14] is the dominant source of *CP* violation in neutral K decays. The analysis of appendix F of ref. [12] tells us immediately that in the six-quark model the effective $\Delta S = 2$ operator is proportional to an integral of the form

$$\int \frac{d^4k}{(2\pi)^4} \left\{\frac{\gamma^\alpha \not{k} \gamma^\beta L}{k^2 - m_W^2}\right\} \left\{\frac{\gamma_\beta \not{k} \gamma_\alpha L}{k^2 - m_W^2}\right\} \left\{\frac{-c_1 s_1 c_3}{k^2 - m_u^2} + \frac{s_1 c_2 (c_1 c_2 c_3 - s_2 s_3 e^{i\delta})}{k^2 - m_c^2} + \frac{s_1 s_2 (c_1 s_2 c_3 + c_2 s_3 e^{i\delta})}{k^2 - m_t^2}\right\}^2 . \tag{2.7}$$

The integral (2.7) is safely convergent even in the absence of the W propagators, so we assume $m_{u,c,t}^2 \ll m_W^2$ and set the W propagators $\approx 1/m_W^2$. We will also assume $m_{c,t}^2 \gg m_u^2$ and set $m_u^2 \simeq 0$. The pole terms in the integrand can then be written as

$$\left\{\frac{s_1 c_1 c_3}{k^2}\left[\frac{c_2^2 m_c^2}{k^2 - m_c^2} + \frac{s_2^2 m_t^2}{k^2 - m_t^2}\right] + \frac{s_1 s_2 c_2 s_3 e^{i\delta} (m_t^2 - m_c^2)}{(k^2 - m_c^2)(k^2 - m_t^2)}\right\}^2 . \tag{2.8}$$

* J. Prentki, private communication.

Then the ratio $\mathrm{Im}\, M_{12}/\Delta m$ of the *CP* violating and *CP* conserving $\Delta S = 2$ terms in the mass matrix is just the ratio of the integral

$$2\int \mathrm{d}^4k \left\{ \frac{s_1 s_2 c_2 s_3 \sin\delta (m_\mathrm{t}^2 - m_\mathrm{c}^2)}{(k^2 - m_\mathrm{c}^2)(k^2 - m_\mathrm{t}^2)} \right\} \left\{ \frac{s_1 c_1 c_3 c_2^2 m_\mathrm{c}^2}{k^2 - m_\mathrm{c}^2} + \frac{s_1 c_1 c_3 s_2^2 m_\mathrm{t}^2}{k^2 - m_\mathrm{t}^2} + \frac{k^2 s_1 s_2 c_2 s_3 \cos\delta (m_\mathrm{t}^2 - m_\mathrm{c}^2)}{(k^2 - m_\mathrm{c}^2)(k^2 - m_\mathrm{t}^2)} \right\}, \tag{2.9}$$

and the integral

$$\int \mathrm{d}^4k \left\{ \frac{s_1^2 c_1^2 c_3^2}{k^2} \left[\frac{c_2^2 m_\mathrm{c}^2}{k^2 - m_\mathrm{c}^2} + \frac{s_2^2 m_\mathrm{t}^2}{k^2 - m_\mathrm{t}^2} \right]^2 + \frac{2 s_1^2 c_1 s_2 c_2 s_3 c_3 \cos\delta (m_\mathrm{c}^2 - m_\mathrm{c}^2)}{(k^2 - m_\mathrm{c}^2)(k^2 - m_\mathrm{t}^2)} \times \left[\frac{c_2^2 m_\mathrm{c}^2}{k^2 - m_\mathrm{c}^2} + \frac{s_2^2 m_\mathrm{t}^2}{k^2 - m_\mathrm{t}^2} \right] + \frac{k^2 s_1^2 s_2^2 c_2^2 s_3^2 \cos 2\delta (m_\mathrm{t}^2 - m_\mathrm{c}^2)^2}{(k^2 - m_\mathrm{c}^2)^2 (k^2 - m_\mathrm{t}^2)^2} \right\}. \tag{2.10}$$

Thus we find

$$\left| \frac{\mathrm{Im}\, M_{12}}{\Delta m} \right| \approx \frac{2 s_1^2 c_1 s_2 c_2 s_3 c_3 \sin\delta (m_\mathrm{t}^2 - m_\mathrm{c}^2) \left\{ c_2^2 m_\mathrm{c}^2 \left[\frac{m_\mathrm{t}^2}{(m_\mathrm{t}^2 - m_\mathrm{c}^2)^2} \ln\left(\frac{m_\mathrm{t}^2}{m_\mathrm{c}^2}\right) - \frac{1}{m_\mathrm{t}^2 - m_\mathrm{c}^2} \right] + s_2^2 m_\mathrm{t}^2 \left[\frac{m_\mathrm{c}^2}{(m_\mathrm{t}^2 - m_\mathrm{c}^2)^2} \ln\left(\frac{m_\mathrm{c}^2}{m_\mathrm{t}^2}\right) + \frac{1}{m_\mathrm{t}^2 - m_\mathrm{c}^2} \right] \right\}}{s_1^2 c_1^2 c_3^2 \left[c_2^4 m_\mathrm{c}^2 + s_2^4 m_\mathrm{t}^2 + \frac{2 s_2^2 c_2^2 m_\mathrm{t}^2 m_\mathrm{c}^2}{(m_\mathrm{t}^2 - m_\mathrm{c}^2)} \ln\left(\frac{m_\mathrm{t}^2}{m_\mathrm{c}^2}\right) \right]}, \tag{2.11}$$

where we have dropped from both the numerator and denominator terms which are relatively small if $s_3^2 \ll c_3^2$ as argued in subsect. 2.1.

We will return to the analysis of (2.11) in subsect. 2.4: for now we just make two observations:

(i) The observed magnitude of *CP* violation effects in the neutral K mass matrix clearly implies $|s_2 s_3 \sin\delta| = \mathrm{O}(10^{-3})$. We leave the derivation of this very small number in a natural way as an exercise for ourselves and our readers.

(ii) The formulae for $|\mathrm{Im}\, M_{12}/\Delta m| \propto (m_\mathrm{t} - m_\mathrm{c})$ quoted in refs. [8,9] are good approximations to (2.11) only when $|m_t - m_\mathrm{c}| \ll m_\mathrm{c}$.

2.3. Direct CP violating decays of K mesons

Lowest order weak decay amplitudes for K^0 mesons will violate *CP* when they involve charmed or top quarks. It is convenient to introduce the idea of "top-spin" which interchanges t and c quarks. It is easy to see from (2.5) that the effective lowest order four-quark *CP* violating operator is a top-spin vector:

$$\mathcal{H}_{\text{eff}}^{\text{CPX}} = \frac{G_F}{\sqrt{2}} \bar{s}\gamma_\mu L(t\bar{t} - c\bar{c})\gamma^\mu L d \,, \tag{2.12}$$

and so its matrix elements between topless * and charmless states would vanish in a top-spin invariant worls. In fact, top-spin is broken by the top and charm quark mass difference, and matrix elements of (2.12) are in general non-zero. We analyse the strength of this *CP* violation in two ways.

Feynman diagrams

In this traditional technique, one inspects all Feynman diagrams (fig. 2) which take the initial valence quarks to the final ones *via* $\mathcal{H}_{\text{eff}}^{\text{CPX}}$ including all possible gluon exchanges **. The *CP* violating contribution of fig. 2a in the 't Hooft-Feynman [20] gauge is given by

$$s_1 s_2 s_3 c_2 \sin\delta g^2 \int \frac{d^4 p}{(2\pi)^4} \frac{\gamma_\mu L(\not{p} + m_X)\gamma^\mu L}{(p^2 - m_X^2)((p-q)^2 - m_W^2)}$$

$$= s_1 s_2 s_3 c_2 \sin\delta g^2 \int \frac{d^4 p}{(2\pi)^4} \frac{-2\not{p} L}{(p^2 - m_X^2)((p-q)^2 - m_W^2)} \,.$$

Its contribution to *CP* violation is the difference between c and t exchanges:

$$s_1 s_2 s_3 c_2 \sin\delta \int \frac{d^4 p}{(2\pi)^4} \frac{-2\not{p} L(m_c^2 - m_t^2)}{((p-q)^2 - m_W^2)(p^2 - m_c^2)(p^2 - m_t^2)}$$

which is finite and proportional to ***

$$\frac{G_F}{\pi^2} s_1 s_2 s_3 c_2 \sin\delta \not{q} L(m_t^2 - m_c^2) \left(1 + O\left(\frac{q^2, m_s^2}{m_W^2}\right)\right) . \tag{2.13}$$

The first term in (2.13) is absorbed in the quark wave function renormalization, so

* Ordinary hadrons are assumed to be topless and bottomless, which is presumably why their degrees of freedom are excited at low energies.

** Graphs with gluons emitted and absorbed by the same fermion line can be discussed similarly.

*** In this and forthcoming formulae, m_s should probably be interpreted as $O(M_K)$.

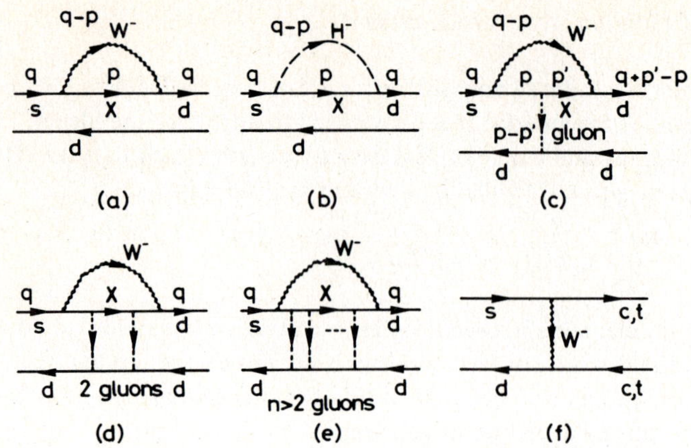

Fig. 2. Feynman diagrams leading to CP violation in amplitudes for K^0, $\bar{K}^0$ decays.

that the final contribution of fig. 2a to the CP violating amplitude is

$$\frac{G_F}{\pi^2} s_1 s_2 s_3 c_2 \sin\delta \; m_s(m_t^2 - m_c^2) O\left(\frac{q^2, m_s^2}{m_W^2}\right) . \tag{2.14}$$

The analysis of figs. 2b and 2c is completely analogous; terms which cánnot be absorbed into quark wave funtion, mass, or coupling renormalization are finite and of the order of (2.14).

Turning now to fig. 2d, we follow intuition based on the $K_L \to \gamma\gamma$ analysis of ref. [12]. When X is a u quark, the quark mass and the digluon mass are similar, and we expect the graph to be given in order of magnitude by the triangle anomaly. When X is a heavy c or t quark, the anomaly contribution is expected to be cancelled as $\beta \equiv m_X^2/m_{\rm digluon}^2 \to \infty$ like $F(\beta)$ in fig. 8 of ref. [12]. We therefore expect a very small contribution to the amplitude, suppressed relative to the u quark diagram by $O(m_u^2/m_c^2 - m_u^2/m_t^2)$ *. As for fig. 2e, our intuition suggests a suppression by a factor $\sim 1/m_X^2$ for each heavy quark propagator, so that the sum of such diagrams should be suppressed relative to the u quark piece by

$$\left(\frac{m_u^2}{m_c^2, m_t^2}\right)^{\alpha} , \tag{2.15}$$

with α an integer $\geqslant 2$. This is consistent with top-spin, which says that the CP violating amplitude should vanish as $m_c \to m_t$, and suggests a contribution of the graphs

* In this formula m_u should probably be interpreted as a typical non-strange meson mass.

of fig. 2e which is

$$s_1 s_2 s_3 c_2 \sin\delta\, G_F\, O\left((m_K)^{1+\alpha}\left(\frac{1}{m_c^\alpha} - \frac{1}{m_t^\alpha}\right)\right), \qquad \alpha \text{ integer} \geqslant 2\,. \qquad (2.16)$$

Finally we turn to the simplest graph of fig. 2f. As observed by Pakvasa and Sugawara [8], its contribution to the amplitude is suppressed by Zweig's rule and we may take its ratio to the allowed diagram with a ($\bar{u}u$) pair in the final state to be *

$$\propto (Z_c^s - Z_t^s)\frac{s_2 c_2 t_3 \sin\delta}{c_1}\,, \qquad (2.17)$$

where $(Z_{c,t}^s)^2$ is the coupling for an $I = 0, J^P = 0^+\pi\pi$ state with a mass m_K to a ($\bar{c}c$), ($\bar{t}t$) pair. The experimental suppression of the cascade decay $\psi'(3700) \to J/\psi(3100) + \pi\pi$ relative to the allowed decay $\rho'(1600) \to \rho(770) + \pi\pi$, where the $\pi\pi$ system has similar isospin, spin and mass, suggests

$$|Z_c^s|^2 = O(10^{-2}) \qquad \text{or} \qquad Z_c^s = O(1/10)\,. \qquad (2.18)$$

We cannot estimate Z_t^s very well, except to believe that as $m_t > m_c$, probably $Z_c^s > Z_t^s$.

Wilson expansion

More modern [21] than the Feynman graph analysis just described is to expand the effective *CP* violating Hamiltonian as a series in local operators of increasing dimension:

$$\mathcal{H}_{\text{eff}}^{\text{CPX}} = G_F\{(\bar{s}d)F_1 + (\bar{s}\not{D}d)F_2 + (\bar{s}D^2 d)F_3 + \bar{s}\gamma_\mu L(c\bar{c} - t\bar{t})\gamma^\mu L d\, F_4$$
$$+ \text{operators of higher dimension}\,, \qquad (2.19)$$

where $D_\mu \equiv \partial_\mu + igB_\mu$, B_μ a gluon field. Dimensional analysis indicates that

$$F_1 = O(m_Q m_W^2)\,, \quad F_2 = O(m_W^2)\,, \quad F_3 = O(m_Q)\,, \quad F_4 = O(1)\,, \qquad (2.20)$$

and higher order terms are suppressed by powers of $1/m_W$. The first two terms of (2.19) are absorbed into mass, wave function and coupling constant renormalizations. The third operator is not a top-spin vector, whereas the underlying interaction was. Therefore the coefficient function $F_3 = O[m_s(m_t^2 - m_c^2)/m_W^2]$. The fourth term is a top-spin vector operator whose matrix elements are given by diagrams like those of

* Hard gluon exchange in fig. 2f is expected to give a short distance enhancement. This may be less [22] than the similar enhancement for the ($\bar{u}u$) diagram [23], but the order of magnitude should be similar.

figs. 2c, d, e and f discussed above *. We expect its matrix element to be proportional to some factor $(X_c - X_t)$ which vanishes as $m_t - m_c \to 0$, and from our Feynman graph discussion may either behave as

$$\left(\frac{m_K}{m_c}\right)^\alpha - \left(\frac{m_K}{m_t}\right)^\alpha \tag{2.21}$$

for some $\alpha > 1$, or like the Zweig factors $(Z_c^s - Z_t^s)$ of (2.17), or both.

2.4. *CP violation in neutral K decays*

We discuss *CP* violation in the $K^0 - \bar{K}^0$ system using the standard definitions and conventions set out by Kleinknecht [14]. There is the mixing parameter

$$\epsilon \equiv \frac{\frac{1}{2}\mathrm{Im}\,\Gamma_{12} + i\,\mathrm{Im}\,M_{12}}{\frac{1}{2}i(\Gamma_S - \Gamma_L) - \Delta m}\,, \tag{2.22}$$

where $\Gamma_{12} = 2\pi\,\Sigma_F\,\rho_F\,\langle \bar{K}^0|\mathcal{H}_{wk}|F\rangle\langle F|\mathcal{H}_{wk}|K^0\rangle$, $\Gamma_S - \Gamma_L$ is the difference between the K_S and K_L lifetimes, and we estimated Im M_{12} and Δm in subsect. 2.2. Choosing a phase convention so that the $K \to 2\pi\ (I = 0)$ amplitude is real apart from the strong interaction phase shift δ_0:

$$\langle \pi\pi(I=0)|K^0\rangle = e^{i\delta_0}A_0\,, \qquad A_0 \text{ real}\,, \tag{2.23a}$$

then

$$\langle \pi\pi(I=2)|K^0\rangle = e^{i\delta_2}A_2\,, \tag{2.23b}$$

and the standard *CP* violation parameters η_{+-} and η_{00} obey

$$\begin{aligned} \eta_{+-} &= \epsilon + \epsilon'\,, \\ \eta_{00} &= \epsilon - 2\epsilon'\,, \end{aligned} \tag{2.24}$$

where

$$\epsilon' \equiv \frac{i}{\sqrt{2}}\, e^{i(\delta_2 - \delta_2)}\, \frac{\mathrm{Im}\,A_2}{A_0}\,. \tag{2.25}$$

From eq. (2.22) it follows that

$$\mathrm{Arg}\,\epsilon = \tan^{-1}\left(\frac{2\Delta m}{\Gamma_S - \Gamma_L}\right) + \phi_D\,,$$

* Note that multigluon operators do not appear in the series (2.19): their coefficients F_i would be suppressed by powers of m_W^{-1}. However, soft gluons in the hadronic wave functions are relevant in taking matrix elements of local operators.

where

$$\phi_D = -\tan^{-1}\left(\frac{\mathrm{Im}\,\Gamma_{12}}{2\,\mathrm{Im}\,M_{12}}\right). \tag{2.26}$$

In the superweak theory where all *CP* violation arises from the $K^0 - \bar{K}^0$ mixing matrix we would have $\epsilon' = 0 = \phi_D$. In milliweak theories ϵ' arises from the *CP* violating phase difference $\xi_{2\pi}$ between $\langle\pi\pi(I=0)|K^0\rangle$ and $\langle\pi\pi(I=2)|K^0\rangle$, while ϕ_D comes from the *CP* violating phase difference between $\langle\pi\pi(I=0)|K^0\rangle$ and the amplitude for all other $(K^0, \bar{K}^0)$ decays.

In the KM model the *CP* violating interaction (2.12) does not contribute to the $|\Delta I| = \frac{3}{2}$ transition (2.23b), so the phase convention defined by (2.23a) implies a redefinition of fields by the phase transformation $e^{-i\xi_{2\pi}S}$, where S is the strangeness operator. The phase of the $K^0 \to \bar{K}^0$ transition is then redefined by

$$\frac{\mathrm{Im}\,M_{12}}{\Delta m} \to \epsilon_m + 2\xi_{2\pi} ,$$

where ϵ_m is the quantity displayed in eq. (2.11). However, we shall see below that $|\xi_{2\pi}| \ll |\epsilon_m|$. Turning now to the specifics of the KM model, we first observe that the successful estimate [12] of m_c from Δm in the four quark model [11] implies that the denominator of (2.11) $\approx s_1^2 m_c^2$. We therefore have

$$\epsilon_m \approx 2c_1 s_2 c_2 s_3 c_3 \sin\delta\left[\cos 2\theta_2\,\frac{m_t^2}{m_t^2 - m_c^2}\ln\left(\frac{m_t^2}{m_c^2}\right) + s_2^2\left(\frac{m_t^2}{m_c^2}\right) - c_2^2\right]. \tag{2.27}$$

By comparison, the *CP* violating phase difference $\xi_{2\pi}$ is determined by the relative strength ξ of the general *CP* violating amplitude estimated in subsect. 2.3. From figs. 2a, b, c we use (2.14) and the approximation

$$\langle\pi\pi(I=0)|K^0\rangle \approx s_1 c_1 c_2\, G_F\, m_K^3 \tag{2.28}$$

to estimate $|\xi_{2\pi}| \simeq |\xi|$ with

$$|\xi| \simeq \left|\frac{s_2 c_2 t_3 \sin\delta}{c_1}\right| \frac{(m_t^2 - m_c^2)}{\pi^2 m_W^2} , \tag{2.29a}$$

where $t_i = \tan\theta_i$. From figs. 2d, e we get

$$|\xi| \simeq \left|\frac{s_2 c_2 t_3 \sin\delta}{c_1}\right| \left(\frac{m_u^2}{m_c^2} - \frac{m_u^2}{m_t^2}\right), \tag{2.29b}$$

where m_u probably $\leqslant 100$ MeV. Finally from fig. 2f we get

$$|\xi| \simeq \left|\frac{s_2 c_2 t_3 \sin\delta}{c_1}\right| (Z_c^s - Z_t^s) , \tag{2.29c}$$

where $Z_c^s \sim \frac{1}{10} \gtrsim Z_t^s$. Comparing the different estimates (2.29) we see that for $m_t^2 \ll m_W^2$ the largest is (2.29c), while (2.29a) may become competitive only for $m_t^2 \simeq m_W^2$. Comparing (2.27) and (2.29c) we see that *CP* violation in the mass mixing matrix is dominant whatever the value of m_t. For example, for $m_t - m_c \ll m_c$ eq. (2.27) becomes

$$|\epsilon_m| \approx 2|c_1 c_2 c_3 s_2 s_3 \sin\delta| \left(\frac{m_t - m_c}{m_c}\right) , \tag{2.30}$$

which reproduces the results of Pakvasa and Sugawara [8] and of Maiani [9]. If we suppose that

$$|Z_c^s - Z_t^s| \approx \frac{1}{10}\left(\frac{m_t - m_c}{m_c}\right) \tag{2.31}$$

for $m_t - m_c \ll m_c$ then we get

$$|\xi| \simeq \frac{1}{10}\left|\frac{c_2 s_2 s_3 \sin\delta}{c_1 c_3}\right| \left(\frac{m_t - m_c}{m_c}\right) , \tag{2.32}$$

and we know from the subsect. 2.1 that $|c_1| \simeq |c_3| \simeq 1$ so that $|\epsilon_m| \gg |\xi|$. For $m_t \gg m_c$,

$$|\epsilon_m| \approx 2|c_1 c_2 c_3 s_2 s_3 \sin\delta| \left[c_2^2\left(\ln\left(\frac{m_t^2}{m_c^2}\right) - 1\right) + s_2^2 \frac{m_t^2}{m_c^2}\right] , \tag{2.33}$$

and whether we take

$$Z_t^s/Z_c^s \approx 0 \qquad \text{or} \qquad 1\Big/\ln\left(\frac{m_t^2}{m_c^2}\right) , \tag{2.34}$$

which seem to bracket the range of possibilities, it is clear that for $|c_2| = O(1)$, again $|\epsilon_m| \gg |\xi|$.

The KM model therefore approximates superweak results whatever the value of m_t. By virtue of the phase convention (2.23a), the amplitude A_2 defined in (2.23b) has the phase $-\xi_{2\pi}$, and we obtain

$$|\epsilon'| = \frac{1}{\sqrt{2}}\left|\frac{\mathrm{Im}\, A_2}{A_0}\right| \simeq \frac{1}{\sqrt{2}} |\xi| \left|\frac{A_2}{A_0}\right| \approx \tfrac{1}{30}|\xi| , \tag{2.35}$$

whereas

$$|\phi_D| \approx \frac{|\xi|}{2}\left[\frac{\Gamma(\mathrm{K}^0 \to 2\pi(I=2)) + \Gamma(\mathrm{K}^0 \to 3\pi)}{\Gamma(\mathrm{K}^0 \to 2\pi(I=0))}\right]\Big/\left|\frac{\mathrm{Im}\, M_{12}}{\Delta m}\right|$$

$$\approx 2 \times 10^{-3}\, |\xi| \Big/ \left|\frac{\mathrm{Im}\, M_{12}}{\Delta m}\right| , \tag{2.36}$$

$$|\epsilon| \simeq \frac{1}{\sqrt{2}}\left|\frac{\mathrm{Im}\, M_{12}}{\Delta m}\right| . \tag{2.37}$$

From eqs. (2.30), (2.32), (2.33) (2.34) and (2.37) we get $|\xi|/|\epsilon| \lesssim \frac{1}{15}$ which gives,

$$|\epsilon'| \lesssim \tfrac{1}{450}\,|\epsilon|\,, \tag{2.38}$$

and from eq. (2.36)

$$|\phi_D| \lesssim 10^{-4}\,. \tag{2.39}$$

By comparison, the present experimental limits [14] are

$$|\epsilon'| \lesssim \tfrac{1}{50}\,|\epsilon|\,, \qquad |\phi_D| \lesssim \tfrac{1}{20}\,.$$

Consider now the decays $K \to 3\pi$. For neutral kaon decay to an arbitrary *CP* eigenstate f, the deviation from the superweak prediction is determined by the parameter ϵ_f' which is defined by

$$\epsilon_f' = \frac{ia(K_{2(1)} \to f)}{a(K_{1(2)} \to f)}\,,$$

yielding

$$\frac{\langle f|K_{L(S)}\rangle}{\langle f|K_{S(L)}\rangle} = \epsilon + \epsilon_f' + O(\epsilon^2)\,, \tag{2.40}$$

if f is *CP* even (odd), so that ϵ_f' vanishes if the phase difference between the amplitudes for $K^0 \to f$ and $K^0 \to 2\pi\,(I = 0)$ is the same.

The amplitude for $K^0 \to 3\pi$ is completely determined in terms of the $K^0 \to 2\pi$ amplitude in the limit of exact chiral symmetry and in the approximation that the matrix element is linear in energy. Thus the phase difference between the dominant $\Delta I = \frac{1}{2}$ amplitudes for $K^0 \to 3\pi$ and $K^0 \to 2\pi$ vanish in this approximation, which is probably good to within 10%. Thus we expect

$$|\epsilon'_{3\pi(\Delta I=\frac{1}{2})}| = \left|\frac{\xi}{10}\right| \lesssim \frac{|\epsilon|}{150}\,,$$

$$|\epsilon'_{3\pi(\Delta I=3/2)}| \simeq |\epsilon'| \simeq \left|\frac{\xi}{20}\right| \lesssim \frac{|\epsilon|}{300}\,.$$

For $K^\pm$ decay, the only observable effects are differences in rates and spectra which are further suppressed by the lack of final state interactions. For semi-leptonic decays, the model predicts no *CP* violation to lowest order in the weak interaction. We conclude that the KM model is completely consistent with existing data on *CP* violation in common neutral K decays, and likely to remain so for some time.

2.5. *CP violation in rare K decays*

We saw in the previous section that the KM model predicts very small deviations from superweak results for the dominant neutral K decays. Rare K decays such as $K \to 2\gamma$, $\mu^+\mu^-$, πe^+e^- and $\pi\nu\bar{\nu}$ occur only in higher order, and at rates generally governed by the GIM mechanism involving cancellations between light and heavy quarks. Therefore, such decays might be expected to show larger deviations from superweak predictions, and we now analyse them in terms of the parameters ϵ_f' defined in eq. (2.40). As we shall see, the relevant rates are so small as to be beyond detection at present. Nevertheless, these effects may eventually (*via* kaon factories?) provide the only hope for distinguishing the KM from the superweak model. Also, our calculation can serve as a starting point for parallel analyses in other models of *CP* violation.

$K \to 2\gamma$

There are two possible states for the photon pair: $2\gamma(+) \equiv F_{\mu\nu}F^{\mu\nu}$ with $CP = +1$, $2\gamma(-) \equiv \epsilon_{\mu\nu\alpha\beta}F^{\mu\nu}F^{\alpha\beta}$ with $CP = -1$. In a *CP* invariant world $K_1 \to 2\gamma(+)$ and $K_2 \to 2\gamma(-)$. The analysis of Gaillard and Lee [12] suggests that the dominant contribution to $K_1 \to 2\gamma(+)$ comes from the usual long distance mechanism of the 2π intermediate state shown in fig. 3. The short distance contribution of fig. 4 is expected to be relatively suppressed by $O(m_q^2/m_W^2)$. On the other hand the $K_2 \to 2\gamma(-)$ decay should go *via* the diagram of fig. 4, with the usual anomalous triangle diagram. However, as discussed in Gaillard and Lee [12], and in subsect. 2.3 of this paper, the triangle diagrams with heavy quarks are suppressed for small $m(\gamma\gamma)$, ostensibly by $O(m_u^2/m_c^2 - m_u^2/m_t^2)$. A more phenomenological way of estimating the contribution of the triangle diagram of fig. 4 is to approximate it by pseudoscalar meson poles [12] as indicated in fig. 5. The dominant contributions then come from the η and η' intermediate states, and the contributions of charmed (top) quarks are suppressed by a Zweig factor $Z^p_{c(t)}$. The observed rate of the decay $\psi'(3700) \to J/\psi(3100) + \eta$ enables us to estimate

$$|Z_c^p| \leqslant \tfrac{1}{10} , \tag{2.41}$$

and again presumably $|Z_t^p| \leqslant |Z_c^p|$. The relative strength of the η_c contribution is $O(m_\eta^2, m_{\eta'}^2/m_{\eta_c}^2)$ and may be comparable.

In the presence of *CP* violation, since the dominant mechanism for producing the $\gamma\gamma(+)$ state is the 2π intermediate state, the pattern of $K_{S,L} \to \gamma\gamma(+)$ decays will be the same as that for $\pi\pi$ final states discussed in subsect. 2.4. Superweak theory will be approximately valid, with deviations characterized by ϵ': $|\epsilon'|/|\epsilon| \leqslant \frac{1}{450}$ from eq. (2.38), and ϕ_D as always given by eq. (2.39). In the case of the $\gamma\gamma(-)$ state, the above discussion suggests:

$$\epsilon'_{\gamma\gamma} \equiv \epsilon'_{\gamma\gamma(-)} \simeq (Z_c^p - Z_t^p)\frac{s_2 c_2 t_3 \sin\delta}{c_1} , \tag{2.42}$$

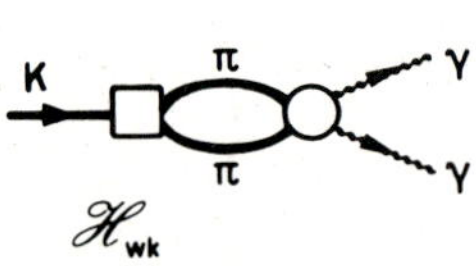

Fig. 3. The 2π intermediate state in $K \to 2\gamma(+)$.

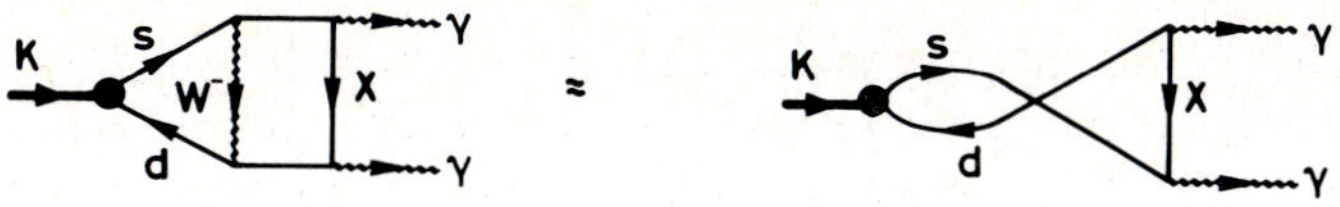

Fig. 4. Quark loop contribution to $K \to 2\gamma(\pm)$.

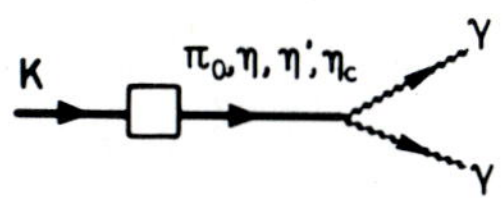

Fig. 5. Phenomenological model for $K \to 2\gamma(-)$.

so that (*cf.* eq. (2.29))

$$\frac{|\epsilon'_{\gamma\gamma}|}{|\epsilon|} \approx \frac{|\xi|}{|\epsilon|} \lesssim \tfrac{1}{15}\,, \tag{2.43}$$

if we follow the analysis of subsect. 2.4. In the case of $\gamma\gamma(-)$ decays the deviations from superweak are larger than for $\pi\pi$ or $\gamma\gamma(+)$, but since even $K_s \to \gamma\gamma$ has not yet been seen, the prospects for experiments able to see $\epsilon'_{\gamma\gamma}$ as given by eq. (2.43) are hardly immediate.

<u>$K \to 2\mu$</u>

Again there are two possible states for the μ pair: $2\mu(+) \equiv \bar{\mu}\mu$ with $CP = +1$, $2\mu(-) \equiv \bar{\mu}\gamma_5\mu$ with $CP = -1$. The analysis of Gaillard and Lee [12] suggests that the dominant contribution to $2\mu(+)$ comes from the $\gamma\gamma(+)$ intermediate state, and hence *via* the $\pi\pi$ intermediate state as in fig. 6. Hence as in $K_{S,L} \to 2\gamma(+)$, the pattern of *CP* violation in $K_{S,L} \to 2\mu(+)$ should be the same as in $K_{S,L} \to \pi\pi$. Gaillard and Lee found that for $2\mu(-)$ there were three relevant diagrams, those shown in fig. 7, with the long distance diagram of fig. 7a dominating the short distance diagrams of fig. 7b, c by a factor of (5 to 10). We then find

$$\epsilon'_{\mu\mu(-)} \simeq \epsilon'_{\gamma\gamma} + (\tfrac{1}{5} \leftrightarrow \tfrac{1}{10})\epsilon'_{\mu\mu}\,, \tag{2.43}$$

where $\epsilon'_{\mu\mu}$ is the ratio of the *CP* violating to *CP* conserving parts of the figs. 7b, c. Taking the Gaillard and Lee calculations of these graphs and extending them to top

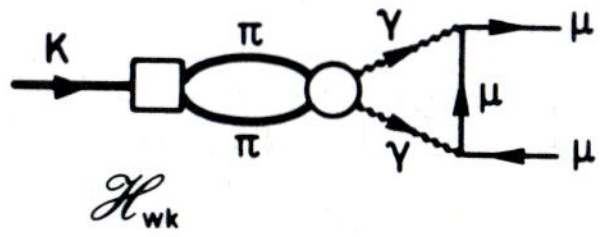

Fig. 6. Intermediate states in $K \to 2\mu(+)$.

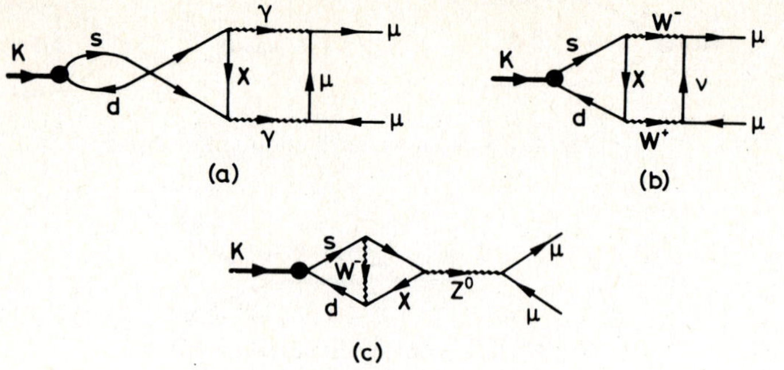

Fig. 7. Intermediate states in $K \to 2\mu(-)$.

quarks we find

$$|\epsilon'_{\mu\mu}| \approx \left(\frac{m_t^2 - m_c^2}{m_c^2}\right) \frac{s_2 c_2 t_3 \sin\delta}{c_1}, \tag{2.44}$$

and hence

$$\frac{|\epsilon'_{\mu\mu}|}{|\epsilon|} \approx \frac{1}{\sqrt{2}\, c_1^2 c_3^2} \frac{\dfrac{m_t^2 - m_c^2}{m_c^2}}{[c_2^2(\ln(m_t^2/m_c^2) - 1) + s_2^2(m_t^2/m_c^2)]}, \tag{2.45}$$

in the case $m_t \gg m_c$. If we assume $s_2^2 \ll c_2^2 \simeq 1 \simeq c_1^2 \simeq c_3^2$, and that the top degree of freedom has not yet been excited at SPEAR II or DORIS, so that $m_t \geqslant 4$ GeV, then we get

$$|\epsilon'_{\mu\mu}|/|\epsilon| \geqslant 4\,. \tag{2.46}$$

Hence the deviations from superweak theory in $K_s \to 2\mu(-)$ are in principle substantial: but they are presumably unobservable.

$K_L \to \pi^0 e^+ e^-$

As pointed out by Gaillard and Lee, the decay $K_2 \to \pi^0 e^+ e^-$ is forbidden by *CP* to proceed *via* γ or Z exchange, and the W^+W^- contribution is expected to be strongly suppressed. Therefore the decay $K_L \to \pi^0 e^+ e^-$ can proceed *via* superweak state mixing or *via CP* violation in the $K^0 \to \pi^0 e^+ e^-$ amplitude itself. Extending the calculations of appendix D of Gaillard and Lee gives

$$|\epsilon'_{\pi ee}| \approx \frac{\ln(m_t^2/m_c^2)}{\ln(m_c^2/m_h^2)} \left|\frac{s_2 c_2 t_3 \sin\delta}{c_1}\right|, \tag{2.47}$$

where m_h is a typical charmless and topless hadronic mass, probably $\sim m_\rho$. It is then easy to get

$$|\epsilon'_{\pi ee}|/|\epsilon| \gtrsim 1\,, \tag{2.48}$$

with $m_t \gtrsim 4$ GeV. Again a substantial deviation from superweak theory in principle, but even so

$$\frac{\Gamma(K_L \to \pi^0 e^+ e^-)}{\Gamma(K_L \to \text{all})} \approx 5 \times 10^{-12}\,. \tag{2.49}$$

The decay $K_L \to \pi^0 e^+ e^-$ can also proceed *via CP* conserving two-photon exchange [24], giving a contribution to the branching ratio of order

$$\left.\frac{\Gamma(K_L \to \pi^0 e^+ e^-)}{\Gamma(K_L \to \text{all})}\right|_{\gamma\gamma} \sim \alpha^2 \frac{\Gamma(K_L \to \pi^0 \gamma\gamma)}{\Gamma(K_L \to \text{all})} \lesssim 10^{-11}\,,$$

where the $\pi\gamma\gamma$ branching ratio estimate is taken from Gaillard and Lee [12]. The two contributions may be comparable, so a K_L, K_S interference measurement would be necessary to determine $\epsilon'_{\pi ee}$.

$\underline{K_L \to \pi^0 \nu\bar{\nu}}$

The decay $K_2 \to \pi^0 \nu\bar{\nu}$ is also strongly suppressed in a *CP* invariant world. By reference to sect. 2 of Gaillard and Lee, we find

$$|\epsilon'_{\pi\nu\bar{\nu}}| \approx \frac{m_t^2 \ln(m_W^2/m_t^2) - m_c^2 \ln(m_W^2/m_c^2)}{m_c^2 \ln(m_W^2/m_c^2)} \left|\frac{s_2 c_2 t_3 \sin\delta}{c_1}\right|, \tag{2.50}$$

and with $m_t \gtrsim 4$ GeV we can get

$$|\epsilon'_{\pi\nu\bar{\nu}}|/|\epsilon| \gtrsim 4\,, \tag{2.51}$$

but still

$$\frac{\Gamma(K_L \to \pi^0 \nu\bar{\nu})}{\Gamma(K_L \to \text{all})} \simeq 10^{-13}\,. \tag{2.52}$$

2.6. *CP violation in charmed particle decays*

The discussion here parallels that for K decays, but with differences because the lifetimes of charmed particles are presumably very short, and because the difference between the lifetimes of the two components of the $D^0 - \bar{D}^0$ system is relatively small. As in the case of strange decays, the KM model produces no detectable *CP* violating effects in amplitudes for semileptonic decays, and *CP* violating effects in $D^\pm$ and $F^\pm$ decays are expected at the same level as in D^0, $\bar{D}^0$ decays, which we now estimate.

Following Pais and Treiman [25], we introduce two components D_S and D_L of the $(D^0, \bar{D}^0)$ system which have definite lifetimes:

$$D_{S,L} = \frac{1}{\sqrt{2(1 + |\epsilon_c|^2)}} [(1 + \epsilon_c)D^0 \pm (1 - \epsilon_c)\bar{D}^0] , \qquad (2.53)$$

where ϵ_c is a *CP* violating mixing parameter, defined analogously to ϵ for the K system:

$$\epsilon_c \equiv \frac{\frac{1}{2} \operatorname{Im} \Gamma^c_{12} + i \operatorname{Im} M^c_{12}}{\frac{1}{2}i(\Gamma^c_S - \Gamma^c_L) - (m^c_S - m^c_L)} \qquad (2.54)$$

where all quantities in eq. (2.54) are defined by analogy with the K parameters in eq. (2.22). We define D_S to be the state whose $S = 0$ decay products mainly have $CP = +$: it is not obvious that $\Gamma^c_S > \Gamma^c_L$. As argued in subsect. 2.1 the dominant decays of charmed states in the KM model are still to strange states:

$$\langle S = -1|D\rangle \propto (c_1 c_2 c_3 - s_2 s_3 e^{i\delta}) ,$$

and decays to non-strange states are still suppressed

$$\langle S = 0|D\rangle \propto (s_1 c_2) .$$

So that we expect (since $|s_3| \ll 1$)

$$\frac{\Gamma(D \to S = 0)}{\Gamma(D \to S = -1)} \approx \frac{s_1^2}{c_1^2 c_3^2} \approx \sin^2 \theta_C .$$

Moreover in a *CP* invariant world the $\Delta C = 1, \Delta S = 0$ four-quark operator is predominantly a *U*-spin vector (see eq. (2.60) below). Hence the parameters $\Delta\Gamma^c = \Gamma^c_S - \Gamma^c_L$ and $\Delta m^c = m^c_S - m^c_L$ are the absorptive and dispersive parts, respectively, of the second order weak amplitude $\langle D^0|\mathcal{H}^2_w|D^0\rangle$. Again $\mathcal{H}^2_w(\Delta C = 2)$ is predominantly an effective *U*-spin vector, and $D^0, \bar{D}^0$ are singlets, so we expect:

$$\left|\frac{\Delta\Gamma^c}{\Gamma^c}\right| \sim \left|\frac{\Delta m^c}{\Gamma^c}\right| \lesssim \sin^2\theta_C \left(\frac{m_s^2 - m_d^2}{m_c^2}\right) \sim 10^{-2} , \qquad (2.55)$$

where $\Gamma^c = \frac{1}{2}(\Gamma^c_L + \Gamma^c_S)$. We also assume for simplicity $|\operatorname{Im} \Gamma^c_{12}| \lesssim |\operatorname{Im} M^c_{12}|$, so that an order of magnitude estimate of $|\epsilon_c|$ can be got from

$$|\epsilon_c| \approx \left|\frac{\operatorname{Im} M^c_{12}}{\Delta m^c}\right| . \qquad (2.56)$$

The calculation of eq. (2.56) then follows precisely the analysis of subsect. 2.2. *CP* violation arises from imperfect cancellation between loops containing b and s quarks, and would vanish in the limit of exact "bottom-spin" ($m_b = m_s$) *. However this limit is not physically relevant as presumably $m_b \gtrsim 1.5$ GeV ** $\gg m_s$, in which case

$$\left|\frac{\text{Im}\, M_{12}^c}{\Delta m^c}\right| \approx \frac{c_3 s_2 s_3 \sin\delta\,[\cos 2\theta_3\, m_s^2 \ln(m_b^2/m_s^2) - c_3^2 m_s^2 + s_3^2 m_b^2]}{c_1 c_2\,[c_3^4 m_s^2 + s_3^4 m_b^2]} \,. \tag{2.57}$$

Comparing eq. (2.57) with eq. (2.11) we see that unless $m_b \gg m_t$, then $|\epsilon_c| \approx |\epsilon|$ in order of magnitude, and we expect small CP violation in the $D^0 - \bar{D}^0$ mass matrix.

As observed by Pais and Treiman [25], the parameters $\Delta m^c/\Gamma^c$, $\Delta\Gamma^c/\Gamma^c$ and Re ϵ_c can in principle be measured in semileptonic decays of states prepared as pure D^0 or $\bar{D}^0$:

$$\frac{\Gamma(D^0 \to \ell^- + \nu + X)}{\Gamma(D^0 \to \ell^+ + \nu + X)}\,\frac{\Gamma(\bar{D}^0 \to \ell^+ + \nu + X)}{\Gamma(\bar{D}^0 \to \ell^- + \nu + X)} \approx \frac{1}{4}\left[\left(\frac{\Delta m^c}{\Gamma^c}\right)^2 + \frac{1}{4}\left(\frac{\Delta\Gamma^c}{\Gamma^c}\right)^2\right]^2 (1 + \ldots)\,, \tag{2.58}$$

which measures the (squared) rate of "wrong sign" semileptonic decays. Also:

$$\frac{\Gamma(D^0 \to \ell^- + \nu + X)}{\Gamma(D^0 \to \ell^+ + \nu + X)} \Bigg/ \frac{\Gamma(\bar{D}^0 \to \ell^+ + \nu + X)}{\Gamma(\bar{D}^0 \to \ell^- + \nu + X)} \approx 1 - 8\,\text{Re}\,\epsilon_c + \ldots\,, \tag{2.59}$$

which may differ from 1 by $O(10^{-2})$ if the estimate (2.56) is accepted. Using eqs. (2.55) and (2.58), the rate of "wrong sign" semileptonic decays may be relatively suppressed by $O(10^{-4})$, so one is looking for small effects in very weak processes!

The most interesting place to look for CP violating effects is presumably in $\Delta S = 0$ hadronic decays. In addition to the superweak mixing effect (2.57) there is direct CP violation in decay amplitudes. From the KM model (2.5) we find (omitting γ matrices and a factor of $G_F/\sqrt{2}$).

$$\begin{aligned}\mathcal{H}_{\substack{\Delta C=-1\\ \Delta S=0}} = \tfrac{1}{2}(\bar{c}u)\,\{(\bar{d}d - \bar{s}s)\,[c_1 s_1 c_2 + s_1 c_3^2 c_1 c_2 - s_1 c_3 s_2 s_3 \cos\delta - i s_1 c_3 s_2 s_3 \sin\delta]\\ + (\bar{d}d + \bar{s}s)\,[c_1 s_1 c_2 s_3^2 + s_1 c_3 s_2 s_3 \cos\delta + i s_1 s_2 s_3 c_3 \sin\delta]\}\,.\end{aligned} \tag{2.60}$$

Introducing $\mathcal{H}_1 \propto (\bar{c}u)(\bar{d}d - \bar{s}s)$, $\mathcal{H}_2 \propto (\bar{c}u)(\bar{d}d + \bar{s}s)$ and extracting the CP violating phases $e^{i\alpha_1}$, $e^{i\alpha_2}$ explicitly, we get

$$\begin{aligned}\mathcal{H}_{\substack{\Delta C=-1\\ \Delta S=0}} &= e^{i\alpha_1}\,\mathcal{H}_1 + e^{i\alpha_2}\,\mathcal{H}_2\,,\\ \mathcal{H}_{\substack{\Delta C=+1\\ \Delta S=0}} &= e^{-i\alpha_1}\,\mathcal{H}_1^\dagger + e^{-i\alpha_2}\,\mathcal{H}_2^\dagger\,,\end{aligned} \tag{2.61}$$

* Eventually it may be necessary to introduce a spin exchanging top and bottom quarks. This we term "side-spin": American readers will recognize this as "english".

** Note that the lifetime considerations of subsect. 2.1 and the SPEAR limits on narrow resonances may allow the $\bar{b}b$ vector meson mass spectrum to start as low as 6–7 GeV, in which case m_b could be as low as 3–4 GeV.

where

$$\tan\alpha_1 = -\frac{c_3 s_2 s_3 \sin\delta}{c_1 c_2 + c_3^2 c_1 c_2 - c_3 s_2 s_3 \cos\delta}\,,$$
$$\tan\alpha_2 = \frac{s_2 s_3 c_3 \sin\delta}{c_1 c_2 s_3^2 + c_3 s_2 s_3 \cos\delta}\,. \tag{2.62}$$

It is apparent from eq. (2.62) that $|\tan\alpha_1|$ is expected to be small, whereas $|\tan\alpha_2|$ is not necessarily small.

We parametrize the matrix elements of $\mathcal{H}$ as follows:

$$\langle \mathrm{f}|\mathcal{H}_i|\mathrm{D}\rangle = \langle \bar{\mathrm{f}}|\mathcal{H}_i^\dagger|\bar{\mathrm{D}}\rangle = T_i\, \mathrm{e}^{i(\beta_i + \delta_\mathrm{s})}, \qquad i = 1, 2\,, \tag{2.63}$$

where δ_s is the elastic phase shift, β_i a residual phase related to inelasticity, and T_i is real. Combining eqs. (2.60), (2.61) and (2.63) we then get, dropping the common phase δ_s

$$\langle \mathrm{f}|\mathcal{H}_{\substack{\Delta C=-1\\ \Delta S=0}}|\mathrm{D}^0\rangle = T_1\, \mathrm{e}^{i(\alpha_1+\beta_1)} + T_2\, \mathrm{e}^{i(\alpha_2+\beta_2)}\,,$$
$$\langle \bar{\mathrm{f}}|\mathcal{H}_{\substack{\Delta C=+1\\ \Delta S=0}}|\bar{\mathrm{D}}^0\rangle = T_1\, \mathrm{e}^{i(\beta_1-\alpha_1)} + T_2\, \mathrm{e}^{i(\beta_2-\alpha_2)}\,. \tag{2.64}$$

In order to use the *CP* violating amplitudes (2.64), we first go back to the decomposition (2.53) of D_S and D_L, from which we find the following expressions for D^0 and $\bar{D}^0$ states:

$$D^0 = \frac{(D_\mathrm{S} + D_\mathrm{L})}{\sqrt{2}} \frac{\sqrt{1+|\epsilon_\mathrm{c}|^2}}{(1+\epsilon_\mathrm{c})}\,, \qquad \bar{D}^0 = \frac{(D_\mathrm{S} - D_\mathrm{L})}{\sqrt{2}} \frac{\sqrt{1+|\epsilon_\mathrm{c}|^2}}{(1-\epsilon_\mathrm{c})}\,.$$

States prepared initially as pure D^0 or $\bar{D}^0$ will then have their D_S, D_L decomposition change in the normal way with time:

$$D^0_\mathrm{op}(t) \equiv \eta[D_\mathrm{S}\, \mathrm{e}^{-\frac{1}{2}\Gamma^\mathrm{c}_\mathrm{S} t} + D_\mathrm{L}\, \mathrm{e}^{-\frac{1}{2}\Gamma^\mathrm{c}_\mathrm{L} t}\, e^{-i\Delta m^\mathrm{c} t}]\,,$$
$$\bar{D}^0_\mathrm{op}(t) \equiv \tilde{\eta}[D_\mathrm{S}\, \mathrm{e}^{-\frac{1}{2}\Gamma^\mathrm{c}_\mathrm{S} t} - D_\mathrm{L}\, \mathrm{e}^{-\frac{1}{2}\Gamma^\mathrm{c}_\mathrm{L} t}\, e^{-i\Delta m^\mathrm{c} t}]\,, \tag{2.65}$$

where

$$\eta \equiv \frac{\sqrt{1+|\epsilon_\mathrm{c}|^2}}{\sqrt{2}(1+\epsilon_\mathrm{c})}\,, \qquad \tilde{\eta} \equiv \frac{\sqrt{1+|\epsilon_\mathrm{c}|^2}}{\sqrt{2}(1-\epsilon_\mathrm{c})}\,.$$

Substituting eqs. (2.64) into (2.65) and neglecting various terms second order in the small quantities ϵ_c, $\Delta m^\mathrm{c}/\Gamma^\mathrm{c}$, $\Delta\Gamma^\mathrm{c}/\Gamma^\mathrm{c}$, T_2 and $\sin\alpha_1$ we find that for states f with $CP = \pm 1$:

$$|\langle \mathrm{f}|\mathrm{D}^0_\mathrm{op}(t)\rangle|^2 = \frac{4T_1^2}{\Gamma_\mathrm{S}} \left\{ |\eta|^2 + 2\,\mathrm{Re}\left[\frac{\epsilon_\mathrm{c} + i(T_2/T_1)\, \mathrm{e}^{i\beta} \sin\alpha_2 + i \sin\alpha_1}{\Gamma^\mathrm{c} \pm i\Delta m^\mathrm{c}}\right]\right\} + \ldots\,,$$

where $\beta \equiv \beta_2 - \beta_1$. From this we get

$$\frac{\Gamma(D^0 \to f)}{\Gamma(\bar{D}^0 \to f)} \simeq \frac{|\eta|^2 + 2(\mathrm{Re}\,\epsilon_c - \frac{T_2}{T_1} \sin\alpha_2 \sin\beta)}{|\tilde{\eta}|^2 - 2(\mathrm{Re}\,\epsilon_c - \frac{T_2}{T_1} \sin\alpha_2 \sin\beta)}$$

$$\approx 1 - 4\frac{T_2}{T_1} \sin\alpha_2 \sin\beta + \ldots \tag{2.66}$$

both for states f with $CP = \pm 1$.

We observe that the result (2.66) is non-zero only if the CP violating phase $\sin\alpha_2$ (*cf.* eq. (2.62)) is non-zero. This agrees with the result of Pais and Treiman [25] in the limit ϵ_c, $\Delta m^c/\Gamma^c$, $\Delta\Gamma^c/\Gamma^c$ small. The phase difference β is not in general small * therefore using eqs. (2.60) and (2.67) we can write

$$\frac{\Gamma(D^0 \to f)}{\Gamma(\bar{D}^0 \to f)} \approx 1 + X\, s_2 s_3 \sin\delta\,, \tag{2.67}$$

where we have set $|c_i| \simeq 1$ and X is a factor O(1) unless there is some special reason favouring or disfavouring matrix elements $\langle f|\mathcal{H}_2|D\rangle$. Since $\mathcal{H}_2$ is a U-spin scalar, whereas $\mathcal{H}_1$ is a U-spin vector, we expect that for transitions to states with definite U-spin

$$\frac{\Gamma(D^0 \to (U=1))}{\Gamma(\bar{D}^0 \to (U=1))} \approx 1 + U\, s_2 s_3 \sin\delta\,,$$

$$\frac{\Gamma(D^0 \to (U=0))}{\Gamma(\bar{D}^0 \to (U=0))} \approx 1 + \frac{1}{U}\, s_2 s_3 \sin\delta\,,$$

where

$$U \approx \frac{m_s^2 - m_d^2}{m_c^2} \quad (\approx \tfrac{1}{10}\,?)\,,$$

is a measure of U-spin symmetry violation. If $|s_2 s_3 \sin\delta| \approx 10^{-3}$ as suggested by the calculation of $|\epsilon|$ in subsect. 2.2, then

$$\frac{\Gamma(D^0 \to (U=1))}{\Gamma(\bar{D}^0 \to (U=1))} - 1 = O(10^{-4})\,,$$

$$\frac{\Gamma(D^0 \to (U=0))}{\Gamma(\bar{D}^0 \to (U=0))} - 1 = O(10^{-2})\,. \tag{2.68}$$

* β need not be small provided that there are several open channels which can communicate strongly with the state f.

The estimates (2.68) seem the least undistinctive consequences of the KM model for *CP* violating effects in charmed particle decays. We re-emphasize that effects of the size (2.68) would not be expected in a "superweak" theory of *CP* violation in the $D^0-\bar{D}^0$ system. The simplest example of a possible $U = 0$ final state is $K^0\bar{K}^0$, and $K^0\bar{K}^0$ scattering at an invariant energy $s \approx (2\text{ GeV})^2$ is presumably sufficiently inelastic so that there will be no *a priori* suppression of the *CP* violating effect ($\sin\beta \sim 1$).

2.7. The electric dipole moment of the neutron

In this section we estimate the dipole moment of the physical neutron in terms of the quark dipole moment. In general [26], the effects of weak interactions are expected to induce electromagnetic form factors which violate parity and/or time reversal invariance, *via* diagrams of the types illustrated in fig. 8. In particular, an effective $\bar{q}q\gamma$ coupling of the form

$$E^D_\mu(\bar{q}q\gamma) = ef_D(k^2)\, i\sigma_{\mu\nu}k^\nu\gamma_5\,, \tag{2.69}$$

violates both P and T, since hermiticity of the electromagnetic current requires that $f_D(k^2)$ be purely imaginary. The strength of the electric dipole moment is given by

$$|D| = e|f_D(0)|\,. \tag{2.70}$$

On general symmetry grounds, the quark dipole moment must vanish in the limits

(a) of vanishing external quark mass, because eq. (2.69) is a helicity flip operator, and

(b) when any two quarks of the same charge are degenerate in mass [9].

To estimate the strength of the dipole moment of the quark q_α, we first separate the contributions of the effective $\bar{q}_\alpha q_\alpha\gamma$ vertex into diagrams which are one-particle irreducible and reducible, respectively, as illustrated in fig. 9. The parity violating part of the self-energy insertion is of the form

$$\Sigma^\alpha(p) = [\not{p}A^\alpha(p^2) + m_\alpha B^\alpha(p^2)]\gamma_5\,, \tag{2.71}$$

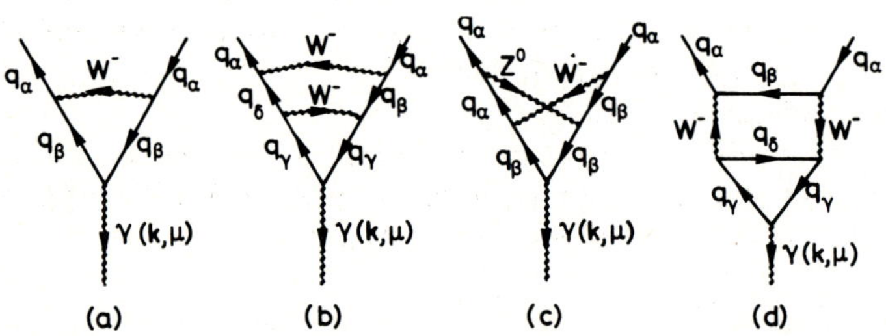

Fig. 8. Contributions to quark electromagnetic form factors.

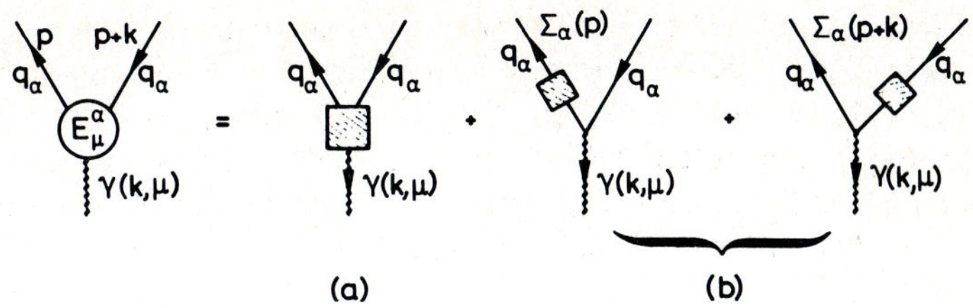

Fig. 9. The effective $\bar{q}_\alpha q_\alpha \gamma$ vertex separated into (a) irreducible and (b) one-particle reducible parts.

and evaluating fig. 9b with external lines on-shell gives

$$E^\alpha_\mu(9b) = eQ_\alpha A^\alpha \gamma_\mu \gamma_5$$

and there is no contribution to the dipole moment.

Next we consider the one particle irreducible vertex of fig. 9a. As noted by Maiani [9], no quark dipole moment arises in lowest order in the weak interaction since the corresponding diagrams (e.g. fig. 8a) are real:

$$E_\mu(8a) \propto U_{\alpha\beta} U^\dagger_{\beta\alpha} F(m_\beta) = |U_{\alpha\beta}|^2 F(m_\beta) , \tag{2.72}$$

where U is the unitary weak coupling matrix defined in eqs. (2.1) and (2.5). More generally, in order to generate a complex amplitude, at least two charged current transitions not related by complex conjugation must occur. Thus the diagram of fig 8b can contribute

$$E_\mu(8b) \propto U_{\alpha\beta} U^\dagger_{\beta\gamma} U_{\gamma\delta} U^\dagger_{\delta\alpha} F(m_\beta, m_\gamma, m_\delta) , \tag{2.73}$$

while the graph of fig. 8c cannot: since Z^0 couplings are real and diagonal, $E_\mu(8c)$ is of the form (2.72). Finally, diagrams involving a closed fermion loop, such as that of fig. 8d also induce real amplitudes:

$$E_\mu(8d) \propto |U_{\alpha\beta}|^2 \, |U_{\gamma\delta}|^2 F(m_\beta, m_\gamma, m_\delta) . \tag{2.74}$$

In summary, the dipole moment arises in lowest order from diagrams in which two charged bosons are exchanged along a single quark line. Since the quark charges are such that the two charged boson lines cannot cross, the lowest order contributions to fig. 9a are those of fig. 10, where $E^{\gamma\beta}_\mu$ is the effective $\bar{q}_\gamma q_\beta \gamma$ vertex and $\Sigma^{\gamma\beta}$ is the off-diagonal self-energy (figs. 11–13).

In the 't Hooft-Feynman gauge [20] which we adopt hereafter, the diagram of 10a in fact gives no contribution to the dipole moment. The W-propagator is proportional to $g_{\mu\nu}$, so the contribution of fig. 10a takes the form

$$E^\alpha_\mu(10a) \propto \gamma_\nu L O_\mu \gamma^\nu L = \gamma_\nu L O_\mu R \gamma^\nu , \tag{2.75}$$

where $L, R = \frac{1}{2}(1 \mp \gamma_5)$ are the usual helicity projection operators. The electric

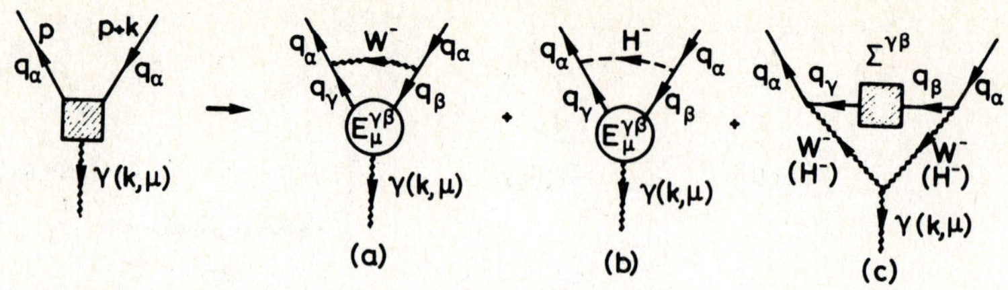

Fig. 10. Diagrams contributing to the electric dipole moment of q_α: $E_\mu^{\gamma\beta}$ is the effective $E_\mu^{\gamma\beta}$ vertex.

dipole moment is determined from E_μ by

$$ef_D(k^2)k^\rho = \tfrac{1}{8}\,\mathrm{Tr}(\sigma^{\mu\rho}\gamma_5 E_\mu^\alpha)\,,$$

and vanishes identically for a term of the form (2.75)

$$ef_D^{(10a)}(k^2)k^\rho \sim -\mathrm{Tr}\,[\gamma_\nu\sigma^{\mu\rho}\gamma^\nu\gamma_5 LO_\mu R]\,, \qquad \gamma_\nu\sigma_{\mu\alpha}\gamma^\nu = 0\,.$$

The diagrams of fig. 10c also give no contribution. The off-diagonal self-energy is of the form

$$\Sigma^{\gamma\beta}(p) = A(p^2)\not{p}L + B(p^2)m_\gamma m_\beta \not{p}R + C(p^2)(m_\gamma L + m_\beta R)\,, \tag{2.76}$$

where to lowest order in the weak coupling, A, B and C are independent of the external masses $m_{\gamma,\beta}$. Inspection of the couplings in fig. 10c shows that a term proportional to $\sigma_{\mu\alpha}\gamma_5$ can arise only from the exchange of one W and one H (unphysical Higgs), but this contribution is symmetric in $m_{\beta,\gamma}$:

$$E_\mu^\alpha(10\mathrm{c}) \sim \int \frac{\mathrm{d}^4 r}{(2\pi)^4} f(r^2, (r-p)^2, (r-p-k)^2)\, m_\alpha [\not{r}, \gamma_\mu]\gamma_5$$

$$\times [r^2 A(r^2) + m_\beta^2 m_\gamma^2 B(r^2) + (m_\beta^2 + m_\gamma^2)\, C(r^2)]\,. \tag{2.77}$$

The imaginary part needed for CP violation is antisymmetric under the exchange $\beta \leftrightarrow \gamma$, and hence vanishes in the sum over intermediate states (β,γ).

Finally we are left with the diagram of fig. 10b, which we study by first examining the effective $\bar{q}_\gamma q_\beta \gamma$ vertex of fig. 11. If $\Gamma_\mu^{\gamma\beta}$ is the one-particle irreducible part

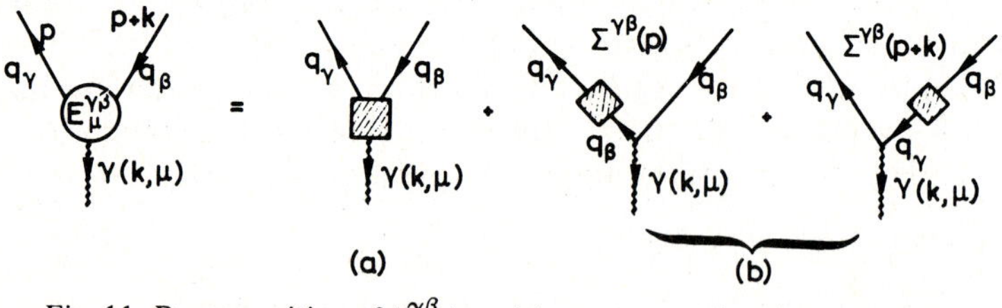

Fig. 11. Decomposition of $E_\mu^{\gamma\beta}$ into (a) vertex and (b) self-energy parts.

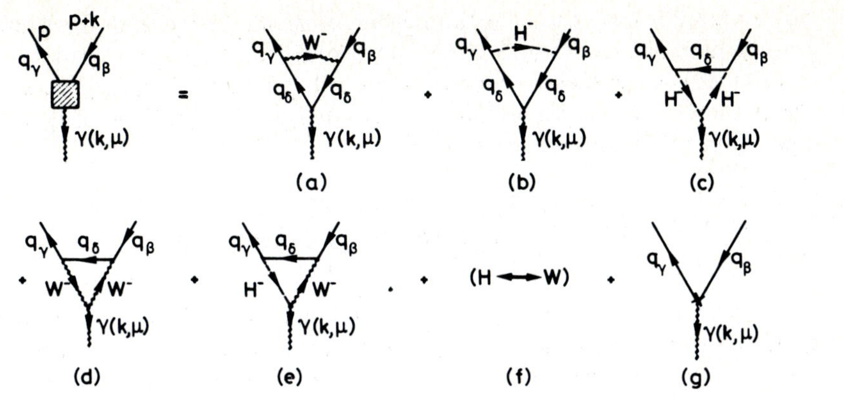

Fig. 12. Contributions to the vertex part of $E_\mu^{\gamma\beta}$.

in fig. 11a, we may use the Ward-Takahashi identity to write

$$k^\mu \Gamma_\mu^{\gamma\beta}(p,k) = eQ_\gamma [\Sigma^{\gamma\beta}(p+k) - \Sigma^{\gamma\beta}(p)] ,$$

$$\Gamma_\mu^{\gamma\beta}(p,k) = eQ_\gamma \frac{\partial}{\partial p^\mu} \Sigma^{\gamma\beta}(p) + O(k) , \tag{2.78}$$

where Q_γ is the charge of q_γ. So the effective $\bar{q}_\gamma q_\beta \gamma$ vertex may be separated into a part determined by the self-energy, eq. (2.76), and the k-dependent parts of the diagrams of fig. 12. Fig. 12g can be ignored, and the other diagrams are all convergent. Furthermore, GIM [11] tells us that they vanish in the limit of degeneracy of exchanged quarks q_δ. As discussed in ref. [12], the GIM cancellation is equivalent to a subtraction of the q_δ propagator. Then the relevant O(k) contributions are still convergent after removal of one W propagator:

$$\Delta_{\mu\nu}^{W} \to ig_{\mu\nu}/m_W^2 .$$

The diagrams of figs. 12c to 12 f contain an additional boson propagator in the loop integration, and fig. 12b has an implicit factor m_W^{-2} for the Higgs coupling. Therefore the diagrams of figs. 12b to 12f must be O(m_W^{-4}). Dropping γ-matrices we have

$$E_\mu^{\gamma\beta}(12\text{b to } 12\text{f}) \sim \frac{eg^2}{16\pi^2} m_\delta^2 \frac{kp}{m_W^2 - p^2} \left[\frac{1}{m_W^2} \text{ or } \frac{1}{m_W^2 - p^2}\right] , \tag{2.79}$$

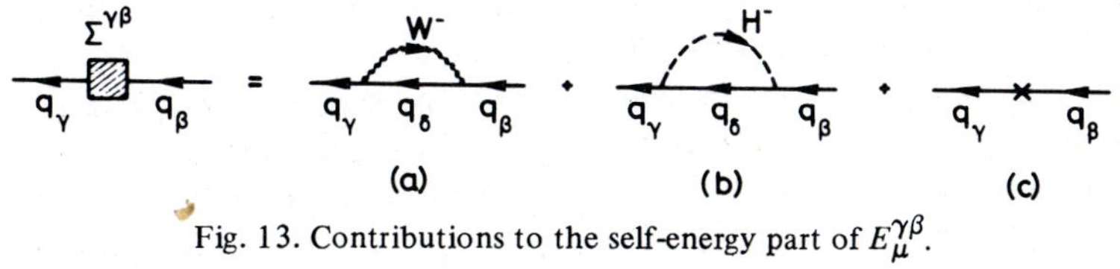

Fig. 13. Contributions to the self-energy part of $E_\mu^{\gamma\beta}$.

up to a possible logarithm. For the diagram of fig. 12a, removal of the W propagator leaves a fermion loop integral with no additional factor of m_W^{-2}. However, the term linear in k then vanishes by symmetric integration over the loop momentum. To obtain a non-vanishing contribution we must integrate over the W propagator and we again obtain an additional suppression:

$$E_\mu^{\gamma\beta}(12a) \sim \frac{eg^2}{16\pi^2} \frac{m_\delta^2}{(m_W^2 - p^2)} kp \ln\left(\frac{m_\delta^2}{m_W^2 - p^2}\right) . \tag{2.80}$$

The result is completely analogous to that found in ref. [12] for the (quark mass-shell) $\bar{s}u\gamma$ vertex: while the charge radius is $O(g^2 \ln m_q^2)$, the induced tensor and pseudotensor terms were found to be suppressed by an extra factor of m_q^2/m_W^2.

Inserting eqs. (2.80) and (2.79) into the diagram of fig. 10b, we obtain for the leading contribution to the dipole moment

$$E_\mu^\alpha(12) \sim \frac{eg^4}{16\pi^2} m_\alpha \frac{m_{\beta,\gamma}^2}{m_W^2} \int \frac{d^4p}{(2\pi)^4} \left[\frac{1}{m_W^2} \text{ or } \frac{1}{m_W^2 - p^2}\right]$$

$$\times \frac{[\not{k}, \gamma_\mu] \ln[(p^2 - m_W^2)/m_\delta^2] p^2}{(p^2 - m_W^2)^2 (p^2 - m_\beta^2)(p^2 - m_\gamma^2)} .$$

The convergence is such that the propagator in brackets may again be replaced by m_W^{-2}. The remaining integral gives another factor of m_W^{-2} with a possible extra factor of $\ln^2 (m_W^2/m_q^2)$. Thus when we put in the appropriate angle factors (assuming $|c_i^2| \simeq 1$) we get a contribution to the dipole moment which is at most

$$\left|\frac{D}{e}\right|_n = \left|\frac{D}{e}\right|_{u,d} \sim m_{u,d} \frac{G_F \alpha}{\pi^3} \frac{(m_t^2 - m_c^2)(m_b^2 - m_s^2)}{m_W^4}$$

$$\times \ln^2\left(\frac{m_W^2}{m_t^2, m_b^2}\right) s_1^2 s_2 s_3 \sin\delta$$

$$\sim (10^{-26} \text{ to } 10^{-27}) \frac{m_u}{m_n} \frac{m_t^2 m_b^2}{m_W^4} \ln^2\left(\frac{m_W^2}{m_t^2, m_b^2}\right) \text{cm} . \tag{2.81}$$

The analysis of the part of $E_\mu^{\gamma\beta}$ which depends on the self-energy intersections $\Sigma^{\gamma\beta}$ shown in fig. 13 does not alter the estimate (2.81).

In formula (2.81) $m_{u,d}$ is a measure of the chiral symmetry breaking fraction of the neutron mass, which is probably $O(m_n/10)$. In the above discussion we have implicitly assumed that all quark masses are small compared with W-masses. If the top and bottom were actually of order m_W (and why not?) the mass-dependent factor in eq. (2.81) would be replaced by a complicated function which we suspect to be O(1). In that case formula (2.81) suggests

$$\left|\frac{D}{e}\right|_{n} \sim (10^{-27} - 10^{-28})\ \text{cm} \quad \text{for} \quad m_{t,b} \sim m_{W}\,. \tag{2.82a}$$

On the other hand, if we take $m_t \sim m_b$ near the present experiment lower limit, we get

$$\left|\frac{D}{e}\right|_{n} \sim (10^{-30} - 10^{-31})\ \text{cm} \quad \text{for} \quad m_{t,b} \sim 5\ \text{GeV}\,. \tag{2.82b}$$

Since the superweak [10] estimate is [14] O(10^{-29}) cm, the values (2.82) mean the KM model prediction is not very distinctive. Within the context of the model, knowledge of m_t and m_b would enable a better prediction of the neutron electric dipole moment, and *vice versa*.

3. Conclusions and discussion

We like the KM model because it introduces *CP* violation in a natural way, as a result of weak mixing between the quarks analogous to the Cabibbo angle in the GIM model. Thus two puzzles are reduced to one. Looked at another way, it gives a *raison d'être* to the fifth and sixth quarks, needed in the simplest Weinberg-Salam theory to cancel anomalies and restore renormalizability if there is a heavy lepton. Unfortunately it is a very well concealed theory, at least as long as the top and bottom degrees of freedom are hidden, and naked top and bottom states are not produced in e^+e^- annihilation or elsewhere. We found:

(a) That superweak results for common K decays are reproduced to high accuracy (*cf.* eqs. (2.38) and (2.39)) whatever the masses of the top and bottom quarks. This is because the "superweak violating" parameter ϵ' is suppressed relative to ϵ both by the Zweig rule (a factor $\lesssim \frac{1}{10}$) and also by the fact that the *CP* violation satisfies a strict $\Delta I = \frac{1}{2}$ rule. The latter suppresses ϵ' by an additional factor $\sim \frac{1}{20}$, characteristic of deviations from the $\Delta I = \frac{1}{2}$ rule in *CP* conserving amplitudes. For non-leptonic decays outside the neutral kaon system, only the first factor is relevant, so one expects *CP* violating effects to be at most of order $\frac{1}{10}\epsilon \sim 10^{-4}$ in amplitude.

(b) That in rare K decays, direct *CP* violating effects in the amplitudes may be comparable with superweak *CP* violating effects, as they (like the parameter ϵ) are determined by higher order graphs involving heavy quark exchange, and the Zweig rule plays no rôle. Unfortunately the branching ratios for these decays are so small (*cf.* eqs. (2.49) and (2.52)) that these effects seem essentially unobservable.

(c) There are in principle observable deviations from superweak predictions for charmed particle decays (*cf.* eq. (2.68)). Since no discovery of a charmed particle has yet been confirmed, checking results like eq. (2.68) will clearly not be done in the near future.

(d) The neutron dipole moment is suppressed by the appearance of explicit quark mass factors which reflect the symmetries under which it vanishes: chiral symmetry ($m_{u,d} \to 0$) imposing a factor m_u/m_n; top spin ($m_t - m_c \to 0$) and bottom spin ($m_b - m_s \to 0$), each imposing a factor $\Delta m_q^2/m_W^2$. Its value is very sensitive to the masses of the top and bottom quarks:

$$\left|\frac{D}{e}\right|_n \approx \begin{Bmatrix} (10^{-27} - 10^{-28})\ \text{cm} \\ (10^{-30} - 10^{-31})\ \text{cm} \end{Bmatrix} \quad \text{for} \quad m_{t,b} \sim \begin{Bmatrix} m_W \\ 5\ \text{GeV} \end{Bmatrix} .$$

Hence a measurement of this quantity would help us tell experimentalists where to look for top and bottom particles. Conversely, if such states were discovered, the above expression for the dipole moment would serve as a test of the validity of the KM model.

In principle the model also predicts small deviations from the GIM model and Cabibbo universality for couplings of the u, d, s and c quarks. This is because there is some "leakage" of weak charge to the t and b quarks, characterized by the angle factors s_2 and s_3 in eq. (2.5). A search for such deviations would be valuable: unfortunately deviations from Cabibbo predictions for the u quark are experimentally no greater than the uncertainties due to radiative corrections and nuclear uncertainties. Checks on the dominance of strange particles in the decays of charmed particles would be important. Unfortunately the decay amplitudes are subject to large SU(3) or SU(4) symmetry violations. A good place to look might be the vector form factors in $D \to K + \ell + \nu$, $D \to \pi + \ell + \nu$ decays. At $m(\ell\nu) \approx 0$ these deviate from symmetry predictions only in second order in symmetry breaking, and a check of the GIM prediction

$$\frac{\langle \pi | J^{wk} | D \rangle}{\langle K | J^{wk} | D \rangle} \approx \tan \theta_C \quad \text{at} \quad q^2 \simeq 0 ,$$

would be least unreliable theoretically.

Probably this model cannot be confirmed or refuted by looking at topless and bottomless states. Unfortunately we have found no constraints on the masses of the top and bottom quarks. Our ignorance does not exclude their being anywhere between lower limits of (3–4) GeV and $\sim m_W$ or m_Z, or beyond. Hence it may prove difficult to expose top and bottom particles. Needless to say, there is no reason known why 8 or more quarks should not exist.

We would like to thank E. de Rafael, V. Glaser, L. Maiani, and J. Prentki for enjoyable and useful discussions about this work.

Appendix

In this appendix we discuss Higgs structures which could give *CP* violation as in the KM model. One method would clearly be to follow Lee [27] and introduce two or more Higgs multiplets, whose vacuum expectation values had "clashing" phases, introducing the requisite complex phases into the weak interaction matrix U of eq. (2.1). However, we are motivated to restrict to a minimum the spectrum of unobserved (and almost unobservable) Higgs bosons, and introduce *CP* violation with just the original [6] doublet Higgs fields.

The neutral Higgs couplings to quarks can be written in the general form

$$\left(\frac{H+v}{v}\right)[\bar{P}_C^R G_P P_C^L + \bar{P}_C^L G_P^\dagger P_C^R] + \left(\frac{H+v}{v}\right)[\bar{N}_C^R G_N N_C^L + \bar{N}_C^L G_N^\dagger N_C^R] , \quad \text{(A.1)}$$

where $P^{L(R)}$ and $N^{L(R)}$ are the multiplets of left-handed (right-handed) quarks of charges $\frac{2}{3}$ and $-\frac{1}{3}$ respectively, v is the vacuum expectation value of the Higgs field, G_P and G_N are coupling matrices which are not necessarily real, and the subscripts C indicate that the quark fields are Cabibbo rotated to the basis where charged weak currents are diagonal. The Higgs couplings (A.1) can be diagonalized by introducing transformations to quark states with well-defined masses:

$$P_C^L = U_P P^L , \qquad N_C^L = U_N N^L ,$$
$$P_C^R = V_P P^R , \qquad N_C^R = V_N N^R . \quad \text{(A.2)}$$

The interactions (A.1) then become

$$\left(\frac{H+v}{v}\right)[\bar{P}^R V_P^\dagger G_P U_P P^L + \bar{P}^L U_P^\dagger G_P^\dagger V_P P^R]$$
$$+ \left(\frac{H+v}{v}\right)[\bar{N}^R V_N^\dagger G_N U_N N^L + \bar{N}^L U_N^\dagger G_N^\dagger V_N N^R] , \quad \text{(A.3)}$$

which is supposed to be diagonal; we can choose the phases of $U_{P,N}$ and $V_{P,N}$ such that

$$V_P^\dagger G_P U_P = M_P , \qquad V_N^\dagger G_N U_N = M_N , \quad \text{(A.4)}$$

where $M_{P,N}$ are the diagonal mass matrices for P,N type quarks. Going to the W^+ coupling which starts off life proportional to

$$W_\mu^+ \bar{P}_C^L \gamma^\mu N_C^L ,$$

this transforms under eq. (A.2) to become

$$W_\mu^+ \bar{P}^L \gamma^\mu U_P^\dagger U_N N_C^L , \quad \text{(A.5)}$$

so that the matrix U of eq. (2.1) is given by

$$U = U_P^\dagger U_N . \quad \text{(A.6)}$$

Consider finally the (unphysical) charged Higgs couplings. From eq. (A.3) we see that

H^+ has couplings proportional to

$$H^+(\bar{P}^R_C G_P N^L_C) - H^+(\bar{P}^L_C G^\dagger_N N^R_C) , \tag{A.7}$$

which become

$$H^+[\bar{P}^R(V^\dagger_P G_P U_N) N^L - \bar{P}^L(U^\dagger_P G^\dagger_N V_N) N^R] , \tag{A.8}$$

under the rotations (A.2). Using (A.4) and (A.6) the expression (A.8) can be rewritten as

$$H^+[\bar{P}^R M_P U N^L - \bar{P}^L U M_N N^R] . \tag{A.9}$$

It is easy to see from the above analysis that we can introduce *CP* violation as in the KM model with just the usual Higgs doublet, and without any extra *CP* violations due to Higgs couplings. For example, if we set

$$U_P = V_P = V_N = I, \qquad G_P = M_P, \qquad G_N = M_N U , \tag{A.10}$$

with U an arbitrary unitary matrix, then the conditions (A.4), (A.6) relating the Higgs coupling matrices $G_{P,N}$ and the weak interaction matrix U are clearly satisfied, and the couplings (A.9) of the (unphysical) Higgs bosons have the expected relation to the W couplings.

We should re-emphasize that *a fortiori* the *CP* violation scheme of Lee [27], using more than one Higgs multiplet, could be employed for the KM model. In that case there would be extra *CP* violating effects in the Higgs sector of the model. These we have ignored, for example in the discussion of the neutron electric dipole moment in subsect. 2.7.

References

[1] R.M. Barnett, Phys. Rev. Letters 34 (1975) 41 and Fermilab-Conf. 75-71 THY, Invited talk presented at 6th Hawaii Topical Conf. on particle physics, Honolulu 1975, and references therein.
[2] H. Harari, Physics Letters 57B (1975) 265; Ann. of Phys. 94 (1975) 391;
Proc. Int. Symposium on lepton and photon interactions at high energies, Stanford 1975, ed. W.T. Kirk, p. 317 (SLAC, Stanford, 1975) and references therein;
S. Pakvasa, W.A. Simmons and S.F. Tuan, Phys. Rev. Letters 35 (1975) 702
[3] G.J. Feldman, Proc. Stanford Symposium, p. 39 and references therein;
J. Heintze, ibid. p. 97;
B.H. Wiik, ibid. p. 69.
[4] M.L. Perl et al., Phys. Rev. Letters 35 (1975) 1489.
[5] C. Bouchiat, J. Iliopoulos and Ph. Meyer, Phys. Letters 38B (1972) 519.
[6] S. Weinberg, Phys. Rev. Letters 19 (1967) 1264;
A. Salam, Proc. of the 8th Nobel Symposium, Stockholm 1968, ed. N. Svartholm, (Almqvist and Wiksells, Stockholm, 1968) p. 367.
[7] M. Kobayashi and K. Maskawa, Progr. Theor. Phys. 49 (1973) 652.
[8] S. Pakvasa and H. Sugawara, University of Hawaii preprint UH-511-204-75 (1975).
[9] L. Maiani, Istituto Superiore di Sanita, Roma, preprint ISS P 75/10 (1975).

[10] L. Wolfenstein, Phys. Rev. Letters 13 (1964) 562.
[11] S.L. Glashow, J. Iliopoulos and L. Maiani, Phys. Rev. D2 (1970) 1285;
M.K. Gaillard, B.W. Lee and J.L. Rosner, Rev. Mod. Phys. 47 (1975) 277.
[12] M.K. Gaillard and B.W. Lee, Phys. Rev. D10 (1974) 897.
[13] M.K. Gaillard, B.W. Lee and R.E. Schrock, CERN preprint TH-2066 (1975) and references therein.
[14] K. Kleinknecht, Proc. 17th Int. Conf. on high-energy physics, London 1974, ed. J.R. Smith, p. III-23 (S.R.C., Rutherford Laboratory, 1974).
[15] C. Jarlskog et al., Nucl. Phys. B109 (1976) 1.
[16] A. de Rujula, H. Georgi and S.L. Glashow, Phys. Rev. Letters 35 (1975) 69;
F.A. Wilczek, A. Zee, R.L. Kingsley and S.B. Treiman, Phys. Rev. D12 (1975) 2768;
H. Fritzsch, M. Gell-Mann and P. Minkowski, Phys. Letters 59B (1975) 293;
G. Branco, T. Hagiwara and R.N. Mohapatra, Phys. Rev. D13 (1976) 104.
[17] F.A. Wilczek, A. Zee, R.L. Kingsley and S.B. Treiman, ref. [16];
E. Golowich and B.R. Holstein, Phys. Rev. Letters 35 (1975) 831;
G. Branco, R.N. Mohapatra and T. Hagiwara, Rockefeller preprint COO-2114-24 (1975);
M.A. Ahmed and G.G. Ross, Phys. Letters 59B (1975) 256.
[18] R.E. Marshak, Riazuddin and C.P. Ryan, Theory of weak interactions in particle physics, (Wiley, New York, 1969) p. 271.
[19] K. Fujikawa and N. Kawamoto, Institute for Nuclear Study, Tokyo preprint INS-report-248.
[20] G. 't Hooft, Nucl. Phys. B35 (1971) 167.
[21] K.G. Wilson, Phys. Rev. 179 (1969) 1499.
[22] J. Ellis, M.K. Gaillard and D.V. Nanopoulos, Nucl. Phys. B100 (1975) 313.
[23] M.K. Gaillard and B.W. Lee, Phys. Rev. Letters 33 (1974) 108;
G. Altarelli and L. Maiani, Phys. Letters 52B (1974) 351.
[24] A. Pais and S.B. Treiman, Phys. Rev. 176 (1968) 1974.
[25] A. Pais and S.B. Treiman, Phys. Rev. D12 (1975) 2744.
[26] A. Pais and J.R. Primack, Phys. Rev. D8 (1973) 3063.
[27] T.D. Lee, Phys. Reports 9 (1973) 143.

THE $\Delta I = 1/2$ RULE AND VIOLATION OF *CP* IN THE SIX-QUARK MODEL ☆

Frederick J. GILMAN and Mark B. WISE [1]
Stanford Linear Accelerator Center, Stanford University, Stanford, CA 94305, USA

Received 2 January 1979
Revised manuscript received 3 February 1979

The consequences for K^0 decay of the six-quark model with its natural possibility of incorporating a *CP* violating phase, δ, are investigated when a particular mechanism to give the $\Delta I = 1/2$ rule is operative. The most important result is that the *CP* violation parameter ϵ' is much larger than in previous analyses, and hence the theory has predictions for $K^0 \to \pi\pi$ decay which are experimentally distinguishable from those of the superweak model.

In models of the weak interactions employing the gauge group [1] SU(2) × U(1) and with three or more left-handed doublets, *CP* violation arises automatically: in addition to real Cabibbo-like mixing angles, there are one or more complex phases which cannot be swept into the fermion fields and which violate *CP*. Such is the case in a six-quark model, proposed for precisely this reason by Kobayashi and Maskawa [2], where there are three real angles, θ_i with $i = 1, 2, 3$ and one *CP* violating phase, δ. In this model the left-handed doublets are

$$\begin{pmatrix} u \\ d' \end{pmatrix}_L , \quad \begin{pmatrix} c \\ s' \end{pmatrix}_L , \quad \begin{pmatrix} t \\ b' \end{pmatrix}_L ,$$

where, with a standard choice of the quark fields [2],

$$\begin{pmatrix} d' \\ s' \\ b' \end{pmatrix} \begin{pmatrix} c_1 & -s_1c_3 & -s_1s_3 \\ s_1c_2 & c_1c_2c_3 - s_2s_3e^{i\delta} & c_1c_2s_3 + s_2c_3e^{i\delta} \\ s_1s_2 & c_1s_2c_3 + c_2s_3e^{i\delta} & c_1s_2s_3 - c_2c_3e^{i\delta} \end{pmatrix} \begin{pmatrix} d \\ s \\ b \end{pmatrix}, \tag{1}$$

and $c_i = \cos\theta_i$, $s_i = \sin\theta_i$.

Such a model has become popular with the increasing evidence for a fifth lepton [3], τ, and more recently for a fifth quark, b, as a constituent for the upsilon family [4] of particles. A sixth quark, t, is expected, being necessary for the (generalized) GIM mechanism [5] as well as the cancellation of anomalies [6]. The six-quark model with $\delta \neq 0$ has been systematically studied by Ellis et al. [7] as the origin of *CP* violating effects in K^0 decay and elsewhere. They found that the contribution to the K^0–$\bar{K}^0$ mass matrix corresponding to fig. 1, with the mixing expressed in eq. (1), leads to *CP* violation in K^0 decay in accord with experiment and in close approximation to predictions of the superweak model [8] for common K decays.

There is another feature of strange-particle decays, and in particular of K decays, which has detectable consequences for the pattern of *CP* violation, but whose effects so far have not been taken adequately into account. This is the $\Delta I = 1/2$ rule. It had been hoped that strong interaction effects at short distances would sufficiently enhance the $\Delta I = 1/2$ portion of the non-leptonic weak interaction to explain the $\Delta I = 1/2$ rule. However, detailed calcula-

☆ Work supported in part by the Department of Energy under contract number EY-76-C-03-0515.
[1] Supported in part by the National Research Council of Canada and Imperial Oil Ltd.

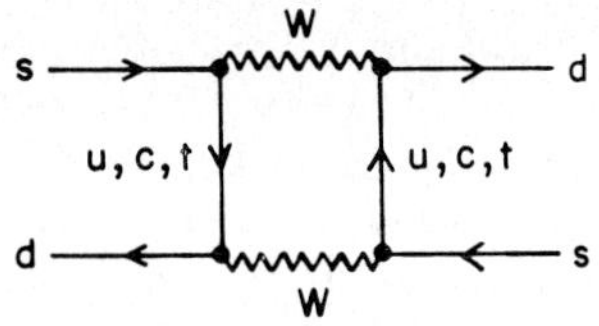

Fig. 1. Diagram contributing to the K^0–$\bar{K}^0$ mass matrix.

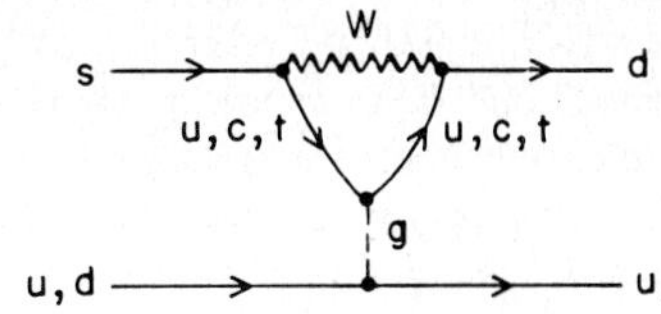

Fig. 2. "Penguin diagram" contributing to weak decays.

tions using quantum chromodynamics (QCD) do not lead to a large enough enhancement [9].

It is claimed [10] that the answer lies in the amplitudes shown in fig. 2, sometimes called "Penguin diagrams". Since the gluon (g) carries no isospin, these amplitudes are pure $\Delta I = 1/2$. Although they appear at first glance to give a smaller contribution to the effective weak hamiltonian than the lowest-order current–current term, when their matrix elements in strange-particle decays are evaluated using light current-quark masses, very large contributions to decay amplitudes result. An extensive analysis [10–12] of both strange-meson and baryon decays supports the hypothesis that the magnitude of the amplitudes and the $\Delta I = 1/2$ rule are understandable on this basis.

However, when the full six-quark model weak current with the mixing of eq. (1) is used to calculate the "Penguin diagrams" in fig. 2, they too will give a *CP* violating amplitude. In this paper we evaluate this contribution to K^0 decay and show it is of the same order as that arising from the standard analysis of fig. 1 for the *CP* violation parameter ϵ. More importantly, the parameter ϵ' is predicted to be much larger than previously in the six-quark model – sufficiently large so as to permit distinguishing the predictions from those of the superweak model.

To be quantitative we first recall the contribution to the mass matrix given by fig. 1. Using eq. (1) with s_1 and s_3 treated as small quantities ‡1, direct calculation gives [7]

$$\mathrm{Im}\,M_{12}/\mathrm{Re}\,M_{12} = 2s_2c_2s_3\sin\delta\,P(\theta_2,\eta) = \epsilon_\mathrm{m}\,, \tag{2a}$$

with

$$P(\theta_2,\eta) = \frac{s_2^2(1+[\eta/(1-\eta)]\ln\eta) - c_2^2(\eta+[\eta/(1-\eta)]\ln\eta)}{c_2^4\eta + s_2^4 - 2s_2^2c_2^2[\eta/(1-\eta)]\ln\eta}\,, \tag{2b}$$

where $\eta = m_\mathrm{c}^2/m_\mathrm{t}^2$ and M_{12} is the element of the K^0–$\bar{K}^0$ mass matrix defined by ‡2

$$M_{12} = \langle K^0|H_\mathrm{w}|\bar{K}^0\rangle + \sum_n \frac{\langle K^0|H_\mathrm{w}|n\rangle\langle n|H_\mathrm{w}|\bar{K}^0\rangle}{m_{K^0} - m_n} + \ldots\,. \tag{3}$$

On the other hand, direct calculation of fig. 2 gives a contribution to the effective weak hamiltonian density of the form

$$\mathcal{H}_\mathrm{W}^{(\mathrm{Penguin})} = 2^{-1/2}G(\alpha_s/12\pi)$$
$$\times\,[s_1c_2^2\ln(m_\mathrm{c}^2/\mu^2) + \mathrm{i}s_1s_2c_2s_3\sin\delta\ln(m_\mathrm{c}^2/\mu^2) + s_1s_2^2\ln(m_\mathrm{t}^2/\mu^2) - \mathrm{i}s_1s_2c_2s_3\sin\delta\ln(m_\mathrm{t}^2/\mu^2)]$$
$$\times\,(\bar{s}\gamma_\mu(1-\gamma_5)\lambda^\alpha d)(\bar{u}\gamma^\mu\lambda^\alpha u + \bar{d}\gamma^\mu\lambda^\alpha d + \ldots) + \mathrm{h.c.}\,. \tag{4}$$

In eq. (4), μ is a typical hadronic mass, α_s the strong interaction fine-structure constant, and u, d, ... represent the corresponding quark fields in both ordinary space–time and in color space, where the matrices λ^α operate. The first two terms in square brackets arise from fig. 2 with a c-quark loop, whereas the last two involve a t-quark. Terms

‡1 The signs of the quark fields may be chosen so that θ_1, θ_2, and θ_3 (but not necessarily δ) lie in the first quadrant and all sines and cosines of them are positive.

‡2 We assume *CPT* invariance and adopt the conventions of Lee and Wu [13], except that we label the K^0 eigenstates by S and L.

involving m_s and m_d, as well as higher powers of the sines of the small Cabibbo-like angles θ_1 and θ_3 have been neglected. Some effects due to higher-order QCD diagrams may be treated by renormalization group techniques. In as much as we will only be interested in the relative size of the real and imaginary terms in eq. (4), only differences in the magnitude of these multiplicative corrections for different terms need concern us. In the following we shall ignore differences between the leading logarithmic corrections to the terms in square brackets in eq. (4), just as they have also been ignored for the terms in the numerator and denominator of eq. (2). These are not expected to effect the character of our results.

When $\delta = 0$ the effective "Penguin" hamiltonian is responsible for a fraction f of the real decay amplitude $A_0^{(\delta=0)}$ for $K^0 \to 2\pi(I=0)$ defined by

$$\langle 2\pi(I=0)|H_w(\delta=0)|K^0\rangle = A_0^{(\delta=0)} e^{i\delta_0}\,, \tag{5}$$

where δ_0 is the $I=0$ strong interaction $\pi\pi$ phase shift. The claim [10–12] from detailed analysis of K^0 decays is that f is not far from unity. The total amplitude (apart from final-state $\pi\pi$ strong interactions) when $\delta \neq 0$ is then

$$A_0 = A_0^{(\delta=0)} - \mathrm{i} f A_0^{(\delta=0)} s_2 c_2 s_3 \sin\delta \, \frac{\ln(m_t^2/\mu^2) - \ln(m_c^2/\mu^2)}{c_2^2 \ln(m_c^2/\mu^2) + s_2^2 \ln(m_t^2/\mu^2)} \approx A_0^{(\delta=0)} e^{i\xi}\,, \tag{6}$$

with the small quantity

$$\xi = f s_2 c_2 s_3 \sin\delta \, \frac{\ln\eta}{\ln(m_c^2/\mu^2) - s_2^2 \ln\eta}\,. \tag{7}$$

However, the standard convention in analyzing CP violation in K^0 decays is to take A_0 real [13]. This may be accomplished by redefining the phases of the K^0 and $\bar{K}^0$ states [‡3]: $|K^0\rangle \to e^{-i\xi}|K^0\rangle$, $|\bar{K}^0\rangle \to e^{i\xi}|\bar{K}^0\rangle$, so that $A_0^{(\delta=0)} e^{i\xi} \to A_0^{(\delta=0)}$ and $\mathrm{Im}\langle K^0|M|\bar{K}^0\rangle \to \mathrm{Im}\{e^{2i\xi}\langle K^0|M|\bar{K}^0\rangle\} \approx \mathrm{Im}\langle K^0|M|\bar{K}^0\rangle + 2\xi\,\mathrm{Re}\langle K^0|M|\bar{K}^0\rangle$. Therefore as a result

$$\mathrm{Im}M_{12}/\mathrm{Re}M_{12} \to \mathrm{Im}M_{12}/\mathrm{Re}M_{12} + 2\xi = \epsilon_m + 2\xi\,. \tag{8}$$

The presence of the 2ξ term is not new – even the notation here is that of ref. [7]. What is new is that ξ as given in eq. (7) is not negligible compared to ϵ_m, as given in eq. (2). The whole "Penguin" contribution previously was neglected because it was guessed that the $\ln(m^2/\mu^2)$ terms in eq. (4) were instead of the form m^2/M_W^2. Indeed, the "Penguin diagram" appears as fig. 2c in ref. [7], but its contribution to ξ is taken as small compared to ϵ_m.

How does all this effect the parameters of CP violation in $K^0 \to \pi\pi$ decay? In standard notation [13]

$$\epsilon = \mathrm{i}(\mathrm{Im}\Gamma_{12} + \mathrm{i}\,\mathrm{Im}M_{12})/(\tfrac{1}{2}(\Gamma_S - \Gamma_L) + \mathrm{i}(m_S - m_L))\,, \tag{9}$$

and since experimentally [14] $|\frac{1}{2}(\Gamma_S - \Gamma_L)| \approx |m_S - m_L|$ and $|\mathrm{Im}\Gamma_{12}/\mathrm{Im}M_{12}| \lesssim 1/10$, we neglect $\mathrm{Im}\Gamma_{12}$ (due to CP violating phase differences between the $K^0 \to \pi\pi(I=0)$ decay and other modes) to obtain

$$|\epsilon| \approx |\mathrm{Im}M_{12}|/2^{1/2}|m_S - m_L|\,. \tag{10}$$

Recalling the result of our phase redefinition, and using $2\,\mathrm{Re}M_{12} = m_S - m_L$, it follows that [‡4]

$$|\epsilon| = 2^{-3/2}|\epsilon_m + 2\xi| = |2^{-1/2} s_2 c_2 s_3 \sin\delta| \cdot |P(\theta_2,\eta) + f\ln\eta/(\ln(m_c^2/\mu^2) - s_2^2\ln\eta)|\,. \tag{11}$$

The CP violation parameter [13]

$$\epsilon' \equiv 2^{-1/2}\,\mathrm{i}\,e^{i(\delta_2-\delta_0)} A_0^{-1}\,\mathrm{Im}A_2\,, \tag{12}$$

where A_0 and A_2 are the $K^0 \to \pi\pi$ decay amplitudes and δ_0 and δ_2 the strong interaction $\pi\pi$ phase shifts for final

‡3 We have also checked explicitly that if the phase convention on A_0 is dropped, the predictions for physical observables in $K^0 \to 2\pi$ decay remain exactly the same.

‡4 Our expression for ϵ (in the absence of "Penguin diagrams") is a factor of two smaller than that in ref. [7], and consequently ϵ'/ϵ is a factor of two bigger than they would have.

isospin 0 and 2, respectively. Neither the *CP* violation arising from the mass matrix (eq. (2)) nor that from the "Penguin diagram" (eq. (4)) can contribute to the $\Delta I = 3/2$ transition involved in the amplitude A_2. However, by virtue of the redefinition of the K^0 and $\bar{K}^0$ phases to make A_0 real, A_2 picks up a phase $e^{-i\xi}$ relative to A_0. Hence,

$$|\epsilon'| \approx 2^{-1/2}|\xi|\,|A_2/A_0|\,. \tag{13}$$

Experimental information on K^+ and K^0 decays yields [13] $|A_2/A_0| \approx 1/20$.

Numerically, we find that the two contributions to ϵ from the mass matrix and "Penguin diagrams" are comparable. For example, with $m_c = 1.5$ GeV, $m_t = 15$ GeV, $\mu = 1$ GeV, $\theta_2 = 15°$, and $f = 0.75$, the two terms in the second set of absolute value signs on the right-hand side of eq. (11) are 5.2 (from ϵ_m) and -3.1 (from 2ξ). The difference in sign holds for all reasonable quark masses, mixing angles, and values of f. Since $\sin\delta$ is adjustable to fit the experimental value of $|\epsilon|$, there is no critical test of the theory at this point.

What is critical, and changed from previous estimates [7], is ϵ'/ϵ, which is independent of δ and θ_3. From eqs. (11) and (13),

$$|\epsilon'/\epsilon| \approx \tfrac{1}{20}\,|2\xi/(\epsilon_m + 2\xi)|\,. \tag{14}$$

With the quark masses and other parameters used above, $|\epsilon'/\epsilon| \approx 1/13$. Bigger values of θ_2, m_c/μ, or m_t can lead to smaller values of $|\epsilon'/\epsilon|$: for example, with $m_t = 30$ GeV and all other parameters left the same $|\epsilon'/\epsilon| \approx 1/30$, while changing μ to 0.2 GeV but leaving $m_t = 15$ GeV gives $|\epsilon'/\epsilon| \approx 1/100$. The phases of ϵ and ϵ' are calculable from eqs. (9) and (12), respectively. With $\sin\delta$ positive [‡1] and the experimental values of $m_S - m_L$ and $\Gamma_S - \Gamma_L$, the phase of ϵ is $43.8 \pm 0.2°$, in agreement with experiment [14]. Using the sign of ξ and the experimentally determined δ_0, δ_2, and the sign of $\mathrm{Re}\,A_2/A_0$, the phase of ϵ' is $37 \pm 6°$ [14]. Our predictions for ϵ'/ϵ are therefore almost real and positive. With ϵ and ϵ' nearly parallel in the complex plane, the experimental [14] value $|\eta_{00}/\eta_{+-}| = |(\epsilon - 2\epsilon')/(\epsilon + \epsilon')| = 1.008 \pm 0.041$ translates to $\epsilon'/\epsilon = -0.003 \pm 0.014$.

Given the uncertainties in θ_2, m_t, μ, and the fraction, f, of the $K \to 2\pi$ amplitude given by the "Penguin diagram", we conclude that the model is not yet ruled out by experiment. However, a quantitative increase in the accuracy of experimental determination of the *CP* violating $K^0 \to \pi\pi$ decay parameter ϵ', which now seems possible [15], would be capable of distinguishing the six-quark model with a *CP* violating phase in the mixing matrix plus any sizable fraction of the $K^0 \to 2\pi$ amplitude being due to "Penguin diagrams" from the superweak model [8] where all the *CP* violation in the K^0 system arises from the mass matrix and $\epsilon' = 0$.

We are grateful to H. Harari and B. Winstein for useful and stimulating discussions. We thank L. Wolfenstein for constructive suggestions.

[1] S. Weinberg, Phys. Rev. Lett. 19 (1967) 1264;
A. Salam, in: Elementary particle physics: relativistic groups and analyticity, Nobel Symp. No. 8, ed. N. Svartholm Almquist and Wiksell, Stockholm, 1968) p. 367.
[2] M. Kobayashi and T. Maskawa, Prog. Theor. Phys. 49 (1973) 652.
[3] G.J. Feldman, rapporteur talk at the XIXth Intern. Conf. on High energy physics (Tokyo, Japan, 1978) unpublished; and SLAC-PUB-2224 contains a complete review, compilation, and references to original experiments on the tau lepton.
[4] For a review of experiments and references to the original work see: L. Lederman, rapporteur talk at the XIXth Intern. Conf. on High energy physics (Tokyo, Japan, 1978), unpublished.
[5] S.L. Glashow, J. Iliopoulos and L. Maiani, Phys. Rev. D2 (1970) 1285.
[6] C. Bouchiat et al., Phys. Lett. 38B (1972) 519;
D.J. Gross and R. Jackiw, Phys. Rev. D6 (1972) 477.
[7] J. Ellis et al., Nucl. Phys. B109 (1976) 213.
[8] L. Wolfenstein, Phys. Rev. Lett. 13 (1964) 562.
[9] M.K. Gaillard and B.W. Lee, Phys. Rev. lett. 33 (1974) 108.
[10] M.A. Shifman et al., JETP Lett. 22 (1975) 55;
M.A. Shifman et al., Nucl. Phys. B120 (1977) 316.
[11] M.A. Shifman et al., ITEP preprints ITEP-63, ITEP-64 (1976), unpublished.
[12] M.K. Gaillard, Lecture at the SLAC Summer Institute on Particle physics (July 10-21, 1978) and Fermilab preprint FERMILAB-Conf-78/64-THY (1978), unpublished.
[13] T.D. Lee and C.S. Wu, Ann. Rev. Nucl. Sci. 16 (1966) 511.
[14] See, for example, K. Kleinknecht, in: Proc. XVIIth Intern. Conf. on High energy physics (London, 1974), ed. J.R. Smith (Science Research Council, Rutherford Lab., 1974) p. III-23.
[15] B. Winstein, private communication.

Nuclear Physics B163 (1980) 289–311

QUANTUM CHROMODYNAMIC EFFECTS AND *CP* VIOLATION IN THE KOBAYASHI-MASKAWA MODEL

B. GUBERINA * and R.D. PECCEI
Max-Planck-Institut für Physik und Astrophysik, Munich, Fed. Rep. Germany

Received 31 July 1979
(Revised 26 October 1979)

We study, in the context of the Kobayashi-Maskawa model, the effects that quantum chromodynamics has on the predictions for *CP* violation in kaon decays. We derive an effective hamiltonian for $\Delta S = 1$ transitions in which all-order corrections in the strong coupling constant α_s are incorporated, including those arising from penguin diagrams. We examine, in particular, the predictions of this hamiltonian for the ratio of the *CP*-violating parameters $|\epsilon'/\epsilon|$. Our results for this ratio are well within the present experimental bound, but are considerably smaller than the recent calculation of Gilman and Wise. The discrepancy between our results and those of Gilman and Wise arises because of the different way we have estimated the amplitude for K^0 decay. A discussion of the reliability of both our estimate and that of Gilman and Wise is included.

1. Introduction

The Kobayashi-Maskawa [1] SU(2) × U(1) model, with three generations of quarks arranged in left-handed doublets and right-handed singlets, is perhaps the simplest gauge-theory model which incorporates the possibility of *CP* violation. The pioneering investigations of Pakvasa and Sugawara [2], Maiani [3] and especially Ellis, Gaillard and Nanopoulos [4] indicated that for strange-particle decays the predictions of this model reduced to the superweak [5] form of *CP* violation. This conclusion has been challenged recently in an interesting letter by Gilman and Wise [6]. These latter authors argued that quantum chromodynamic (QCD) gluonic effects can produce important additional contributions, which do not fit in the superweak pattern. Most interestingly, if the estimates of Gilman and Wise are correct, it might be possible, with a moderate increase in the accuracy of the relevant experiments, to test the combined predictions of the Kobayashi-Maskawa model and QCD.

* Alexander von Humboldt Fellow, on leave of absence from the Rudjer Bošković Institute, Zagreb, Croatia, Yugoslavia.

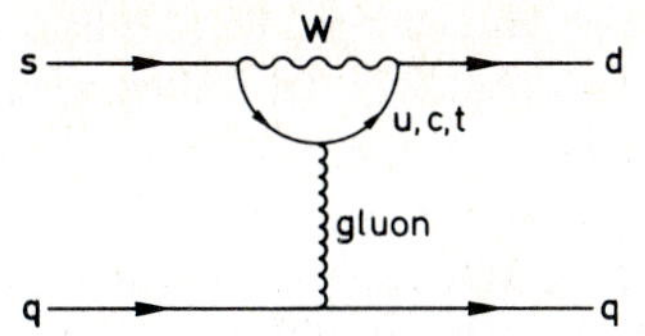

Fig. 1. Penguin diagrams that contribute for $\Delta S = 1$ processes.

The observation of Gilman and Wise is based on the work of Shifman, Vainshtein and Zakharov [7] on the $\Delta I = \frac{1}{2}$ rule in non-leptonic weak decays. Shifman et al., argued that the contribution of the pure $\Delta I = \frac{1}{2}$ penguin diagrams, shown in fig. 1, although of order α_s, accounts for the bulk of the observed amplitudes in non-leptonic interactions. In the Kobayashi-Maskawa model the penguin diagrams are not purely real and thus can in principle give rise to a direct *CP*-violating contribution to the process $K^0 \to 2\pi$. If this penguin contribution is large it can cause additional *CP*-violating effects above those arising from the kaon mass matrix, in contrast to the usual assumptions of the superweak theory. Gilman and Wise estimated the effects that the penguin contribution induces on the ratio of the *CP*-violating parameters [8] $|\epsilon'/\epsilon|$, which vanishes in the superweak theory. They found that although $|\epsilon'/\epsilon|$ is small, it is perilously close to the present experimental limit [8]: $|\epsilon'/\epsilon| \lesssim \frac{1}{50}$ and for some choices of parameters it even exceeds this limit.

The original calculation of Gilman and Wise was only to lowest order in α_s. Furthermore, they took as a phenomenological input the fraction, f, of the total amplitude in $K^0 \to 2\pi$ which Shifman et al., attributed to the penguin contribution. In view of how close to the present bound on $|\epsilon'/\epsilon|$ the result of Gilman and Wise came, it seemed worthwhile to us to re-examine this question in more detail by considering the effects of QCD to all orders in α_s ★. There are two good physical reasons that spurred us to undertake this improved calculation. (i) Higher-order QCD corrections in general enhance the contribution of the penguin diagrams in a different way for the real and imaginary parts. Roughly speaking, the imaginary part receives most of its enhancement from the region in k^2 space going from m_t^2 to m_c^2, while the real part receives its enhancement from the region going from m_c^2 to μ^2, with μ a typical hadronic mass. (ii) The estimate of Shifman et al., of the importance of the penguin contribution is essentially concerned with how much the real part gets enhanced and is rather parameter dependent since it involves the region from m_c^2 to μ^2. In contrast, the imaginary part enhancement arising from the region going from m_t^2 to m_c^2 is more reliable. Thus, a direct calculation of the imaginary

★ After this paper was submitted we became aware of a similar all-order calculation of the $\Delta S = 1$ effective weak hamiltonian by Gilman and Wise [13]. We have incorporated in the text, where appropriate, a comparison of our work with their recent calculation. See also ref. [14].

part of the effective penguin hamiltonian, to all orders in α_s, avoids two problems at once. One need not assume that the real and imaginary parts get the same enhancement, and one need not rely on an estimate of the importance of the real penguin contributions to the total amplitude.

The plan of this paper is as follows. In sect. 2 we derive the QCD corrected effective hamiltonian for the Kobayashi-Maskawa model for $\Delta S = 1$ weak processes. Our derivation of this hamiltonian is along the lines developed by Shifman et al., but is somewhat more complicated since we must consider operator mixing in two regions now, instead of one. Having the effective hamiltonian in hand, we proceed in sect. 3 to examine *CP* violation in the $K^0 \to 2\pi$ decay. We obtain an expression for the ratio of the *CP*-violating parameters $|\epsilon'/\epsilon|$ which, in contrast to the work of Gilman and Wise, is essentially independent of the magnitude of the contribution of the real penguins to the decay $K^0 \to 2\pi$. Our results for this ratio, for a broad range of parameters, are well within the experimental bound. In this section we also discuss the reason why our results are so much smaller than those obtained by Gilman and Wise [6,13]. We point out that this discrepancy is principally caused by differing assumptions on how to evaluate the relevant K^0 amplitudes once the effective hamiltonian is in hand. Our conclusions, as well as some final observations on the reliability of our results are contained in sect. 4. Some technical details are left to the appendix.

2. Effective hamiltonian for $\Delta S = 1$ processes

We want to derive an effective hamiltonian for $\Delta S = 1$ processes in the context of the Kobayashi-Maskawa model. To lowest order in the weak interactions, for processes where the momentum transfer can be neglected with respect to M_W^2, the $\Delta S = 1$ weak hamiltonian takes the usual current-current form

$$H = -\sqrt{\tfrac{1}{2}} G_F J_\mu^+ J^{-\mu} , \tag{1}$$

where the currents $J_\mu^\pm$ are given by

$$J_\mu^+ = \bar{\psi}\gamma_\mu(1-\gamma_5)\, U\psi , \tag{2a}$$

$$J_\mu^- = \bar{\psi}\gamma_\mu(1-\gamma_5)\, U^\dagger\psi . \tag{2b}$$

Here we have suppressed color indices and for each color the field $\bar{\psi}$ is a six-component field given by

$$\bar{\psi} = (\bar{u}, \bar{c}, \bar{t}, \bar{d}, \bar{s}, \bar{b}) . \tag{3}$$

The 6×6 matrix U has the block form

$$U = \begin{bmatrix} 0 & C \\ 0 & 0 \end{bmatrix} , \tag{4}$$

where the 3 X 3 matrix C is the usual Kobayashi-Maskawa matrix

$$C = \begin{bmatrix} c_1 & -s_1c_3 & -s_1s_3 \\ s_1c_2 & c_1c_2c_3 - s_2s_3\,e^{i\delta} & c_1c_2s_3 + s_2c_3\,e^{i\delta} \\ s_1s_2 & c_1s_2c_3 + c_2s_3\,e^{i\delta} & c_1s_2s_3 - c_2c_3\,e^{i\delta} \end{bmatrix}, \qquad (5)$$

with $c_i = \cos\theta_i$, $s_i = \sin\theta_i$. For $\Delta S = 1$ processes we need to retain only terms involving an s-quark in (1). Furthermore, since we are only interested in processes where $\Delta C = \Delta B = \Delta T = 0$, we may further drop terms that violate these selection rules in (1). Notice further that we have not included in (1) any contribution from neutral currents, since in the Kobayashi-Maskawa model these currents are flavor diagonal and hence cannot contribute to $\Delta S = 1$ processes. Effectively then, the $\Delta S = 1$ weak hamiltonian can be taken to be

$$H^{(0)} = -\sqrt{\tfrac{1}{2}}G_F\,[a_1\bar{d}\gamma_\mu(1-\gamma_5)\,u\bar{u}\gamma^\mu(1-\gamma_5)\,s$$

$$+\, a_2\bar{d}\gamma_\mu(1-\gamma_5)\,c\bar{c}\gamma^\mu(1-\gamma_5)\,s + a_3\bar{d}\gamma_\mu(1-\gamma_5)\,t\bar{t}\gamma^\mu(1-\gamma_5)\,s]\,, \qquad (6)$$

where

$$a_1 = -c_1s_1c_3\,, \qquad (7a)$$

$$a_2 = s_1c_2\,[c_1c_2c_3 - s_2s_3\,e^{i\delta}]\,, \qquad (7b)$$

$$a_3 = s_1s_2\,[c_1s_2c_3 + c_2s_3\,e^{i\delta}]\,. \qquad (7c)$$

Note that since the Glashow-Illiopoulos-Maiani mechanism [9] is incorporated in the Kobayashi-Maskawa model one has

$$a_1 + a_2 + a_3 = 0\,. \qquad (8)$$

Using eq. (6) as our starting point, we may now consider what correction QCD induces on this structure. To lowest order in α_s one encounters the diagrams of fig. 2. These diagrams are all logarithmically divergent because of our approximation of neglecting the weak propagator. If the weak propagator were to be restored,

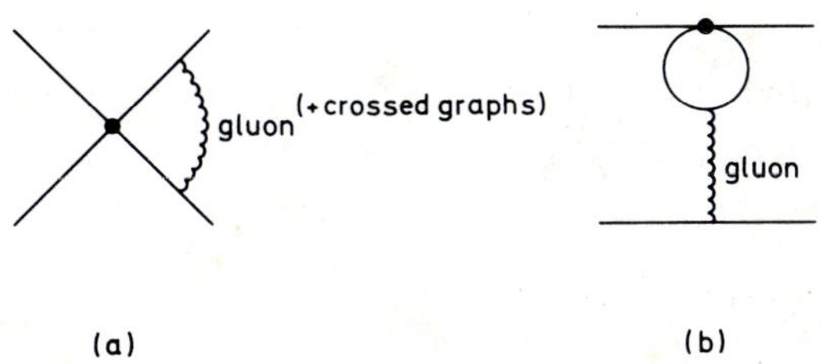

Fig. 2. Diagrams which give the lowest-order QCD corrections to the hamiltonian $H^{(0)}$. The diagrams of fig. 2b are the penguin contributions.

one would find that the relevant cut-off is provided by M_W^2. Of particular interest are the diagrams of fig. 2b which are essentially the penguin diagrams of fig. 1 in the limit as $M_W^2 \to \infty$. In the limit in which the gluon momentum transfer k^2 is much larger than all the quark masses entering in the quark loop, it is easy to check that the penguin diagrams vanish, since their contribution is proportional to $a_1 + a_2 + a_3$. However, for problems involving strange-particle decays $k^2 \ll m_c^2, m_t^2$ and the cancellation is not complete. This is essentially the crucial observation of Shifman et al. [7], which was exploited by Gilman and Wise [6].

An explicit evaluation of the penguin diagrams, dropping terms of order m_i^2/M_W^2, with m_i a quark mass, leads to the following effective $\Delta S = 1$ hamiltonian arising from the penguin diagrams:

$$H^{\text{penguin}} = \sqrt{\tfrac{1}{2}} G_F \frac{\alpha_s}{12\pi} [\bar{d}\gamma_\mu(1-\gamma_5)\lambda_a s \sum_f \bar{q}_f \gamma^\mu \lambda_a q_f]$$

$$\times \left\{ a_3 \ln \frac{M_W^2}{m_t^2} + a_2 \ln \frac{M_W^2}{m_c^2} + a_1 \ln \frac{M_W^2}{k^2} \right\} . \tag{9}$$

The above reflects the fact that for the c- and t-quark loops of fig. 1, the effective scale of the diagrams is set by the masses of the quarks instead of the momentum transfer. Using eq. (8) we may equally well write the quantity in the curly bracket in eq. (9) as

$$\{\ \} = (a_1 + a_2) \ln \frac{m_t^2}{m_c^2} + a_1 \ln \frac{m_c^2}{k^2} . \tag{10}$$

This rearrangement allows one to interpret the penguin hamiltonian in more physical terms. For momentum transfer scales ranging above m_t^2 there is no penguin contribution. For effective momentum transfers between m_t^2 and m_c^2 the t-quark loop can be neglected, but the c- and u-quark loops are active. Finally, for momentum transfers below m_c^2 only the u-quark loop is important. Of course the true hamiltonian is the sum of all these contributions.

The contribution of the diagrams of fig. 2a can also be readily evaluated, remembering that the effective cut-off in the logarithms is to be taken as $\Lambda^2 \simeq M_W^2$. For our future development it is convenient to write $H^{(0)}$ as the sum of two distinct terms:

$$H^{(0)} = -\sqrt{\tfrac{1}{2}} G_F [O_{189} + O_{405}] , \tag{11}$$

where the operators O_{189} and O_{405} for $\Delta S = 1, \Delta C = \Delta B = \Delta T = 0$ processes are given by

$$O_{189} = \sum_{i=1}^{3} [\tfrac{1}{3} a_i \bar{d}\gamma_\mu(1-\gamma_5) q_i \bar{q}_i \gamma^\mu (1-\gamma_5) s$$

$$- \tfrac{1}{4} a_i \bar{d}\gamma_\mu(1-\gamma_5)\lambda_a q_i \bar{q}_i \gamma^\mu(1-\gamma_5)\lambda_a s] , \tag{12a}$$

$$O_{405} = \sum_{i=1}^{3} [\tfrac{2}{3} a_i \bar{d}\gamma_\mu(1-\gamma_5)\, q_i \bar{q}_i \gamma^\mu(1-\gamma_5)\, s$$

$$+ \tfrac{1}{4} a_i \bar{d}\gamma_\mu(1-\gamma_5)\,\lambda_a q_i \bar{q}_i \gamma^\mu(1-\gamma_5)\,\lambda_a s] \,. \tag{12b}$$

Here $q_1 = u, q_2 = c, q_3 = t$. These operators correspond precisely to the $\Delta S = 1$, $\Delta C = \Delta B = \Delta T = 0$ projection of operators which transform as the 189- and 405-dimensional representations of SU(6). A simple calculation shows that the diagrams of fig. 2a again yield an effective hamiltonian which just involves these same operators

$$H^{2a} = -\sqrt{\tfrac{1}{2}}G_F \left[\frac{\alpha_s}{8\pi} \ln\frac{M_W^2}{\mu^2}\,(8O_{189} - 4O_{405})\right] . \tag{13}$$

Here μ^2 is a typical hadronic scale which should be identified with the momentum transfer involved in the relevant processes to be studied.

The total hamiltonian to $O(\alpha_s)$ is given by the sum of $H^{(0)}$, H^{penguin} and H^{2a}. There are three distinct types of logarithms that enter in this hamiltonian $\ln(M_W^2/m_t^2)$, $\ln(m_t^2/m_c^2)$ and $\ln(m_c^2/\mu^2)$. Associated with these logarithms there are corresponding operators. For the $\ln(M_W^2/m_t^2)$ terms only the operators O_{189} and O_{405} enter, while for the rest the penguin operators of eq. (9) are also relevant. However, we should comment that associated with the logarithms of the type $\ln(m_t^2/m_c^2)$ it is sensible to neglect the contributions of the t-quarks in the operators, since effectively these quarks do not play an active role. Similarly, for the $\ln(m_c^2/\mu^2)$ terms it is permissible to drop both the t- and c-quark contributions. Hence the effective hamiltonian to $O(\alpha_s)$ splits into three separate contribution which we can associate with momentum transfer regions ranging from M_W^2 to m_t^2, from m_t^2 to m_c^2 and from m_c^2 to μ^2.

One can proceed to analyze higher-order QCD corrections to the weak hamiltonian in the same vein, by studying more complicated Feynman diagrams. However, this approach is cumbersome and can be effectively bypassed by making use of renormalization group methods [10]. Indeed we could have phrased our discussion from the beginning in this more elegant way. Our insistence in looking at Feynman graphs was motivated only by trying to bring to the fore, in a perhaps more direct way, the existence of three relevant regions of interest. If one writes the weak hamiltonian in the form

$$H^{(0)} = -\sqrt{\tfrac{1}{2}}G_F \sum_i C_i O_i \,, \tag{14}$$

with the C_i being c-number coefficients and O_i being the relevant operators, then a renormalization group analysis of the effects of QCD informs us that the final effective hamiltonian will have an analogous form, with perhaps more operators entering, but where the coefficient functions now depend on M_W^2/μ^2, μ^2 being a relevant scale for the processes studied. In the particular case in question, where there are different scales entering, the correct procedure for arriving at the final hamiltonian consist of deriving at each scale an effective hamiltonian which then provides the starting point hamiltonian for the next stage. This is the procedure adopted for the 4-quark model by Shifman et al. [7]. A justification for this rather pragmatic approach can be found in the interesting paper of Witten [11].

Let us begin by considering QCD corrections to the $\Delta S = 1$ weak hamiltonian

$$H^{(0)} = -\sqrt{\tfrac{1}{2}} G_F [O_{189} + O_{405}] \tag{15}$$

in the region ranging from M_W^2 to m_t^2. In this region the penguin operators are ineffective and the only relevant operators which enter are again just O_{189} and O_{405}. To calculate the QCD corrections to the coefficients

$$C_{189} = 1 \,, \tag{16a}$$

$$C_{405} = 1 \,, \tag{16b}$$

it suffices to compute the renormalization matrix appropriate to these operators. Because O_{189} and O_{405} transform differently under SU(6), these operators will not mix, so that under renormalization

$$\begin{bmatrix} O_{189} \\ O_{405} \end{bmatrix} \to \begin{bmatrix} Z_{189} & 0 \\ 0 & Z_{405} \end{bmatrix} \begin{bmatrix} O_{189} \\ O_{405} \end{bmatrix} . \tag{17}$$

The renormalization factors Z_{189} and Z_{405} can be readily extracted from an analysis of the diagrams of fig. 2a. Evaluating these diagrams with a cut-off Λ, and remembering to include appropriate factors associated with the renormalization of the individual external quark lines, one finds

$$Z_{189} = 8 \left[\frac{\alpha_s}{4\pi} \ln \Lambda \right] , \tag{18a}$$

$$Z_{405} = -4 \left[\frac{\alpha_s}{4\pi} \ln \Lambda \right] . \tag{18b}$$

Defining, as usual, the anomalous dimension by

$$\gamma_n = -\frac{\partial \ln Z}{\partial \ln \Lambda} \equiv -\frac{\alpha_s}{4\pi} \bar{\gamma}_n \,, \tag{19}$$

a straightforward application of the renormalization group [10] yields the QCD corrected coefficients, for the region ranging to m_t:

$$C_{189}(m_t) = C_{189}(K_t)^{\bar{\gamma}_{189}/2b} \,, \tag{20a}$$

$$C_{405}(m_t) = C_{405}(K_t)^{\bar{\gamma}_{405}/2b} . \tag{20b}$$

Here

$$K_t = \frac{\alpha_s(m_t^2)}{\alpha_s(M_W^2)} \,, \tag{21}$$

and the QCD running coupling constant $\alpha_s(q^2)$ is given by the expression

$$\alpha_s(q^2) = \frac{\alpha_s(\mu^2)}{1 + \alpha_s(\mu^2)(b/4\pi)\ln(q^2/\mu^2)}, \tag{22}$$

with

$$b = 11 - \tfrac{2}{3}n_f, \tag{23}$$

where n_f is the number of flavors. For the region in question, since all 6 quarks are active, $b = 7$.

Using eqs. (16) and (18) one finds

$$C_{189}(m_t) = K_t^{4/7}, \tag{24a}$$

$$C_{405}(m_t) = K_t^{-2/7}. \tag{24b}$$

We note parenthetically that the anomalous dimensions of the operators O_{189} and O_{405} are precisely the same as the ones found in ref. [12] for the case of SU(4). This is not surprising, since these operators contain, respectively, the operators that transform as the 20- and 84-dimensional representation of SU(4). The only difference, at this stage between the 4-quark and 6-quark model resides in the different value of b for the two cases.

The effective QCD corrected hamiltonian at m_t, which is to be used as the starting point for computing the QCD corrections to m_c is then *

$$H(m_t) = -\sqrt{\tfrac{1}{2}}G_F\,[K_t^{4/7}O_{189} + K_t^{-2/7}O_{405}]. \tag{25}$$

However, as we mentioned earlier, since we are now propagating to regions below m_t we may drop effectively from the definitions of O_{189} and O_{405} of eqs. (12) all terms containing t-quarks. In the region from m_t^2 to m_c^2 it is no longer true that the operators O_{189} and O_{405} just reproduce themselves, since as our lowest-order calculation has indicated in this region the penguin operators appear with non-vanishing coefficients. Let us define the penguin operators, appropriate to the 4-quark case, as

$$O_P = (a_1 + a_2)[\bar{d}\gamma_\mu(1-\gamma_5)\lambda_a s \cdot \{\bar{u}\gamma^\mu\lambda_a u + \bar{c}\gamma^\mu\lambda_a c + \bar{d}\gamma^\mu\lambda_a d + \bar{s}\gamma^\mu\lambda_a s\}]. \tag{26}$$

An analysis of the diagrams of fig. 2, where now the four-fermion operator is not H_{weak} but O_P, indicates that this operator in turn produces, in higher order in α_s,

* In principle we should also split the region going from m_t to m_c into regions going from m_t to m_b and then from m_b to m_c. The only effect of this further splitting is to slightly change the anomalous dimension factors, since b changes when less flavors are active. We have neglected this rather unimportant effect. This effect is included in ref. [13].

other structures besides itself. Let us define three additional operators

$$A_{\rm P} = (a_1 + a_2)[\bar{d}\gamma_\mu(1-\gamma_5)\lambda_a s \cdot \{\bar{u}\gamma^\mu\gamma_5\lambda_a u + \bar{c}\gamma^\mu\gamma_5\lambda_a c$$

$$+ \bar{d}\gamma^\mu\gamma_5\lambda_a d + \bar{s}\gamma^\mu\gamma_5\lambda_a s\}] , \tag{27}$$

$$\tilde{O}_{\rm P} = (a_1 + a_2)[\bar{d}\gamma_\mu(1-\gamma_5) s \cdot \{\bar{u}\gamma^\mu u + \bar{c}\gamma^\mu c + \bar{d}\gamma^\mu d + \bar{s}\gamma^\mu s\}] , \tag{28}$$

$$\tilde{A}_{\rm P} = (a_1 + a_2)[\bar{d}\gamma_\mu(1-\gamma_5) s \cdot \{\bar{u}\gamma^\mu\gamma_5 u + \bar{c}\gamma^\mu\gamma_5 c$$

$$+ \bar{d}\gamma^\mu\gamma_5 d + \bar{s}\gamma^\mu\gamma_5 s\}] . \tag{29}$$

Then a straightforward calculation shows that O_{189}, O_{405}, $O_{\rm P}$, $A_{\rm P}$, $\tilde{O}_{\rm P}$ and $\tilde{A}_{\rm P}$ form a closed set of operators which mix among themselves. The renormalization matrix for these operators is readily computed and is displayed in detail in the appendix. We have

$$Z = \frac{\alpha_{\rm s}}{4\pi} \ln \Lambda\, \bar{\Gamma} . \tag{30}$$

The matrix $\bar{\Gamma}$ can be diagonalized by a similarity transformation

$$S^{-1}\bar{\Gamma}S = \Lambda , \tag{31}$$

where

$$\Lambda_{ij} = \delta_{ij}\lambda_j . \tag{32}$$

The effective hamiltonian at $m_{\rm c}$ is then readily calculated from that at $m_{\rm t}$. One has

$$H(m_{\rm c}) = -\sqrt{\tfrac{1}{2}}G_{\rm F}\{K_{\rm t}^{4/7} \sum_{ij} S_{1i}(K_{\rm c})^{\lambda_i/2b}(S^{-1})_{ij}O_j$$

$$+ K_{\rm t}^{-2/7} \sum_{ij} S_{2i}(K_{\rm c})^{\lambda_i/2b}(S^{-1})_{ij}O_j\} . \tag{33}$$

Here

$$K_{\rm c} = \frac{\alpha_{\rm s}(m_{\rm c}^2)}{\alpha_{\rm s}(m_{\rm t}^2)} , \tag{34}$$

and the set of operators O_j, $j = 1 \ldots 6$, are given by

$$O_j = \{O_{189}, O_{405}, O_{\rm P}, A_{\rm P}, \tilde{O}_{\rm P}, \tilde{A}_{\rm P}\} . \tag{35}$$

Note that in (33) b should be given by the value appropriate for 4 flavors, $b = \frac{25}{3}$.

Eventually, to complete the calculation we must propagate the hamiltonian (33) down to μ^2. In this last region only the u-, d- and s-quarks are active. If we drop the c-quark contributions, it is possible to rewrite the operators appearing in eq. (35) in terms of the standard set of operators used by Shifman et al. [7]. Defining

$$q_{\rm L} = \tfrac{1}{2}(1-\gamma_5)q , \tag{36a}$$

$$q_R = \tfrac{1}{2}(1+\gamma_5)\,q\,, \tag{36b}$$

we consider the following set of operators

$$O_1 = \bar{d}_L\gamma_\mu s_L \bar{u}_L\gamma^\mu u_L - \bar{d}_L\gamma_\mu u_L \bar{u}_L\gamma^\mu s_L\,, \tag{37}$$

$$O_2 = \bar{d}_L\gamma_\mu s_L \bar{u}_L\gamma^\mu u_L + \bar{d}_L\gamma_\mu u_L \bar{u}_L\gamma^\mu s_L + 2\bar{d}_L\gamma_\mu s_L \bar{d}_L\gamma^\mu d_L + 2\bar{d}_L\gamma_\mu s_L \bar{s}_L\gamma^\mu s_L\,, \tag{38}$$

$$O_3 = \bar{d}_L\gamma_\mu s_L \bar{u}_L\gamma^\mu u_L + \bar{d}_L\gamma_\mu u_L \bar{u}_L\gamma^\mu s_L + 2\bar{d}_L\gamma_\mu s_L \bar{d}_L\gamma^\mu d_L - 3\bar{d}_L\gamma_\mu s_L \bar{s}_L\gamma^\mu s_L\,, \tag{39}$$

$$O_4 = \bar{d}_L\gamma_\mu s_L \bar{u}_L\gamma^\mu u_L + \bar{d}_L\gamma_\mu u_L \bar{u}_L\gamma^\mu s_L - \bar{d}_L\gamma_\mu s_L \bar{d}_L\gamma^\mu d_L\,, \tag{40}$$

$$O_5 = \bar{d}_L\gamma_\mu\lambda_a s_L \cdot [\bar{u}_R\gamma^\mu\lambda_a u_R + \bar{d}_R\gamma^\mu\lambda_a d_R + \bar{s}_R\gamma^\mu\lambda_a s_R]\,, \tag{41}$$

$$O_6 = \bar{d}_L\gamma_\mu s_L \cdot [\bar{u}_R\gamma^\mu u_R + \bar{d}_R\gamma^\mu d_R + \bar{s}_R\gamma^\mu s_R]\,. \tag{42}$$

It is easy to rewrite the set (35) in terms of the above operators. One finds

$$O_{189} = -2O_1\,, \tag{43a}$$

$$O_{405} = \tfrac{2}{5}O_2 + \tfrac{4}{15}O_3 + \tfrac{4}{3}O_4\,, \tag{43b}$$

$$O_P = 2O_5 + \tfrac{4}{3}O_2 - \tfrac{8}{3}O_1\,, \tag{43c}$$

$$A_P = 2O_5 - \tfrac{4}{3}O_2 + \tfrac{8}{3}O_1\,, \tag{43d}$$

$$\tilde{O}_P = 2O_6 + O_1 + O_2\,, \tag{43e}$$

$$\tilde{A}_P = 2O_6 - O_1 - O_2\,. \tag{43f}$$

Using eqs. (33) and (43) we may rewrite the hamiltonian at m_c in the form

$$H(m_c) = -\sqrt{2}G_F \sum_{i=1}^{6} A_i O_i\,. \tag{44}$$

The coefficient functions A_i are displayed explicitly in the appendix in eqs. (A.7)–(A.12).

Our final task is to use the hamiltonian (44) as a starting point to compute the QCD corrections from the scale set by m_c^2 down to μ^2. The operators O_1, O_2, O_5 and O_6 get mixed in this last zone. However, both O_3 and O_4 do not mix since they carry distinct quantum numbers. The renormalization matrix for the operators O_1,

O_2, O_5 and O_6 is given in the appendix:

$$Z = \frac{\alpha_s}{4\pi} \ln \Lambda \bar{\Gamma}' , \tag{45}$$

while for both O_3 and O_4 one finds

$$Z_{3,4} = \frac{\alpha_s}{4\pi} \ln \Lambda(-4) . \tag{46}$$

The matrix $\bar{\Gamma}'$ can be diagonalized by a similarity transformation

$$T^{-1}\bar{\Gamma}'T = \Lambda' \tag{47}$$

and its eigenvalues are given in the appendix. Armed with these eigenvalues and the matrices T^{-1} and T, it is now straightforward to evolve $H(m_c)$ down to μ^2, by following the same procedure as outlined before. The final hamiltonian for $\Delta S = 1$ processes, which includes corrections to all orders in α_s, is given by the formula

$$H(\mu) = -\sqrt{2}G_F \sum_i B_i O_i , \tag{48}$$

where the coefficients B_i are given by sum of products of the coefficients A_i with new coefficients obtained from the last zone of interest. Specifically we have

$$B_i = \sum_j C_{ij} A_j . \tag{49}$$

The new coefficients C_{ij} are also given explicitly in the appendix in eqs. (A.22)–(A.39). We should comment that the anomalous exponents d_i to which

$$K = \frac{\alpha(\mu^2)}{\alpha(m_c^2)} \tag{50}$$

in the coefficients C_{ij} is raised, are given by

$$d_i = \lambda_i'/2b , \tag{51}$$

where b here is that appropriate for the case of 3 flavors, $b = 9$.

The final hamiltonian we have obtained is rather cumbersome because we have had to evolve from M_W^2 down to μ^2 through three distinct regions, and in the last two regions there was considerable operator mixing. As a check in our hamiltonian we note that if $K_t = K_c = K = 1$ we recover the original weak hamiltonian, as far as the u-, d- and s-quarks are concerned. If $K = K_c = 1$ we recover $H(m_t)$, and if $K = 1$ we recover $H(m_c)$. In sect. 3 we shall use this hamiltonian to study *CP* violation in the decay $K^0 \to 2\pi$. An additional comment is in order. We have checked that the effective hamiltonian (48), which we have derived, is in agreement with the recent all-order computation of Gilman and Wise [13]. Their final expression is given in terms of a different set of operators but, on changing to our basis, similar results to ours obtain.

3. *CP* violation in K-decays

We want to study now the implication of the hamiltonian (48) for *CP* violation in strange particle decays. Because the resulting hamiltonian is apparently so complicated, it is extremely useful first to try to focus on what are the dominant terms. Eventually, for a final analysis, one may also want to include the effects of the terms dropped in the first instance. The operators O_1 through O_4 all have a (V – A) × (V – A) structure, while O_5 and O_6 have a (V – A) × (V + A) structure. While the first type of structure is invariant under a Fierz transformation, the latter produces an (S + P) × (S – P) structure. When considering matrix elements of the operators O_i between pseudoscalar states, there will be a considerable enhancement of the (S + P) × (S – P) operators relative to the (V – A) × (V – A) operators. This is the physical reason why the penguin operator, which includes the operator O_5, although multiplied by coefficients which are of $O(\alpha_s)$ can give important contributions to non-leptonic decays [7].

We are going to be principally interested in the imaginary part of $H(\mu)$. Even though the coefficients that multiply the operators O_1, O_2, O_5 and O_6 for the imaginary part are comparable, in view of the above considerations it is permissible to focus only on the terms connected with O_5 and O_6. Furthermore, it is easy to show that between color singlet states

$$O_6 = \tfrac{3}{16} O_5 \,. \tag{52}$$

Given the fact that the coefficients that multiply O_6 are typically more than a factor of two smaller than those that multiply O_5, it is also sensible to neglect, as a first approximation, all terms connected with O_6. We shall focus then on the approximate hamiltonian

$$H_5(\mu) = -\sqrt{2} G_F B_5 O_5 \,. \tag{53}$$

For comparison with the work of Gilman and Wise [6,13], it is also useful to retain in B_5 terms which are purely real, although these terms play no role in *our* evaluation of *CP*-violating effects.

There are a few more reasonable approximations which we can effect on B_5. Recall that

$$B_5 = \sum_j C_{5j} A_j \,. \tag{54}$$

The dominant terms in the above expression will be those in which at least one of the coefficients in C_{5j} or A_j do not vanish when K_c, K_t, K go to unity. Furthermore, given that all the Cabibbo-like angles θ_i are small, we may also drop terms which involve non-leading products of the various small angles. In this way we arrive at the following approximate expression for $H_5(\mu)$:

$$H_5(\mu) = \sqrt{2} G_F \{c_1 s_1 c_3 a_R + i s_1 c_2 s_2 s_3 (\sin\delta) a_I\} O_5 \,, \tag{55}$$

where the coefficients a_R and a_I of the real and imaginary parts, respectively, are given by:

$$a_R = -K_t^{0.57}K_c^{0.48}[0.0390K^{0.80} - 0.0334K^{0.42} - 0.0026K^{-0.12} - 0.0030K^{-0.30}]$$

$$- \tfrac{1}{5}K_t^{-0.29}K_c^{-0.24}[0.0681K^{0.80} + 0.0071K^{0.42} + 0.0708K^{-0.12} - 0.1460K^{-0.30}] , \tag{56a}$$

$$a_I = -\{K_t^{0.57}[0.0370K_c^{0.85} - 0.0308K_c^{0.42} - 0.0024K_c^{-0.13} - 0.0038K_c^{-0.35}]$$

$$+ K_t^{-0.29}[0.0124K_c^{0.85} + 0.0028K_c^{0.42} + 0.0140K_c^{-0.13} - 0.0292K_c^{-0.35}]\}$$

$$\times \{0.8509K^{0.80} + 0.0091K^{0.42} + 0.1222K^{-0.12} + 0.0178K^{-0.30}\} . \tag{56b}$$

We note that the real part of the above hamiltonian is essentially that given by Shifman et al. [7], except that we have broken down the contribution coming from M_W^2 to m_c^2 into two distinct terms proportional to $K_t^{0.57}$ and $K_c^{0.48}$, respectively. The magnitude of the real part of $H_5(\mu)$ is rather sensitive to what values one takes for K, since the terms in (56a) proportional to K vanish as $K \to 1$. In contrast the imaginary part is much less sensitive to the value of K assumed. In table 1, we display the values taken by a_R and a_I for a broad range of parameters. In calculating these values we have used for K the formula

$$K = \frac{\alpha_s(\mu^2)}{\alpha_s(m_c^2)} = 1 + \frac{9}{4\pi}\alpha_s(\mu^2)\ln\frac{m_c^2}{\mu^2} , \tag{57}$$

and have varied both μ and m_c, taking $\alpha_s(\mu^2) = 1$. Although one should perhaps worry about trusting the QCD enhancements down to values where $\alpha_s = 1$, this is precisely the approach followed by Shifman et al. [7]. Once K is determined, K_c and K_t follow once values of m_t and M_W are given. In table 1 we have taken m_t as a variable parameter and fixed $M_W = 80$ GeV.

We note that the values of a_R are essentially independent on what we assume for m_t but are crucially dependent on the values of μ and m_c taken. In fact, we see that the values of a_R, for different choices of parameters can vary by an order of magnitude. In contrast a_I is much less parameter dependent, ranging by at most a factor of two over the range of parameters chosen. One should note that the parameters which make a_R largest (small μ, large m_c) make a_I the smallest. We shall comment again below on this point.

To give a quantitative evaluation of *CP* violation we need to know what is the matrix element of the operator O_5 for the decay $K^0 \to 2\pi$. Shifman, Vainshtein and Zakharov [7] have evaluated this matrix element in a valence quark approximation. It proves convenient to quote their answer in terms of the experimental amplitude for this decay. They find, approximately

$$\sqrt{2}G_F c_1 s_1 c_3 \langle\pi\pi(I=0)|O_5|K^0\rangle \simeq -3.4A[K^0 \to \pi\pi(I=0)] . \tag{58}$$

Table 1
Value of the coefficients a_R and a_I for a range of parameters

μ(GeV)	m_c(GeV)	m_t(GeV)	a_R	a_I
1	1.2	15	−0.0138	−0.1140
1	2	15	−0.0441	−0.0831
1	1.2	50	−0.0138	−0.1517
1	2	50	−0.0439	−0.1203
0.5	1.2	15	−0.0550	−0.0992
0.5	2	15	−0.0789	−0.0764
0.5	1.2	50	−0.0549	−0.1341
0.5	2	50	−0.0787	−0.1118
0.14	1.2	15	−0.1146	−0.0867
0.14	2	15	−0.1334	−0.0687
0.14	1.2	50	−0.1146	−0.1206
0.14	2	50	−0.1334	−0.1028

Thus the real part of the hamiltonian $H_5(\mu)$ gives a contribution to the decay amplitude

$$\langle\pi\pi(I=0)|H_5^{\mathrm{Real}}(\mu)|\mathrm{K}^0\rangle \simeq -3.4a_\mathrm{R}A[\mathrm{K}^0\to\pi\pi(I=0)]\ . \tag{59}$$

To obtain the result of Shifman et al., in which the penguin diagrams give approximately 85% of the total amplitude, requires $a_\mathrm{R}\simeq -0.25$. A glance at table 1 indicates that this is a rather unrealistic expectation. In fact, to be fair, this is also the conclusion of Shifman et al. The rather large value of a_R used by them is not really a QCD prediction, but comes from a fit of hyperon decays. In view of the crucial dependence of a_R on K, these authors preferred to take a_R as a free parameter.

Fortunately to find how much $H_5(\mu)$ contributes to the imaginary part of the decay amplitude, we are much less sensitive to the parameters used. Using eq. (58) we have that

$$\langle\pi\pi(I=0)|H_5^{\mathrm{Im}}(\mu)|\mathrm{K}^0\rangle = a_\mathrm{I}\, i s_1 c_2 s_2 s_3 \sin\delta\sqrt{2}G_\mathrm{F}\langle\pi\pi(I=0)|O_5|\mathrm{K}^0\rangle$$

$$\simeq -i\left[\frac{c_2 s_2 s_3 \sin\delta}{c_1 c_3}(3.4a_\mathrm{I})\right]A[\mathrm{K}^0\to 2\pi(I=0)]\ . \tag{60}$$

The quantity in the square brackets above is precisely the parameter ξ, defined by Gilman and Wise [6], which enters in the analysis of *CP* violation. We obtain, thus,

$$\xi = \frac{c_2 s_2 s_3 \sin\delta}{c_1 c_3}(3.4a_\mathrm{I}) \simeq c_2 s_2 s_3 \sin\delta(3.4a_\mathrm{I})\ , \tag{61}$$

dropping small-angle corrections. For the broad ranges of parameters displayed in table 1 the coefficient of $c_2 s_2 s_3 \sin\delta$ in eq. (61) ranges between −0.234 and

Table 2
Values for $\xi/s_2c_2c_3 \sin\delta$ for a range of parameters evaluated according to our formula, eq. (61), and the formula of Gilman and Wise, eq. (62), with $f = 0.75$

μ(GeV)	m_c(GeV)	m_t(GeV)	Present determination	Gilman and Wise ($f = 0.75$)
1	1.2	15	−0.388	−6.196
1	2	15	−0.283	−1.413
1	1.2	50	−0.516	−8.245
1	2	50	−0.409	−2.055
0.5	1.2	15	−0.337	−1.353
0.5	2	15	−0.260	−0.745
0.5	1.2	50	−0.456	−1.832
0.5	2	50	−0.380	−1.065
0.14	1.2	15	−0.295	−0.567
0.14	2	15	−0.234	−0.386
0.14	1.2	50	−0.410	−0.789
0.14	2	50	−0.350	−0.577

−0.516, with the lowest value corresponding to parameters for which a_R is maximal, $a_R = -0.1333$.

It is important to compare our evaluation of ξ with that of Gilman and Wise [6,13]. These authors preferred to rely not on the estimate of Shifman et al. [7] for the matrix element of the operator O_5 in the $K^0 \to 2\pi$ process but rather on the total contribution that the penguin diagrams make for $A(K^0 \to \pi\pi(I = 0))$. Let f be the fraction of this total amplitude that can be attributed to penguins, then Gilman and Wise estimate ξ as

$$\xi_{GW} = -c_2 s_2 s_3 \sin\delta \left[f \frac{a_I}{a_R} \right] . \tag{62}$$

From ref. [7] one has that $f \simeq 75$–85%. Adopting for f the value $f = 0.75$ and using our results * for a_I and a_R from table 1, we compare in table 2 the determination of ξ given by eq. (61) and that given by eq. (62). It is clear from the table that the values of ξ obtained by Gilman and Wise are typically much larger than ours and are rather strongly dependent on the scale parameter μ. Such a dependence is not surprising since a_R, as we have seen, is strongly μ-dependent. Roughly speaking, using a

* We have mentioned earlier that as μ decreases so does $a_I(\alpha)$. In ref. [13] the opposite behaviour obtains. However, this fact is an artifact of the ways one calculates K, K_c and K_t. Here we have kept $\alpha_s(\mu) = 1$ and varied μ in eq. (57) to get K and then K_c and K_t. What Gilman and Wise do is to calculate their results for varying $\alpha_s(\mu) = 1.25, 1, 0.75$ and employ the formula $\alpha_s(Q^2) = 12\pi/(33 - 2n_f)\ln(Q^2/\Lambda^2)$ with $\Lambda^2 = 0.1\ \mathrm{GeV}^2$. For the purposes of table 2 we have used our results for a_I and a_R rather than those of Gilman and Wise. However, we believe that the qualitative trend is rather similar.

lowest-order estimate, one has

$$\frac{a_{\rm I}}{a_{\rm R}} \simeq \frac{\ln(m_{\rm t}/m_{\rm c})}{\ln(m_{\rm c}/\mu)}. \tag{63}$$

Hence for values of μ near $m_{\rm c}$ one gets large values of ξ.

We believe that our estimate of ξ is more reliable since it is much less dependent on μ. Physically, one expects that measurable quantities should not depend on the normalization scale μ. In our way of calculating ξ, we have implicitly assumed that the valence-quark estimate for the matrix element of O_5 is independent of μ and hence our only μ-dependence arises from that in $a_{\rm I}$. In as much as $a_{\rm I}$ is not terribly μ-dependent, our result satisfies the physical criteria discussed above. However, in fairness, we should point out that a similar assumption for the matrix element of O_5, coupled with our evaluation of $a_{\rm R}(\mu)$ would give a rather large variation with μ for the real part of the $K^0 \to 2\pi$ amplitude. It is true, however, that there are other contributions to the real part of the $K^0 \to 2\pi$ amplitude besides those arising from the penguins. But typically the coefficients multiplying the other operators are not that rapidly varying with μ and it appears difficult to argue that, for the real amplitude, the μ-dependence of the penguin contribution is cancelled by that of the remaining contributions. We think it is perhaps more sensible to argue that the rapid dependence on μ of $a_{\rm R}$ is perhaps an artifact of trusting a leading log calculation in a region where $\alpha_{\rm s}$ is large. After all the principal contribution for $a_{\rm R}$ comes precisely from the region going from $m_{\rm c}^2$ to μ^2 and here non-leading contributions may be important. For $a_{\rm I}$, on the other hand, one is dealing with the region from $m_{\rm t}^2$ to $m_{\rm c}^2$ and a leading log calculation may be more sensible.

As we discussed above, the result of Gilman and Wise for ξ is rather μ-dependent. However, this μ-dependence should perhaps not be too severely criticized. Basically, Gilman and Wise have taken a different approach to calculate ξ. They trust that the penguin contributions dominate in $K^0 \to 2\pi$, thus $f \simeq 0.75$. However, they do not commit themselves to what value of μ such a result applies and hence present a broad range of values for $\xi(\mu)$. In contrast, we do not assume that the penguin contributions explain the $\Delta I = \frac{1}{2}$ rule but try to calculate directly at least what is their contributions to the phase ξ.

Having obtained an expression for ξ, the value of the measured *CP*-violating parameters ϵ and ϵ' [8] can now be estimated straightforwardly. Using the notation of Gilman and Wise one has

$$|\epsilon| = \frac{|\epsilon_{\rm m} + 2\xi|}{2\sqrt{2}}, \tag{64}$$

$$|\epsilon'| = \sqrt{\tfrac{1}{2}}\,|\xi| \left|\frac{A_2}{A_0}\right| . \tag{65}$$

In the above $\epsilon_{\rm m}$ is given by the ratio of the imaginary to real part of the kaon mass matrix. This parameter has been evaluated in the Kobayashi-Maskawa model by

Table 3
Values of $|\epsilon'/\epsilon|$ for a range of parameters

m_c(GeV)	m_t(GeV)	$\|\epsilon'/\epsilon\|$ ($\theta_2 = 5°$)	$\|\epsilon'/\epsilon\|$ ($\theta_2 = 15°$)
1.2	15	0.0039	0.0030
1.5	15	0.0042	0.0033
2	15	0.0044	0.0037
1.2	30	0.0025	0.0022
1.5	30	0.0028	0.0022
2	30	0.0032	0.0024
1.2	50	0.0015	0.0020
1.5	50	0.0018	0.0019
2	50	0.0023	0.0020

Here we have taken μ = 0.7 GeV. For μ = 1 GeV all values would increase by roughly 10%, while for μ = 0.14 GeV all values would decrease by roughly 20%.

Ellis et al. [4], with the result *

$$\epsilon_m = 2s_2c_2s_3(\sin\delta)P(\theta_2,\eta)\,, \tag{66}$$

where

$$P(\theta_2,\eta) = \frac{s_2^2\left[1+\dfrac{\eta\ln\eta}{1-\eta}\right] - c_2^2\left[\eta+\dfrac{\eta\ln\eta}{1-\eta}\right]}{c_2^4\eta + s_2^4 - 2s_2^2c_2^2\dfrac{\eta\ln\eta}{1-\eta}}\,, \tag{67}$$

with $\eta = m_c^2/m_t^2$. The ratio $|A_2/A_0|$ in eq. (65) is the ratio of the decay amplitudes for $K^0 \to 2\pi$ in which the final pions are in a state of isospin 2 or 0. Experimentally, $|A_2/A_0| \simeq \frac{1}{20}$.

Using our results for ξ we may write

$$|\epsilon| = \left|\frac{s_2c_2s_3\sin\delta}{\sqrt{2}}\right| |P(\theta_2,\eta) + 3.4a_I|\,, \tag{68}$$

$$\left|\frac{\epsilon'}{\epsilon}\right| = \frac{1}{20}\left|\frac{3.4a_I}{P(\theta_2,\eta)+3.4a_I}\right|. \tag{69}$$

As Gilman and Wise correctly pointed out, there is not much that can be said about $|\epsilon|$ since both θ_2 and δ are adjustable parameters. However, in contrast to the results of Gilman and Wise, $3.4a_I$ is not comparable to $P(\theta_2,\eta)$ but typically an order of magnitude smaller than it. Thus we find that $|\epsilon'/\epsilon|$ is well below the present experimental bound [8] $|\epsilon'/\epsilon| \lesssim \frac{1}{50}$. Our results for this ratio, for a variety of parameters, are displayed in table 3.

* A complete calculation should also include QCD enhancements for ϵ_m. For our purposes it probably suffices to rely on the lowest-order estimate since $|\epsilon_m| >> |\xi|$.

4. Concluding remarks

We have evaluated QCD corrections to the $\Delta S = 1$ weak hamiltonian in the Kobayashi-Maskawa model with a view of seeing whether the effects of *CP* violation in this model can be distinguished from the superweak model. Our results for the ratio $|\epsilon'/\epsilon|$, although non-vanishing, are quite small and well below the present level of experimental uncertainty. It is unclear to us whether experimental limits can be pushed down to test the predictions of this model for $|\epsilon'/\epsilon|$, although it certainly would be of importance to obtain confirmation of a departure from the superweak theory.

Our results for $|\epsilon'/\epsilon|$ are at variance with those of Gilman and Wise but we believe we have understood the reason for this discrepancy and feel that our results are reliable. As can be seen from table 3, our predictions for $|\epsilon'/\epsilon|$ vary by about a factor of 2 depending on different choices of parameters. There are, however, two further causes of uncertainty in our estimates which ought to be mentioned. Our results depend on the value assumed for the matrix element $\langle\pi\pi|O_5|K^0\rangle$. In this paper we have relied on the estimate of this matrix element by Shifman, Vainshtein and Zakharov, who used a valence-quark approximation. It may well be that using other approaches the value of this matrix element may change. However, we would be surprised if the change were more than a factor of 2 or so *. Finally, we have dropped from our calculations both subdominant terms in $H_5(\mu)$, as well as all terms in which c- and t-quarks enter. We have checked that the retention of the contribution of other operators besides O_5 in $H_5(\mu)$ does not affect our results for ξ. Of more importance, perhaps, is the inclusion of operators in which $c\bar{c}$ and $t\bar{t}$ structures enter. The effect of these kind of terms is difficult to estimate. However, since these contributions to the amplitude are Zweig suppressed, a reasonable assumption [2,4] is that they should contribute to ξ an amount of the order

$$\xi \sim \tfrac{1}{10}\, |c_2 s_2 s_3 \sin\delta|\,. \tag{70}$$

If this is indeed the case, our results would not be markedly affected.

In conclusion we feel that the Kobayashi-Maskawa model remains a viable model for *CP* violation. The inclusion of QCD effects gives non-vanishing contributions to $|\epsilon'/\epsilon|$ which, unfortunately, appear to be tantalizingly small.

We are both grateful to Hector Rubinstein for some useful conversations. One of us (B.G.) would like to thank the Max-Planck-Institute for its hospitality.

* Dr. J. Donoghue has kindly informed one of us (RDP) that a bag model evaluation of the matrix element of O_5 in kaon decay differs by about a factor of two from that in the valence-quark approximation.

Appendix

Operator mixing and effective $\Delta S = 1$ hamiltonian for the Kobayashi-Maskawa model

The operators $O_{189}, O_{405}, O_P, A_P, \tilde{O}_P$ and $\tilde{A}_P$ defined in the text in general mix when one considers QCD corrections. Under renormalization one finds

$$\begin{bmatrix} O_{189} \\ O_{405} \\ O_P \\ A_P \\ \tilde{O}_P \\ \tilde{A}_P \end{bmatrix} \to \frac{\alpha_s}{4\pi} \ln \Lambda \begin{bmatrix} 8 & 0 & -\frac{1}{3} & 0 & 0 & 0 \\ 0 & -4 & -\frac{1}{3} & 0 & 0 & 0 \\ 0 & 0 & \frac{37}{9} & 5 & 0 & \frac{32}{3} \\ 0 & 0 & \frac{41}{9} & 9 & \frac{32}{3} & 0 \\ 0 & 0 & -\frac{2}{3} & 3 & 0 & 0 \\ 0 & 0 & \frac{11}{3} & 0 & 0 & 0 \end{bmatrix} \begin{bmatrix} O_{189} \\ O_{405} \\ O_P \\ A_P \\ \tilde{O}_P \\ \tilde{A}_P \end{bmatrix} . \tag{A.1}$$

The eigenvalues of the above matrix, which we shall denote by $\overline{\Gamma}$, are readily found numerically:

$$\lambda_i = \{14.0856; 8; 7.0020; -2.1948; -4; -5.7817\} . \tag{A.2}$$

The matrix $\overline{\Gamma}$ can be diagonalized by a similarity transformation

$$S^{-1}\overline{\Gamma}S = \Lambda . \tag{A.3}$$

The required matrices S and S^{-1} which diagonalize $\overline{\Gamma}$ are given by

$$S = \begin{bmatrix} -0.21042 & 1 & 0.63782 & -0.01957 & 1 & -0.03814 \\ -0.07080 & 0 & -0.05786 & 0.11052 & 0 & -0.29500 \\ 3.84154 & 0 & 1.90966 & -0.59858 & 0 & -1.57684 \\ 5.53018 & 0 & -1.02972 & -1.37842 & 0 & 0.98656 \\ 0.99601 & 0 & -0.62279 & 1.70231 & 0 & -0.69372 \\ 1 & 0 & 1 & 1 & 0 & 1 \end{bmatrix} , \tag{A.4}$$

$$S^{-1} = \begin{bmatrix} 0 & 0 & 0.06343 & 0.11271 & 0.08535 & 0.04803 \\ 1 & 0 & -0.12500 & 0.12500 & 0.16667 & -0.16667 \\ 0 & 0 & 0.20248 & -0.15414 & -0.23481 & 0.30845 \\ 0 & 0 & -0.05072 & -0.07492 & 0.36411 & 0.24652 \\ 0 & 1 & -0.04167 & 0.04167 & -0.11111 & 0.11111 \\ 0 & 0 & -0.21518 & 0.11635 & -0.21466 & 0.39699 \end{bmatrix} . \tag{A.5}$$

Using these matrices and the procedure indicated in the text one can derive the effective hamiltonian $H(m_c)$. Expanding this hamiltonian in terms of the set of operators O_i, $i = 1 \ldots 6$, defined by Shifman et al. [7] and given explicitly in the text, one has

$$H(m_c) = -\sqrt{2}G_F \sum_{i=1}^{6} A_i O_i , \tag{A.6}$$

where the coefficient functions A_i are given by the following formulas:

$$A_1 = (a_1 + a_2)\{K_t^{0.57}[-0.0178K_c^{0.85} + 0.5000K_c^{0.48} - 0.4764K_c^{0.42} - 0.0006K_c^{-0.13} - 0.0052K_c^{-0.35}] + K_t^{-0.29}[-0.0060K_c^{0.85} + 0.0432K_c^{0.42} + 0.0030K_c^{-0.13} - 0.0402K_c^{-0.35}]\} + a_1\{-K_t^{0.57}K_c^{0.48}\}, \quad (A.7)$$

$$A_2 = (a_1 + a_2)\{K_t^{0.57}[0.0030K_c^{0.85} - 0.0216K_c^{0.42} - 0.0014K_c^{-0.13} + 0.0200K_c^{-0.35}] + K_t^{-0.29}[0.0010K_c^{0.85} + 0.0020K_c^{0.42} + 0.0082K_c^{-0.13} - 0.1666K_c^{-0.24} + 0.1554K_c^{-0.35}]\} + a_1\{\tfrac{1}{5}K_t^{-0.29}K_c^{-0.24}\}, \quad (A.8)$$

$$A_3 = a_1\{\tfrac{2}{15}K_t^{-0.29}K_c^{-0.24}\}, \quad (A.9)$$

$$A_4 = a_1\{\tfrac{2}{3}K_t^{-0.29}K_c^{-0.24}\}, \quad (A.10)$$

$$A_5 = (a_1 + a_2)\{K_t^{0.57}[-0.0370K_c^{0.85} + 0.0308K_c^{0.42} + 0.0024K_c^{-0.13} + 0.0038K_c^{-0.35}] + K_t^{-0.29}[-0.0124K_c^{0.85} - 0.0028K_c^{0.42} - 0.0140K_c^{-0.13} + 0.0292K_c^{-0.35}]\}, \quad (A.11)$$

$$A_6 = (a_1 + a_2)\{K_t^{0.57}[-0.0280K_c^{0.85} + 0.0470K_c^{0.42} - 0.0120K_c^{-0.13} - 0.0070K_c^{-0.35}] + K_t^{-0.29}[-0.0094K_c^{0.85} - 0.0042K_c^{0.42} + 0.0674K_c^{-0.13} - 0.0538K_c^{-0.35}]\}. \quad (A.12)$$

To obtain the final hamiltonian we must consider how the operators O_1 through O_6 mix in the final region from m_c^2 to μ^2. Because O_3 and O_4 transform like a 27-dimensional representation of SU(3) they will not mix with the other operators. They also do not mix with each other since one of them is a $\Delta I = \frac{1}{2}$ operator and the other is a $\Delta I = \frac{3}{2}$ operator. For the operators O_1, O_2, O_5 and O_6 one finds that under renormalization

$$\begin{bmatrix} O_1 \\ O_2 \\ O_5 \\ O_6 \end{bmatrix} \to \frac{\alpha_s}{4\pi} \ln \Lambda \begin{bmatrix} \frac{68}{9} & \frac{2}{9} & \frac{1}{3} & 0 \\ \frac{20}{9} & -\frac{46}{9} & -\frac{5}{3} & 0 \\ \frac{8}{3} & -\frac{4}{3} & 12 & \frac{32}{3} \\ 0 & 0 & 3 & 0 \end{bmatrix} \begin{bmatrix} O_1 \\ O_2 \\ O_5 \\ O_6 \end{bmatrix}, \quad (A.13)$$

a result found already by Shifman et al. [7]. Denoting this matrix by $\overline{\Gamma}'$, one finds the eigenvalues

$$\lambda_i' = \{14.4443; 7.5118; -2.1522; -5.3594\} . \qquad \text{(A.14)}$$

$\bar{\Gamma}'$ can be diagonalized by the similarity transformation

$$T^{-1}\bar{\Gamma}'T = \Lambda' . \qquad \text{(A.15)}$$

The relevant matrices T and T^{-1} are given by

$$T = \begin{bmatrix} 0.22055 & -9.18839 & 0.01512 & 0.29840 \\ -0.38529 & -1.94819 & 0.41546 & -14.66226 \\ 4.81475 & 2.50395 & -0.71742 & -1.78647 \\ 1 & 1 & 1 & 1 \end{bmatrix} , \qquad \text{(A.16)}$$

$$T^{-1} = \begin{bmatrix} 0.06476 & -0.01131 & 0.17672 & 0.13050 \\ -0.10680 & -0.00226 & 0.00364 & 0.00517 \\ 0.02874 & 0.07894 & -0.17040 & 0.84452 \\ 0.01330 & -0.06537 & -0.00996 & 0.01981 \end{bmatrix} . \qquad \text{(A.17)}$$

For completeness we remark that under renormalization

$$O_3 \to \frac{\alpha_s}{4\pi} \ln \Lambda(-4)\, O_3 , \qquad \text{(A.18)}$$

$$O_4 \to \frac{\alpha_s}{4\pi} \ln \Lambda(-4)\, O_4 . \qquad \text{(A.19)}$$

Using the procedure described in the text one can construct the final hamiltonian for $\Delta S = 1$ processes, which included all order corrections in α_s. The structure of the hamiltonian is given by the formula

$$H = -\sqrt{2}G_F \sum_{i=1}^{6} B_i O_i , \qquad \text{(A.20)}$$

where the coefficient functions B_i are given by

$$B_i = \sum_{j=1}^{6} C_{ij} A_j . \qquad \text{(A.21)}$$

Direct computation, using the above formulas, yields the following expressions for the coefficients C_{ij}, which are non-vanishing:

$$C_{11} = 0.0143K^{0.80} + 0.9813K^{0.42} + 0.0004K^{-0.12} + 0.0040K^{-0.30} , \qquad \text{(A.22)}$$

$$C_{12} = -0.0250K^{0.80} + 0.2081K^{0.42} + 0.0119K^{-0.12} - 0.1950K^{-0.30} , \qquad \text{(A.23)}$$

$$C_{15} = 0.3118K^{0.80} - 0.2674K^{0.42} - 0.0206K^{-0.12} - 0.0238K^{-0.30} , \qquad \text{(A.24)}$$

$$C_{16} = 0.0648K^{0.80} - 0.1068K^{0.42} + 0.0287K^{-0.12} + 0.0133K^{-0.30}, \quad (A.25)$$

$$C_{21} = -0.0025K^{0.80} + 0.0208K^{0.42} + 0.0012K^{-0.12} - 0.0195K^{-0.30}, \quad (A.26)$$

$$C_{22} = 0.0044K^{0.80} + 0.0044K^{0.42} + 0.0328K^{-0.12} + 0.9584K^{-0.30}, \quad (A.27)$$

$$C_{25} = -0.0545K^{0.80} - 0.0057K^{0.42} - 0.0566K^{-0.12} + 0.1168K^{-0.30}, \quad (A.28)$$

$$C_{26} = -0.0113K^{0.80} - 0.0022K^{0.42} + 0.0789K^{-0.12} - 0.0654K^{-0.30}, \quad (A.29)$$

$$C_{33} = K^{-0.22}, \quad (A.30)$$

$$C_{44} = K^{-0.22}, \quad (A.31)$$

$$C_{51} = 0.0390K^{0.80} - 0.0334K^{0.42} - 0.0026K^{-0.12} - 0.0030K^{-0.30}, \quad (A.32)$$

$$C_{52} = -0.0681K^{0.80} - 0.0071K^{0.42} - 0.0708K^{-0.12} + 0.1460K^{-0.30}, \quad (A.33)$$

$$C_{55} = 0.8509K^{0.80} + 0.0091K^{0.42} + 0.1222K^{-0.12} + 0.0178K^{-0.30}, \quad (A.34)$$

$$C_{56} = 0.1767K^{0.80} + 0.0037K^{0.42} - 0.1704K^{-0.12} - 0.0100K^{-0.30}, \quad (A.35)$$

$$C_{61} = 0.0288K^{0.80} - 0.0475K^{0.42} + 0.0128K^{-0.12} + 0.0059K^{-0.30}, \quad (A.36)$$

$$C_{62} = -0.0503K^{0.80} - 0.0101K^{0.42} + 0.3509K^{-0.12} - 0.2905K^{-0.30}, \quad (A.37)$$

$$C_{65} = 0.6283K^{0.80} + 0.0130K^{0.42} - 0.6059K^{-0.12} - 0.0354K^{-0.30}, \quad (A.38)$$

$$C_{66} = 0.1305K^{0.80} + 0.0052K^{0.42} + 0.8445K^{-0.12} + 0.0198K^{-0.30}. \quad (A.39)$$

References

[1] M. Kobayashi and T. Maskawa, Prog. Theor. Phys. 49 (1973) 652.
[2] S. Pakvasa and H. Sugawara, Phys. Rev. D14 (1976) 305.
[3] L. Maiani, Phys. Lett. 62B (1976) 183.
[4] J. Ellis, M.K. Gaillard and D.V. Nanopoulos, Nucl. Phys. B109 (1976) 213.
[5] L. Wolfenstein, Phys. Rev. Lett. 13 (1964) 562.
[6] F. Gilman and M. Wise, Phys. Lett. 83B (1979) 83.
[7] M.A. Shifman, A.I. Vainshtein and V.I. Zakharov, JETP Lett. 22 (1975) 55; Nucl. Phys. B120 (1977) 316; JETP (Sov. Phys.) 45 (1977) 670.
[8] K. Kleinknecht, Proc. 17th Int. Conf. on High-energy physics, London 1974, ed. J.R. Smith (Science Research Council, Rutherford Laboratory 1974), vol. 3, p. 23.
[9] S.L. Glashow, J. Iliopoulos and L. Maiani, Phys. Rev. D2 (1970) 1285.

[10] G. Altarelli, R.K. Ellis, L. Maiani and R. Petronzio, Nucl. Phys. B88 (1975) 215.
[11] E. Witten, Nucl. Phys. B120 (1977) 387.
[12] B.W. Lee and M.K. Gaillard, Phys. Rev. Lett. 33 (1974) 108;
G. Altarelli and L. Maiani, Phys. Lett. 52B (1976) 351.
[13] F. Gilman and M. Wise, Phys. Rev. D, to be published.
[14] Y. Prokhorov, JETP Lett., to be published.

Nuclear Physics B307 (1988) 19–33
North-Holland, Amsterdam

PREDICTIONS OF THE CKM MODEL FOR *CP* ASYMMETRIES IN B DECAY

P. KRAWCZYK*, D. LONDON, R.D. PECCEI and H. STEGER

Deutsches Elektronen Synchrotron (DESY), Hamburg, FRG

Received 29 December 1987

We discuss present-day uncertainties for the value of the CP-violating phase δ in the CKM matrix and point out how a knowledge of m_t and/or x_s could substantially reduce this uncertainty. A model-independent measurement of δ is, in principle, possible by studying certain CP-violating asymmetries, involving B^0 mesons decaying into CP-conjugate hadronic final states. There exist three different classes of these asymmetries and we give estimates for their values, based on our present knowledge of the CKM matrix. Some comments on the experimental requirements for detecting these asymmetries are also presented.

1. Introduction

In the standard electroweak model, with three generations of quarks and leptons, CP-violating phenomena arise simply from the presence of a nontrivial phase δ in the Cabibbo–Kobayashi–Maskawa (CKM) mixing matrix V_{ij}. Although the standard model cannot explain the deeper origin for this phase, it is obviously very important to know whether or not the observed CP violation in the kaon system arises from the phase δ. All the evidence we have at present, including the recent positive signal of a nonvanishing value for ε'/ε [1], is consistent with this hypothesis. However, the evidence for CP violation being due only to the CKM phase δ is weak, and this phase itself is badly determined.

In the coming years this situation is likely to be improved by new experimental observations. Of particular importance would be a determination of the top quark mass and of the value of the mixing parameter x_s in the B_s–$\overline{B}_s$ system. It may well be that, with these measurements in hand, one will find an inconsistency with the simple CKM mixing scheme. However, even if this turns out not to be the case, there is likely to remain considerable uncertainty attached to the value of δ. This is because we are still unable to calculate the hadronic matrix elements of weak operators reliably and this, obviously, directly affects δ. The purpose of this note is

* Alexander Von Humboldt fellow. On leave of absence from the Institute of Theoretical Physics, Warsaw University, Warsaw, Poland.

0550-3213/88/$03.50 © Elsevier Science Publishers B.V.
(North-Holland Physics Publishing Division)

to discuss critically what future experimental information is most likely to provide clear tests of the CKM scheme and, in particular, will allow for a reliable determination of the *CP*-violating phase δ.

The plan of this paper is as follows. We shall begin by briefly reviewing the present uncertainties in determining the *CP*-violating phase in the CKM matrix, which stem both from our ignorance of a precise value for m_t and of the value of the ratio $|V_{ub}/V_{cb}|$, as well as from the unreliability of hadronic-matrix-element calculations. We shall show, next, that these latter uncertainties, however, can be largely obviated by studying certain classes of *CP*-violating asymmetries in B decays, involving decays of neutral B's into *CP*-conjugate hadronic final states. The non-negligible asymmetries of this type, either time integrated or time dependent, measure one of three possible combinations of phases of the matrix elements V_{ij} (in a convenient parametrization); each of them, of course, being a function of δ. Using the best information available at present on the CKM matrix, we then present ranges of predictions for these important *CP*-violating asymmetries in the B system and draw some conclusions on their likely observability.

2. The mixing matrix

For our purposes, it is convenient to parametrize the CKM matrix in the form suggested by Maiani [2]

$$V=\begin{pmatrix} c_1c_3 & s_1c_3 & s_3e^{i\delta} \\ -s_1c_2-c_1s_2s_3e^{-i\delta} & c_1c_2-s_1s_2s_3e^{-i\delta} & s_2c_3 \\ s_1s_2-c_1c_2s_3e^{-i\delta} & -c_1s_2-s_1c_2s_3e^{-i\delta} & c_2c_3 \end{pmatrix}, \tag{1}$$

where $s_1 \equiv \sin\theta_1$, $c_1 \equiv \cos\theta_1$, etc. Since the angles θ_i are known to have a hierarchical pattern, it is useful to write, following Wolfenstein [3]

$$s_1=\lambda, \qquad s_2=A\lambda^2, \qquad s_3=A\rho\lambda^3, \tag{2}$$

with λ corresponding essentially to the Cabibbo angle, $\lambda \simeq 0.22$. Then, to $O(\lambda^4)$, but keeping for the moment the phase information for each of the elements of the CKM matrix, one has

$$V=\begin{pmatrix} 1-\frac{1}{2}\lambda^2 & \lambda & A\rho\lambda^3e^{i\delta} \\ -\lambda(1+A^2\lambda^4\rho\,e^{-i\delta}) & 1-\frac{1}{2}\lambda^2-A^2\rho\lambda^6e^{-i\delta} & A\lambda^2 \\ A\lambda^3(1-\rho\,e^{-i\delta}) & -A\lambda^2(1+\lambda^2\rho\,e^{-i\delta}) & 1 \end{pmatrix}. \tag{3}$$

One sees immediately that, with this parametrization, only two elements of V can have a significant imaginary part: V_{ub} and V_{td}. It is these phases which will play a crucial role in the B-decay *CP* asymmetries and which should give rise to rather substantial experimental signals. In evaluating the *CP*-violating ε and ε' parameters in the K-sector, however, one needs also to keep track of the "small" phase in V_{cd}. Of course, all *CP*-violating phenomena disappear if the CKM phase $\delta = 0$ or π.

The parameters A and ρ are, in principle, obtainable from B decay. For the estimate of these quantities, we rely (as is commonly done) on the quark picture supplemented by some hadronization model. The ratio*

$$R = \frac{\Gamma(\mathrm{b} \to \mathrm{u})}{\Gamma(\mathrm{b} \to \mathrm{c})} \simeq 2\frac{|V_{bu}|^2}{|V_{bc}|^2} \simeq 2(\lambda\rho)^2, \tag{4}$$

measures ρ, while the B lifetime fixes A. A recent analysis by Altarelli and Franzini [5] gives

$$|V_{bc}|^2 = \lambda^4 A^2 = \frac{(2.9 \pm 0.6) \times 10^{-3}}{\tau_B(10^{-12}\,\mathrm{s})}. \tag{5}$$

Using $\tau_B = (1.11 \pm 0.16) \times 10^{-12}$ s [6], the above implies

$$A = 1.05 \pm 0.17. \tag{6}$$

In the analysis which follows, we shall take A equal to its central value. The ratio R is subject to more theoretical uncertainty which is related to problems with the interpretation of the lepton spectra from semileptonic B decays. From a study of these decays [7], one infers the rather conservative upper bound $R \leq 0.08$, which implies $\rho \leq 0.9$**. On the other hand, the recent observation of charmless B decays by the ARGUS collaboration [9] shows that $V_{ub} \neq 0$ and so provides a lower bound for ρ. Another very conservative analysis [9] gives $\rho \geq 0.3$, so that ρ lies in the range

$$0.3 \leq \rho \leq 0.9. \tag{7}$$

The phase δ is directly, but far from uniquely, determined by the *CP*-violating parameter ε in the K system. A further constraint on δ is also provided by the recent observation of B_d–$\bar{B}_d$ oscillations. The mixing parameter x_d is proportional to $|V_{td}|^2$ and hence it is sensitive to the combination $(1 + \rho^2 - 2\rho\cos\delta)$. In principle, the new data on ε'/ε [1] provide a further constraint on δ. However, the theoretical uncertainties are such that this measurement does not restrict δ beyond the range allowed by ε and x_d [10]. Since m_t is not known, and ρ is only fixed to be in the interval of eq. (7), we will display, below, the allowed values of δ as a function of

* The factor 2 in eq. (4) arises from phase space and QCD correctors. See, for example, ref. [4].

** Theoretical analysis (for a discussion, see ref. [5]), along with the CLEO result [8], indicate that $\rho \leq 0.6$. However, in light of uncertainties in extracting R from experiments, and in eqs. (4) and (5), we will use the conservative upper bound $\rho \leq 0.9$. Since we will plot the allowed region in ρ–δ space, the CLEO bound is easily seen.

these two parameters, indicating, furthermore, the effects of the uncertainties induced by the hadronic matrix elements. Similar analyses have been carried out by a number of different groups recently [11–14].

The standard analysis of Buras et al. [15] gives for $|\varepsilon|$ the formula

$$|\varepsilon| = \frac{G_F^2 f_K^2 M_K M_W^2}{6\sqrt{2}\,\pi^2 \Delta M_K} B_K \left(A^2 \rho \lambda^6 \sin\delta\right)\left(y_c\{\eta_3 f_3(y_c, y_t) - \eta_1\}\right.$$
$$\left. + \eta_2 y_t f_2(y_t) A^2 \lambda^2 (1 - \rho\cos\delta)\right) + \sqrt{\tfrac{1}{2}}\,\xi. \quad (8)$$

Here, $y_i = m_i^2/M_W^2$ and f_2 and f_3 are weakly dependent functions of the top and charm masses

$$f_2(y_t) = 1 - \frac{3y_t(1+y_t)}{4(1-y_t)^2}\left(1 + \frac{2y_t}{1-y_t^2}\ln y_t\right), \quad (9a)$$

$$f_3(y_c, y_t) = \ln\frac{y_t}{y_c} - \frac{3y_t}{4(1-y_t)}\left(1 + \frac{y_t}{1-y_t}\ln y_t\right). \quad (9b)$$

The η_i are QCD correction factors ($\eta_1 \simeq 0.7$, $\eta_2 \simeq 0.6$, $\eta_3 \simeq 0.4$ [16]), while B_K encapsulates our present ignorance of the matrix element of $(\bar{d}\gamma^\mu(1-\gamma_5)s)^2$ between K^0 and $\bar{K}^0$, with $B_K = 1$ corresponding to the vacuum insertion approximation. Finally, ξ is the phase parameter of the $(I = 0)$ $K \to 2\pi$ weak amplitudes ($\xi = \mathrm{Im}\,A_0/\mathrm{Re}\,A_0$). In the quark phase convention which we are using, its presence guarantees that $|\varepsilon|$ is actually convention independent. ξ is directly related to $|\varepsilon'|$ [15] and one has

$$|\varepsilon'| = \frac{1}{\sqrt{2}}\frac{\mathrm{Re}\,A_2}{\mathrm{Re}\,A_0}|\xi| \simeq 0.035|\xi|, \quad (10)$$

where the numerical value above uses experimental information for the kaon amplitudes. The recent determination of $|\varepsilon'/\varepsilon|$ [1] implies that $|\xi| \simeq 0.1|\varepsilon|$. In view of this, and of the other uncertainties in eq. (8), we shall neglect ξ altogether in our analysis.

3. B_d–$\bar{B}_d$ mixing

The observation of B_d–$\bar{B}_d$ mixing by the ARGUS collaboration [17] has provided an independent constraint on the parameters of the Cabibbo–Kobayashi–Maskawa matrix. Since for the B_d system one expects [15] $\Delta\Gamma \ll \Delta M$ and the magnitude of the $\Delta B = 2$ *CP*-violation to be small ($|(1-\varepsilon_d)/(1+\varepsilon_d)| \simeq 1$) the measured ratio,

$$r_d = \frac{\Gamma(B_d \to \ell^- X)}{\Gamma(B_d \to \ell^+ X)} = 0.21 \pm 0.08\,, \tag{11}$$

directly fixes the mixing parameter $x_d = \Delta M/\Gamma$. Using the expression

$$r_d \simeq \frac{x_d^2}{2 + x_d^2} \tag{12}$$

and the ARGUS measurement [17] yields $x_d = 0.73 \pm 0.18$. In our analysis, following Ali [11], we shall use for x_d the 90% confidence limit provided jointly by this measurement and the upper bound of the CLEO collaboration [18]

$$0.78 \geq x_d \geq 0.44\,. \tag{13}$$

Theoretically, x_d receives its dominant contribution by the presence of top quarks in the box diagram and one finds [15, 19]

$$x_d = \tau_B \frac{G_F^2}{6\pi^2} M_B M_W^2 \left(f_{B_d}^2 B_{B_d}\right) \eta_B y_t f_2(y_t) \left\{A^2 \lambda^6 (1 + \rho^2 - 2\rho \cos\delta)\right\}. \tag{14}$$

Here, the hadronic uncertainty is hidden in the factor $f_{B_d}^2 B_{B_d}$, whose meaning is analogous to that of the corresponding quantities in the kaon system, except that here also f_B is not measured. The parameter η_B is a QCD correction factor, which in refs. [15, 19] has been estimated to be $\eta_B \simeq 0.85$. A recent calculation [20], however, including certain higher order QCD effects, obtains a lower value, $\eta_B \simeq 0.63$. We shall adopt this value here, but we note that since $f_{B_d}^2 B_{B_d}$ is quite uncertain, one cannot really tell the difference between these two assumptions. Being rather conservative, we shall allow, for $(f_{B_d}^2 B_{B_d})^{1/2}$, the range 100–200 MeV, which is slightly larger than that used in ref. [5].

We have not included in our analysis the MARK II [21] or UA1 [22] results on B–$\bar{B}$ oscillations, since these experiments cannot distinguish B_s–$\bar{B}_s$ from B_d–$\bar{B}_d$ oscillations. As Ali [11] has pointed out, these results can powerfully constrain the CKM matrix, given a knowledge of the relative amount of B_d and B_s produced. However, these constraints are very dependent on the B_d/B_s production ratio. The experiments measure the quantity χ, which is

$$\chi = P_d \frac{x_d^2}{2(1 + x_d^2)} + P_s \frac{x_s^2}{2(1 + x_s^2)}\,, \tag{15}$$

where P_d (P_s) is the probability that a B_d (B_s) is produced. The MARK II

TABLE 1
The upper bound on x_d from the MARK II data [21], assuming a value of $x_s = 7$, is shown here as a function of the production probabilities P_d and P_s.

P_d	P_s	$(x_d)_{max}$
0.4	0.2	0.35
	0.15	0.55
	0.1	0.74
0.375	0.2	0.37
	0.15	0.57
	0.1	0.78

experiment gives a 90% confidence level upper limit on χ.

$$\chi \leq 0.12. \tag{16}$$

Thus, for a given P_d, P_s and x_s, this gives an upper limit on x_d. In table 1, we have shown this upper limit, for $x_s = 7$ (a typical value), as a function of P_d and P_s. As is evident from the table and eq. (13), depending on the values of P_d and P_s taken, this data can either rule out the standard model (e.g. $P_d = 0.4$, $P_s = 0.2$) or give no bounds whatsoever (e.g. $P_d = 0.375$, $P_s = 0.1$). Given the uncertainty in the information which these experiments provide, we have preferred to be conservative and ignore this information altogether.

Assuming some (typical) values for B_K and $f_{B_d}^2 B_{B_d}$, eqs. (8) and (14) determine δ as a function of m_t and ρ*. For example, taking $B_K = 1$ and $f_{B_d}^2 B_{B_d} = (150 \text{ MeV})^2$ and letting m_t vary from 40 to 180 GeV** gives the "moon-shaped" allowed region in the ρ–δ plane of fig. 1 (a similar analysis has been carried out in ref. [13]). Low values of m_t require that the phase δ be near π [11–14], so as to enhance the $(1 + \rho^2 - 2\rho \cos\delta)$ factor in eq. (14). A substantial portion of the allowed δ values is eliminated when one imposes the lower bound of $\rho > 0.3$. For this choice of theoretical parameters, the observation of charmless B decays [9] cuts the moon in half, eliminating all δ-values below $\delta \leq 2.7$! This effect is still present, although less sharply so, when one allows variations in the theoretically uncertain parameters over sensible ranges: $1/3 \leq B_K \leq 1$; $(100 \text{ MeV})^2 \leq f_{B_d}^2 B_{B_d} \leq (200 \text{ MeV})^2$***. This is demonstrated in fig. 2. We see that for $B_K = 1$, but ranging over $f_{B_d}^2 B_{B_d}$, the moon shaped region of fig. 1 expands. This expansion grows further as B_K is lowered to $\frac{2}{3}$ and a considerable portion of the ρ–δ plane is filled if $B_K = \frac{1}{3}$. In the analysis of

* The measurement of ε and the allowed range for ρ constrain $\sin\delta$ to be positive.

** While the lower bound on m_t follows from the internal consistency of the presented analysis and refs. [11–14], the upper bound results from the study of radiative corrections within the standard model (see ref. [23a]); a bound of $m_t \leq 180$ GeV at 90% confidence limit is obtained in ref. [23b].

*** These ranges are extensive enough that they compensate any reasonable variation in A.

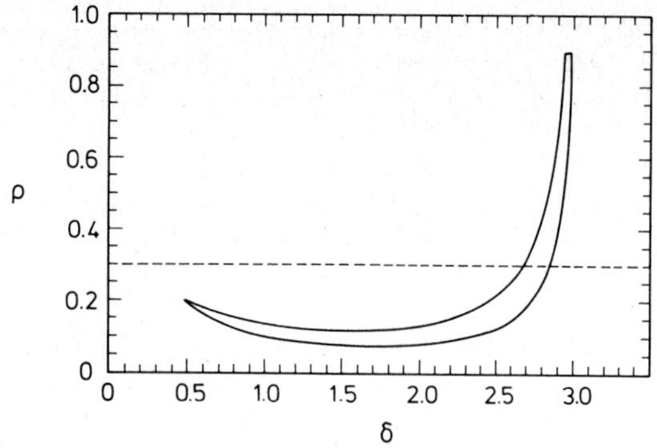

Fig. 1. The domain in ρ–δ space (δ in radians), within which the standard model is compatible with the measurements of ε and x_d (90% confidence limit). We vary m_t between 40 GeV and 180 GeV and use $(f_{B_d}^2 B_{B_d})^{1/2} = 150$ MeV and $B_K = 1$. The dashed line represents the lower bound $\rho \geq 0.3$ inferred from the observation of charmless B decays by ARGUS [9].

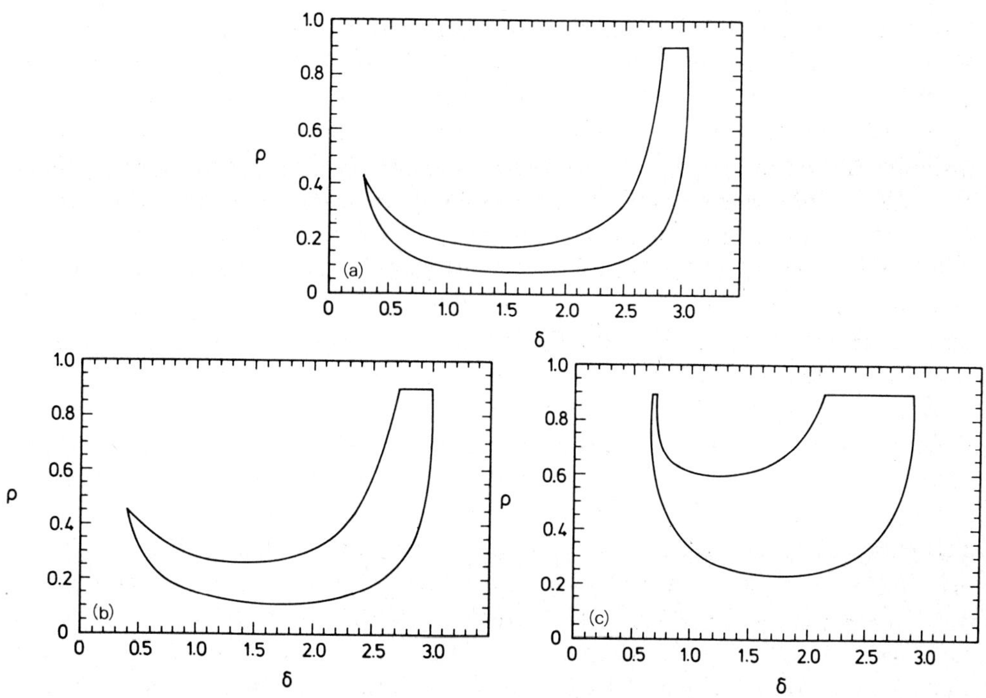

Fig. 2. As in fig. 1, but now, in addition, $(f_{B_d}^2 B_{B_d})^{1/2}$ is varied between 100 MeV and 200 MeV: (a) $B_K = 1$, (b) $B_K = \frac{2}{3}$, (c) $B_K = \frac{1}{3}$.

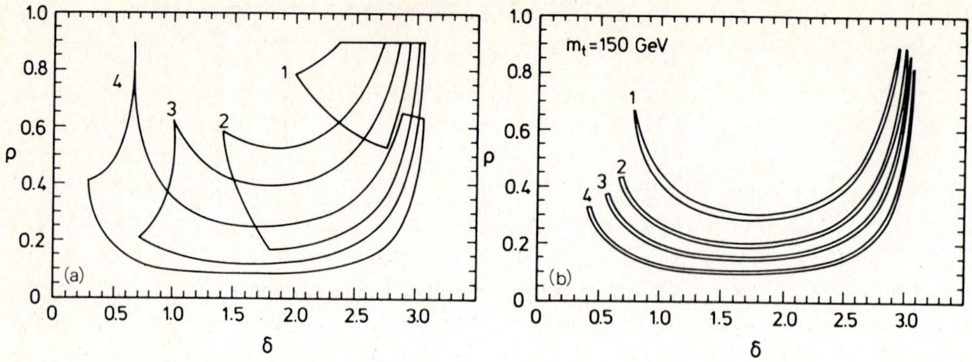

Fig. 3. As in fig. 1, but now m_t is kept fixed while B_K and $(f_{B_d}^2 B_{B_d})^{1/2}$ are allowed to vary within the ranges $\frac{1}{3} \le B_K \le 1$ and 100 MeV $\le (f_{B_d}^2 B_{B_d})^{1/2} \le$ 200 MeV. (a) The areas 1, 2, 3, 4 correspond to fixed values m_t = 60, 90, 120, 180 GeV, respectively. (b) Taking m_t = 150 GeV, the strips 1, 2, 3, 4 correspond to $B_K = \frac{1}{3}, \frac{1}{2}, \frac{2}{3}, 1$, respectively.

CP-violating phenomena in the B-system, we shall take, for definitiveness, $B_K = \frac{2}{3}$, but shall continue to allow m_t and $f_{B_d}^2 B_{B_d}$ to vary over the ranges indicated above.

A direct measurement of m_t and/or of B_s–$\bar{B}_s$ oscillations would do much to clarify the above situation, even if ρ cannot be restricted better than eq. (7). This is illustrated in fig. 3a, where we show the allowed ρ–δ ranges for four values of m_t (m_t = 60, 90, 120, 180 GeV), for the full uncertain theoretical ranges. This uncertainty comes mainly from our lack of knowledge of B_K, as is demonstrated in fig. 3b. There, we plot the allowed areas corresponding to fixed values of $B_K = \frac{1}{3}, \frac{1}{2}, \frac{2}{3}$ and 1, for m_t = 150 GeV. In the SU(3) limit we have $\tau_{B_d} M_{B_d} \eta_{B_d} f_{B_d}^2 B_{B_d} =$

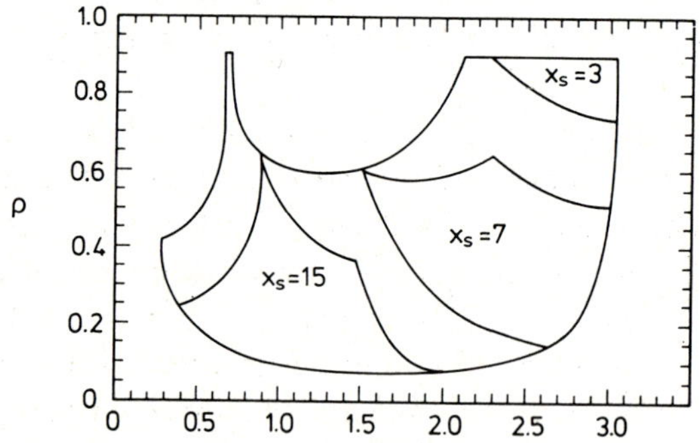

Fig. 4. As in fig. 1, but now all three parameters m_t, $(f_{B_d}^2 B_{B_d})^{1/2}$ and B_K are varied: 40 GeV $\le m_t \le$ 180 GeV, 100 MeV $\le (f_{B_d}^2 B_{B_d})^{1/2} \le$ 200 MeV and $\frac{1}{3} \le B_K \le 1$. In addition, the further restrictions given by fixed x_s = 3, 7, 15 are shown.

$\tau_{B_s} M_{B_s} \eta_{B_s} f_{B_s}^2 B_{B_s}$ and thus the ratio x_d/x_s is given by

$$\frac{x_d}{x_s} = \lambda^2 (1 + \rho^2 - 2\rho \cos \delta) . \tag{17}$$

Even if $f_{B_s}^2 B_{B_s}$ should turn out to differ from $f_{B_d}^2 B_{B_d}$ by a factor of 2, a measurement of x_s would still give substantial restrictions in the ρ–δ plane. To illustrate this point, in fig. 4, using eq. (17), we plot the constraints on the allowed ρ–δ range, for three values of x_s ($x_s = 3, 7, 15$), using the range of eq. (13) for x_d. Because fixing x_s gives correlated ranges for $f_{B_d}^2 B_{B_d}$ and m_t, the interval (13) for x_d is not always fully allowed. This has been taken into account in fig. 4. Thus, a measurement of x_s is particularly constraining for the Cabibbo–Kobayashi–Maskawa model and we very much hope it will be attempted at Cornell (or DESY?).

4. *CP*-violating asymmetries

Because B mesons possess many more decay channels than K mesons, there exist a considerable variety of *CP*-violating phenomena that one can search for experimentally. This subject, naturally, has generated intense theoretical interest and has, in some sense, been fully explored [24–27]. However, most of the investigations to date have been "broad band", concentrating on the totality of the phenomena, without looking at any one decay, or class of decays, in detail. Furthermore, in many instances, predictions are given only for what the *maximum* signal of *CP* violation could be. Thus, many optimistic dynamical assumptions are made and the Cabibbo–Kobayashi–Maskawa parameters are stretched to their limits [26, 28]. Here, we would like to take a rather more "narrow band" approach, by concentrating on *CP* asymmetries which are essentially independent of theoretical assumptions. Furthermore, we want to predict what are reasonable expectations for these asymmetries, based on the constraints which we know today already exist for the CKM matrix elements.

To observe *CP*-violating effects in B decays requires that there should be interference between two amplitudes with different phases. Because the B and the $\overline{B}$ states mix, if one looks at decays of B mesons into a final state f which can be reached by both B^0 and $\overline{B}^0$ decays, then the required interference exists. The interesting asymmetry to consider is the difference between the decay probability of a state which at $t = 0$ started as B^0 – denoted here by $B^0(t)$ – into f, compared to the decay probability of $\overline{B}^0(t)$ into $\bar{f}$ (for a recent discussion, see [25, 26]). One finds the time integrated asymmetry

$$A_f = \frac{\Gamma(B^0(t) \to f) - \Gamma(\overline{B}^0(t) \to \bar{f})}{\Gamma(B^0(t) \to f) + \Gamma(\overline{B}^0(t) \to \bar{f})} = -\frac{2x \,\mathrm{Im}\, \lambda_f}{2 + x^2 + x^2 |\rho_f|^2} . \tag{18}$$

Here, x is the mixing parameter of the B meson, while

$$\rho_f = \frac{A(\overline{B}^0 \to f)}{A(B^0 \to f)}, \tag{19}$$

$$\lambda_f = \frac{1-\varepsilon_B}{1+\varepsilon_B}\rho_f, \tag{20}$$

where ε_B is the analogue of ε for the B system. This asymmetry becomes *independent* of strong interaction effects if f is a *CP* eigenstate ($\bar{f} = \pm f$), and the weak-decay process is dominated by just *one* amplitude* [27]. In this case $|\rho_f| = 1$ and eq. (18) reduces to

$$A_f = -\frac{x}{1+x^2}\,\mathrm{Im}\,\lambda_f. \tag{21}$$

Observe that the sign of A_f depends on the *CP* eigenvalue of f; thus, final states with opposite *CP* properties give rise to asymmetries of opposite sign [25]. Various comments are in order:

(i) The asymmetry A_f vanishes, either in the case of no mixing ($x \to 0$) or full mixing ($x \to \infty$). For B_d, eq. (13) puts one almost in an ideal situation since $x_d/(1+x_d^2) \simeq 0.5$. For B_s, on the other hand, the situation is less favourable. Using eq. (17), our analysis suggests x_s extends over the range: $3 \leq x_s \leq 20$, so that $0.3 \geq x_s/(1+x_s^2) \geq 0.05$.

(ii) The magnitude of the factor $(1-\varepsilon_B)/(1+\varepsilon_B)$ in eq. (20) is very nearly unity [15]. However, one must be careful about its phase, since only by including this phase information will $\mathrm{Im}\,\lambda_f$ be independent of the phase convention adopted. In the quark phase convention we are using, since the top-quark graph totally dominates, one finds simply that

$$\frac{1-\varepsilon_B}{1+\varepsilon_B} = \begin{cases} \dfrac{V_{tb}^*}{V_{tb}}\dfrac{V_{td}}{V_{td}^*} = \dfrac{V_{td}}{V_{td}^*} \equiv e^{2i\phi}, & (B_d), \\[2ex] \dfrac{V_{tb}^*}{V_{tb}}\dfrac{V_{ts}}{V_{ts}^*} \simeq 1, & (B_s), \end{cases} \tag{22}$$

where the second line follows from the form of our CKM matrix, eq. (3), in which only two elements have non-negligible phases, V_{ub} and V_{td}.

(iii) Since $|\rho_f| = 1$, ρ_f itself is also a pure phase. In fact, since only one weak-decay amplitude enters by assumption, ρ_f is a ratio of two Cabibbo–Kobayashi–Maskawa

* This will happen, in general, if the quark subprocesses in the decay do not contain both a u and a $\bar{c}$ quark. Decays where the quark subprocess involves, for instance, $b \to u\bar{u}d$ or $b \to c\bar{c}s$ (e.g. $B_d \to \pi^+\pi^-$, $B_d \to \Psi K_S$) are examples where one expects $|\rho_f| = 1$.

matrix elements (times a CP-sign, which we take to be positive in what follows). The ratio of CKM matrix elements in ρ_f containing only light quarks (u, d, s, c), with the convention of eq. (3), is essentially real and unity. Thus, it can be ignored in the following. Since V_{cb} is real, only for Cabibbo suppressed decays will ρ_f involve a phase.

$$\rho_f \simeq \begin{cases} \dfrac{V_{ub}}{V_{ub}^*} \equiv e^{2i\delta} & \text{(Cabibbo suppressed)}, \\ \dfrac{V_{cb}}{V_{cb}^*} = 1, & \text{(Cabibbo allowed)}. \end{cases} \tag{23}$$

The above simple considerations tell us that, in B decays, there are three classes of *model-independent* asymmetries which can be sizeable. Each of these classes measures a *different* combination of phases of the CKM matrix elements – all, of course, related ultimately to δ:

(1) Cabibbo-allowed B_d decays (e.g. $B_d \to \Psi K_S$ [29]),

$$\operatorname{Im} \lambda_1 \simeq \operatorname{Im} \frac{V_{td}}{V_{td}^*} = \sin 2\phi = \frac{2\rho \sin\delta(1-\rho\cos\delta)}{1+\rho^2-2\rho\cos\delta}; \tag{24}$$

(2) Cabibbo-suppressed B_d decays (e.g. $B_d \to \pi^+\pi^-$ [25]),

$$\operatorname{Im} \lambda_2 \simeq \operatorname{Im} \frac{V_{td}}{V_{td}^*}\frac{V_{ub}}{V_{ub}^*} = \sin 2(\phi+\delta) = \frac{2\sin\delta(\cos\delta-\rho)}{1+\rho^2-2\rho\cos\delta}; \tag{25}$$

(3) Cabibbo-suppressed B_s decays (e.g. $B_s \to \rho^0 K_S$ [14]),

$$\operatorname{Im} \lambda_3 \simeq \operatorname{Im} \frac{V_{ub}}{V_{ub}^*} = \sin 2\delta = 2\sin\delta\cos\delta. \tag{26}$$

It is obviously very interesting to know what ranges of $\operatorname{Im}(\lambda_i)$ are allowed by present data. The relevant plots are presented in fig. 5, where $\operatorname{Im}(\lambda_i)$ $(i = 1, 2, 3)$ is plotted against ρ, in the range $0.3 \le \rho \le 0.9$. In these graphs we have let m_t range from 40 to 180 GeV, have fixed $B_K = \frac{2}{3}$ and let 100 MeV $\le (f_{B_d}^2 B_{B_d})^{1/2} \le$ 200 MeV. Fig. 6 presents the same quantities but now for specific m_t values (m_t = 60, 90, 120, 150, 180 GeV).

One sees from figs. 5b, c that $\operatorname{Im}(\lambda_2)$ and $\operatorname{Im}(\lambda_3)$, for $0.3 \le \rho \le 0.9$, can take on rather large values. For $\operatorname{Im}(\lambda_1)$, on the other hand, values greater than ~ 0.4 appear to be excluded. The actual measured asymmetries are, however, reduced by the mixing factor $x/(1+x^2)$. This is, at least, a factor of 2 for B_d and could be near a factor of 10 for B_s. From this viewpoint, therefore, the most promising processes

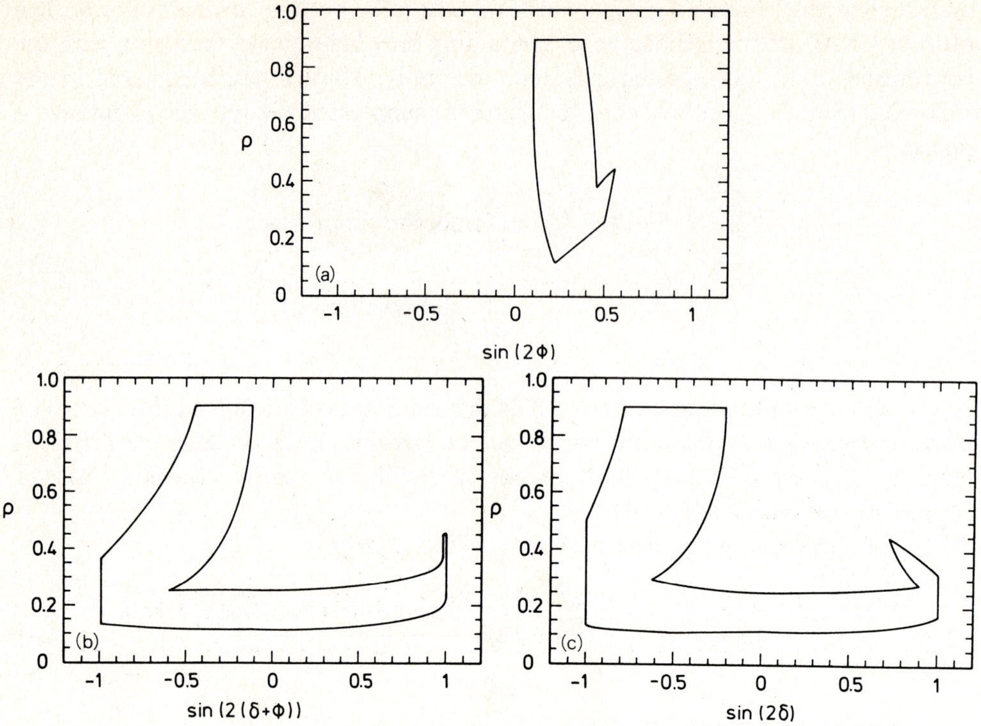

Fig. 5. Varying $(f_{B_d}^2 B_{B_d})^{1/2}$ and m_t, 100 MeV $\leq (f_{B_d}^2 B_{B_d})^{1/2} \leq$ 200 MeV and 40 GeV $\leq m_t \leq$ 180 GeV, and fixing $B_K = \frac{2}{3}$, the areas within which the standard model is compatible with the measurements of ε and x_d (90% confidence limit) are shown for the following parameter spaces: (a) $(\sin 2\phi, \rho)$, (b) $(\sin 2(\delta + \phi), \rho)$, (c) $(\sin 2\delta, \rho)$.

appear to involve $\mathrm{Im}(\lambda_2)$. However, since these asymmetries concern Cabibbo-suppressed B_d decays, this overall rate is going to be considerably smaller. For instance, the branching ratio $B_d \to \pi^+\pi^-$ is probably of $O(10^{-5})$, while we know that $BR(B_d \to \Psi K_S)$ is of $O(10^{-3})$. If it is possible to follow the time development of the B decays [24, 27, 30], then one gets rid of the reduction factor $x/(1+x^2)$, since the probability of obtaining a state f at time t, for a beam which at $t = 0$ was pure B^0, is simply

$$N_f(t) = N_f(0) e^{-\gamma t} [1 - \mathrm{Im}\, \lambda_f \sin \Delta m t]. \quad (27)$$

If one has a large mixing parameter, $x = \Delta m / \gamma$, as is likely to be the case for B_s, then the non-exponential behaviour of eq. (27) should be visible, provided of course that one can track the decay at all.

It is difficult to estimate the number of B^0 decays needed to perform the *CP*-violation tests we have discussed. First of all, these asymmetries A_f require that one know if the decaying B was originally a B^0 or a $\overline{B}^0$. To determine this, perhaps

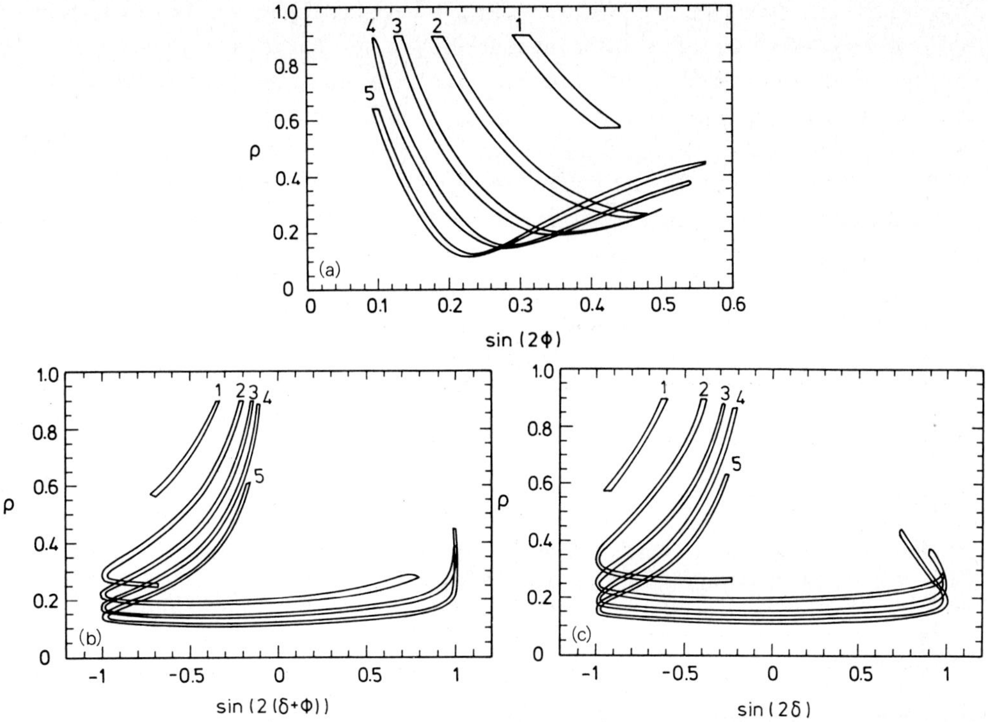

Fig. 6. As in figs. 5a, b, c, but now the strips 1, 2, 3, 4, 5 correspond to fixed values of m_t: 60, 90, 120, 150, 180 GeV, respectively.

the best method is to try to establish the charge of the associated B [27]. This requires looking for another secondary vertex, besides that of the original decaying B. Even being optimistic, this should cost at least a factor of 10. Consider the decay $B_d \to \pi^+\pi^-$ and imagine $\mathrm{Im}(\lambda_2) \simeq -0.5$, so that $A_{\pi^+\pi^-} \simeq +0.25$. Establishing this asymmetry at the 3σ level requires approximately 150 tagged $B_d(t) \to \pi^+\pi^-$ events, which, with $\mathrm{BR}(B_d \to \pi^+\pi^-) \simeq 10^{-5}$ and a tagging efficiency of 10%, calls for 10^8 B decays. This number is quite typical and appears discouraging*. Perhaps it is more important, therefore, to look for final states with clear experimental signals. The decay $B_d \to p\bar{p}$**, for instance, whose branching ratio should also be of $O(10^{-5})$, appears very interesting. However, a cautionary remark is in order. Since the $p\bar{p}$ pair in the final state can either be in a *p*- or an *s*-wave configuration, which have opposite *CP* eigenvalues, one may expect a large cancellation of the asymmetry,

* One should, of course, do a detailed study for any given process, before quoting a definitive number of B mesons needed. However, we are skeptical of optimistic statements in the literature [31].

** This decay was suggested by Harari [32a], but see also ref. [32b]. The relevance of B decays into baryonic final states has also been stressed (although not in the context of *CP* violation) by Stech [32c].

unless there is a dynamical suppression of the p-wave configuration. In this respect, the final state $\pi^+\pi^-$ is safe since it involves only spin-0 particles.

Although the number of 10^8 B's is unpleasant to countenance, perhaps we should point out that the situation would be much worse if the predicted asymmetries were below 10%. Fortunately, as fig. 5 shows, $\mathrm{Im}(\lambda_i)$ in the standard model seems to be well away from this unfortunate region. Obviously, as fig. 6 shows, a knowledge of m_t would allow a much more restricted prediction for these CP-violating asymmetries in B decays.

We would like to thank A. Ali and P. Langacker for helpful conversations.

References

[1] I. Mannelli, Nucl. Phys. B (Proc. Suppl.) 3 (1988) 367
[2] L. Maiani, Phys. Lett. B62 (1976) 183
[3] L. Wolfenstein, Phys. Rev. Lett. 51 (1983) 1945
[4] G. Altarelli, N. Cabibbo, G. Corbo, L. Maiani and G. Martinelli, Nucl. Phys. B208 (1982) 365
[5] G. Altarelli and P. Franzini, CERN preprint CERN TH4745/87
[6] D.M. Ritson, *in* Proc. XXIII International Conference on High energy physics (Berkeley, USA, 1986)
[7] M. Gilchriese, *in* Proc. XXIII International Conference on High energy physics (Berkeley, USA, 1986)
[8] A. Chen et al. (CLEO Collaboration), Phys. Rev. Lett. 52 (1984) 1084
[9] W. Schmidt-Parzefall, Nucl. Phys. B (Proc. Suppl.) 3 (1988) 257
[10] G. Altarelli and P. Franzini, CERN preprint CERN TH4914/87;
J. Ellis, J. Hagelin, S. Rudaz and D.D. Wu, CERN preprint CERN TH4816/87
[11] A. Ali, DESY preprint DESY 87-083
[12] G. Altarelli and P. Franzini, ref. [5];
J. Donoghue, T. Nakada, E. Paschos and D. Wyler, Phys. Lett B195 (1987) 285;
J. Ellis, J. Hagelin and S. Rudaz, Phys. Lett. B192 (1987) 201;
I.I. Bigi and A.I. Sanda, Phys. Lett. B194 (1987) 307;
J. Maalampi and M. Roos, Phys. Lett. B195 (1987) 489;
V. Barger, T. Han, D.V. Nanopoulos and R.J.N. Phillips, Phys. Lett. B194 (1987) 312
[13] H. Harari and Y. Nir, Phys. Lett. B195 (1987) 586;
Y. Nir, SLAC-PUB-4368, July 1987
[14] V.A. Khoze and N.G. Uraltsev, Leningrad preprint 1290, June (1987)
[15] A.J. Buras, W. Slominski and H. Steger, Nucl. Phys. B238 (1984) 529; B245 (1984) 369
[16] F.J. Gilman and M.B. Wise, Phys. Rev. D27 (1983) 1128
[17] H. Albrecht et al. (ARGUS Collaboration), Phys. Lett. B192 (1987) 245
[18] A. Bean et al. (CLEO Collaboration) Phys. Rev. Lett. 58 (1987) 183
[19] J. Hagelin, Nucl. Phys. B193 (1981) 123
[20] W.A. Kaufman, H. Steger and Y.P. Yao, University of Michigan preprint UM-TH-87-13, July (1987)
[21] T. Schaad et al. (MARK II Collaboration), Phys. Lett. B160 (1985) 188
[22] C. Albajar et al. (UAI Collaboration), Phys. Lett. B186 ((87) 427
[23] (a) M. Veltman, Nucl. Phys. B123 (1977) 89;
(b) U. Amaldi et al., Phys. Rev. D36 (1987) 1385
[24] Y. Azimov, V.A. Khoze and N.G. Uraltsev, Yad. Fiz. 45 (1987) 1412
[25] I.I. Bigi and A.I. Sanda, Nucl. Phys. B281 (1987) 41
[26] D. Du, I. Dunietz and D. Wu, Phys. Rev. D34 (1986) 3414

[27] I. Dunietz and J. Rosner, Phys. Rev. D34 (1986) 1404; I.I. Bigi and A.I. Sanda, ref. [25]
[28] L.L. Chau and H.Y. Cheng, Phys. Rev. Lett. 59 (1987) 958
[29] I.I. Bigi and A.I. Sanda, Nucl. Phys. B193 (1981) 85
[30] I.I. Bigi and A.I. Sanda, Comments Nucl. Part. Phys. 14 (1985) 149
[31] D. Du and Z. Zhao, Phys. Rev. Lett. 59 (1987) 1072
[32] (a) H. Harari, *in* 1987 Int. Symposium on Lepton and photon interactions at high energy, (Hamburg, 1987), unpublished;
(b) R.D. Peccei, *in* Proc. Workshop on the experimental program at UNK (Protvino, September, 1987);
(c) B. Stech, Heidelberg preprint HD-THEP-87-19

Erratum

There is an inconsistency in the definitions of the phases δ and ϕ and the result presented in the figures. All the figures are correct if one replaces in eqs. (1) and (3) everywhere δ by $-\delta$ and if one replaces in eq. (22) ϕ by $-\phi$.

Nuclear Physics B193 (1981) 85–108

NOTES ON THE OBSERVABILITY OF *CP* VIOLATIONS IN B DECAYS

I.I. BIGI

Institut für Theor. Physik der RWTH Aachen, D-5100 Aachen, FR Germany

A.I. SANDA[1]

Rockefeller University, New York 10021, USA

Received 16 June 1981

We describe a general method of exposing *CP* violations in on-shell transitions of B mesons. Such *CP* asymmetries can reach values of the order of up to 10% within the Kobayashi–Maskawa model for plausible values of the model parameters. Our discussion focuses on those (mainly non-leptonic) decay modes which carry the promise of exhibiting clean and relatively large *CP* asymmetries at the expense of a reduction in counting rates. Accordingly we address the complexities encountered when performing *CP* tests with a high statistics B meson factory like the Z^0 (and a toponium) resonance.

1. Introduction

Our knowledge on *CP* violations is quite unsatisfactory – both experimentally, since they have so far been detected only in the decays of neutral kaons, and theoretically. It is of crucial importance for our understanding of this fundamental phenomenon to observe it also in other systems. It had been realized some time ago that the decays of bottom (=B) mesons might provide ample opportunities to study *CP* violations [1, 2]. Yet a careful analysis yields a somewhat sobering result: when one looks for *CP* violations in B^0 decays in the same way as one did in K^0 decays [3], namely via searching for *CP* impurities in B^0-$\bar{B}^0$ mixing, one has to chase after very tiny effects. Implementing *CP* violation, for example, via a complex Kobayashi-Maskawa (KM) mixing matrix [4] one predicts asymmetries of the order of 10^{-4}–10^{-3} [2]. While such a minute effect has been established for K^0 mesons in a celebrated experiment [5], the prospects appear rather dim for repeating this remarkable feat in B decays. In any case, considering the importance of the question under discussion, we feel that it should be approached from many different angles.

In ref. [6] a new method was introduced to search for *CP* violations in *non-leptonic* B meson decays. The basic idea was to study a final state which can be fed both from B^0 and $\bar{B}^0$ mesons. Assuming sizeable B^0-$\bar{B}^0$ mixing (as is commonly done),

[1] Supported in part by the US Department of Energy under contract grant number DE-ACO2-81ER4003.B00.

one obtains interference effects which can exhibit a *CP*-violating phase. Unfortunately the numbers presented in ref. [6] were derived for a somewhat unrealistic choice of hadronic final states. Including other hadronic states to make the experimental test feasible leads to some cancellations in the asymmetry. The reduction factor depends on the branching ratios of various exclusive final states and we cannot compute it at present. The asymmetry can, however, be reduced by as much as a factor of ten.

Therefore, we feel that this analysis should be extended in three respects:

(i) In ref. [6] the decays

$$\begin{array}{l} B_d^0 \to D^0 + \pi\text{'s} \searrow \\ \qquad\qquad\qquad\qquad K_S + \pi\text{'s} \\ \bar{B}_d^0 \to \bar{D}^0 + \pi\text{'s} \nearrow \end{array}$$

were discussed. We will emphasize that transitions like

$$\begin{array}{l} B_d^0 \searrow \\ \qquad \psi K_S + X,\ D\bar{D}[F^+F^-]K_S + X\,, \\ \bar{B}_d^0 \nearrow \end{array}$$

$$\begin{array}{l} B_s^0 \searrow \\ \qquad \psi + X,\ D\bar{D}[F^+F^-] + X\,, \\ \bar{B}_s^0 \nearrow \end{array}$$

should be studied where it is *not* necessary to reconstruct the D and F meson decays.

(ii) We will describe how the general method can be applied not only near bottom threshold, but also at energies well above it, in particular on the eagerly awaited Z^0 resonance in e^+e^- annihilation.

(iii) Large mixing being an essential ingredient of the method is expected for B^0-$\bar{B}^0$ systems, but not for T^0-$\bar{T}^0$ top mesons. Nevertheless, top meson decays anticipated as leading mainly to bottom states can quite conveniently be employed in our analysis.

The paper will be organized as follows: in sect. 2 we recapitulate briefly the mechanism underlying the *CP* asymmetry described in ref. [6]. In sect. 3 we make our discussion more quantitative by using more specific approaches towards *CP* violation, namely the Kobayashi–Maskawa construction, an ansatz based on Higgs exchanges and the superweak model. In sect. 4 we show how the modes $B \to \psi + X$ and $B \to D\bar{D} + X$ can profitably be employed in our enterprise. In sect. 5 we discuss direct B production in e^+e^- annihilation on the Z^0 resonance while in sect. 6 we include effects generated by top mesons serving as a B meson factory. Sect. 7 finally contains our conclusions. In the appendix we add some comments on the relationship between mixing and dilepton events.

2. The general method

In this section we will briefly recapitulate how violations of *CP* invariance can be exhibited in decay reactions. The arduous student of this fascinating subject can find more details in many excellent reviews [2, 3, 6, 7].

CP invariance is violated on the lagrangian level if one (or several) coupling constants are complex leading to non-trivial phases in amplitudes which describe for example decay processes. Yet phases are elusive quantities. Only relative phases being observable, one has to look for reactions to which (at least) two different amplitudes contribute, which are comparable in size.

We list three possibilities to perform *CP* tests:

(i) One searches for a decay channel which has different strength for particle and antiparticle:

$$\Gamma(\mathrm{i} \to \mathrm{f}) \neq \Gamma((\mathrm{i})_{CP} \to (\mathrm{f})_{CP}) . \tag{2.1}$$

Two conditions have to be met to make it a useful *CP* test:

(a) As said above one needs two different decay modes leading to the same final state.

(b) In order to exhibit the *CP*-violating phase stemming for example from the weak mixing matrix (see sect. 3) one has to rely on strong final-state interactions. Otherwise *CPT* already enforces equality of these exclusive rates.

While condition (b) makes it very hard to give quantitative predictions (see, however, ref. [8]), it follows from condition (a) that one has to deal with transitions suppressed by presumably small mixing angles [6].

(ii) Neutral mesons can show the phenomenon of mixing. The "classical", i.e. so far only example of it is provided by the K^0-$\bar{K}^0$ system. Since mixing requires two amplitudes to interfere the mass eigenstates, like K_L and K_S, can contain *CP* impurities. These can be sampled by searching for $K_L \to \pi\pi$ or $B^0\bar{B}^0 \to \ell^+\ell^+ + X$, $\ell^-\ell^- + X$ with the two like-sign dilepton yields being unequal. Unfortunately this dilepton signal is expected to be very small, of the order of 10^{-4}–10^{-3} [2, 6].

(iii) To avoid the shortcomings of methods (i) and (ii), Carter and Sanda put forward a new suggestion [6] which we will describe in some detail since our subsequent analysis depends on it.

It is also based on mixing as an essential ingredient, yet it does *not* depend on *CP* impurities in it. *CP* violation originates in the *non*-leptonic decays of the mixed states, while mixing is required only to have interference between two different amplitudes. We will generically phrase our discussion for B mesons since it is obvious how to apply it to other cases.

By B^0 [$\bar{B}^0$] we denote mesons containing a bottom b quark [$\bar{b}$ antiquark] which are eigenstates under the strong interactions. In general they are *not* the asymptotic mass eigenstates and their time development is described by a non-diagonal hamiltonian matrix:

$$H\begin{pmatrix} B^0 \\ \bar{B}^0 \end{pmatrix} = \begin{pmatrix} M - \frac{1}{2}i\Gamma & M_{12} - \frac{1}{2}i\Gamma_{12} \\ M_{12}^* - \frac{1}{2}i\Gamma_{12}^* & M - \frac{1}{2}i\Gamma \end{pmatrix} \begin{pmatrix} B^0 \\ \bar{B}^0 \end{pmatrix} . \tag{2.2}$$

The mass eigenstates B_1 and B_2 are found by diagonalizing the mass matrix (2.2),

$$B_{\frac{1}{2}} = \frac{(1+\varepsilon)B^0 \pm (1-\varepsilon)\bar{B}^0}{\sqrt{2(1+|\varepsilon|^2)}} \equiv pB^0 \pm q\bar{B}^0 , \tag{2.3}$$

where

$$\frac{p}{q} = \frac{1+\varepsilon}{1-\varepsilon} = \sqrt{\frac{M_{12} - \frac{1}{2}i\Gamma_{12}}{M_{12}^* - \frac{1}{2}i\Gamma_{12}^*}} . \tag{2.4}$$

A state of pure B^0 or $\bar{B}^0$ prepared by the strong interactions at $t = 0$ evolves into a superposition at later times t:

$$|B^0(t)\rangle = f_+(t)|B^0\rangle + \frac{q}{p} f_-(t)|\bar{B}^0\rangle , \tag{2.5a}$$

$$|\bar{B}^0(t)\rangle = \frac{p}{q} f_-(t)|B^0\rangle + f_+(t)|\bar{B}^0\rangle , \tag{2.5b}$$

with

$$f_\pm(t) = \tfrac{1}{2}\{\exp[it(-m_1 + \tfrac{1}{2}i\Gamma_1)] \pm \exp[it(-m_2 + \tfrac{1}{2}i\Gamma_2)]\} , \tag{2.6}$$

$$\Gamma = \Gamma_1 + \Gamma_2 ,$$

$$\Delta\Gamma = \Gamma_1 - \Gamma_2 = -4 \,\mathrm{Im}\, \sqrt{(M_{12} - \tfrac{1}{2}i\Gamma_{12})(M_{12}^* - \tfrac{1}{2}i\Gamma_{12}^*)} , \tag{2.7}$$

$$\Delta m = m_1 - m_2 = 2 \,\mathrm{Re}\, \sqrt{(M_{12} - \tfrac{1}{2}i\Gamma_{12})(M_{12}^* - \tfrac{1}{2}i\Gamma_{12}^*)} .$$

One picks a final state f into which both B^0 and $\bar{B}^0$ can decay. Later we will discuss examples of it. For now we write:

$$M = \langle f|H|B^0\rangle , \tag{2.8a}$$

$$\bar{M} = \langle f|H|\bar{B}^0\rangle . \tag{2.8b}$$

We choose the phase of the *CP* operator such that $CP|B^0\rangle = -|\bar{B}^0\rangle$ and $CP|f\rangle = |f\rangle$. If *CP* were a good symmetry, we would have $M = -\bar{M}$. In sect. 3 we will describe interactions where this does not hold.

The amplitudes for states *initially pure* B^0 and $\bar{B}^0$ to decay to f at time t are

$$\langle f|H|B^0(t)\rangle = f_+(t)M + f_-(t)\frac{q}{p}\bar{M} \equiv \frac{q}{p}\bar{M}(\lambda f_+(t) + f_-(t)) , \tag{2.9a}$$

$$\langle f|H|\bar{B}^0(t)\rangle = f_-(t)\frac{p}{q}M + f_+(t)\bar{M} \equiv \bar{M}(\lambda f_-(t) + f_+(t)) , \tag{2.9b}$$

where

$$\lambda = pM/q\bar{M}.$$

The lifetimes and mixing times of B mesons are so short that experimental measurements are performed essentially at $t \sim \infty$. Therefore, we have to integrate over time up to infinity to obtain a useful, since accessible, quantity.

The widths Γ, $\bar{\Gamma}$ for initially pure states B^0, $\bar{B}^0$ to decay into the final state f are obtained by squaring (2.9a, b) and integrating from $t = 0$ to ∞:

$$\begin{Bmatrix} \bar{\Gamma} \\ \Gamma \end{Bmatrix} = \text{const} \cdot \begin{Bmatrix} |p|^2 \\ |q|^2 \end{Bmatrix} [1 \pm a + |\lambda|^2(1 \mp a) - 2 \,\mathrm{Re}\, \lambda (y \pm ixa)], \tag{2.10}$$

where

$$y = \frac{\Delta\Gamma}{\Gamma}, \qquad x = \frac{2\Delta m}{\Gamma}, \qquad a = \frac{1-y^2}{1+x^2}.$$

Thus, we can display a CP asymmetry between the conjugate decay widths $\Gamma = \Gamma(B^0 \to f)$ and $\bar{\Gamma} = \Gamma(\bar{B}^0 \to f)$:

$$\frac{\Gamma - \bar{\Gamma}}{\Gamma + \bar{\Gamma}} = -\frac{xa \sin 2\phi}{1 + y \cos 2\phi}, \tag{2.11}$$

with $\lambda \equiv -e^{-2i\phi}$

Since we think it unlikely that such tests can be performed in any reaction but e^+e^- annihilation, we cannot content ourselves to deal only with single B^0 or $\bar{B}^0$ production, but we have to consider $B\bar{B}$ pair production.

Attributing an appreciable production probability to $B_u = (b\bar{u})$, $B_d = (b\bar{d})$ and on a somewhat reduced level to $B_s = (b\bar{s})$ mesons, but not to $B_c = (b\bar{c})$, η_b etc. states we have to consider four different cases at energies sufficiently above bottom threshold:

(i) $e^+e^- \to B_u\bar{B}_u + X$;

(ii) $e^+e^- \to \begin{matrix} B_d\bar{B}_d \\ B_s\bar{B}_s \end{matrix} + X$;

(iii) $e^+e^- \to \begin{matrix} B_u\bar{B}_d,\ \bar{B}_uB_d + X \\ B_u\bar{B}_s,\ \bar{B}_uB_s + X \end{matrix}$;

(iv) $e^+e^- \to B_d\bar{B}_s, \bar{B}_dB_s + X$.

In *case (i)* there can be no mixing and thus *no CP* asymmetry of the kind discussed here. (See, however, a comment in sect. 7.)

For *case (ii)* a new element comes into the game, namely that both B mesons produced can mix into each other. To deal with this complication we will invoke a "*CP taster*", namely a decay mode which can tell us whether the meson which has decayed was a B^0 or $\bar{B}^0$ at *that* moment. The mode which we consider best suited to do this job – the reasons will later become apparent – is the semileptonic transition $B^0 \to \ell^- + X^+$ versus $\bar{B}^0 \to \ell^+ + X^-$. Production of $\ell^-[\ell^+]$ at time t_0 shows that a $B^0[\bar{B}^0]$ has decayed into it at t_0. This in turn determines the other meson as a particular linear combination of B^0 and $\bar{B}^0$ at that time. With (2.5a, b) the wave function of this second meson is then known for all previous and subsequent times and we may ask for the probability that it decays into the final state f accessible both from B^0 and $\bar{B}^0$.

To be more specific we consider the time dependence of a $B_d\bar{B}_d$ or a $B_s\bar{B}_s$ state with charge parity C in its rest frame:

$$|B(t), \boldsymbol{k}; \bar{B}(t), -\boldsymbol{k}\rangle + (-1)^C |\bar{B}(t), \boldsymbol{k}; B(t), -\boldsymbol{k}\rangle . \tag{2.12}$$

Thus the amplitude for a meson with momentum $\boldsymbol{k}$ to decay to $\ell^- + X^+$ at time t_0, while the other meson decays to f at time t is

$$\begin{aligned} A[(\ell^-)_{t_0}, (f)_t] = {} & \langle \ell^- X^+|H|B^0(t_0), \boldsymbol{k}\rangle\langle f|H|\bar{B}^0(t), -\boldsymbol{k}\rangle \\ & + (-1)^C \langle \ell^- X^+|H|\bar{B}^0(t_0), \boldsymbol{k}\rangle\langle f|H|B^0(t), -\boldsymbol{k}\rangle . \end{aligned} \tag{2.13}$$

Employing eq. (2.6), we can express the dependence on t and t_0 explicitly:

$$\begin{aligned} A[(\ell^-)_{t_0}, (f)_t] = {} & \langle \ell^- X^+|H|B^0, \boldsymbol{k}\rangle \\ & \times \left\{ \frac{p}{q} [f_+(t_0) f_-(t) + (-1)^C f_-(t_0) f_+(t)] \langle f|H|B^0, -\boldsymbol{k}\rangle \right. \\ & \left. + [f_+(t_0) f_+(t) + (-1)^C f_-(t_0) f_-(t)] \langle f|H|\bar{B}^0, -\boldsymbol{k}\rangle \right\} . \end{aligned} \tag{2.14a}$$

The *CP*-conjugate situation is described by an amplitude

$$\begin{aligned} A[(\ell^+)_{t_0}, (f)_t] = {} & \langle \ell^+ X^-|H|\bar{B}^0, \boldsymbol{k}\rangle \\ & \times \left\{ [f_+(t_0) f_+(t) + (-1)^C f_-(t_0) f_-(t)] \langle f|H|B^0, -\boldsymbol{k}\rangle \right. \\ & \left. + \frac{q}{p} [f_-(t_0) f_+(t) + (-1)^C f_+(t_0) f_-(t)] \langle f|H|\bar{B}^0, -\boldsymbol{k}\rangle \right\} . \end{aligned} \tag{2.14b}$$

For convenience we define

$$M(\boldsymbol{k}) = \langle f|H|B^0, \boldsymbol{k}\rangle, \qquad \bar{M}(\boldsymbol{k}) = \langle f|H|\bar{B}^0, \boldsymbol{k}\rangle , \tag{2.15}$$

$$F(\boldsymbol{k}) = \langle \ell^- X^+|H|B^0, \boldsymbol{k}\rangle , \qquad \bar{F}(\boldsymbol{k}) = \langle \ell^+ X^-|H|\bar{B}^0, \boldsymbol{k}\rangle , \tag{2.16}$$

where we would have, if *CP* were a good symmetry, $M(\boldsymbol{k}) = -\bar{M}(-\boldsymbol{k})$, $F(\boldsymbol{k}) = -\bar{F}(-\boldsymbol{k})$. For an apparatus with equal detection efficiency for $\boldsymbol{k}$ and $-\boldsymbol{k}$ (no sweat for a theorist) we can omit the momentum dependence, obtaining

$$A[\ell^- f] = F\bar{M}\{\lambda[f_+(t_0) f_-(t) + (-1)^C f_-(t_0) f_+(t)] + [f_+(t_0) f_+(t) + (-1)^C f_-(t_0) f_-(t)]\} , \tag{2.17a}$$

$$A[\ell^+ f] = \frac{q}{p}\bar{F}\bar{M}\{\lambda[f_+(t_0)f_+(t) + (-1)^C f_-(t_0)f_-(t)] + [f_-(t_0)f_+(t) + (-1)^C f_+(t_0)f_-(t)]\}\,. \tag{2.17b}$$

As before, one can hope to have experimental access to time-integrated rates only. Integrating over the reference times t_0 and t from zero to infinity, we find

$$\begin{Bmatrix}\sigma(\ell^+ f)\\ \sigma(\ell^- f)\end{Bmatrix} = \text{const}\cdot\begin{Bmatrix}|p|^2|F|^2\\ |q|^2|\bar{F}|^2\end{Bmatrix}$$
$$\times[|\lambda|^2(1-a^2+y^2+x^2a^2)+(1+a^2+y^2-x^2a^2)-2\,\text{Re}\,(y\mp ixa^2)\lambda]\,, \tag{2.18}$$

$$C = \text{even}\,.$$

The CP asymmetry between the conjugate channels takes the form

$$A^{(\text{ii})} \equiv \frac{\sigma(\ell^+ f)-\sigma(\ell^- f)}{\sigma(\ell^+ f)+\sigma(\ell^- f)} = \begin{cases} -\dfrac{2xa^2\sin 2\phi}{1+y^2+y\cos 2\phi}\,, & \text{for } C=\text{even}\,, \quad (2.19\text{a})\\ 0\,, & \text{for } C=\text{odd}\,, \quad (2.19\text{b})\end{cases}$$

where we have put, for definiteness, $|p|^2 \simeq |q|^2$, $|F|^2 \simeq |\bar{F}|^2$, $|\lambda|^2 \simeq 1$.

The *cases (iii) and (iv)* listed above have to be treated differently, since the two mesons $B_i\bar{B}_j$ do not form a particle–antiparticle pair for $i \neq j$ and, thus, do not mix into each other.

Before passing on we want to clarify an important point: the *final state f* was introduced as one into which both B^0 and $\bar{B}^0$ can decay. $B_u = B^-$ and $\bar{B}_u = B^+$ can obviously not decay into such a common channel. Therefore, case (iii) is basically treated like single B^0 or $\bar{B}^0$ production and eq. (2.11) is applied (with a minor modification as explained later).

The situation is more complex for case (iv). Although there are no common decay channels f with $\overset{(-)}{B}_d$, $\overset{(-)}{B}_s \to f$ that can interfere, one has to keep in mind that B_d and B_s can mix into their corresponding antiparticles. We also take into account that the system $B_i\bar{B}_j + X$ ($i, j = s, d$) can be produced in a state of specified charge parity C. Accordingly we write down for the wave function at time t:

$$|B_d(t), \boldsymbol{k}; \bar{B}_s(t), \boldsymbol{k}'; \boldsymbol{k}_X\rangle + (-1)^C|\bar{B}_d(t), \boldsymbol{k}; B_s(t), \boldsymbol{k}'; \bar{X}, \boldsymbol{k}_X\rangle\,. \tag{2.20}$$

The state X will contain (at least) one strange particle, mostly a kaon. If it is a charged strange particle, then $X \neq \bar{X}$ and no interference between the two terms in eq. (2.20) can take place. If, however, a neutral kaon is produced instead then X and $\bar{X}$ can interfere because of the large K^0-$\bar{K}^0$ mixing. Yet as long as one does not identify the neutral kaon as K_L or K_S, but sums over them, the interference term drops out again.

We use the notation

$$\begin{matrix} B_d & & & B_s & & \\ & \searrow & & & \searrow & \\ & & f_d, & & & f_s\,, \\ & \nearrow & & & \nearrow & \\ \bar{B}_d & & & \bar{B}_s & & \end{matrix}$$

and arrive after some lengthy algebra at

$$A_{km}^{(\mathrm{iv})} = \frac{\sigma(\ell^+ f_k) - \sigma(\ell^- f_k)}{\sigma(\ell^+ f_k) + \sigma(\ell^- f_k)} = -\frac{1}{1+x_m^2}\,\frac{x_k}{1+x_k^2}\sin 2\phi_k\,, \tag{2.21}$$

where

$$k, m = \mathrm{d}, \mathrm{s}\,, \qquad \lambda_k = -\mathrm{e}^{-2i\phi_k}\,.$$

In deriving eq. (2.21) we have assumed, as before,

$$\begin{aligned} |\langle \ell^- X|H|B_d\rangle|^2 \simeq |\langle \ell^- X|H|B_s\rangle|^2 \simeq |\langle \ell^+ X|H|\bar{B}_d\rangle|^2 \simeq |\langle \ell^+ X|H|\bar{B}_s\rangle|^2\,, \\ |\langle f_d|H|B_d\rangle|^2 \simeq |\langle f_d|H|\bar{B}_d\rangle|^2 \simeq |\langle f_s|H|B_s\rangle|^2 \simeq |\langle f_s|H|\bar{B}_s\rangle|^2\,. \end{aligned} \tag{2.2}$$

To simplify matters further we have also put

$$y_d,\ y_s \simeq 0\,, \tag{2.23}$$

an approximation to be justified in sect. 3.

If an additional K_S is identified in X, say in

$$e^+e^- \to B_d\bar{B}_sK_S,\ \bar{B}_dB_sK_S\,,$$

then there will be an additional interference term and the asymmetry will be modified. This modification can be computed in the same way.

Of course, well above bottom threshold this is not a very practical proposition and we have to apply eq. (2.21) there.

This concludes the discussion of the general principles and the features behind the kind of *CP* asymmetry we study here. In sect. 3 we will rely on somewhat complementary models of *CP* violations to give estimates of the quantities x, y, a and ϕ.

3. Models on *CP* violations

3.1. THE KOBAYASHI–MASKAWA CONSTRUCTION

Most theorists adhere these days to a parametrization of *CP* violation first suggested by Kobayashi and Maskawa [4] which is based on the mixing of three $SU(2)_L$ quark doublets. The representation of the KM matrix we shall use is

$$U \equiv \begin{pmatrix} U_{du} & U_{su} & U_{bu} \\ U_{dc} & U_{sc} & U_{bc} \\ U_{dt} & U_{st} & U_{bt} \end{pmatrix} = \begin{pmatrix} c_1 & -s_1c_3 & -s_1s_3 \\ s_1c_2 & c_1c_2c_3 - s_2s_3\,e^{i\delta} & c_1c_2s_3 + s_2c_3\,e^{i\delta} \\ s_1s_2 & c_1s_2c_3 + c_2s_3\,e^{i\delta} & c_1s_2s_3 - c_2c_3\,e^{i\delta} \end{pmatrix}, \tag{3.1}$$

where $s_i = \sin\theta_i$, $c_i = \cos\theta_i$. The matrix U relates the $I_3 = -\frac{1}{2}$ members d′, s′, b′ of the $SU(2)_L$ doublets to the eigenstates d, s, b of strong flavour

$$\begin{pmatrix} d' \\ s' \\ b' \end{pmatrix} = U \begin{pmatrix} d \\ s \\ b \end{pmatrix} . \tag{3.2}$$

Now consider one of these doublets consisting of a lighter charge $-\frac{1}{3}$ quark L and its heavier charge $+\frac{2}{3}$ partner H:

$$\begin{pmatrix} H \\ L \end{pmatrix} = \begin{pmatrix} c \\ s \end{pmatrix}, \begin{pmatrix} t \\ b \end{pmatrix} .$$

The L quarks, in contrast to the H quarks, must decay *out* of their doublets, so their widths are suppressed both by phasespace *and* by Cabibbo-like mixing angles, which we assume to be small. Thus in general

$$\Gamma(L) < \Gamma(H) . \tag{3.3}$$

Conventional wisdom approximates the mass difference Δm arising due to mixing of conjugate neutral mesons by the contribution from the familiar box diagram (fig. 1). If the quark mixing angles are small compared to the ratio of quark masses in succeeding doublets, then the resulting Δm is proportional to the mass squared of the exchanged quark from the *same* doublet:

$$(\Delta m)_{L\bar{q}} \propto M_H^2 , \tag{3.4a}$$

$$(\Delta m)_{H\bar{q}} \propto M_L^2 . \tag{3.4b}$$

Thus one concludes as a general expectation

$$\left(\frac{\Delta m}{\Gamma}\right)_{L\bar{q}} > \left(\frac{\Delta m}{\Gamma}\right)_{H\bar{q}} . \tag{3.5}$$

Mixing in the B^0-$\bar{B}^0$ system might therefore be very strong, as it actually is for K^0-$\bar{K}^0$, while similar prospects for D^0-$\bar{D}^0$ and eventually T^0-$\bar{T}^0$ systems seem correspondingly dimmer.

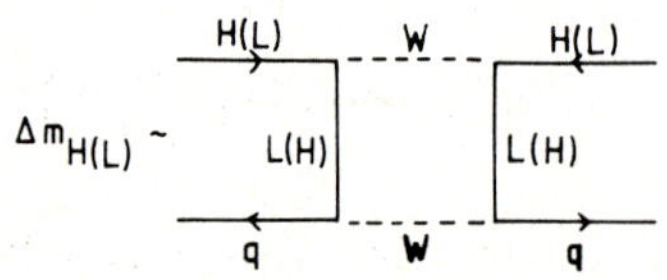

Fig. 1. Box diagram leading to Δm.

More specifically one obtains – as described in more detail in refs. [1, 2] – for the mixing parameters x and y:

$$x_d = \frac{2\Delta m(B_d)}{\Gamma_d} \simeq \frac{m^2(T)}{700\ \mathrm{GeV}^2}\left(\frac{1}{(1+\xi\cos\delta)^2(1+\tau^2)}\right) \gtrsim 0.5\,,$$

$$\xi = \frac{s_3}{s_2}\,,\qquad \tau = \frac{\xi\sin\delta}{1+\xi\cos\delta}\,,\tag{3.6}$$

$$y_d = \frac{\Delta\Gamma_d}{\Gamma_d} \simeq \frac{1}{12}\,,$$

for B_d mesons and

$$x_s = \frac{2\Delta m(B_s)}{\Gamma_s} \simeq \frac{m^2(T)}{37\ \mathrm{GeV}^2}\left(\frac{1}{(1+\xi\cos\delta)^2(1+\tau^2)}\right) \gtrsim 10\,,$$

$$y_s = \frac{\Delta\Gamma_s}{\Gamma_s} \ll 1\,,\tag{3.7}$$

for B_s mesons.

Looking at the expressions (2.11), (2.19), (2.21) given for *CP* asymmetries and inserting the numbers listed above we can already conclude that such *CP* asymmetries will presumably be smaller for B_s than for B_d mesons. The reason is that the mixing time – $x = \tau_B/\tau_{mix}$ – is so short that it tends to wash out the *CP* asymmetry for B_s decays [6].

This expectation is actually further strengthened when one computes the angle ϕ in the KM scheme. While we obtain for B_d mesons

$$\lambda_d = -\frac{(U^*_{bt}U_{bc}U^*_{sc})^2}{|U_{bt}U_{bc}U_{sc}|^2} \equiv -e^{-2i\phi_d}\,,\tag{3.8}$$

and thus

$$\sin\phi_d \simeq \left\{\frac{s_3}{\sqrt{s_3^2+s_2^2+2s_2s_3\cos\delta}}\right\}\sin\delta\,,\tag{3.9}$$

we find for B_s mesons that

$$\lambda_s = -\frac{(U^*_{bt}U_{bc}U^*_{ts})^2}{|U_{bt}U_{bc}U_{ts}|^2}\tag{3.10}$$

is *real* and therefore

$$\sin\phi_s = 0 + O\left(\frac{m_c}{m_t}, s^2\right).\tag{3.11}$$

Therefore, *within the KM ansatz* one gets relatively sizeable numbers for various *CP* asymmetries in B_d decays, as exhibited in fig. 2, while for all practical purposes

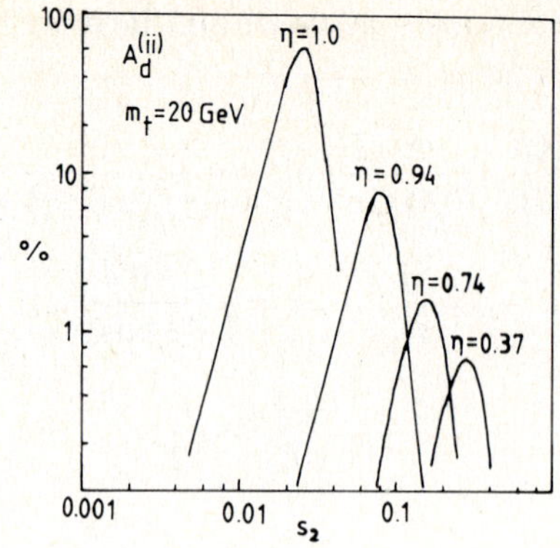

Fig. 2. The *CP* asymmetry $A_d^{(ii)}$, eq. (2.19a), plotted for C = even and for various values of the KM parameters; s_3 and $\sin\delta$ were eliminated in favour of m_t = mass of the top quark and of η, which is related to $\langle K^0|H|\bar{K}^0\rangle$. $\eta = 1.0$ and 0.4 correspond to the vacuum saturation approximation and the MIT bag model computation of $\langle K^0|H|\bar{K}^0\rangle$, respectively.

one does *not* expect to see any in B_s decays. It will not be lost on the more cynical minded reader to observe that a relatively moderate variation in the unknown parameter η, which is connected to $\langle K^0|H|\bar{K}^0\rangle$, induces a large change in the asymmetry. For us it constitutes another reason to adopt and advocate a strategy aimed at exploring phenomena rather than firmly predicting them.

3.2. *CP* VIOLATION DUE TO HIGGS EXCHANGE

A different way of introducing *CP* violation consists in enlarging the Higgs content of the weak sector of the theory: one adds (at least) two extra Higgs doublets to the minimal Higgs doublet of the standard model. With a mass of ~2–100 GeV they mediate spontaneous *CP* violation, while unwanted flavour-changing neutral current interactions are avoided by imposing a set of discrete symmetries [9].

We realize that this ansatz has lost some appeal with the (apparent) discovery of a fifth, the bottom, quark. Furthermore it not only requires some *ad hoc* postulates, but it also predicts a much larger ε'/ε ratio in K decays than seems to be measured [10]. Nevertheless, even if it cannot reproduce the known features of *CP* violations just on its own, it can serve us as an imagination stretcher – a commodity not to be belittled. In any case we will treat it as an ansatz complementary to the KM model to illustrate certain points in our discussion [2].

B^0-$\bar{B}^0$ mixing is calculated as before via the box diagram with two W boson lines (fig. 1). For an estimate of *CP* violation one computes the box diagram where one W line is replaced by Higgs exchange and finds

$$\sin\phi_s \simeq \sin\phi_d\,, \qquad \text{Higgs exchange}\,, \tag{3.12}$$

in clear contrast to the KM ansatz where we obtained [eqs. (3.9), (3.11)]

$$0 \sim \sin \phi_s \ll \sin \phi_d\,, \qquad \text{KM}\,.$$

Thus, the *CP* asymmetry introduced in sect. 2 [eqs. (2.11), (2.19)] should show up in B_d decays while basically vanishing in B_s decays if the KM description of *CP* violation is correct. If, however, *CP* violations receive significant contributions from Higgs exchange then a *CP* asymmetry as described above should appear both in B_d and B_s decays with comparable strength.

3.3. SUPERWEAK ANSATZ

There is a class of theories which describe the *CP* violation in K decays by introducing interactions which give a non-negligible contribution only to the off-diagonal mass matrix $\langle K^0|H|\bar{K}^0\rangle$ [11]. Such schemes generally predict much smaller *CP* violation in B than in K decays [12] and this grim prospect has cost them a large part of their popularity compared with the KM scheme. We will not comment on it any further.

4. On the actual observation of *CP* asymmetries

So far the discussion on *CP* asymmetries has been quite general. We shall now make a few comments on how to actually observe such effects. Among the major decays of the b quark there are three types:

(a) $b \to e^-, \mu^-, \tau^- + \bar{\nu} + c\,,$

(b) $b \to c\bar{u}d\,,$

(c) $b \to c\bar{c}s\,.$

CP is presumably conserved in the semileptonic transitions (a) due to *CPT*; they can be used only to probe *CP* impurities in B^0-$\bar{B}^0$ mixing which would lead to $\sigma(e^+e^- \to b\bar{b} \to \ell^+\ell^+ + X) \neq \sigma(e^+e^- \to b\bar{b} \to \ell^-\ell^- + X)$. Yet, as already stated, the size of such an effect is estimated to be very small [6].

The *CP* asymmetries (2.11) and (2.19) which were discussed in ref. [6] in connection with decay mode (b) were derived for a final state $B^0, \bar{B}^0 \to f$ which satisfies

$$CP|f\rangle = |f\rangle\,. \tag{4.1}$$

Note that the corresponding asymmetry flips sign for a final state $B^0, \bar{B}^0 \to f'$ which satisfies

$$CP|f'\rangle = -|f'\rangle\,. \tag{4.2}$$

Thus the asymmetry in the processes

$$\begin{array}{l} B_d \to D^0 + \pi\text{'s} \\ \qquad\quad \searrow \\ \qquad\qquad K_S + \pi\text{'s}\,, \\ \qquad\quad \nearrow \\ \bar{B}_d \to \bar{D}^0 + \pi\text{'s} \end{array}$$

which contain *both CP odd and even* final states, will be diluted. Since not very much is understood at present about the final states of weak mesonic decays, it is premature to attempt to compute the diluting effect. Instead, we shall discuss the decay mode (c) whose hadronic final states are, to a first approximation, restricted to have simple properties under *CP* transformations. There are hardly any background sources and very likely only few exclusive channels will make up the inclusive transition (c) thus greatly reducing the danger of mutual cancellations of *CP* asymmetries.

4.1. B_d DECAYS

There are two types of final states into which B_d mesons can decay:

(i) $B_d \to D\bar{D}\bar{K} + \pi\text{'s},\ D\bar{F} + \pi\text{'s},\ F\bar{F}\bar{K} + \pi\text{'s}\,,$

(ii) $B_d \to \psi\bar{K} + \pi\text{'s},\ \psi'\bar{K} + \pi\text{'s},\ \eta_c\bar{K} + \pi\text{'s},\ P_c\bar{K} + \pi\text{'s}\,.$

The channels shared by B_d^0 and $\bar{B}_d^0$ are

(i′) $$\begin{array}{l} B_d \\ \quad \searrow \\ \qquad D\bar{D}K_S + \pi\text{'s},\ F^+F^-K_S + \pi\text{'s}\,, \\ \quad \nearrow \\ \bar{B}_d \end{array}$$

(ii′) $$\begin{array}{l} B_d \\ \quad \searrow \\ \qquad \psi K_S + \pi\text{'s},\ \eta_c K_S + \pi\text{'s},\ P_c K_S + \pi\text{'s}\,. \\ \quad \nearrow \\ \bar{B}_d \end{array}$$

Our present understanding of weak non-leptonic decays does not yet allow us to predict precise branching ratios. To illustrate the general features, however, we will make some educated guesses as to the size of various branching ratios. Our guiding principle in obtaining numbers will be just to count the allowed transitions and to attribute equal probability to each of them if there does not exist a specific reason to assume otherwise. It will be obvious how to sharpen our numerical analysis after more information on inclusive B decays has been obtained, say from CESR. Thus we advocate a strategy aimed at exploring phenomena rather than firmly predicting them.

The decay $b \to c\bar{c}s$ is strongly reduced due to phase space. One can roughly estimate in a free quark model

$$\frac{\Gamma(b \to c\bar{c}s)}{\Gamma(b \to \text{all})} \sim 15\% \,. \tag{4.3}$$

It is more uncertain how much of this is taken up by channels with hidden charm, i.e. with ψ, ψ', η_c and the P_c states. A rough estimate would be

$$\mathrm{BR}(B_d \to (c \cdots)(\bar{c} \cdots) + X) \sim 10\% \,, \tag{4.4a}$$

$$\mathrm{BR}(B_d \to (c\bar{c}) + X) \sim 5\% \,, \tag{4.4b}$$

where $(c \cdots)$ stands for D, D*, F, . . . , and $(c\bar{c})$ stands for ψ, ψ', ... Since D*, . . . decay into D mesons and ψ presumably dominates $(c\bar{c})$, we estimate

$$\mathrm{BR}(B_d \to D\bar{D}K_S + \pi\text{'s}) \sim 3\text{–}5\% \,, \tag{4.5a}$$

$$\mathrm{BR}(B_d \to \psi K_S + \pi\text{'s}) \sim 1\text{–}2\% \,. \tag{4.5b}$$

Recent measurements at CESR on the number of kaons per B decay indicate somewhat larger values than expected in the naive quark model. Taken at face value it would lead us to conclude that BR $(B_d^0 \to (c \cdots)(\bar{c} \cdots))$ could be as large as 25% and the number for $\mathrm{BR}(B_d^0 \to D\bar{D}K_S + \pi\text{'s})$ had to be multiplied by a factor of two to three accordingly!

Such an enhancement of this channel is not that unlikely since large dynamical enhancements are observed in e^+e^- annihilation close to $D\bar{D}$ threshold and therefore they can well appear in the analogous decay $B_d^0 \to D\bar{D} + X$.

Since the decays $\psi \to e^+e^-$, $\mu^+\mu^-$ possess a clear signature, it is tempting to suggest a *CP* test based on this channel with a branching ratio

$$\mathrm{BR}\left(B_d \to \psi K_S + X \to \begin{matrix} e^+e^- \\ \mu^+\mu^- \end{matrix} + K_S + X\right) \sim 0.1\text{–}0.3\% \,. \tag{4.6}$$

There are two possible classes of final states which could signal the occurrence of $D\bar{D}$ mesons:

(a) $B_d \to D\bar{D}K_S + X$
$\quad\quad\quad \hookrightarrow K^+ + X'$,

(b) $B_d \to D\bar{D}K_S + X$
$\quad\quad\quad \hookrightarrow \ell^- + X'$.

A rough guess for the branching ratio of reaction (a) is

$$\mathrm{BR}(B_d \to D\bar{D}K_S + X) \sim 2\% \,.$$
$$\quad\quad \hookrightarrow K^+ + X' \,.$$

Thus the channel $B_d \to K^+K_S + X$ will be swamped by $K\bar{K}$ pair creation through strong interactions, unless the $D\bar{D}K_S$ decay mode is unusually prominent, as some preliminary CESR data might indicate.

Case (b) is a much more promising candidate. Detection of a negatively charged direct secondary lepton ℓ^- shows the occurrence of $\bar{D}$ mesons. It is now clear that a secondary lepton ℓ^- can be separated from a primary lepton which stems from a semileptonic B decay by imposing an energy cut for ℓ^- [13]. The Cabibbo suppressed decay modes $B_d^0 \to c\bar{c}d\bar{d}$, $u\bar{c}s\bar{d}$, $c\bar{c}$ may be background sources for the reaction (b). The data for $B \to K_S\ell + X$ will soon be available from CESR. Our rough guess is

$$BR(B_d \to D\bar{D}[F\bar{F}] + K_S + X \to \ell^- + K_S + X) \sim 0.3\text{–}0.6\% \,, \qquad (4.7)$$

and that the background from Cabibbo suppressed decays can be ignored.

4.2. B_s DECAYS

We include two mechanisms into our discussion of B_s decays, namely quark decay

$$B_s = (b\bar{s}) \to c\bar{c}s\bar{s} \,,$$

and weak annihilation

$$B_s = (b\bar{s}) \to c\bar{c} \,.$$

Again there are two types of final states, namely with open and with hidden charm:

(a) $B_s \to D\bar{D} + \pi$'s, $D\bar{D}K\bar{K} + \pi$'s, $D\bar{D}\phi + \pi$'s, $F^+F^- + \pi$'s,

(b) $B_s \to \psi + \pi$'s, $\psi\phi + \pi$'s, $\psi K\bar{K} + \pi$'s.

Assuming weak annihilation to be significant in B_s^0 decays, say to contribute of the order of 20–30% of $\Gamma(B_s)$ [14] we estimate

$$\text{(a)} \quad BR(B_s^0 \to \text{open charm}) \sim 25\text{–}35\% \,, \qquad (4.8a)$$

$$\text{(b)} \quad BR(B_s^0 \to \psi + X) \sim 2\text{–}3\% \qquad (4.8b)$$

and thus conclude

$$\text{(a)} \quad BR(B_s^0 \to D\bar{D}[F\bar{F}] + X \to \ell^- + \tilde{X}) \sim 4\% \,, \qquad (4.9a)$$

$$\text{(b)} \quad BR\left(B_s^0 \to \psi + X \to \begin{matrix} e^+e^- \\ \mu^+\mu^- \end{matrix} + X\right) \sim 0.3\text{–}0.4\% \,. \qquad (4.9b)$$

Background to reaction (a) due to Cabibbo disfavoured modes can be ignored on this level.

Even dileptons from open charm should occur with not too tiny a branching fraction:

$$(\tilde{a})\quad BR(B_s \to D\bar{D}[F\bar{F}] + X \to \ell^+\ell^- + \tilde{X}) \sim 0.6\text{–}0.7\% . \tag{4.9c}$$

In passing we want to add that the transition $B_s \to \phi + X$ might be useful to be studied in this context be it only as a first step. A ball park guess is $BR(B_s^0 \to \phi + X) \sim$ few %.

To summarize this section: if our crude estimates for the branching ratios of B decays are correct, then it may be difficult to observe *CP* asymmetries just above bottom threshold in e^+e^- annihilation. Experimentalists should *not*, however, be discouraged from searching for such an asymmetry, since, as indicated above, sizeable deviations from naive quark model predictions can be present in B decays and some unexpected effect may allow observation of *CP* asymmetries just above bottom threshold. Otherwise a high luminosity B meson factory is needed. The Z^0 resonance would provide such an entity and so would, to some extent, a toponium resonance. In the next section we describe how B production on the Z^0 can be harnessed for our *CP* analysis.

5. On the Z^0 resonance

Physics on the Z^0 resonance is often portrayed as the promised land where, if not milk and honey, at least huge event rates will flow freely and generously. E.g. with a luminosity of 3×10^{31} cm^{-2} s^{-1} one expects one $Z^0 \to b\bar{b}$ decay per second and one does not have to worry unduly about small branching ratios. Instead one has to face a different complexity already referred to in sect. 2: the b states produced are not necessarily particle–antiparticle combinations any more:

$$Z^0 \to B_u\bar{B}_u + X,\ B_u\bar{B}_d + X,\ B_u\bar{B}_s + X,\ B_d\bar{B}_d + X,\ B_d\bar{B}_s + X, \ldots .$$

Thus an observed asymmetry will in general receive contributions from many channels and it will be a non-trivial problem to disentangle the various channels. Assuming for $q\bar{q}$ excitation from the vacuum a ratio

$$u\bar{u} : d\bar{d} : s\bar{s} = 2:2:1 ,$$

we get for the inclusive production

$$\sigma(B_u\bar{B}_u) : \sigma(B_d\bar{B}_d) : \sigma(B_s\bar{B}_s) : \sigma(B_u\bar{B}_d + B_d\bar{B}_u) : \sigma(B_u\bar{B}_s + B_s\bar{B}_u) : \sigma(B_d\bar{B}_s + B_s\bar{B}_d)$$
$$\sim \tfrac{4}{25} : \tfrac{4}{25} : \tfrac{1}{25} : \tfrac{8}{25} : \tfrac{4}{25} : \tfrac{4}{25}$$

where we have invoked short-range compensation of strangeness.

We also want to state already at this point that a search for a *CP* asymmetry via $\sigma(\ell^\pm K_S)$ as described in ref. [6] might not be that promising since secondary production of $K\bar{K}$ via $s\bar{s}$ excitation could well bury the whole effect. Thus a chase for *CP* violations is better based on channels (a) and (c) of sect. 4.

The six $B\bar{B}$ channels fall into four categories, which have already been discussed in sect. 2:

(i) $Z^0 \to B_u\bar{B}_u + X$;

(ii) $Z^0 \to B_d\bar{B}_d + X$, $B_s\bar{B}_s + X$;

(iii) $Z^0 \to B_u\bar{B}_d + \bar{B}_u B_d + X$, $B_u\bar{B}_s + \bar{B}_u B_s + X$;

(iv) $Z^0 \to B_d\bar{B}_s + \bar{B}_d B_s + X$.

The first question one has to face is: how can one separate those events in e^+e^- annihilation on the Z^0 resonance which do contain $b\bar{b}$ quarks and which one therefore wants to analyse from all the others which do not? In this section we treat a situation where top quarks are *not* produced on the Z^0; they will be dealt with in sect. 6. We have advocated before the use of direct leptons as "*CP* taster". Actually they can at the same time serve as "flavour tasters", too, whose observation in an event narrows the competition down to bottom flavoured versus charm flavoured events. The longitudinal spectrum of direct leptons coming from the *primary* bottom decay and from charm decay should be the same for all practical purposes at these energies: $E = M_{Z^0} \gg m_B$, m_D and thus it will not distinguish between the two cases. But the $p_\perp$ spectra will still allow to separate bottom and charm events [15].

For a *CP* asymmetry based for example on

$$A_\psi = \frac{\sigma(\ell^+\psi X) - \sigma(\ell^-\psi X)}{\sigma(\ell^+\psi X) + \sigma(\ell^-\psi X)}, \tag{5.1}$$

we expect for the four classes of b flavoured events listed above:

$$\text{(i)} \quad A^{(i)} = 0, \qquad \text{(no mixing!)}, \tag{5.2}$$

$$\text{(ii)} \quad A^{(ii)}_{d,s} = -(1 + (-1)^C)\frac{x_{d,s} a^2_{d,s} \sin 2\phi_{d,s}}{1 + y^2_{d,s} + 2y_{d,s}\cos 2\phi_{d,s}}, \tag{5.3}$$

$$\text{(iii)} \quad A^{(iii)}_{d,s} = -\frac{x_{d,s} a_{d,s} \sin 2\phi_{d,s}}{1 + r_{d,s} + y_{d,s}\cos 2\phi_{d,s}},$$

$$r_{d,s} = \frac{\Gamma(B_u \to \psi X)}{\Gamma(B_{d,s} \to \psi X)}, \tag{5.4}$$

$$\text{(iv)} \quad A^{(iv)} = -\frac{\rho_d}{1 + x_s^2}\left[\frac{x_d}{1 + x_d^2}\sin 2\phi_d\right] - \frac{\rho_s}{1 + x_d^2}\left[\frac{x_s}{1 + x_s^2}\sin 2\phi_s\right],$$

$$\rho_{d,s} = \frac{\Gamma(B_{d,s} \to \psi X)}{\Gamma(B_d \to \psi X) + \Gamma(B_s \to \psi X)}. \tag{5.5}$$

The various quantities are defined in sect. 2 [see (2.10, 11, 20)]. Putting $\sin 2\phi_s \simeq 0$, $x_s \gg 1$, $x_s a_s \ll 1$, $y \simeq 0$ as suggested in the KM ansatz (sect. 3) we obtain much simpler expressions:

$$\text{(ii)}\quad A_s^{(ii)} \simeq 0\,, \tag{5.6a}$$

$$A_d^{(ii)} \simeq -(1+(-1)^C) x_d a_d^2 \sin 2\phi_d\,, \tag{5.6b}$$

$$\text{(iii)}\quad A_s^{(iii)} \simeq 0\,, \tag{5.7a}$$

$$A_d^{(iii)} \simeq -\frac{x_d a_d \sin 2\phi_d}{1+r_d}\,, \tag{5.7b}$$

$$\text{(iv)}\quad A^{(iv)} \simeq 0\,. \tag{5.8}$$

Of course, these different classes cannot be observed separately. The experimental situation consists always of a mixture of them with a certain weighting. We advocate to search in all possible channels for a *CP* asymmetry. After such an effect has been found, one can study more finely how the asymmetry is affected by choosing a different final state. Thus it is at least in principle possible to disentangle the various classes and their contributions. The numbers we present in this context are mainly intended to serve as guide lines, not as firm predictions.

A_ψ [see (5.1)] or an asymmetry in the lepton distribution

$$A_{D\bar{D}} = \frac{\sigma(\ell^+ \ell^-_{sec} X) - \sigma(\ell^- \ell^+_{sec} X)}{\sigma(\ell^+ \ell^-_{sec} X) + \sigma(\ell^- \ell^+_{sec} X)}\,, \tag{5.9}$$

which is averaged over the different $B\bar{B}$ production channels, depends obviously on the semileptonic branching ratios of B mesons. It has been suggested that they are not the same for B_u, B_d and B_s mesons – they might differ by up to ~50% [14] due to weak annihilation. Also the branching fractions $BR(B_d \to D\bar{D}+X)$ and $BR(B_s \to D\bar{D}+X)$, or $BR(B_d \to \psi + X)$ and $BR(B_s \to \psi + X)$, might well be quite different because of weak annihilation. We have included that possibility in our discussion of sect. 4; now, however, we want to ignore such differences in order not to overburden our presentation. It will be obvious how to improve on our estimates once the various branching ratios have actually been measured. Using $b_{SL}(B_u) \sim b_{SL}(B_d) \sim b_{SL}(B_s)$, $BR(B_{d,u} \to D\bar{D}+X) \sim BR(B_s \to D\bar{D}+X)$ and $BR(B_{d,u} \to \psi + X) \sim BR(B_s \to \psi + X)$, we find

$$\left.\begin{matrix} A_\psi \\ A_{D\bar{D}} \end{matrix}\right\} = \tfrac{4}{25} A_d^{(ii)} + \tfrac{1}{25} A_s^{(ii)} + \tfrac{8}{25} A_d^{(iii)} + \tfrac{4}{25} A_s^{(iii)} + \tfrac{4}{25} A^{(iv)}\,. \tag{5.10}$$

A further simplification arises in the KM ansatz:

$$\left.\begin{matrix} A_\psi \\ A_{D\bar{D}} \end{matrix}\right\} \simeq \tfrac{4}{25} A_d^{(ii)} + \tfrac{8}{25} A_d^{(iii)}\,. \tag{5.11}$$

One reads off from (5.11) that by averaging over the various $B\bar{B}$ channels one dilutes an expected *CP* asymmetry, but only by a factor of roughly two (depending on the parameters, see fig. 2). Such a moderate reduction of the effect is a small price to pay for the gain of large counting rates on the Z^0 resonance. One can also demand identification of a K_S in the same hemisphere as ψ or the secondary lepton, i.e. one studies

$$A_{\psi K_S} = \frac{\sigma(\ell^+ \psi K_S X) - \sigma(\ell^- \psi K_S X)}{\sigma(\ell^+ \psi K_S X) + \sigma(\ell^- \psi K_S X)}, \tag{5.12}$$

$$A_{D\bar{D}K_S} = \frac{\sigma(\ell^+ \ell^-_{sec} K_S X) - \sigma(\ell^- \ell^+_{sec} K_S X)}{\sigma(\ell^+ \ell^+_{sec} K_S X) + \sigma(\ell^- \ell^+_{sec} K_S X)}. \tag{5.13}$$

On the same crude level of approximation we have mostly been using here, these two asymmetries are also given by eq. (5.11). Yet on a more refined level one can anticipate that $A_{\psi K_S}$ and $A_{D\bar{D}K_S}$ are weighted more favourably towards B_d decays than are A_ψ and $A_{D\bar{D}}$, while it is the reverse for B_s decays. Therefore by comparing $A_{\psi K_S}$ with A_ψ and $A_{D\bar{D}K_S}$ with $A_{D\bar{D}}$ one can presumably gain information on whether the *CP* asymmetry is fed from just B_d decays – as predicted in the KM scheme – or also from B_s decays.

6. Impact of toponium and open top production

So far we have omitted the hypothetical top states in our analysis; now they will be included.

Mixing, an essential ingredient in our method can in principle also occur for neutral top mesons T^0 and $\bar{T}^0$. But conventional wisdom, as sketched in sect. 3, has it that T^0-$\bar{T}^0$ mixing should be analogous to D^0-$\bar{D}^0$ mixing, i.e. small, whereas the B^0-$\bar{B}^0$ case resembles more the large K^0-$\bar{K}^0$ mixing. Nevertheless top states can serve useful purposes in the context of our method:

(i) The theoretical bias for simplicity and some admittedly crude estimates [16] make us expect the top quark to decay dominantly into bottom quarks. Thus top states act as a powerful B factory. This feature becomes most evident *off* the Z^0 resonance where we estimate

$$BR(\gamma^* \to b + \bar{b} + X) \sim 30\% \qquad \text{for } E_{\gamma^*} > 2M_T, \tag{6.1a}$$

to be contrasted with

$$BR(\gamma^* \to b + \bar{b} + X) \sim 9\% \qquad \text{for } E_{\gamma^*} < 2M_T. \tag{6.1b}$$

(ii) Top decays can furthermore help us to identify in individual events whether we are dealing with B or with $\bar{B}$ decays. This can happen in different ways:

(a) At energies well above top threshold the decay products of T and of $\bar{T}$ will populate two clearly separable hemispheres. If, e.g., a positively charged lepton, which is sufficiently energetic to come from the primary top decay, is observed in one hemisphere, we can conclude that a B state was produced along with it in the same hemisphere, while in the *opposite* hemisphere a $\bar{\mathrm{B}}$ state was generated. Thus the case of single B or $\bar{\mathrm{B}}$ production is realized, eq. (2.11), which according to our estimates should exhibit larger *CP* violation. Such asymmetries can be represented in different ways, for example by

$$A = \frac{\sigma(\ell^+[\psi]\mathrm{K_S}) - \sigma(\ell^-[\psi]\mathrm{K_S})}{\sigma(\ell^+[\psi]\mathrm{K_S}) + \sigma(\ell^-[\psi]\mathrm{K_S})}, \tag{6.2}$$

where the hard lepton and the ψ plus $\mathrm{K_S}$ are either in the same hemisphere or in opposite hemispheres. These two asymmetries would be equal, but of opposite sign.

(b) On a toponium resonance just above threshold top mesons will decay almost at rest into B states. Even below threshold there is a considerable probability that toponium decays weakly like

$$(\mathrm{t\bar{t}}) \to \mathrm{t} + \bar{\mathrm{b}}\mathrm{W}^-, \ \bar{\mathrm{t}} + \mathrm{bW}^+$$

if $M(\mathrm{t\bar{t}}) \gtrsim 50$ GeV [17]. In both cases one can therefore discuss semileptonic top decays. If one observes a lepton which is sufficiently energetic to come from top decay one knows that due to mainly kinematical constraints the corresponding B state is flying into the opposite direction with some large probability. More precisely, calculating just $\mathrm{t} \to \mathrm{be}^+\nu_\mathrm{e}$, one finds

$$\frac{\mathrm{d}\sigma}{\mathrm{d}z\ \mathrm{d}\cos\theta} \propto \frac{z^2(1-z)^2}{[1-\frac{1}{2}z(1-\cos\theta)]^2}, \tag{6.3}$$

with

$$z = \frac{2E_{\mathrm{e}+}}{M_\mathrm{t}}, \qquad \theta = \sphericalangle(\boldsymbol{p}_{\mathrm{e}+}, \boldsymbol{p}_\mathrm{b}) .$$

Therefore, in the direction opposite to the positively [negatively] charged lepton one has a much higher chance to find a B[$\bar{\mathrm{B}}$] than a $\bar{\mathrm{B}}$[B] state. To a good approximation one can thus treat such a situation like single B or $\bar{\mathrm{B}}$ production, eq. (2.11), where larger *CP* asymmetries are expected.

7. Summary and conclusions

B^0-$\bar{\mathrm{B}}^0$ mixing is, as emphasized before, essential to exhibit *CP* asymmetries as described by us (and others). Therefore, the first task is to observe mixing and determine its parameters; this can be achieved by studying like-sign dileptons in $\mathrm{e^+e^-}$ annihilation as explained in more detail in the appendix.

One can then search for a *CP* violation by comparing like-sign dileptons of positive and negative charge [1, 2]. Estimates on the size of such an effect are, however, sobering as indicated in sect. 1 – they range from 10^{-4} to 10^{-3}. On the other hand one has a fighting chance to observe a difference in $\sigma(\mathrm{e^+e^-} \to \mathrm{K^+K_SX})$

versus $\sigma(e^+e^- \to K^- K_S X)$ etc. just above bottom threshold as described in ref. [6] and sects. 2 and 4 of this paper.

If the effects turn out to be too small for being measured in the threshold region one has to have recourse on a high luminosity B meson factory. The best candidate is obviously the Z^0 resonance; a toponium resonance and open top states can also serve as one.

At such high energies one expects to derive essential information from the emerging leptons. As before, one can compare like-sign dileptons of positive and negative charge where it is understood that both leptons are *primary* leptons coming from B decay: the expected asymmetry is still 10^{-4}–10^{-3}. Or one searches for a difference in

$$\sigma\Big(e^+e^- \to \ell^+ \underset{\hookrightarrow \ell^+\ell^-}{\psi} X\Big) \text{ versus } \sigma\Big(e^+e^- \to \ell^- \underset{\hookrightarrow \ell^+\ell^-}{\psi} X\Big). \tag{7.1}$$

or in

$$\sigma\Big(e^+e^- \to \ell^+ D\underset{\hookrightarrow \ell^-}{\bar{D}} X\Big) \text{ versus } \sigma\Big(e^+e^- \to \ell^- \bar{D}\underset{\hookrightarrow \ell^+}{D} X\Big). \tag{7.2}$$

Here one also studies mainly leptons, but some of them are understood to stem *not* from primary bottom, but from *secondary* charm and ψ decays. As discussed in sect. 4 one gives up in counting rates relative to the case of primary dileptons, but it is not unreasonable to expect *CP* asymmetries of the order of 1–10%.

Finally we remind the reader that there are some *CP*-violating observables which do *not* depend on B^0-$\bar{B}^0$ mixing. An example which has been discussed in ref. [6] is an asymmetry between $\Gamma(B^- \to K_S K_S + X)$ and $\Gamma(B^+ \to K_S K_S + X)$ which arises from interference of the two amplitudes

$$A\left[\begin{matrix} B^+ \to D^0 K_S + X_1 \\ \quad\;\; \hookrightarrow K_S + X_2 \end{matrix}\right] \quad \text{and} \quad A\left[\begin{matrix} B^+ \to \bar{D}^0 K_S + \bar{X}_1 \\ \quad\;\; \hookrightarrow K_S + \bar{X}_2 \end{matrix}\right].$$

As emphasized in sect. 2, this asymmetry depends on the difference between the final state interactions of $(D^0 K_S + X)$ and $(\bar{D}^0 K_S + X)$. If nothing dramatic happens in the final state interactions, we expect the asymmetry to be of the order of 0.1–10%.

Clearly one has to pay a certain price when testing *CP* in B decays on the Z^0 resonance. The numbers so obtained will in general be averaged over many channels and a single measurement cannot determine which specific decay of B_d or B_s mesons contains the finally observed *CP* violation. By demanding, however, that additional particles are identified in the final state X of expressions (7.1), (7.2) – in particular K_S – one can at least in principle disentangle the different contributions – after the main objective has been met: *establishing the first CP violation outside the neutral kaon system.* The experimental task involved is not easy, but the prospective premium justifies a considerable effort.

This collaboration started while the authors were at CERN. One of the authors (A.I.S.) thanks Prof. J. Prentki for the warm hospitality extended to him while he was visiting CERN. The other author gratefully acknowledges helpful discussions with S. Ono, L. Sehgal and P.M. Zerwas.

Appendix

MIXING AND DILEPTONS

Since our *CP* asymmetry depends crucially on B^0-$\bar{B}^0$ mixing, we shall discuss detection of the mixing effect in e^+e^- colliding beam processes.

Let the semileptonic decay modes $B^0 \to \ell^- + X$, $\bar{B}^0 \to \ell^+ + X$ be the normal modes in the absence of B^0-$\bar{B}^0$ mixing. Furthermore let $N^L_{\pm\pm}$, $N^L_{\pm-}$ be the number of events of the type

$$e^+e^- \to \underbrace{(B^0\bar{B}^0)_L}_{} + X$$
$$\hookrightarrow \ell^\pm\ell^\pm + X^{\mp\mp},\ \ell^+\ell^- + X^0,$$

respectively. L is the orbital angular momentum of the B^0-$\bar{B}^0$ pair. $N^L_{\pm\pm}$ and N^L_{+-} can be computed using techniques described in the text*. The result is

$$N^L_{--} \sim \left|\frac{1-\varepsilon_B}{1+\varepsilon_B}\right|^2 \left[\frac{1+(-1)^L y^2}{(1-y^2)^2} - \frac{1-(-1)^L x^2}{(1+x^2)^2}\right], \tag{A.1}$$

$$N^L_{++} \sim \left|\frac{1+\varepsilon_B}{1-\varepsilon_B}\right|^2 \left[\frac{1+(-1)^L y^2}{(1-y^2)^2} - \frac{1-(-1)^L x^2}{(1+x^2)^2}\right], \tag{A.2}$$

$$N^L_{+-} \sim 2\left[\frac{1+(-1)^L y^2}{(1-y^2)^2} + \frac{1-(-1)^L x^2}{(1+x^2)^2}\right], \tag{A.3}$$

where

$$x = \frac{2\Delta m}{\Gamma}, \qquad y = \frac{\Delta\Gamma}{\Gamma}, \qquad \Delta m = m_1 - m_2, \qquad \Gamma = \Gamma_1 + \Gamma_2.$$

The N^L depend on L since the time evolution of the $B\bar{B}$ state must satisfy Bose symmetry. For example a $(B\bar{B})_{L=1}$ state *cannot* be found in a BB or $\bar{B}\bar{B}$ state at any time in the future.

Mixing will lead to like-sign dilepton events where the ratio

$$R^L = \frac{N^L_{++} + N^L_{--}}{N^L_{+-}} \tag{A.4}$$

can be studied. In fig. 3 we have shown R^L in the approximation that $y \simeq 0$.

* The $L = 1$ case was first obtained in ref. [18].

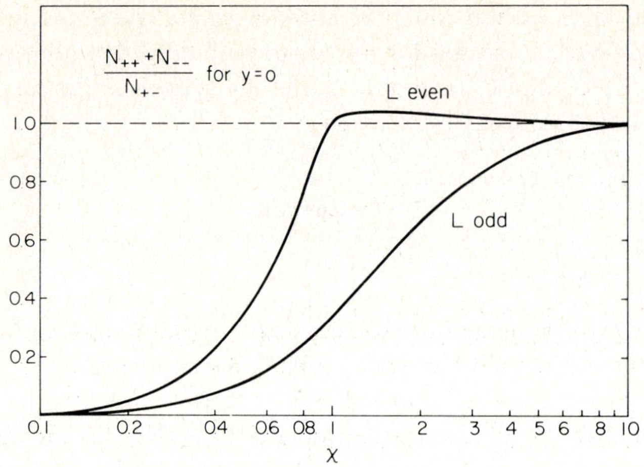

Fig. 3. An effect of B^0-$\bar{B}^0$ mixing in e^+e^- colliding beam experiments. Note that the observable effect depends on the angular momentum L of the B^0-$\bar{B}^0$ state.

Measuring R^L will determine the value of x and this in turn will fix the asymmetries which are discussed in the text.

It is evident from the figure that for the range of interesting values of x, i.e. $x \leqslant 8$, the curves for L even and L odd are quite distinct. If one is just sufficiently above $B\bar{B}$ threshold to produce $B\bar{B}$, but not $B\bar{B}^* + B^*\bar{B}$ (B^* is the corresponding vector state), then one is dealing with a $L = 1$ configuration:

$$e^+e^- \to \gamma^* \to (B\bar{B})_{L=1}\,. \tag{A.5}$$

If also the vector states can be excited which subsequently decay into B plus a photon, then one is dealing with an L = even configuration:

$$e^+e^- \to \gamma^* \to \begin{matrix} B^*\bar{B} \\ B\bar{B}^* \end{matrix} \to (B\bar{B})_{L=\text{even}}\gamma\,. \tag{A.6}$$

On the Υ_{4s} resonance we may observe some mixture of L = odd and L = even $B\bar{B}$ states. The precise mixture depends rather sensitively on $m_{B^*} - m_B$.

References

[1] A. Pais and S.B. Treiman, Phys. Rev. D12 (1975) 2744;
L.B. Okun, V.I. Zakharov and B.M. Pontecorvo, Nuovo Cim. Lett. 13 (1975) 218;
J. Ellis, M.K. Gaillard and D.V. Nanopoulos, Nucl. Phys. B109 (1976) 213
[2] A. Ali and Z.Z. Aydin, Nucl. Phys. B148 (1978) 165
[3] J. Bell and J. Steinberger, Proc. Oxford Int. Conf. on Elem. Part. 1965, vol. I, p. 207ff.
[4] M. Kobayashi and K. Maskawa, Prog. Theor. Phys. 49 (1973) 652
[5] J.H. Christenson et al., Phys. Rev. Lett 13 (1964) 138
[6] A.B. Carter, A.I. Sanda, Phys. Rev. Lett. 45 (1980) 952; Phys. Rev. D23 (1981) 1567;
T.D. Lee and C.S. Wu, Ann. Rev. Nucl. Sci. 16 (1966) 471

[7] M.K. Gaillard, Lecture at 1978 SLAC Summer Inst., ed. M. Zipf;
J. Ellis, Lectures at 7th Int. Meeting on Fundamental physics, Segovia, Spain, 1979
[8] M. Bander, D. Silverman and A. Soni, Phys. Rev. Lett. 43 (1979) 242
[9] S. Weinberg, Phys. Rev. Lett 31 (1976) 657
[10] A.I. Sanda, Rockefeller Report 2232B-217 (1981);
N.G. Deshpande, Oregon preprint OITS 160 (1981)
[11] L. Wolfenstein, Phys. Rev. Lett. 13 (1964) 180
[12] A.I. Sanda, Rockefeller Report 2232B-221 (1981)
[13] A.Ali, Z. Phys. C1 (1979) 25
[14] I.I. Bigi, Z. Phys. C5 (1980) 313;
V. Barger et al., Phys. Rev. D22 (1980) 693;
H. Fritzsch and P. Minkowski, Phys. Lett. 90B (1980) 455;
W. Bernreuther, O. Nachtmann and B. Stech, Z. Phys. C4 (1980) 257
[15] M.J. Puhala et al., Jowa State University preprint IS-J 487
[16] V. Barger, W. F. Long and S. Pakvasa, Phys. Rev. Lett. 42 (1979) 1585;
R.E. Shrock, S.B. Treiman and L.L. Wang, Phys. Rev. Lett. 42 (1979) 1589
[17] J. Ellis and M.K. Gaillard, CERN Yellow Report 76-18 (1976);
K. Fujikawa, Prog. Theor. Phys. 61 (1979) 1186;
G. Goggi and G. Penso, Nucl. Phys. B165 (1980) 429;
I.I. Bigi and H. Krasemann, Z. Phys. C7 (1981) 127
[18] L.B. Okun, V.I. Zakharov and P.M. Pontecorvo, Nuovo Cim. Lett. 13 (1975) 218

Nuclear Physics B246 (1984) 45–51

CP VIOLATION IN B^0–$\bar{B}^0$ MIXING

L. WOLFENSTEIN[1]

Institute for Theoretical Physics, University of California, Santa Barbara, California 93106, USA

Received 13 October 1983
(Final version received 21 June 1984)

A parameter ε_B for the B^0 system analogous to ε for the K^0 system is defined. Recent information on the Kobayashi–Maskawa matrix yields the conclusion that, if the KM matrix explains *CP* violation, $|\varepsilon_B|$ is probably of order unity. The measurement of ε_B requires a class of experiments first analyzed by Bigi and Sanda involving relatively rare exclusive decay channels.

At the present time the only evidence for *CP* violation involve observations on the $K^0-\bar{K}^0$ system. The possibility of finding similar *CP*-violating effects associated with $B^0-\bar{B}^0$ mixing has been widely discussed [1, 2]. It has been concluded that the charge asymmetry of dilepton pairs is expected to be very small. Here we discuss *CP*-violating effects of $B^0-\bar{B}^0$ mixing which are expected to be large and possibly maximal in the Kobayashi–Maskawa (KM) model. These effects, which involve relatively rare exclusive decays, were first analyzed by Bigi and Sanda [3]. Recent information on the KM matrix make it possible to show that the *CP* violation is likely to be large for these decays. The present paper is self-contained, although some of the major results are derivable from ref. [3].

It is useful first to review the $K^0-\bar{K}^0$ system. Using the Wu–Yang phase convention, for which the dominant K^0 decay amplitude is real, one writes

$$K_S=(K_1+\varepsilon K_2)/(1+|\varepsilon|^2)\,,$$
$$K_L=(K_2+\varepsilon K_1)/(1+|\varepsilon|^2)\,, \qquad (1)$$

where K_S, K_L are the decaying eigenstates and K_1, K_2 are the *CP* eigenstates. The complex mixing parameter ε is a measure of *CP* violation. Two types of experiments can be distinguished: (i) the charge asymmetry in $K_L\to\pi\ell\nu$ decay which depends only on Re ε, and (ii) the Fitch–Cronin experiment measuring $K^0\to\pi\pi$, which depends on the magnitude of ε and thus on both Re ε and Im ε.

As is well known, phase conventions (unlike experimental facts) are subject to human whims. A change in the phase convention ($K^0\to K^0\,e^{i\alpha}$) has the consequence of changing the imaginary part of the mixing parameter of eqs. (1). Since ε is experimentally measurable we prefer to use another symbol $\bar{\varepsilon}$ to describe the mixing parameter in an arbitrary phase convention [4]. While Re $\bar{\varepsilon}$ = Re ε, Im $\bar{\varepsilon}$ can be changed at will. The observable Im ε is in general a combination of Im $\bar{\varepsilon}$ and an imaginary decay amplitude. Thus the Fitch–Cronin effect (ii) cannot unambiguously be described as an effect of $K^0-\bar{K}^0$ mixing, although it appears to be so in the Wu–Yang phase convention.

In the case of the $B^0-\bar{B}^0$ system we use a phase convention in which the predominant B^0 decays have real amplitudes. In this phase convention we determine the parameter ε_B, which turns out to be almost purely imaginary with a magnitude of order unity. As a consequence type (i) experiments show little *CP* violation, but type (ii) experiments can show large, possibly maximal, *CP* violation.

It is convenient [5] to expand the KM matrix U in terms of a small parameter $\lambda \approx \sin\Theta_c \approx 0.22$. To order λ^3, which is adequate for our purposes*,

$$U=\begin{pmatrix} 1-\frac{1}{2}\lambda^2 & \lambda & \lambda^3 A(\rho-i\eta) \\ -\lambda & 1-\frac{1}{2}\lambda^2 & \lambda^2 A \\ \lambda^3 A(1-\rho-i\eta) & -\lambda^2 A & 1 \end{pmatrix}. \tag{2}$$

The term $\lambda^2 A$ is determined from the B lifetime to be 0.053 ± 0.010 [6] so that $A=1.1\pm0.2$. The experimental finding that the branching ratio of $b\to u$ relative to $b\to c$ is less than 0.05 [7] yields the constraint

$$\rho^2+\eta^2<0.5\,. \tag{3}$$

There are no other significant constraints on the matrix other than a correlated constraint between η and m_t if the KM mechanism is assumed to produce the *CP* violation in the K^0 system. Our parameters are related to those of Maiani [8] by $\lambda=s_\Theta$, $\lambda^2 A=s_\gamma$, $\lambda^3 A(\rho-i\eta)=s_\beta\, e^{i\delta}$. The phase convention we use emphasizes that *CP*-violating effects enter only at the order λ^3.

We define the *CP* eigenstates

$$B_-=\sqrt{\tfrac{1}{2}}(B^0+\bar{B}^0)\,,\qquad CP=-\,,$$
$$B_+=\sqrt{\tfrac{1}{2}}(B^0-\bar{B}^0)\,,\qquad CP=+\,, \tag{4}$$

with the phase convention defined by the KM matrix form of eq. (2). In the $B_- - B_+$ representation the mass matrix has the form

$$M=\begin{pmatrix} \mathrm{Re}\, M_{12} & i\,\mathrm{Im}\, M_{12} \\ -i\,\mathrm{Im}\, M_{12} & -\mathrm{Re}\, M_{12} \end{pmatrix}, \tag{5}$$

where M_{12} is given by [2]

$$M_{12}=\frac{G_F^2 f_B^2 m_B}{12\pi^2}\left[(U_{tb}U_{td}^*)^2\left(m_t^2+\tfrac{3}{4}m_b^2\ln\frac{m_t^2}{m_b^2}+\cdots\right)\right], \tag{6}$$

and f_B is the B meson decay constant. The omitted terms are of order $(m_b/m_t)^2$ or

* *CP*-violating effects in the B^0 system are of order λ^3. For the K^0 system they are of order λ^5 and additional terms are needed in U [5]. We consider only B ($\equiv b\bar{d}$).

smaller relative to the leading term. Using eq. (2) this becomes

$$M_{12} = \mu(1-\rho+i\eta)^2 , \tag{7}$$

where, neglecting the log term in eq. (6),

$$\mu \approx (4\times10^{-14}\,\text{GeV})\left(\frac{m_t}{40\,\text{GeV}}\right)^2\left(\frac{f_B}{160\,\text{MeV}}\right)^2\left(\frac{A}{1.25}\right)^2 . \tag{8}$$

In the K^0 system it is also important to consider the width matrix Γ_{12}. In the B^0 system [2] Γ_{12} is smaller than M_{12} by a factor $(m_b/m_t)^2$ and will be neglected here, although it is essential for the smaller *CP*-violating effects discussed elsewhere [1, 2].

The mass eigenstates are now given by

$$|B_H\rangle = \cos\tfrac{1}{2}\Theta|B_-\rangle - i\sin\tfrac{1}{2}\Theta|B_+\rangle ,$$

$$|B_L\rangle = -i\sin\tfrac{1}{2}\Theta|B_-\rangle + \cos\tfrac{1}{2}\Theta|B_+\rangle , \tag{9}$$

with a mass difference

$$m_H - m_L = 2\mu[(1-\rho)^2+\eta^2] , \tag{10a}$$

$$\tan\tfrac{1}{2}\Theta = \eta/(1-\rho) . \tag{10b}$$

The condition $\eta = (1-\rho)$ yields maximum *CP* violation in the mixing matrix in the sense that the mass eigenstates are equal mixtures of the *CP* eigenstates. This is meaningful, however, only if the phase convention has some physical meaning, since by changing the phase convention ($B \to B\,e^{i\alpha}$) we can effectively rotate Θ to zero.

We consider now decays to final states which are eigenstates of *CP*. Three classes are shown in table 1, each defined by the quark decay indicated. Examples of final *CP* eigenstates are given for each class. The last column of table 1 contains estimates of the branching ratio for all decays of the class; the exclusive decays of interest

TABLE 1

CP eigenstates for B^0 decay

Quark decay	Final state	*CP*	Amplitude factor	Total BR
$b\to c+\bar{c}+s$	$\psi+K^0_s$ $D+\bar{D}+K^0_s$	— + [a])	$A\lambda^2$	12%
$b\to c+\bar{c}+d$	$D+\bar{D}$ $\psi+\pi^0$	+ +	$A\lambda^3$	1%
$b\to c+\bar{u}+d$	$D^0_-+\pi^0$	+ [b])	$A\lambda^2$	54%

[a]) Because of angular momentum barriers this is mainly +, but $CP = -$ is not absolutely forbidden.

[b]) D^0_- is the *CP*-odd state identifiable by the decay $K_s+\pi^0$.

may have a much smaller branching ratio. The first class is suppressed by phase space, while the second is suppressed both by phase space and mixing angles. For all the decays considered the quark decay amplitude is real (neglecting corrections of order λ^2); this is the crucial significance of the phase convention for our purposes. Thus B_- decays only to *CP*-odd states and B_+ to *CP*-even. This conclusion depends on the dynamical assumption that the indicated quark decay is the only significant contribution to the indicated B^0 decay. For the first two classes the contributions of other quark decays are severely suppressed by the necessity of creating a $\bar{c}c$ pair. For the last class there is a contribution from the quark diagram $b \to \bar{c} + u + \bar{d}$ but this is suppressed by a factor λ^2.

An analogue of the Fitch–Cronin parameter ε (or η_{+-}) is now defined as

$$\varepsilon_B \equiv \frac{A(B_L \to CP \text{ odd})}{A(B_H \to CP \text{ odd})} = \frac{A(B_H \to CP \text{ even})}{A(B_L \to CP \text{ even})}$$

$$= i \tan \tfrac{1}{2}\Theta = \frac{i\eta}{1-\rho}, \qquad (11)$$

where the *CP*-even or -odd states belong to the classes illustrated in table 1. This definition is appropriate for $|\Theta| < 90°$, which follows from eq. (3). In order to explain quantitatively the *CP* violation in the K^0 system m_t must be greater than 40 GeV according to the standard analysis [9, 10]. For m_t between 40 and 60 GeV, η must be in the neighborhood of 0.5 and ρ between -0.5 and 0.5 with the corresponding range of $|\varepsilon_B|$ between $\frac{1}{4}$ and unity*. In contrast to the K^0 case the value of ε_B depends only on the elements of the KM matrix with the very minimal dynamical assumption discussed above.

In order to measure ε_B it is necessary to exploit some difference between B_H and B_L. Since, as previously stated, the width difference is very small we must make use of the mass difference. Thus, even if ε_B is near unity, experimental measures may be suppressed by powers of $(m_H - m_L)/\Gamma$, which is expected to be below unity.

As an example we consider a B^0 version of the standard K^0 oscillation experiment; because of the relatively long B lifetime this may indeed become possible. In high-energy collisions, either hadronic or e^+e^-, typically a $(B^0, \bar{B}^0)$ pair is produced. If a $\bar{B}^0$ decay is observed at sufficiently short times then the companion is identified as a B^0. The B_0 has a time-dependent form given by

$$|B_0 t\rangle = \sqrt{\tfrac{1}{2}}\, e^{+i\Theta/2}\, e^{-im_L t}\, e^{-\Gamma t/2} \{B_H\, e^{-i(m_H - m_L)t} + B_L\}.$$

The probability for decay into one of the *CP* eigenstates listed in table 1 has a probability given by

$$\text{Prob}\,(B_0 t \to \pm) = \text{const}\,\{(1 + |\varepsilon_B|^2 \pm 2(i\varepsilon_B) \sin(m_H - m_L)t\}\, e^{-\Gamma t}$$

$$= \text{const}\,\{1 \mp \sin\Theta \sin(m_H - m_L)t\}\, e^{-\Gamma t}, \qquad (12)$$

* We have used the equations of ref. [10] for ε with the matrix element parameter $B = \frac{1}{3}$. If B is raised to $\frac{2}{3}$ our numerical reslts for ε_B and $r_\pm$ hold for 30 GeV $< m_t <$ 45 GeV.

where $\pm$ refers to the CP eigenvalue and $\Gamma = \Gamma_1 = \Gamma_2$. The existence of the oscillatory interference term is a definite sign of CP violation [4]. The CP violation is also shown by the reversal of the $\pm$ sign in eq. (12) for $\bar{B}^0$ decays. Without seeing the time dependence a measure of CP violation is given by the total rate asymmetry [3]

$$r_{\pm} = \frac{\Gamma(B^0 \to \pm) - \Gamma(\bar{B}^0 \to \pm)}{\Gamma(B^0 \to \pm) + \Gamma(\bar{B}^0 \to \pm)}$$

$$= \mp \frac{(m_H - m_L)/\Gamma}{1 + (m_H - m_L)^2/\Gamma^2} \sin \Theta . \quad (13)$$

Unlike some other asymmetries [1, 2] this has a linear dependence on $(m_H - m_L)/\Gamma$, for small values of which we can rewrite eq. (13):

$$r_{\pm} = \mp 4\eta(1-\rho)\mu/\Gamma , \quad (14a)$$

using eqs. (10). For $f_B = 160$ MeV and m_t between 40 and 60 GeV values of η and ρ determined from the standard analysis of the K^0 system inserted in eqs. (13) and (10) yield values of $|r_{\pm}|$ near 0.25. Thus in the KM model we expect $|r_{\pm}|$ to be near half its maximum possible value.

The measurement of $r_{\pm}$ requires that a B^0 (or $\bar{B}^0$) "beam" has been tagged by the observation at a sufficiently early time of a $\bar{B}^0$ (or B^0). An experiment that requires no time determinations is the total asymmetry between the final states ($\bar{B}^0$, $\pm$) and (B^0, $\pm$). As emphasized in ref. [3] this asymmetry depends upon the correlation between the B^0 and $\bar{B}^0$ in the initial state and indeed vanishes for the simple reaction $e^+ + e^- \to B^0 + \bar{B}^0$. In collisions far above the $B^0\bar{B}^0$ threshold, with many particles in the final state, we expect B^0 and $\bar{B}^0$ to be uncorrelated. In this case the total asymmetry A is found [3] to be given by

$$A = r_{\pm}[1 + (m_H - m_L)^2/\Gamma^2]^{-1} . \quad (14b)$$

In a realistic experiment one might expect detector 1 to see a charged lepton so as to identify a b or $\bar{b}$ decay and detector 2 to see a characteristic CP eigenstate. In high-energy collisions it is likely that a $B^0(\bar{B}^0)$ will be accompanied by some other bound $\bar{b}$(b) state such as the charged B or Λ_b so that much of the time the decay detected by detector 1 will not be a B^0 or $\bar{B}^0$. In this case eq. (14b) will not apply. As long as the initial state has CP symmetry, as e^+e^- or $\bar{p}p$, this asymmetry is still a sign of CP violation and the value of A will be somewhat greater than that of eq. (14b), closer to $r_{\pm}$. However the total asymmetry is not a useful test of CP violation in the case of pp collisions because of the intrinsic asymmetry of the initial state*.

A CP-violating effect more closely related to the original Fitch–Cronin discovery can be found in $e^+ - e^-$ annihilation. This is primarily of didactic interest. By Bose statistics the process $e^+ + e^- \to B^0 + \bar{B}^0$ necessarily produces one B_H and one B_L. Thus

* I am indebted to Steve Ellis for this comment.

if both B^0 decay into a positive (or negative) *CP* eigenstate we have direct evidence of *CP* violation. An example would be the sequence

$$e^+ + e^- \to B_H + B_L \to (\psi + K_s) + (\psi + K_s)\,. \tag{15}$$

At first glance it might seem that this effect depends only on ε_B and not on $(m_H - m_L)$. However, this is clearly untrue, as can be seen by noting that the original process can also be described as producing one B_- and one B_+. Since only B_- decays into $\psi + K_s$, the process (15) depends on an oscillation of B_+ into B_-. If the branching ratio for B_- to decay into $\psi + K_s$ is r, then the probability for process (15) is given by

$$\tfrac{1}{2}r^2 \sin^2\Theta \int_0^\infty \Gamma\, \mathrm{d}t\, \mathrm{e}^{-\Gamma t} \sin^2[\tfrac{1}{2}(m_H - m_L)t] = \tfrac{1}{4}r^2 \sin^2\Theta \frac{(m_H - m_L)^2}{(m_H - m_L)^2 + \Gamma^2}\,. \tag{16}$$

When $(m_H - m_L)/\Gamma \gg 1$ we obtain the naive result $\frac{1}{4}r^2 \sin^2\Theta$.

Another class of final *CP* eigenstates, exemplified by $\pi^+\pi^-$ or $\pi^0\pi^0$, is associated with the quark decay $b \to u + \bar{u} + d$. With our phase convention this decay amplitude is complex. The corresponding measure of *CP* violation given by

$$\varepsilon'_B = \frac{A(B_L \to \pi^+\pi^-)}{A(B_H + \pi^+\pi^-)}$$

will not be equal to ε_B. It is of course possible to change the phase convention so that ε'_B is due to mixing only

$$\varepsilon'_B = \tan\tfrac{1}{2}\Theta'\,, \tag{17}$$

with

$$\Theta' = \Theta + \alpha\,,$$

$$\tan\tfrac{1}{2}\alpha = \frac{\eta}{\rho}\,. \tag{18}$$

The difference between ε'_B and ε_B is a sign that the *CP* violation is not a superweak effect; that is, all *CP* violation cannot be blamed on the mass matrix. This is analogous to the difference between η_{+-} and η_{00} in K^0 decay. The neglect of other quark decays contributing via non-spectator diagrams in deriving eq. (18) may be somewhat less justified than for the calculation of ε_B.

It is possible that *CP* violation in the K^0 system cannot be explained by the KM mechanism; that is, the parameter η might be zero or very small. Even if this is the case it is still probable that B decay is described to a good approximation by the KM matrix. The *CP* violation in the K^0 system and presumably also in the B^0 system is then mainly due to new contributions to the imaginary part of the mass matrix. These may be due to superweak $\Delta S = 2$ tree diagrams or to new box diagrams as in the $SU(2)_L \times SU(2)_R \times U(1)$ model [11]. The *CP*-violating effects that may be significant for the B^0 system in these models are exactly those discussed in this paper. In the limit that η goes to zero we have $\varepsilon_B = \varepsilon'_B$. Whereas in the KM model we expect $|\varepsilon_B|$ of order unity in spite of the small value of ε for the K^0 system, there is no reason to expect such a difference in the alternative models, although it is not impossible. It is also possible with additional contributions to the mass matrix to generate a larger value for $(m_H - m_L)$ and so somewhat increase the experimental effects described by eq. (13).

In conclusion we have shown that our present knowledge of the KM matrix leads to a good probability that *CP* violation is intrinsically large in the $B^0 - \bar{B}^0$ system. The parameter $|\varepsilon_B|$ is expressed unambiguously in terms of KM matrix elements and may well be of order unity. Unfortunately the necessary experiments to measure $|\varepsilon_B|$ involve relatively rare exclusive decays such as those listed in table 1.

I am indebted to discussions with J.S. Hagelin and E.A. Paschos. This research was initiated at the Aspen Center for Physics and was supported in part by the US Department of Energy, a John Simon Guggenheim Memorial Foundation Fellowship, and the National Science Foundation under grant no. PHY77-27084, supplemented by funds from the National Aeronautics and Space Administration.

References

[1] J. Ellis et al., Nucl. Phys. B131 (1977) 285; B132 (1978) 541;
E. Ma et al., Phys. Rev. D20 (1977) 2888;
A. Ali and Z.Z. Aydin, Nucl. Phys. B148 (1979) 165;
V. Barger et al., Phys. Rev. D21 (1980) 952;
E. Franco et al., Nucl. Phys. B194 (1982) 403;
E.A. Paschos, B. Stech and U. Türke, Phys. Lett. 128B (1983) 240

[2] J.S. Hagelin, Nucl. Phys. B193 (1981) 123

[3] I.I. Bigi and A.I. Sanda, Nucl. Phys. B193 (1981) 85

[4] L. Wolfenstein, in Theory and phenomenology in particle physics, ed. A. Zichichi (Academic Press, 1969) p. 218

[5] L. Wolfenstein, Phys. Rev. Lett. 51 (1983) 1945

[6] N.S. Lockyer et al., Phys. Rev. Lett. 51 (1983) 1316

[7] C. Klopfenstein et al., Phys. Lett. 130B (1983) 444

[8] L. Maiani, in Proc. 1977 Int. Symp. on Lepton and photon interactions at high energies, DESY, Hamburg, 1978, p. 867

[9] P.H. Ginsparg et al., Phys. Rev. Lett. 50 (1983) 1415; erratum 51 (1983) 1395

[10] F. Gilman and J.S. Hagelin, Phys. Lett. 126B (1983) 111

[11] R.N. Mohapatra and J.C. Pati, Phys. Rev. D11 (1975) 566;
D. Chang, Nucl. Phys. 214 (1983) 435

5. Neutron Electric Dipole Moment

Introduction

Of possible experiments to detect directly time-reversal (T) violation, the most sensitive experiments appear to be measurements of electric dipole moments of non-degenerate systems. The greatest effort and the most sensitive results involve the electric dipole moment D_n of the neutron.

The interaction $\boldsymbol{\sigma} \cdot \boldsymbol{E}$ violates both parity (P) and time-reversal (T) invariance. With the discovery of large P violation in the weak interaction, the measurement of electric dipole moments became a way of studying T violation. The pioneering experiments [1] on the neutron were done by Purcell, Ramsey and collaborators. Experiments now being done involve confining neutrons for the significant fraction of a lifetime and aim at finding values down to 10^{-26} e cm. The Soviet group of Lobashev and collaborators have reported [2] a result of $-(1.4 \pm 0.6) \times 10^{-25}$ e cm, but until more work is done this is to be considered as giving an upper limit. The Grenoble group has quite recently found [3] a result of $-(0.3 \pm 0.2) \times 10^{-25}$ e cm, with a systematic error of 0.4×10^{-25} e cm. The upper limit on $|D_n|$ of 2×10^{-25} e cm derived from these results is probably the best limit on possible T violation now available. A very good introduction to the subject of the neutron electric dipole moment was given by Golub and Pendlebury [4].

The order-of-magnitude expected for D_n can be written as

$$\frac{D_n}{\mu_n} = \left(\frac{GM_n^2}{4\pi}\right) \times (\text{time-reversal factor}),$$

where μ_n is the magnetic moment, and the first factor on the right is an estimate of the price paid for parity violation. If the time-reversal factor is set equal to ϵ', then the expected value is 10^{-25} e cm, just the values now being probed experimentally. In the milliweak models discussed in chapter 3 (Weinberg model and left–right model), the theoretical values have this order of magnitude. However, in the standard KM model the value of D_n is zero to lowest order. In fact, the electric dipole moment of a quark vanishes even in second order [5]. However, as pointed out by many authors, the electric dipole moment of the neutron is not zero in second order, but is dominated by long-distance contributions. A reasonable but uncertain calculation [6] gives $D_n = 2 \times 10^{-32}$ e cm, which in any case, is much too small to be detected. A review of calculations of D_n has been given by Shabilin [7].

There is a quite different possible origin for a neutron electric dipole moment. Once CP is violated, there exists the possibility of adding to the Lagrangians of QCD a term involving gluon fields that violates both P and T symmetry. The coefficient of this term is called θ_{QCD}. From an experimental view, the only possible way of detecting this term is from a non-zero value of D_n. The present limit on D_n translates [8] into an upper limit on θ_{QCD} of 10^{-9} to 10^{-10}.

In the standard model, taken literally with CP violation introduced via complex Yukawa couplings, the value of θ_{QCD} becomes an arbitrary renormalization constant. Its very small value then appears to be unexplained; this is often referred to as the "strong CP problem". One resolution [9] is to add a new interaction involving a new particle called the axion. One then obtains a new symmetry which makes θ_{QCD} vanish. While the original axion model can be ruled out experimentally, variations called invisible axion models [10] are possible.

Another possibility is to consider that CP violation occurs spontaneously as discussed by Lee (chapter 3). In this case, θ_{QCD} is calculable. In many models (e.g., the Weinberg model as discussed by Branco in chapter 3) the value of θ_{QCD} turns out much too large. A particularly interesting possibility is that the spontaneous CP violation occurs at a high mass scale and that it results in an effective interaction at the $SU(2) \times U(1)$ scale that is just the standard KM model. A model of this sort has been developed by Nelson [11]. In this particular model, θ_{QCD} is expected [12] to be 10^{-11} or greater. Thus it is possible to have CP violation in the K^0 system explained by the standard KM model and still have a measurable value for D_n. General criteria for this class of models have been given by Barr [13].

It is also possible to search for the electric dipole moment D_e of the electron by studying electric dipole moments of atoms. In general this has not appeared to be as sensitive a test although it may be in some models. Theoretical calculations of D_e in different models has been reviewed by Cheng [14].

References

[1] J.H. Smith, E.M. Purcell and N.F. Ramsey, Phys. Rev. 108 (1957) 120. Paper 30, this volume.

[2] I.S. Altarev et al., JETP Lett. 44 (1986) 460. Paper 31, this volume.

[3] B.R. Heckel, private communication.
For an earlier review of experiments see N.F. Ramsey, Ann. Rev. Nucl. Sci. 32 (1982) 211.

[4] R. Golub and J.M. Pendlebury, Contemp. Phys. 13 (1972) 519.

[5] E.P. Shabilin, Sov. J. Nucl. Phys. 28 (1978) 75.

[6] I.B. Khriplorich and A.R. Zhitnitskii, Phys. Lett. B 109 (1982) 490. Paper 32, this volume.

[7] E.P. Shabilin, Sov. Phys. Usp. 26 (1983) 297.

[8] R.J. Crewther, P.Di Vecchia, G. Veneziano and E. Witten, Phys. Lett. B 88 (1979) 123; Erratum B 91 (1980) 487. Paper 33, this volume.

[9] R.D. Peccei and H.R. Quinn, Phys. Rev. D 16 (1977) 1791. Paper 34, this volume.

[10] J.E. Kim, Phys. Rev. Lett. 43 (1979) 103;
M. Dine, W. Fischler and M. Srednicki, Phys. Lett. B 104 (1981) 199.

[11] A. Nelson, Phys. Lett. B 136 (1984) 387. Paper 35, this volume.

[12] A. Nelson, Phys. Lett. B 143 (1984) 165.

[13] S. Barr, Phys. Rev. D 30 (1984) 1805.

[14] H.Y. Cheng, Phys. Rev. D 28 (1983) 150.

Experimental Limit to the Electric Dipole Moment of the Neutron

J. H. SMITH,* E. M. PURCELL, AND N. F. RAMSEY
Oak Ridge National Laboratory, Oak Ridge, Tennessee, and Harvard University, Cambridge, Massachusetts
(Received May 17, 1957)

An experimental measurement of the electric dipole moment of the neutron by a neutron-beam magnetic resonance method is described. The result of the experiment is that the electric dipole moment of the neutron equals the charge of the electron multiplied by a distance $D=(-0.1\pm2.4)\times10^{-20}$ cm. Consequently, if an electric dipole moment of the neutron exists and is associated with the spin angular momentum, its magnitude almost certainly corresponds to a value of D less than 5×10^{-20} cm.

1. INTRODUCTION

SEVERAL years ago Purcell and Ramsey[1] pointed out that the usual parity arguments for the nonexistence of electric dipole moments for nuclei and elementary particles, although appealing from the point of view of symmetry, were not necessarily valid. In particular they pointed out that the validity of the parity assumption must rest on experimental evidence and that the experimental evidence was not as conclusive as then generally supposed in the case of nuclei and elementary particles, even though there was abundant evidence for the assumption in the case of electromagnetic forces. Analysis of the experimental evidence against the existence of electric dipole moments of both nuclei and elementary particles showed[1] that most experiments would not have revealed an electric dipole moment smaller than 5×10^{-23} esu. Such a moment is equal to the charge of the electron multiplied by a distance D of 10^{-13} cm. Henceforth we shall express the nuclear dipole moment in cm, it being understood that the "dipole length" D is to be multiplied by e, the magnitude of the electron charge. Probably the most sensitive experiments at the time were those to measure the neutron-electron interaction[2,3]; it was found that the observed results of these experiments could result from a neutron electric dipole moment with a D of 3×10^{-18} cm. Purcell and Ramsey[1] proposed a sensitive neutron-beam resonance experiment for the detection of an electric dipole moment.

This experiment was successfully completed several years ago. However, the negative results of the experiment were in accordance with the then widely accepted views on parity so the detailed description[4] of the experiment was not published. The upper limit to the electric dipole moment determined in this experiment has occasionally been quoted in other publications.[5,6]

Lee and Yang[6] have analyzed the effects of parity nonconservation on the angular distributions of beta decay, the angular distributions of π-μ-e decays, and existence of electric dipole moments of particles. The effects of the first two of these have been observed by Wu, Ambler, Hayward, Hoppes, and Hudson,[7] and by Garwin, Lederman, and Weinrich.[8] Since electric moments are primarily determined by the strong forces, Lee and Yang[6] showed that the effect of mixed parity should produce an electric dipole moment even smaller than the upper limit set by the experiment described in the present paper. In their most recent theories, Lee and Yang[9] no longer anticipate the existence of an electric dipole moment for the neutron, and arguments involving time-reversal invariance[9,10] can be advanced against its existence. These arguments, however, like the original ones of parity, can be questioned.

Although the negative results of the experiment described here are fully consistent with the current theories, a brief description of the experiment and its results seems appropriate at the present time since the experiment provides the most sensitive experimental upper limit to an electric dipole moment of any elementary particle or nucleus and since the original parity arguments against the existence of such electric dipole moments are now known to be invalid.

2. METHOD AND APPARATUS

The method used in this experiment was similar to that used to measure the magnetic moment of the neutron.[11–14] We might observe that the neutron is the only particle suitable for such an experiment. Any charged particle exposed to an electric field which is not small in the time average will necessarily suffer a large change in momentum. A schematic diagram of the apparatus is shown in Fig. 1. A beam of neutrons 1-cm high and 0.1-cm wide leaves the pile and is polarized by total reflection from a polished, magnetized iron

* Now at the University of Illinois, Urbana, Illinois.
[1] E. M. Purcell and N. F. Ramsey, Phys. Rev. **78**, 807 (1950).
[2] Havens, Rabi, and Rainwater, Phys. Rev. **72**, 634 (1947).
[3] E. Fermi and L. Marshall, Phys. Rev. **72**, 1139 (1947).
[4] J. H. Smith, Ph.D. thesis, Harvard University, 1951 (unpublished).
[5] N. F. Ramsey, *Molecular Beams* (Oxford University Press, Oxford, 1956).
[6] T. D. Lee and C. N. Yang, Phys. Rev. **104**, 254 (1956).
[7] Wu, Ambler, Hayward, Hoppes, and Hudson, Phys. Rev. **105**, 1413 (1957).
[8] Garwin, Lederman, and Weinrich, Phys. Rev. **105**, 1415 (1957).
[9] T. D. Lee and C. N. Yang, private communication (to be published).
[10] L. Landau (private communication).
[11] L. W. Alvarez and F. Bloch, Phys. Rev. **57**, 111 (1940).
[12] W. R. Arnold and A. Roberts, Phys. Rev. **71**, 878 (1947).
[13] Bloch, Nicodemus, and Staub, Phys. Rev. **74**, 1025 (1948).
[14] Corngold, Cohen, and Ramsey, Phys. Rev. **104**, 283 (1956).

mirror[15] at A. The beam then passes into a region of homogeneous magnetic field produced by the magnet B. The polarization of the beam is analyzed by a magnetically saturated iron isthmus[16] in the magnet at A'. The neutrons traversing the apparatus are counted in a BF_3 proportional counter at D. When a radio-frequency magnetic field is applied to the neutron beam in two separated coils[17] C and C' at the average Larmor frequency of the neutrons in the magnetic field between these coils, transitions may be induced to the opposite spin state. The intensity of the beam passing through the analyzer will consequently decrease. If the neutron had an electric dipole moment oriented along its spin, a steady electric field applied parallel to the homogeneous magnetic field in the region between the coils C and C' would alter the torque on the neutron and hence would change its precession frequency. This would result in a shift in the frequency at which a maximum number of spin transitions were induced. Such an electric field was applied between the plates of a condenser E.

The magnetized iron mirror used as the polarizer was polished flat on a pitch lap to one or two wavelengths of light per inch. The mirror was six inches long and one and one-half inches high, though a beam of neutrons only one centimeter high was used. The magnetic field at the surface of the mirror was nearly 2500 oersteds. The intensity of neutrons reflected from this mirror as a function of mirror angle was compared to that theoretically expected from a flux of neutrons with a 500°K Maxwellian velocity distribution and found to agree within experimental error. The polarization theoretically expected at the mirror angles used in the experiment was approximately 85%. The analyzer was a magnetically saturated piece of cold-rolled steel such as other experimenters have used for both polarizer and analyzer.[11–13] A magnetic field of approximately 10 500 oersteds was maintained in the analyzer block. A second mirror would have been preferable, but limited space prevented its use. With the mirror polarizer and steel block analyzer changes of 40% in intensity were observed in varying the frequency through resonance.

The homogeneous field magnet was constructed by clamping two cold-rolled steel pole faces 71-in. long by 5-in. high by 1-in. thick 1.75-in. apart. Sixteen small Alnico magnets bolted between the pole faces along the bottom supplied the magnetomotive force. This construction provided a convenient trough in which the radio-frequency coils and electric-field plates could be suspended. The magnet provided a field of about 250 oersteds which corresponds to a neutron transition frequency of 750 kc/sec. The split rf coil technique does not require a field of high homogeneity, so the homogenizing was done by rubber-cementing suitably cut

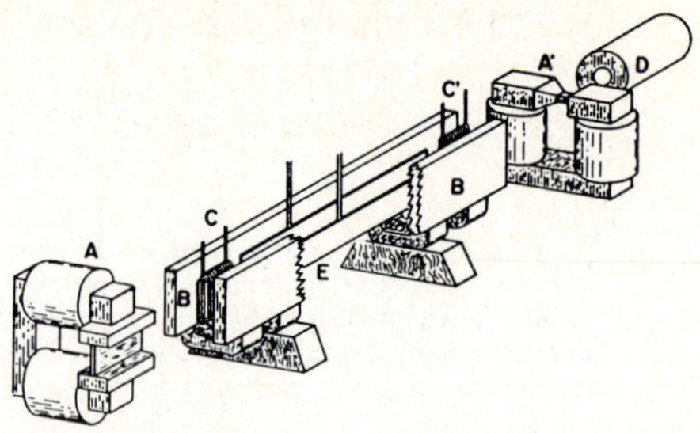

FIG. 1. Schematic diagram of the apparatus. A, the magnetized iron mirror polarizer. A', the magnetized iron transmission analyzer. B, the pole faces of the homogeneous field magnet. Note the horseshoe-like magnets bolted along the bottom. C, C', the coils for the radio-frequency magnetic field. D, the BF_3 neutron counter. The magnetic fields in the polarizing magnet and the homogeneous field magnet are at right angles, and two twisted iron strips were used between them to rotate the neutron spins adiabatically.

steel shims to the magnet pole faces. The final field was uniform to within $\frac{1}{3}$ oersted of its mean value over the relevant region traversed by the neutron beam. All magnetic-field measurements were made with a proton-resonance apparatus. The field was not exceedingly stable. It was necessary to enclose the whole apparatus in a box whose temperature was regulated to about 0.2°C. Large steel objects and electromagnets in neighboring laboratories disturbed the field noticeably. Such day to day disturbances are not particularly objectionable because the comparisons involved in any one measurement extend over a relatively short period. Data were discarded, however, when it was discovered that the field was drifting excessively.

The electric field was produced by applying a voltage of approximately 25 000 volts between highly polished nickel-plated copper plates 0.349-cm apart and 135-cm long. This whole structure was enclosed in a vacuum chamber for insulation and suspended in the magnetic field. The nickel disturbed the magnetic field to some extent, but unfortunately the ferromagnetic substances seem to have the best vacuum sparking characteristics.

The radio-frequency magnetic field was produced in two helical coils 5-cm long with their axes along the neutron beam. The coils were spaced 159 cm apart. The radio-frequency current was supplied by a conventional resistance-stabilized electron-coupled Hartley oscillator driving a 1625 beam power tube through two amplifier-buffer stages. The master oscillator was of very rigid construction and was quite stable. During the experimental runs the frequency was monitored on a Bendix Radio LM-18 frequency meter. The average frequency during a run was known to be better than 2 cps per Mc/sec.

All counts were referred to the counts recorded in a thin BF_3 proportional counter used as a monitor, through which the beam passed as it left the pile.

Typical resonance curves are shown in Fig. 2. As the radio-frequency magnetic field is changed in frequency

[15] D. J. Hughes and M. Burgy, Phys. Rev. **75**, 463 (1949).
[16] Bloch, Hammermesh, Staub, and Condit, Phys. Rev. **64**, 47 (1943); **70**, 972 (1945).
[17] N. F. Ramsey, Phys. Rev. **75**, 996 (1949).

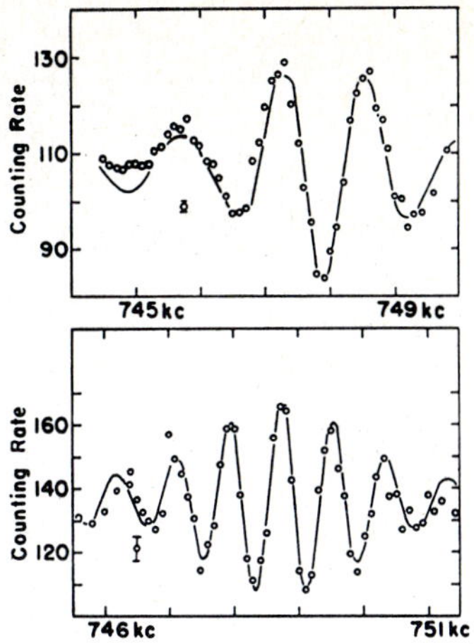

FIG. 2. Resonance curves of the neutron counting rate *versus* the frequency of the radio-frequency magnetic field. The upper curve is for a mirror angle of 3.14×10^{-3} radian, and the lower for 5.77×10^{-3} radian. A typical root-mean-square counting statistical error is shown. The lower curve shows a narrower resonance due to the fact that only slower neutrons are reflected at the larger mirror angles. It also shows more resonance detail since the polarized neutrons are more nearly monochromatic. The central peak is a minimum in the upper curve and a maximum in the lower curve since the phase of one of the coils was reversed.

the intensity varies as the neutrons have their spins flipped by varying amounts. The curves shown are theoretical calculations with the resonant frequency and main peak-to-valley ratio as adjustable parameters. A Maxwellian 500°K spectrum was used for the neutrons. The second curve of Fig. 2 shows the expected narrowing of the resonance curve at large mirror angles where only the slower neutrons are totally reflected. The agreement between theory and experiment is excellent, as shown by the experimental points. A sample probable error is shown for comparison. The upper curve corresponds more closely to the parameters used in the experiment.

3. RESULTS

To find the effect of the electric field on the resonance frequency of the neutrons, the radio-frequency magnetic field was set to a frequency corresponding to a point on one of the steep sides of the central resonance. Counting rates were then recorded as the polarity of the electric field was switched back and forth. The results are shown in Table I. The first column gives the number of the

TABLE I. Summary of results.

Run	Number of counting periods	Average number of counts in each counting period	Difference between averages with field + and −	Total difference between + and − voltages	Slope of resonance curve	Run taken above or below resonance
1	16	17 102.3	1.7	47 000	8.28	above
2	16	16 991.3	−52.5	46 000	10.58	below
3	16	18 217.3	68.1	51 000	12.11	below
4	18	17 920.4	−30.7	51 000	11.86	above
5	8	18 447.6	−17.3	51 000	12.11	below
6[a]	8	18 018.5	−248.0	54 000	11.86	above
7	16	17 350.4	−93.3	56 000	12.90	below
8	16	17 987.0	−16.0	56 000	10.35	above
9	8	17 156.4	6.8	56 000	12.90	below
10	16	17 310.3	69.9	56 000	12.33	above
11	16	17 626.8	−29.5	56 000	11.32	above
12	15	16 223.5	39.8	53 000	13.68	below
13	16	16 613.2	47.8	54 000	13.68	below
14	17	16 581.6	−31.5	54 000	13.68	below
15	16	16 615.5	−20.2	54 000	13.14	above
16	14	16 596.2	−18.4	54 000	13.14	above

[a] Run number 6 was discarded.

run; the second, the number of readings taken; the third, the average number of counts per reading; the fourth, the difference in the average of the counting rates with the field positive and negative; the fifth, the total change in the potential difference between the condenser plates; the sixth, the slope of the resonance curve in counts per cycle; the seventh, whether the data was taken above or below the resonance frequency. In computing the upper limit for the electric dipole moment, run number 6 has been discarded. Although nothing was found wrong with this particular run, the result looks peculiar and indeed the next day the counting equipment failed completely.

From the data of Table I, the value of the neutron electric dipole moment is found to be

$$D=(-0.1\pm2.4)\times10^{-20}\text{ cm},$$

and the indicated error is the root-mean-square deviation from the mean. From this one can conclude that if an electric dipole moment of the neutron exists and is associated with the spin angular momentum it is probably less than the charge of the electron multiplied by 3×10^{-20} cm and it is almost certainly less than 5×10^{-20} cm. As discussed in the first section, this result is consistent both with the dipole moment being very small as would be expected if the only violation of parity conservation were in the weak forces, and with the dipole moment vanishing as would be expected from arguments of time-reversal invariance.

Search for an electric dipole moment of the neutron

I. S. Altarev, Yu. V. Borisov, N. V. Borovikova, A. B. Brandin, A. I. Egorov, S. N. Ivanov, É. A. Kolomenskiĭ, M. S. Lasakov, V. M. Lobashev,[1] A. N. Pirozhkov, A. P. Serebrov, Yu. V. Sobolev, R. R. Tal'daev, and B. V. Shul'gina

B. P. Konstantinov Leningrad Institute of Nuclear Physics, Academy of Sciences of the USSR

(Submitted 21 August 1986)

Pis'ma Zh. Eksp. Teor. Fiz. **44**, No. 8, 360–363 (25 October 1986)

A *CP*-breaking effect of an electric dipole moment of the neutron has been measured by a method of prolonged confinement of ultracold neutrons. The measured value, $d_n = -(1.4 \pm 0.6) \times 10^{-25}$ e·cm, is interpreted as a new upper limit: $|d_n| < 2.6 \times 10^{-25}$ e·cm at a 95% confidence level.

In this letter we report the results of new experiments in a search for an electric dipole moment of the neutron. The experiments were carried out with ultracold neutrons at the VVR-M reactor at the B. P. Konstantinov Leningrad Institute of Nuclear Physics.

Earlier experiments[1,2] had used a magnetic-resonance spectrometer of the so-called continuous-flow type with confinement of the ultracold neutrons for an average time on the order of 5 s. Two chambers were used to confine neutrons with oppositely directed electric fields (this is a differential method for measuring the electric dipole moment).

The results of those measurements yielded a value $d_n = -(2 \pm 1) \times 10^{-25}$ e·cm for the electric dipole moment of the neutron. This result was interpreted as a limitation on the magnitude of the dipole moment: $|d_n| < 4 \times 10^{-25}$ e·cm at a 95% confidence level.

It appeared to be difficult to achieve any further improvement in the sensitivity in the continuous-flow version of the spectrometer, so a modification was developed with prolonged confinement of ultracold neutrons in a closed chamber. In recent experiments by a group from the Laue-Langevin Institute (Grenoble), a confinement time on the order of 80 s was achieved in a storage version with a single chamber, and the result $d_n = (0.3 \pm 4.8) \times 10^{-25}$ e·cm was achieved.[3]

In the present experiments we have gone back to the differential method, using two neutron-confinement chambers with oppositely directed electric fields (Fig. 1)

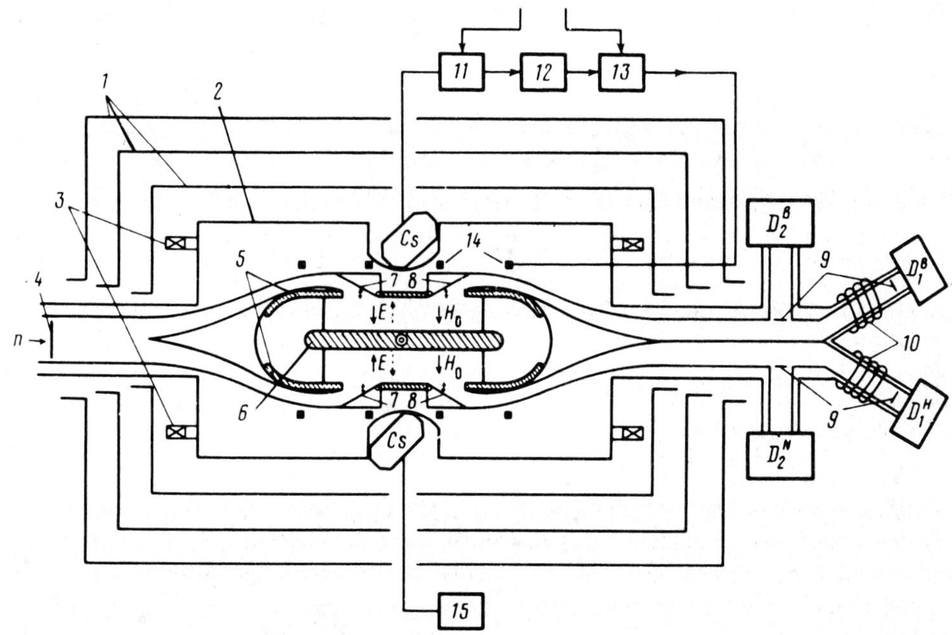

FIG. 1. The magnetic-resonance spectrometer for ultracold neutrons for measuring the electric dipole moment of the neutron. 1—Magnetic shields; 2—vacuum chambers; 3—Helmholtz coils for producing a static magnetic field H_0; 4—polarizer of the ultracold neutrons; 5—grounded electrodes; 6—high-voltage electrode; 7—entrance shutters; 8—exit shutters; 9—analyzers; 10—rf flipper; 11—computer-controlled Cs magnetometer; 12—frequency divider; 13—computer-controlled unit for producing the rf pulses; 14—coils for producing an oscillating field; 15—Cs magnetometer; D—ultracold-neutron detectors ("U" and "L" specify the upper and lower chambers; 1 and 2 specify opposite polarizations of the ultracold neutrons).

and with common static magnetic and rf fields. This approach has made it possible to cancel the fluctuations in the magnetic field and to monitor possible systematic errors. The chamber is designed in such a way that the high-voltage electrode which separates the chambers is between two grounded electrodes (caps), which in a sense form a grounded screen around the high-voltage electrode. The average electric field is $\pm(12-15)$ kV/cm at leakage currents below 30 nA. The chamber walls are quartz rings on which films of BeO and Be_3N_2 have been deposited. A system of shutters in the entrance and exit neutron ducts can be opened and closed in a time $\sim$0.3 s. As in the earlier measurements,[1,2] we use a system for simultaneously detecting ultracold neutrons polarized in opposite directions.

Since the three-layer shield of the Permalloy type used by us does not ensure the stability of the magnetic field which would be required for a confinement time of 50–100 s, the position of the resonance is stabilized by means of self-generating cesium quantum magnetometers, positioned above and below the neutron confinement chamber. Since the ratio of the resonant frequency of the cesium magnetometers to the frequency of the neutron resonance is 119.9—approximately 120—we developed a stabilization system in which the frequency of one magnetometer or the average frequency of two magnetometers is divided by 120. The signal is sent from the divider to an rf coil wound around the neutron confinement chamber.[4] This system makes it possible to maintain resonance conditions with a stability corresponding to a magnetic-field stability better than 2×10^{12} T over 5–10 min. The effective stabilization factor in the single-channel version is $\sim$7, and that in the two-channel version is $\sim$15.

A further suppression of the fluctuations in the resonance conditions is achieved by virtue of the differential measurement method, so that the difference between the mean square scatter and the statistical scatter is no more than 5–10%.

In the early measurements, the source of ultracold neutrons was a liquid-hydrogen converter in a beryllium reflector.[5] In the bulk of the measurements we used a new, universal source of cold and ultracold neutrons.[6]

The three- or fourfold increase in the intensity of the ultracold neutrons from the new source, combined with the confinement system of the spectrometer which was developed, with an ultracold-neutron confinement time $\sim$50 s, made it possible to achieve a maximum sensitivity $\sim(2.5\text{–}2.7)\times10^{-25}$ e·cm per day in the measurement of the electric dipole moment.

The measurement cycle includes filling the chamber with ultracold neutrons (30 s), confinement (50–55 s), and the discharge and counting of the ultracold neutrons (30 s). Radio-frequency signals 1.2 s long are applied at the beginning and end of the confinement time. The polarity of the electric field is switched in a 40-s interval between measurement cycles. Otherwise, the measurement conditions are similar to those in Ref. 1.

For each detector we calculate the magnitude of the electric dipole moment, $d^{u,L}_{1,2}$. The "U" and "L" and the 1 and 2 match the notation used in Fig. 1.

The results of the measurements for the four detectors are used to form the following combinations, by analogy with Ref. 1:

$$d_n = \frac{1}{4}\left[(d_1^U + d_2^U) + (d_1^L + d_2^L)\right]$$

$$R = \frac{1}{4}\left[(d_1^U + d_2^U) - (d_1^L + d_2^L)\right]$$

$$P = \frac{1}{4}\left[(d_1^U - d_2^U) - (d_1^L - d_2^L)\right]$$

$$C = \frac{1}{4}\left[(d_1^U - d_2^U) + (d_1^L - d_2^L)\right]$$

The quantity d_n determines the effect which is being sought: the electric dipole moment of the neutron. The quantity R determines the synchronous displacement of the neutron resonance in the two chambers; P serves as a measure of the effect of the switching of the high voltage on the counting systems; and C is a measure of how well these systematic and random factors are canceled out.

Table I summarizes the results of the various measurements of the electric dipole moment, including the results of Refs. 1 and 2 with a continuous-flow system. We see that the results found on the electric dipole moment by the continuous-flow method and those found by the confinement method agree well. In some series of measurements, the value of R is extremely large, but the accuracy with which the synchronous effects for the two chambers have been canceled—estimated from an analysis of individual measurements with a large value of R—yields a factor ~ 0.05. In other words, the possible contribution is, on the average, smaller than 0.3×10^{-25} e·cm.

According to our estimates, other possible systematic effects are inconsequential. For example, the contribution of the effect of a deviation of the electric and magnetic fields from a parallel alignment and the effect of leakage currents $\langle$at $1 < 30$ nA$)$ do not exceed 10^{-26} e·cm.

The overall result of the measurements is therefore

$$d_n = -(1.4 \pm 0.6) \times 10^{-25}\ \text{e}\cdot\text{cm}.$$

Although the value found for the electric dipole moment differs slightly from zero, we cannot draw the conclusion that the existence of a nonzero electric dipole moment, determined by the magnitude of this error, is a certainty, particularly in view of the fundamental importance of this phenomenon. Accordingly, the most natural interpretation of this result is that it gives us a new upper limit

$$|d_n| < 2.6 \times 10^{-25}\ \text{e}\cdot\text{cm (95\% confidence level)}.$$

This result definitely contradicts the estimates of the electric dipole moment based on the Weinberg model, e.g., $d_n = -9 \times 10^{-25}$ e·cm (Ref. 7) or $d_n \sim 10^{-22}$ e·cm (Ref. 8). Limitations on the parameters which can be used can be extracted from several other models.

We wish to thank V. L. Ryabov for assistance in processing the measurements, V. L. Varentsov for assistance in the measurements, and V. G. Muratov, L. A. Grogor'eva, and V. N. Slyusar, for assistance in adjusting the apparatus. We are

Table I. Results of the measurements of the electric dipole moment (EDM) of the neutron (in units of 10^{-25} e·cm).

Measurement conditions	EDM	Effect of magnetic field	Effect of transients	Cancellation effect
	$d_n\left(\frac{\text{S.D.}}{\text{stat}}\right)$	R (S.D.)	P (S.D.)	C (S.D.)
Continuous-flow	$2.1\left(\frac{2.4}{2.4}\right)$	16.6 (4.8)	5.3 (3.5)	4.8 (2.4)
spectrometer	$-1.0\left(\frac{2.8}{2.7}\right)$	1.5 (5.8)	− 5.0 (4.1)	− 2.6 (2.7)
	$-3.4\left(\frac{1.3}{1.3}\right)$	1.1 (2.8)	0.4 (2.2)	− 3.6 (1.3)
Average	− 2.0 (1.0)	4.5 (2.2)	0.6 (1.7)	− 1.8 (1.0)
Storage spectrometer				
Source of ultracold neutrons in beryllium reflector; single-channel stabilization of the resonance	$0.97\left(\frac{2.60}{2.55}\right)$	− 3.7 (4.5)	1.3 (2.9)	0.8 (2.6)
Universal source of cold and ultracold neutrons				
$H_0\downarrow$; two-channel stabilization	$10.4\left(\frac{4.9}{4.2}\right)$	− 12.5 (8.2)	5.0 (5.7)	2.9 (4.8)
$H_0\downarrow$; two-channel stabilization	$-1.9\left(\frac{1.1}{1.1}\right)$	8.9 (1.7)	− 0.7 (1.6)	0.9 (1.1)
$H_0\downarrow$; single-channel stabilization	$-0.6\ \frac{2.3}{2.3}$	9.4 (4.4)	− 0.8 (3.1)	− 1.2 (2.1)
$H_0\uparrow$; single-channel stabilization	$-3.9\left(\frac{2.2}{2.1}\right)$	5.2 (4.4)	− 0.5 (2.5)	1.0 (2.1)
$H_0\uparrow$; two-channel stabilization	$-0.5\left(\frac{1.4}{1.3}\right)$	− 1.9 (2.3)	5.4 (1.7)	0.3 (1.3)
Average	− 1.1 (0.7)	4.0 (1.2)	1.7 (1.0)	0.4 (0.7)
Overall average	− 1.4 (0.6)	4.1 (1.0)	1.4 (0.9)	0.3 (0.6)

particularly indebted to the staff at the VVR-M reactor, including K. A. Konoplev, R. G. Pikulik, and all the members of the cryogenic service for arranging the conditions required for these experiments.

[1)]Institute of Nuclear Research, Academy of Sciences of the USSR.

[1]I. S. Altarev *et al.*, Phys. Lett. **102B**, 269 (1981).
[2]V. M. Lobashev and A. P. Serebrov, J. Phys. (Paris) **45**, 3-11, 1984.
[3]M. J. Pendlebury *et al.*, Phys. Lett. **136B**, 327 (1984).
[4]I. S. Altarev, Yu. V. Borisov, A. B. Brandin, S. N. Ivanov, V. M. Pobashev, A. P. Serebrov, and P. P. Tal'daev, Preprint LIYaF-1117, Leningrad Institute of Nuclear Physics, Leningrad, 1985.
[5]I. S. Altarev *et al.*, Phys. Lett. **80A**, 413 (1980).
[6]I. S. Altarev *et al.*, Pis'ma Zh. Eksp. Teor. Fiz. **44**, 269 (1986) [JETP Lett. **44**, 344 (1986)].
[7]A. R. Zhitnitskiĭ and I. B. Khriplovich, Yad. Fiz. **34**, 167 (1981) [Sov. J. Nucl. Phys. **34**, 95 (1981)].
[8]A. A. Anselm *et al.*, Phys. Lett. **152B**, 116 (1985).

Translated by Dave Parsons

 0021-3640/86/200465-04$01.00

WHAT IS THE VALUE OF THE NEUTRON ELECTRIC DIPOLE MOMENT IN THE KOBAYASHI–MASKAWA MODEL?

I.B. KHRIPLOVICH and A.R. ZHITNITSKY
Institute of Nuclear Physics, 630090, Novosibirsk 90, USSR

Received 18 November 1981

A new mechanism is considered due to which the neutron electric dipole moment D_n arises in the Kobayashi–Maskawa model. This mechanism leads to the estimate $D_n \sim 2 \times 10^{-32}$ e cm, by two orders of magnitude larger than the contributions considered previously.

The search for the electric dipole moment (edm) of the neutron, being carried out for many years, have allowed to restrict essentially the class of possible models of *CP*-violation. Due to the well-known successes of the renormalizable theory of electromagnetic and weak interactions the attempts to describe *CP*-violation within the same approach appear to be the most natural. Now two schemes of this kind are discussed intensively. In one of them *CP*-violation arises in the Higgs sector of the model [1,2], in the other in the fermion sector [3].

The edm of the neutron in the most popular version of the model of the first type, suggested by Weinberg [2], was considered in refs. [2,4–7]. In ref. [7] the neutron edm D_n was shown to be due in this model mainly to the contribution of the strange quarks. The estimate $D_n \simeq -9 \times 10^{-25}$ e cm [‡1] obtained in this way, formally contradicts already the last experimental data [8] $D_n = (2.3 \pm 2.3) \times 10^{-25}$ $\times\, e$ cm. Due, however, to the ambiguities in the value of the model parameters which are extracted from the $K_L \to 2\pi$ decay, only the order of magnitude of the theoretical estimate can be guaranteed, but not its sign. Nevertheless, the improvement of the accuracy of the experiment on the neutron edm by 3–4 times (which can be expected within the next few months) would finally decide the fate of the Weinberg model of *CP*-violation [2].

In the present work we discuss the value of the neutron edm in the Kobayashi–Maskawa (K–M) model [3]. In the first papers on this topic [9–11] it was realized that D_n does not arise here in the first order in the Fermi constant G_F. Then Shabalin [12] showed that a quark edm vanishes also in the two-loop approximation [‡2]. In ref. [12] it was also noted that D_n can arise due to a weak interaction between quarks constituting the neutron. The investigation of the corresponding diagrams (see fig. 1) has led to the estimate $D_n \sim 10^{-34}$ e cm [15]. Other, larger values for this contribution [16,17] are an overestimation.

In the present work another mechanism leading to the neutron edm and due to the interquark weak interaction is considered. Following the idea used in

‡1 In ref. [7] we have made an error in the sign of the strange quark contribution that has led to the estimate $D_n \simeq +7 \times 10^{-25}$ e cm.

‡2 Even taking into account strong-interaction radiative corrections, one cannot get a quark edm exceeding 10^{-34} $\times\, e$ cm [13]. Larger predictions [14] for D_n obtained in this way are perhaps overestimated.

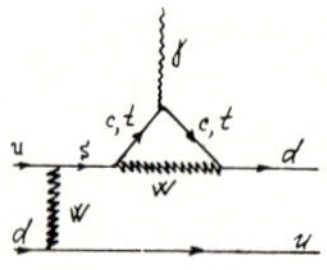

Fig. 1.

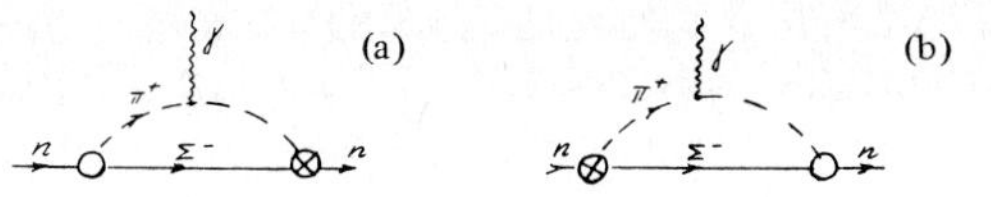

Fig. 2.

ref. [18] to calculate the neutron edm in another model (see also refs. [19,7]), we shall look for the contribution to D_n which is singular in the chiral limit. This contribution, proportional to ln $(1/m_\pi)$, arises from the diagrams 2a,b. The symbol ○ represents here the usual non-leptonic weak interaction described by the phenomenological hamiltonian

$$H = \mathrm{i}G_F m_\pi^2 \bar{u}_n (A + B\gamma_5) u_\Sigma \cdot \varphi_\pi ,$$

$$A = -1.93(1), \quad B = -0.65(7). \tag{1}$$

We follow the convention on the signs of the amplitudes of non-leptonic decays adopted in ref. [20].

As to the *CP*-odd $\pi n \Sigma$ interaction, represented at figs. 2a,b by the symbol ⊗, it is determined by the quark diagram 3 (here the dashed line refers to the gluon) where we only retain the structure

$$H_1 = \mathrm{i}2^{-1/2} G_F \sin\delta\, s_1 s_2 s_3 c_2 [\alpha_s(\bar{m})/12\pi]\,\Delta$$

$$\times \ln(m_t^2/m_c^2)\, \bar{s}\gamma_\mu (1+\gamma_5)\lambda^a \mathrm{d} \sum_{\mathrm{q=u,d}} \bar{\mathrm{q}}\gamma_\mu \lambda^a \mathrm{q},$$

$$m_c < \bar{m} < m_t . \tag{2}$$

In formula (2) we follow the convention on the K–M matrix adopted in ref. [21]. Note that at $\alpha_s(\bar{m}) = 0.2$ the correction factor Δ in the operator (2), arising due to a strong interaction at small distances, constitutes $\Delta \simeq 1.3$. It only slightly deviates from unity since the characteristic virtual momenta are here large enough so that α_s is small. The operator (2) is essentially the well-known "penguin" diagram which was used for the first time in ref. [20] in another connection, for the description of non-leptonic decays of hyperons and kaons.

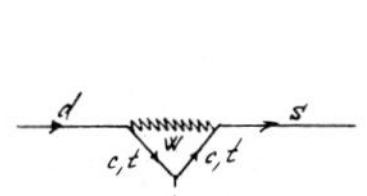

Fig. 3.

Unlike another CP-odd structure, $\bar{s}\sigma_{\mu\nu}(1+\gamma_5) \times \lambda^a \mathrm{d}\, G^a_{\mu\nu}$, arising from this diagram, here the cancellation between the c-quark and t-quark contributions in the limit $m_c = m_t$ does not lead to the factor $(m_t^2 - m_c^2)/m_W^2 \sim 10^{-1}$, but to ln (m_t^2/m_c^2) which at $m_t = 30\,\mathrm{GeV}$, $m_c = 1.5\,\mathrm{GeV}$ is equal to 6.

The matrix element $\langle \Sigma^- \pi^+ | H_1 | \mathrm{n} \rangle$ is calculated in the standard way by factorizing the pion. The result is

$$\langle \Sigma^- \pi^+ | H_1 | \mathrm{n} \rangle = -2^{-1/2} G_F (\alpha_s \cdot \Delta/12\pi) \sin\delta\, s_1 s_2 s_3 c_2$$

$$\times \ln \frac{m_t^2}{m_c^2} \frac{4}{9} \frac{f_\pi (m_\Sigma + m_n) m_\pi^2}{m_s (m_u + m_d)} \tag{3}$$

$$\times \bar{u}_\Sigma \left(\frac{m_\Sigma - m_n}{m_\Sigma + m_n} + (2\alpha - 1) g_A \frac{m_K^2}{m_K^2 - q^2} \gamma_5 \right) u_n .$$

Here $\alpha \simeq 0.64$ is the relative weight of the D-coupling in the interaction of baryon and meson octets; $g_A = 1.25$; q is the pion momentum. One should pay attention to the factor $m_K^2/(m_K^2 - q^2)$ (m_K is the kaon mass) at the γ_5-structure. It is due to the effective pseudoscalar in the axial current.

Using the expressions (1), (3), we find by standard calculations the contribution of the diagrams 2a,b to the neutron edm:

$$D_n = -eG_F^2 \sin\delta\, s_1 s_2 s_3 c_2 [\alpha_s(\bar{m})\Delta/\sqrt{2}\, 27\pi^3]$$

$$\times \ln \frac{m_t^2}{m_c^2} \frac{f_\pi m_\pi^4}{m_s(m_u + m_d)} A (2\alpha - 1) g_A \ln \frac{m_K}{m_\pi} . \tag{4}$$

We neglect in this expression the term $B(m_\Sigma - m_N)/(m_\Sigma + m_N) \simeq -0.08$ in comparison with $A(2\alpha - 1)g_A \simeq -0.7$. The numerical estimate by the formula (4) at $c_2 s_2 s_3 \sin\delta = 10^{-3}$, $\alpha_s(\bar{m}) = 0.2$, $\Delta = 1.3$, $\ln(m_K/m_\pi) = 1$ gives $D_n \simeq 2 \times 10^{-32}\, e$ cm.

Calculating the contribution of the diagrams 2a,b, we have omitted the analogous diagrams with the $\pi^- \Sigma^+$ intermediate state. The reasons are as follows. The s-wave amplitude of the corresponding decay is very small, $A(\Sigma^+ \to n\pi^+) = 0.06(1)$, and can be neglected. But if in the *CP*-even vertex the amplitude B is retained only, in another vertex, the *CP*-odd one, only the scalar structure is essential. The last struc-

Fig. 4.

ture turns to zero both in the limit of zero pion momentum, as can be checked by direct calculation of the corresponding commutator in PCAC technique, and in the pion factorization approach. The proof of the validity of these arguments is that they can be applied as well to the *CP*-even amplitude $A(\Sigma^+ \to n\pi^+)$ which is very small indeed.

It is useful to explain why our estimate for D_n is by two orders of magnitude larger than that obtained in ref. [15]. Firstly, as it was pointed out above, the cancellation of the diagrams 3 with c- and t-quarks in the limit $m_c = m_t$ is guaranteed here by the factor $\ln(m_t^2/m_c^2) \simeq 6$ instead of $(m_t^2 - m_c^2)/m_W^2 \sim 10^{-1}$. Secondly, due to the presence of right-handed currents in the operator (2), the factor $m_\pi^2/f_\pi(m_u + m_d) \sim 12$ arises. And finally, there is a factor of the same order of magnitude in the amplitude A compensating in it the smallness of the Cabibbo angle. Taken together, they overweight with odds α_s and a small numerical factor arising due to the extra loop (fig. 2a,b).

There is a problem of the contribution of the diagrams 4a,b where the symbol ⊗ refers to a *CP*-odd scalar vertex of the second order in G_F, and the dot refers to the usual strong πNN vertex. It contains also the factors $\ln(m_t^2/m_c^2)$ and $m_\pi^2/f_\pi(m_u + m_d)$; the absence of the last large factor is compensated numerically by a large value of the constant $g_{\pi NN} \simeq 13$. However, we do not see any grounds to expect that such a contribution should cancel out that obtained above.

Therefore, we arive at the estimate for the neutron edm in the K–M model:

$$D_n \simeq 2 \times 10^{-32} e \text{ cm} . \qquad (5)$$

References

[1] T.D. Lee, Phys. Rev. D8 (1973) 1226.
[2] S. Weinberg, Phys. Rev. Lett. 37 (1976) 657.
[3] M. Kobayashi and T. Maskawa, Prog. Theor. Phys. 49 (1973) 652.
[4] A.A. Anselm and D.I. D'yakonov, Nucl. Phys. B145 (1978) 271.
[5] A.A. Anselm and N.G. Uraltsev, Yad. Fiz. 30 (1979) 465.
[6] A.R. Zhitnitsky, Yad. Fiz. 33 (1981) 1320.
[7] I.B. Khriplovich and A.R. Zhitnitsky, Yad. Fiz. 34 (1981) 167.
[8] I.S. Altarev et al., Phys. Lett. B102 (1981) 13.
[9] L. Maiani, Phys. Lett. B62 (1976) 183.
[10] J. Ellis, M.K. Gaillard and D.V. Nanopoulos, Nucl. Phys. B109 (1976) 23.
[11] B.W. Lee, Phys. Rev. D15 (1977) 3394.
[12] E.P. Shabalin, Yad. Fiz. 28 (1978) 151.
[13] E.P. Shabalin, preprint ITEP-87 (1979).
[14] J. Ellis and M.K. Gaillard, Nucl. Phys. B150 (1979) 141.
[15] E.P. Shabalin, Yad. Fiz. 32 (1980) 443.
[16] B.F. Morel, Nucl. Phys. B157 (1979) 23.
[17] D.V. Nanopoulos, A.Yildiz and P.H. Cox, Ann. Phys. 127 (1980) 126.
[18] R.J. Crewther, P. Di Vecchia, G. Veneziano and E. Witten, Phys. Lett. 88B (1979) 123; 91B (1980) 487E.
[19] G. Barton and E.D. White, Phys. Rev. 184 (1969) 1660.
[20] A.I. Vainshtein, V.I. Zakharov and M.A. Shifman, Zh. Eksp. Teor. Fiz. 72 (1977) 1275; Sov. Phys. JETP 45 (1977) 670.
[21] L.B. Okun', Leptons and quarks (Nauka, Moscow, 1981).

CHIRAL ESTIMATE OF THE ELECTRIC DIPOLE MOMENT OF THE NEUTRON IN QUANTUM CHROMODYNAMICS

R.J. CREWTHER, P. DI VECCHIA and G. VENEZIANO
CERN, Geneva, Switzerland

and

E. WITTEN
Lyman Laboratory of Physics, Harvard University, Cambridge, MA 02138, USA

Received 7 September 1979

Current algebra for *CP* violating strong interactions is investigated. In particular, the neutron electric dipole moment D_n is shown to behave as $\theta m_\pi^2 \ln m_\pi^2$ for small pion mass m_π and *CP* violating parameter θ. This logarithm is explicitly calculable: it contributes $5.2 \times 10^{-16}\theta$ cm to D_n. This result is somewhat larger than a previous $O(m_\pi^2)$ estimate based on the bag model.

In the last few years, it has been realized [1] that the usual lagrangian of quantum chromodynamics (QCD) [‡1]

$$\mathcal{L} = -\tfrac{1}{4}F^2 + \sum_k \bar{q}_k(i\not{D} - m_k)q_k\ , \qquad (1)$$

can be generalized by including an additional *P* and *CP* violating interaction:

$$\mathcal{L} \to \mathcal{L}_{QCD} = \mathcal{L} - \theta(g^2/32\pi^2)F\cdot F^*\ . \qquad (2)$$

Despite its being a total divergence, this additional interaction modifies the physics of strong interactions.

The purpose of this paper is to investigate the current-algebraic consequences of the *CP* violating parameter θ being small but not zero. Using an effective lagrangian originally derived by Baluni [2], we will show that, to lowest order in chiral symmetry breaking, many θ-dependent QCD amplitudes can be explicitly calculated by means of current algebra.

Although we will also consider processes such as *CP* violating meson decays and pion–nucleon couplings, our most interesting current algebra theorem concerns the electric dipole moment of the neutron, since it is here that a θ dependent effect may be observable. When the up and down quark masses m_u and m_d are very small (but not zero), the contribution of θ to the neutron electric dipole moment D_n can be calculated explicitly, with no undetermined parameters. Instead of being proportional to m_π^2 for small m_u and m_d, as one might expect, D_n actually behaves as $m_\pi^2 \ln m_\pi^2$. The coefficient of the logarithmically enhanced term is uniquely fixed by current algebra, as in similar examples [3] of non-analytic dependence on m_π^2. The logarithm comes from the $p\pi^-$ intermediate states shown in fig. 1.

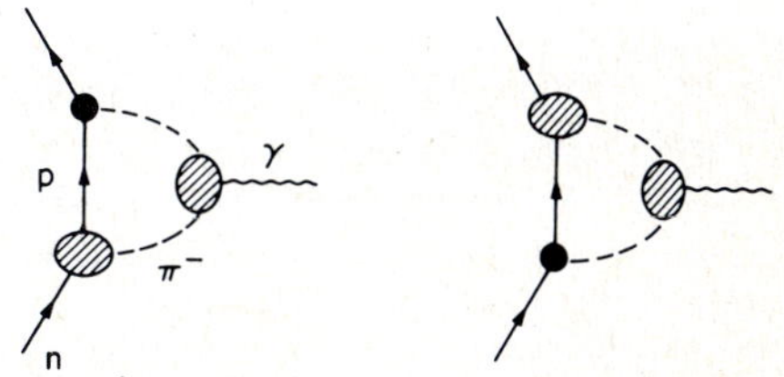

Fig. 1. Mechanism responsible for the $O(m_\pi^2 \ln m_\pi^2)$ contribution to the electric dipole moment of the neutron. The dark blob indicates a *CP* violating πNN interaction induced by the parameter θ.

‡1 Notation: gauge covariant derivative D_μ and field-strength tensor $F^a_{\mu\nu}$, dual tensor $F^*_{\mu\nu} = \frac{1}{2}\epsilon_{\mu\nu\alpha\beta}F^{\alpha\beta}$, with $\epsilon_{0123} = +1$ and Bjorken–Drell conventions for metric and γ-matrices, topological charge [1] $(g^2/32\pi^2)\int d^4x\, F\cdot F^*$, quark fields q_k (or u, d, s, ...) with flavour index k and mass parameters $m_k > 0$.

Our final remark concerns the problem of demonstrating on experimental grounds that in nature, θ is close to zero, not π. Zero and π are values of θ at which *CP* may be conserved (although the status of $\theta = \pi$ is somewhat in doubt). We will try to show that the conventional assumption $|\theta| \ll 1$ is really justified by observation.

Let us briefly review Baluni's effective lagrangian. (Similar considerations have been made by Peskin [4] and Coleman [5].)

For purposes of current algebra, it is convenient to remove the $\theta F \cdot F^*$ term from the lagrangian by introducing an equivalent θ dependence in the quark mass matrix. Then there is a *CP* violating contribution of the form

$$\delta \mathcal{L}_{CP} = \bar{\mathrm{q}} \mathrm{i} \gamma_5 A \mathrm{q} \,, \qquad (3)$$

where A is a three-dimensional hermitean matrix acting in flavour space on the quark fields (u, d, s).

There is an apparent arbitrariness in the choice of A, because A can be redefined by means of an SU(3) chiral rotation. One should use this freedom to choose A such that chiral perturbation theory can be carried out in a straightforward manner. The condition for this is that $\bar{\mathrm{q}}\gamma_5 A \mathrm{q}$ should have zero amplitude to create any of the approximate Goldstone boson states π, K or η:

$$\langle 0 | \bar{\mathrm{q}} \mathrm{i} \gamma_5 A \mathrm{q} | M^a \rangle = 0 \,, \quad (M^a = \pi, \mathrm{K}, \eta) \,. \qquad (4)$$

A perturbation not satisfying this condition would create Goldstone bosons in profusion and cause a shifting of the vacuum [6]; this shift would have to be taken into account before performing chiral perturbation theory, and then one would reach the same conclusions as those obtained more simply by using a perturbation that satisfies (4).

Eq. (4) implies that A must be proportional to the unit 3×3 matrix, in agreement with a theorem of Nuyts [7]. For small θ, the constant of proportionality can be easily determined by writing out the full mass matrix (including both the usual mass terms and the *CP* violating term (3) and noting that the argument of the determinant of the left-handed part must equal θ in order to agree with the standard normalization (2)). Hence the hamiltonian density is perturbed by a chiral asymmetric term

$$\begin{aligned} \mathcal{H}' &= m_\mathrm{u} \bar{\mathrm{u}}\mathrm{u} + m_\mathrm{d} \bar{\mathrm{d}}\mathrm{d} + m_\mathrm{s} \bar{\mathrm{s}}\mathrm{s} \\ &\quad - \mathrm{i}\theta m_\mathrm{u} m_\mathrm{d} m_\mathrm{s} (m_\mathrm{u} m_\mathrm{d} + m_\mathrm{u} m_\mathrm{s} + m_\mathrm{d} m_\mathrm{s})^{-1} \\ &\quad \times (\bar{\mathrm{u}}\gamma_5 \mathrm{u} + \bar{\mathrm{d}}\gamma_5 \mathrm{d} + \bar{\mathrm{s}}\gamma_5 \mathrm{s}) \\ &= m_\mathrm{u} \bar{\mathrm{u}}\mathrm{u} + m_\mathrm{d} \bar{\mathrm{d}}\mathrm{d} + m_\mathrm{s} \bar{\mathrm{s}}\mathrm{s} - \delta \mathcal{L}_{CP} \,. \end{aligned} \qquad (5)$$

Note that we have dropped a factor 3 from Baluni's eq. (16) in order to be consistent with the standard convention (2). Also, our eq. (5) is normalized properly only to first order in θ; $\mathrm{O}(\theta^2)$ effects will be neglected.

Since m_s is much larger than m_u or m_d, we can safely approximate the mass-dependent factor $(\Sigma_k m_k^{-1})^{-1}$ in $\delta \mathcal{L}_{CP}$ by $m_\mathrm{u} m_\mathrm{d} / (m_\mathrm{u} + m_\mathrm{d})$. This simplification should be understood in what follows.

Before turning to the electric dipole moment of the neutron, let us give the simplest example of a physical prediction of eq. (5). The *CP* violating term in (5) permits three of the light pseudoscalar mesons M^a to couple with strength

$$G_{abc} = \langle 0 | \delta \mathcal{L}_{CP} | M^a M^b M^c \rangle \quad (a, b, c = 1, \ldots, 8) \qquad (6)$$

in the zero-momentum limit, so there is a prediction for the θ-dependent contribution to the rate of the decay $\eta \to \pi\pi$. According to the usual current-algebraic technique (previously applied [8] to $\eta \to \pi\pi$ in a different context), each contraction of a meson M^a in eq. (6) yields a commutator $-\mathrm{i}F_\pi^{-1} [Q_5^a, \;]$ involving the corresponding SU(3) $\times$ SU(3) generator Q_5^a and the pion-decay constant F_π ($\approx$93 MeV):

$$\begin{aligned} G_{abc} &= -\theta m_\mathrm{u} m_\mathrm{d} (m_\mathrm{u} + m_\mathrm{d})^{-1} F_\pi^{-3} \\ &\quad \times \langle 0 | [Q_5^a, [Q_5^b, [Q_5^c, \bar{\mathrm{q}}\gamma_5 \mathrm{q}]]] | 0 \rangle \,. \end{aligned} \qquad (7)$$

Evaluating the commutators and using current algebra [9] to express the resulting vacuum expectation value in terms of m_π^2 and F_π, we obtain the following effective lagrangian for the three-meson couplings:

$$\begin{aligned} \mathcal{L}_\mathrm{eff} &= \tfrac{1}{6} G_{abc} M^a M^b M^c \\ &= -\tfrac{1}{6} \theta m_\pi^2 F_\pi^{-1} m_\mathrm{u} m_\mathrm{d} (m_\mathrm{u} + m_\mathrm{d})^{-2} \, \mathrm{Tr} M^3 \,. \end{aligned} \qquad (8)$$

Here M denotes the traceless 3×3 matrix $\lambda^a M^a$ of pseudoscalar boson fields M^a.

Thus for the decay $\eta \to \pi\pi$, we find that the effective lagrangian is

$$\mathcal{L}_\mathrm{eff} = -(\theta m_\pi^2 / \sqrt{3} F_\pi) m_\mathrm{u} m_\mathrm{d} (m_\mathrm{u} + m_\mathrm{d})^{-2} \boldsymbol{\pi}^2 \eta \,. \qquad (9)$$

It is amusing that such an exact formula can be given, but the effect is much too small to be observable. Our analysis of the neutron electric dipole moment will result in the bound $|\theta| \lesssim 10^{-9}$, so the branching ratio for $\eta \to \pi\pi$ implied by eq. (9) is at least 13 orders of magnitude smaller than the present upper bound [10] 1.5×10^{-3}.

The calculation of the neutron electric dipole moment depends on another interesting quantity, the *CP* violating pion–nucleon coupling constant $\bar{g}_{\pi NN}$, for which there is an explicit prediction. Let the full πNN interaction be represented by the effective lagrangian

$$\mathcal{L}_{\pi NN} = \boldsymbol{\pi} \cdot \bar{N}\boldsymbol{\tau}(i\gamma_5 g_{\pi NN} + \bar{g}_{\pi NN})N , \qquad (10)$$

where $\boldsymbol{\tau}$ are isospin Pauli matrices. To compute $\bar{g}_{\pi NN}$, we contract the pion in the amplitude

$$\langle \pi^a N_f | \delta\mathcal{L}_{CP} | N_i \rangle = -\theta F_\pi^{-1} m_u m_d (m_u + m_d)^{-1} \langle N_f | \bar{q}\tau^a q | N_i \rangle , \qquad (11)$$

and observe that $\langle N_f | \bar{q}\tau^a q | N_i \rangle$ is related by ordinary SU(3) symmetry to the *F* coupling which splits Ξ from N in the baryon octet. Thus our result is

$$\bar{g}_{\pi NN} = -\theta (M_\Xi - M_N) m_u m_d / [F_\pi (m_u + m_d) \times (2m_s - m_u - m_d)] , \qquad (12)$$

or numerically

$$|\bar{g}_{\pi NN}| \approx 0.038 |\theta| , \qquad (13)$$

where the known quark mass ratios [11] have been substituted. Of course, this is minute compared with the usual *CP* conserving coupling $|g_{\pi NN}| \approx 13.4$.

To first order in θ, the neutron electric dipole moment D_n is given by

$$T\langle n(p_f) | J_\mu(0) i \int d^4x \, \delta\mathcal{L}_{CP}(x) | n(p_i) \rangle = -D_n \bar{u}_f \sigma_{\mu\nu} k^\nu \gamma_5 u_i + O(k^2) , \qquad (14)$$

where $k = p_f - p_i$ is the momentum carried by the hadronic electromagnetic current J_μ. This matrix element can be written as a sum over intermediate states $|X\rangle$ inserted between the two operators. We have to evaluate products of the type

$$\langle n(p_f) | J_\mu | X \rangle \langle X | \delta\mathcal{L}_{CP} | n(p_i) \rangle , \qquad (15)$$

plus terms with J_μ and $\delta\mathcal{L}_{CP}$ exchanged.

Baluni [2] estimated (14) by assuming the sum over $|X\rangle$ to be saturated by the lowest-lying $1/2^-$ resonances N(1535) and N(1700), and applying the MIT bag model to the resulting matrix elements. Our intention is to avoid such model dependence by establishing a theorem which rigorously determines D_n in the chiral limit $m_\pi \to 0$. However, we shall see that Baluni's numerical estimate is not very different from ours.

To establish an exact result, we have to treat the intermediate states $|X\rangle$ as multiparticle states of the stable hadrons:

$$|X\rangle = |N\rangle, |N\pi\rangle, |N\pi\pi\rangle, ..., |NN\bar{N}\rangle . \qquad (16)$$

Note that it would be double counting to include a resonance state $|N^*\rangle$ or a state like $|N^*\pi\rangle$ on this list.

The next step is to show that the most singular contribution to D_n for $m_\pi \to 0$ arises solely from the intermediate state $|X\rangle = |N\pi\rangle$. This state contributes a term $O(m_\pi^2 \ln m_\pi^2)$: the factor m_π^2 comes from the explicit factor $m_u m_d/(m_u + m_d)$ in $\delta\mathcal{L}_{CP}$, while the non-analytic factor $\ln m_\pi^2$ is generated by an infra-red divergence of the loop integral at $m_\pi = 0$.

The other states in the complete set (16) produce analytic contributions $O(m_\pi^2)$ or non-analytic terms $O(m_\pi^3)$ or $O(m_\pi^4 \ln m_\pi^2)$, all of which are less singular than $m_\pi^2 \ln m_\pi^2$. The only states capable of generating a logarithmic dependence are those which become degenerate ‡2 in energy with the external neutron states in the soft-pion limit, so states like $|NN\bar{N}\rangle$ can be excluded. Non-analytic contributions due to states $|X\rangle$ with a nucleon and two or more pions are at least $O(m_\pi^2)$ compared with the contribution of $|\pi N\rangle$ because of lack of phase space at threshold. Finally, the contribution of the single nucleon state $|X\rangle = |N\rangle$ is $O(m_\pi^2)$: it is proportional to the neutron magnetic moment μ_n which is finite in the limit $m_\pi \to 0$ (with $O(m_\pi)$ corrections [3]).

It is therefore sufficient to restrict our attention to the states ‡3

$$|X\rangle = |N\pi_{soft}\rangle , \qquad (17)$$

in order to determine the leading $O(m_\pi^2 \ln m_\pi^2)$ term completely. One of the factors in (15) becomes

‡2 We neglect the proton–neutron mass difference.

‡3 Considered many years ago [12] as a contribution in the saturation of sideways dispersion relations.

$\langle N\pi_{soft}|\delta\mathcal{L}_{CP}|n\rangle$, which has already been evaluated in eqs. (10)–(13). The other factor $\langle n|J_\mu|N\pi_{soft}\rangle$ is known from current algebra [13] to be given by the Born terms plus a term proportional to $\gamma_5\sigma_{\mu\nu}k^\nu$. Explicit calculation shows that this last term and the Born terms in which the photon couples to the nucleon give $O(m_\pi^2)$ contributions to D_n. The photon does not couple to π^0 at low momenta, so the entire $O(m_\pi^2 \ln m_\pi^2)$ contribution arises from the Born term in $\langle n|J_\mu|p\pi^-\rangle$ which corresponds to the photon coupling to π^-.

This justifies our claim that the coefficient of $m_\pi^2 \ln m_\pi^2$ in the expansion of D_n at small m_π is given rigorously by the diagrams of fig. 1.

It is also intuitively reasonable that these diagrams provide the most important contribution to D_n. The neutron acquires an electric dipole moment by dissociating into charged constituents. The largest contribution comes from the lightest constituent (the pion) because it attains the greatest distance from the centre of the charge distribution.

The calculation is now reduced to a simple one-loop integral with the πNN vertices given by (10) and the $\gamma\pi\pi$ vertex by

$$\langle\pi^-(p+k)|J_\mu|\pi^-(p)\rangle = -(2p+k)_\mu + O(k^2). \qquad (18)$$

Our result for the logarithmic contribution to the electric dipole moment of the neutron (in units of elementary charge e) is thus

$$D_n = g_{\pi NN}\bar{g}_{\pi NN} \ln(M_N/m_\pi)/4\pi^2 M_N . \qquad (19)$$

Numerically, we find in centimeter units

$$D_n = 5.2 \times 10^{-16}\theta \text{ cm} , \qquad (20)$$

where eq. (13) for $\bar{g}_{\pi NN}$, the experimental value of $g_{\pi NN}$, and $\ln(M_N/m_\pi) = 1.9$ have been substituted in (19). This is our main result.

A comparison with Baluni's result depends on allowing for the factor of 3 difference between our eq. (5) and his eq. (16). When θ is normalized in the conventional way, Baluni's bag-model estimate of the $1/2^-$ resonance contributions is

$$D_n(\text{Baluni}) = 2.7 \times 10^{-16}\theta \text{ cm} . \qquad (21)$$

Although our calculation becomes rigorous for $m_\pi \to 0$, we cannot guarantee that the logarithm dominates in the real world with $m_\pi \neq 0$. In particular, our result suffers from a difficulty common to all estimates [3] of non-analytic terms: we do not know whether to write the logarithm in (19) as $\ln(M_N/m_\pi)$ or $\ln(m_\rho/m_\pi)$ (for instance), because the difference remains finite as m_π vanishes and finite terms are not determined by current algebra. Therefore our numerical value (20) should be treated as a rough order-of-magnitude estimate.

The theoretical importance of our result is that it provides a model-independent demonstration that the parameter θ is responsible for a non-zero contribution to D_n. It is a current-algebraic theorem, so there can be no unrecognized mechanism that forbids or strongly suppresses the effect.

The contribution of θ to D_n is of considerable interest at present because, according to the popular Kobayashi–Maskawa theory [14], the weak interaction contribution to D_n is very small, 10^{-30} cm or less [15]. Therefore, an experimental observation of D_n at a significantly higher level could be interpreted as a θ dependent effect and so yield a measurement of θ.

The present experimental limit [16] $|D_n| \lesssim 10^{-24}$ cm indicates that θ cannot be much greater than 10^{-9}. Little is known of how large θ should be, but it is unlikely to be exactly zero in the presence of the *CP* violation observed in nature. If θ is not much less than about 10^{-11}, it may be measured in the near future; it is hoped [17] that D_n will soon be detectable at the level of 10^{-26} or 10^{-27} cm.

Throughout this discussion and in common with most of the literature, we have supposed that θ must be close to zero because *CP* is almost conserved. Could θ be close to π instead? It appears that $\theta = \pi$ describes a *CP* conserving world, because *CP* changes $\theta = \pi$ to $\theta = -\pi$ which, in view of the periodicity $\theta \to \theta + 2\pi$, should be equivalent to $\theta = \pi$. (Actually, the periodicity is not totally beyond doubt, but here we assume it to be 2π.)

We would like to show that the choice $|\theta| \ll 1$ instead of $\theta \approx \pi$ is not arbitrary but can be justified from observed facts. To do this, we must explain what it means to have $\theta = \pi$ in QCD.

The theory with $\theta = \pi$ in (2) is equivalent to a theory with a *CP* conserving (γ_5 free) quark mass matrix with negative determinant. This means that some of the quark mass terms appear with the "wrong" sign. In principle, it does not matter which quark mass terms are reversed in sign, because the various choices are related by SU(3) chiral rotations. However, the most convenient choice for calculations is to reverse the sign of the mass term of the quark of *smallest* mass – the up quark. This corresponds to a chiral asymmetric term

$$\mathcal{H}' = -m_u \bar{u}u + m_d \bar{d}d + m_s \bar{s}s \,, \qquad (22)$$

in the hamiltonian density.

The point here is that our substitution $m_u \to -m_u$ is not accompanied by a shift of the vacuum through a chiral SU(3) rotation: the values of $\langle \bar{u}u \rangle$, $\langle \bar{d}d \rangle$ and $\langle \bar{s}s \rangle$ remain negative and approximately equal. This can be seen from the formulas [9,11] which (for $\theta = 0$) express the squares of pseudoscalar octet masses m^2 as positive linear combinations of m_u, m_d and m_s. Since m_u is the smallest quark mass, changing its sign preserves the positivity of m^2 and hence the stability of the vacuum.

(But this simple procedure would not work if we were to reverse the sign of m_s, for example. Then the $\theta = 0$ formula $(m_d + m_s)\mu$ for $m^2_{K^0}$ would become negative, $(m_d - m_s)\mu < 0$, indicating an instability due to a wrong choice of vacuum. Indeed, the correct vacuum minimizing $\langle \mathcal{H}' \rangle$ would give $\langle \bar{s}s \rangle$ a positive value, and the usual formulas of current algebra would have to be systematically changed. That is an inconvenience which we prefer to avoid.)

According to the above argument, the correct answer is obtained by simply substituting $-m_u$ for m_u in all formulas of current algebra. In particular, the $\theta = 0$ formula [11] ‡4 for the ratio

$$(m^2_{K^0} - m^2_{K^0} - m^2_{\pi^0} + m^2_{\pi^+})/m^2_\pi = \frac{m_d - m_u}{m_d + m_u} \,, \qquad (23)$$

becomes at $\theta = \pi$:

$$(m^2_{K^0} - m^2_{K^+} - m^2_{\pi^0} + m^2_{\pi^+})/m^2_\pi = \frac{m_d + m_u}{m_d - m_u} \,. \qquad (24)$$

Thus this ratio is predicted to be less than 1 for $\theta = 0$ and more than 1 for $\theta = \pi$. The experimental value of the ratio is about 0.3, so we learn that θ is near zero, not near π.

The above argument assumes that $(m_d + m_u)/m_s$ is much smaller than $(m_d - m_u)/(m_d + m_u)$, which is the case for the conventional values [11] of the quark mass ratios. If m_u is much closer to m_d, there is a large spontaneous *CP* violation at $\theta = \pi$, as first pointed out by Dashen [6]. This case is also ruled out experimentally [8].

Further discussion of θ dependent effects in current algebra, including comments form the point of view of the $1/N$ expansion [19], will be given in future papers [19,20].

‡4 As is usual, we have subtracted off the pion mass difference so that, according to Dashen's theorem [18], purely electromagnetic contributions to $m^2_{K^0} - m^2_{K^+}$ are cancelled.

We thank S. Coleman for interesting conversations. E.W. acknowledges the hospitality of the CERN Theory Division and the support of NSF grant Phy-77-22864.

Note added. The result (9) for $\eta \to \pi\pi$ decay has been independently derived by Shifman et al. [21].

[1] A.A. Belavin, A.M. Polyakov, A.S. Schwartz and Yu.S. Tyupkin, Phys. Lett. 59B (1975) 85;
V.N. Gribov, unpublished;
G. 't Hooft, Phys. Rev. Lett. 37 (1976) 8;
R. Jackiw and C. Rebbi, Phys. Rev. Lett. 37 (1976) 172;
C.G. Callan Jr., R.F. Dashen and D.J. Gross, Phys. Lett. 63B (1976) 334.
[2] V. Baluni, Phys. Rev. D19 (1979) 2227.
[3] L.-F. Li and H. Pagels, Phys. Rev. Lett. 26 (1971) 1204; 27 (1971) 1089; Phys. Rev. D5 (1972) 1509;
M.A.B. Bég and A. Zepeda, Phys. Rev. D6 (1972) 2912;
D.G. Caldi and H. Pagels, Phys. Rev. D10 (1974) 3739;
H. Pagels, Phys. Rep. 16C (1975) 219.
[4] M. Peskin, unpublished.
[5] S. Coleman, unpublished.
[6] R. Dashen, Phys. Rev. D3 (1971) 1879.
[7] J. Nuyts, Phys. Rev. Lett. 26 (1971) 1604; 27 (1971) 361(E).
[8] M.A.B. Bég, Phys. Rev. D4 (1971) 3810.
[9] M. Gell-Mann, R.J. Oakes and B. Renner, Phys. Rev. 175 (1968) 2195.
[10] J.J. Thaler et al., Phys. Rev. D7 (1973) 2569.
[11] S. Weinberg, in: Festschrift for I.I. Rabi, ed. L. Motz (N.Y. Academy of Sciences, New York, 1977);
P. Langacker and H. Pagels, Phys. Rev. D19 (1979) 2070.
[12] G. Barton and E.D. White, Phys. Rev. 184 (1969) 1660.
[13] S. Fubini, G. Furlan and C. Rosetti, Nuovo Cimento 40 (1965) 1171;
S.L. Adler and R.F. Dashen, Current algebras (Benjamin, New York, 1968).
[14] M. Kobayashi and K. Maskawa, Prog. Theor. Phys. 49 (1973) 652.
[15] J. Ellis and M.K. Gaillard, Nucl. Phys. B150 (1979) 141;
B.F. Morel, Harvard Univ. preprint HUTP-79/A009 (1979);
D.V. Nanopoulos, A. Yildiz and P.H. Cox, Phys. Lett. 87B (1979) 53; Harvard Univ. preprint HUTP-79/A048 (1979).
[16] W.B. Dress et al., Phys. Rev. D15 (1977) 9.
[17] N.F. Ramsey, Phys. Rep. 43C (1977) 409.
[18] R.F. Dashen, Phys. Rev. 183 (1969) 1245.
[19] P. Di Vecchia, G. Veneziano and E. Witten, to be published.
[20] R.J. Crewther, Lectures at 1979 Kaiserslautern Summer Institute, Field theoretic methods in elementary particle physics, to be published.
[21] M.A. Shifman, A.I. Vainshtein and V.I. Zacharov, preprint ITEP-64 (1979).

CP Conservation in the Presence of Pseudoparticles*

R. D. Peccei and Helen R. Quinn†
Institute of Theoretical Physics, Department of Physics, Stanford University, Stanford, California 94305
(Received 31 March 1977)

We give an explanation of the *CP* conservation of strong interactions which includes the effects of pseudoparticles. We find it is a natural result for any theory where at least one flavor of fermion acquires its mass through a Yukawa coupling to a scalar field which has nonvanishing vacuum expectation value.

It is experimentally obvious that we live in a world where *P* and *CP* are good symmetries at the level of strong interactions. In the context of quantum chromodynamics the strong interactions are believed to be due to non-Abelian vector gluons coupled to massive quarks. In such a theory, when the effects of gluon configurations of nonzero pseudoparticle number are included, *CP* invariance requires a very special choice of parameters. We will show, however, that *CP* invariance of the strong interactions is, in fact, a natural consequence, provided at least one flavor of quark acquires its mass from a Yukawa coupling to a scalar field which has a nonzero vacuum expectation value, and the Lagrangian originally possesses a U(1) invariance involving all Yukawa couplings.

The physical importance of gauge field configurations with nontrivial topology has been stressed by 't Hooft.[1] He has reminded us that the physics of such theories involves a parameter θ which does not appear in the original Lagrangian.[2] This parameter defines the choice of vacuum[3] among an infinity of possible distinct and generally inequivalent vacua. Each θ represents a possible true vacuum and there are in general an infinity of distinct theories arising from any given Lagrangian.

If all fermions which couple to the non-Abelian gauge fields are massless then the various θ choices give equivalent theories.[1,3] This is most clearly seen by remarking that a change in the effective value of θ can be induced by making an $\exp[i\gamma_5\eta]$ rotation of the fermion fields. We define the effective Euclidean action in the qth sector to be

$$S_{\rm eff}{}^q = \int d^4x \mathcal{L} + i\theta q, \qquad (1)$$

where

$$q = (g^2/32\pi^2)\int d^4x\, F_{\mu\nu}{}^a \tilde{F}^{a\mu\nu}. \qquad (2)$$

The rotation of a fermion field by $\exp[i\gamma_5\eta]$ induces a change in the effective action given by

$$\delta S_{\rm eff}{}^q = -i\int(\partial^\mu j_\mu{}^5)\eta = -2iq\eta \qquad (3)$$

since

$$\partial^\mu j_\mu{}^5 = (g^2/16\pi^2) F_{\mu\nu}{}^a \tilde{F}_\mu{}^{\mu\nu a}. \qquad (4)$$

Thus in such a theory the net effect of such a rotation is

$$\theta \to \theta' = \theta - 2\eta. \qquad (5)$$

If, however, all fermions are massive such a rotation will also change the fermion mass term in $\mathcal{L}$. Hence one can define inequivalent theories which have the same mass term and various choices of θ. There are of course also classes of equivalent theories related by $\exp[i\gamma_5\eta]$ transformations. Only one such class of theories, those in which $\theta \to 0$ when all fermion masses have been made real by a suitable $\text{ex}[i\gamma_5\eta]$ rotation, yield a CP- and P-invariant theory of the strong interactions.[1,3] We will present a mechanism which explains why we live in such a world.

We consider theories in which at least one fermion flavor acquires its mass from a Yukawa coupling to a color-singlet scalar field which has a nonzero vacuum expectation value. The effective potential for the scalar fields is θ dependent and does not have the same symmetry as the scalar polynomial $U(\varphi)$ appearing in the Lagrangian. This reduction in symmetry has been used by 't Hooft[1] in explaining away the U(1) problem. Here it means that the minimum of the potential corresponds to a particular choice of phases for the various scalar vacuum expectation values. These phases appear in the fermion mass terms and we find they are always such that when all fermion masses are made real by $\exp[i\gamma_5\eta]$ rotations of the fermion fields the resulting θ is zero.

We will illustrate this result first for a toy model of the strong interactions in which there is only a single fermion flavor and a single color-singlet complex scalar field. Introducing additional fermion flavors and scalar multiplets will not change our result. Finally we will discuss the effect of adding weak interactions to the theory.

Let us begin by examining our toy model. The Lagrangian is

$$\mathcal{L} = -\tfrac{1}{4}F_{\mu\nu}{}^{a}F^{\mu\nu a} + i\bar{\psi}D_\mu\gamma^\mu\psi + \bar{\psi}[G\varphi\tfrac{1}{2}(1+\gamma_5) + G^*\varphi^*\tfrac{1}{2}(1-\gamma_5)]\psi - |\partial_\mu\varphi|^2 - \mu^2|\varphi|^2 - h|\varphi|^4; \quad \mu^2<0. \tag{6}$$

We are here thinking of, for example, a color SU(3) gauge theory where the fermion belongs to the fundamental triplet representation. We can write the generating functional for Green's functions in the θ vacuum as

$$Z_\theta(J,J^*) = \sum_q e^{i\theta q}\int (dA_\mu)_q \int d\psi \int d\bar{\psi}\int d\varphi \int \varphi^* \exp[\int d^4x\, \mathcal{L}(\varphi,\psi,A) + J^*\varphi^*]. \tag{7}$$

The scalar vacuum expectation value is then defined by

$$\frac{1}{Z_\theta}\frac{\delta Z_\theta}{\delta J}\bigg|_{J=J^*=0} = \langle\varphi\rangle = \lambda e^{i\beta}, \tag{8}$$

where λ and β are real constants to be determined.

To proceed further we need to know some properties of the fermion eigenmodes in the $q=n$ sector. The essential points for our argument are the following: (i) The number of orthogonal eigenmodes with eigenvalue zero is n. (This is true when the fermions are in the fundamental representation of the gauge group.[4]) (ii) The zero eigenmodes for $n>0$ are purely right handed.[1] (iii) The zero eigenmodes for $q=-n$ are of opposite chirality to those for $q=+n$ but otherwise identical. (iv) The nonzero eigenmodes are the same for $q=n$ and $q=-n$,[5] and are admixtures of both left- and right-handed parts.

These four facts allow us to schematically express $Z_\theta(J,J^*)$ as it appears after integrating out both the vector and the fermion fields. We introduce new scalar variables ρ and σ by the definition

$$\varphi = e^{i\beta}(\lambda+\rho+i\sigma). \tag{9}$$

Then we find

$$Z(J,J^*) = \int d\rho \int d\sigma \{A_0(\rho,\sigma^2) + \sum_n A_n(\varphi\varphi^*)[Ge^{i\theta}\varphi]^n + A_n^*(\varphi\varphi^*)[G^*e^{-i\theta}\varphi^*]^n\}$$
$$\times \exp[Je^{i\beta}(\lambda+\rho+i\sigma) + J^*e^{-i\beta}(\lambda+\rho-i\sigma)]. \tag{10}$$

In this equation the A_n are polynomials of the form

$$A_n(\varphi\varphi^*) = \sum_m \prod_{i=1}^{m} [\int dx_i \int dy_i\, \varphi(x_i)\varphi^*(y_i)]\, c_m{}^n(x_i,y_i). \tag{11}$$

The $c_m{}^n$ are real functions which depend on $|G|^2$, μ, h, and λ but are β independent. The terms proportional to G^n arise from the fermion zero eigenmodes in the $q=n$ sector, while those with $(G^*)^n$ are from the $q=-n$ sector. Now we define

$$\alpha = \arg[Ge^{i(\theta+\beta)}]$$

and rewrite (10) as

$$Z(J,J^*)=\int d\rho \int d\sigma\{A_0(\rho,\sigma^2)+\sum_n[F_n(\rho,\sigma^2)\cos n\alpha-\sigma G_n(\rho,\sigma^2)\sin n\alpha]\} \times \exp[Je^{i\beta}(\lambda+\rho+i\sigma)+J^*e^{-i\beta}(\lambda+\rho-i\sigma)], \quad (12)$$

where F_n and σG_n are the real and imaginary parts of $A_n|G|^n(\lambda+\rho+i\sigma)^n$, respectively. We note that F_n and G_n are even functions of σ and that $\int d\sigma\, \sigma^{2n+1}$ vanishes identically. Then by definition λ and β must be chosen so that

$$\langle\rho\rangle=0=\int d\rho\int d\sigma\, \rho[A_0+\sum_n F_n\cos n\alpha],$$
$$\langle\sigma\rangle=0=\int d\rho\int d\sigma\, \sigma^2\sum_n G_n \sin n\alpha. \quad (13)$$

The first of these equations is satisfied for arbitrary α by appropriately choosing $\lambda=\lambda(\alpha)$. The second equation can then in general only be satisfied for $\alpha=0,\ \pi$. These values are both stationary points of the potential. To find which is the true minimum we must examine $V(\varphi)$. This we can only do to leading order in G and h, for which we find

$$V(\varphi)=\mu^2\varphi^*\varphi+h(\varphi^*\varphi)^2-K|\varphi|\cos\alpha,$$

where K is a real positive constant. Thus if G and h are sufficiently small the minimum occurs at

$$\alpha=0. \quad (14)$$

In terms of the shifted scalar variables the fermions acquire a mass term

$$\lambda\bar\psi[Ge^{i\beta}\tfrac{1}{2}(1+\gamma_5)+G^*e^{-i\beta}\tfrac{1}{2}(1-\gamma_5)]\psi. \quad (15)$$

This mass can be made real by rotating the fermion fields by $\exp[i\gamma_5\theta/2]$. Such a rotation gives

$$\theta\to\theta'=\theta-\theta=0 \quad (16)$$

(Q.E.D.).

The generalization to multiple fermion flavors and numerous scalars follows the same pattern. The pseudoparticles now have a zero mode for each fermion flavor. Provided the original scalar Lagrangian possesses a U(1) invariance, which is violated by the additional terms arising from the fermion determinant. there will always exist a range of parameters for which the CP-conserving phase choices correspond to the true minimum of the scalar potential. The requirement $\alpha=0$ generalizes in these cases to

$$\arg\left\{\prod_{i=1}^{m}(G_i\exp[i\beta_{j_i}])\exp[i\theta]\right\}=0, \quad (17)$$

where G_i is the Yukawa coupling of the ith fermion flavor to the j_ith neutral scalar. Just as in the single-flavor case the resulting fermion masses can all be made real by a rotation the ith fermion field by $\gamma_5\eta_i$ where, from (17),

$$\sum_i\eta_i=\theta/2. \quad (18)$$

The resulting theory again has $\theta_{\rm eff}=0$ when all fermion masses are real. This point will be discussed, with examples, in a subsequent paper.[6]

If some fermion masses appear in the original Lagrangian the expression (17) will be modified by $e^{i\theta}\to e^{i\theta}\Pi_k m_k$. When all masses including the m_k are made real this gives $\theta_{\rm eff}=0$.

The introduction of electromagnetic and weak interactions in the form of gauge fields coupling to fermion flavor can be treated as a perturbative correction to this theory. P and CP noninvariance can be introduced at the level of the weak interactions (or CP noninvariance may occur in the scalar self-couplings). There are two features of such a theory which could invalidate our earlier arguments. The first is the fact that the there are also $q_{\rm weak}\neq 0$ sectors in the integration over weak-field configurations. 't Hooft has argued[1] that the contribution of these sectors is extremely small, being approximately of order $\exp[-(2\pi/\alpha)q]$. The remaining possible problem for standard phenomenology is the occurrence of parity-nonconserving order-α corrections to the fermion masses.[7] These do not alter our conclusions. In the presence of such terms it is important to recognize that the quantity G appearing in (16) must be corrected to include all orders of weak interactions—it is the Yukawa coupling renormalized at zero scalar four-momentum. The argument then proceeds as before. The vacuum expectation value of the scalar field picks up a phase such that the effective θ will be zero when the fermion fields are redefined so that all renormalized fermion masses are real.

We have thus shown that the observed CP invariance of the strong interactions is a natural feature of a theory such as quantum chromodynamics provided only that at least one fermion flavor acquires its mass through a Yukawa coupling to a scalar field which has a nonvanishing vacuum expectation value. The physical mechanism that forces this result is the same asymmetry of the scalar potential in the θ vacuum which was used by 't Hooft to explain away the U(1) problem.

We wish to thank S. Weinberg and J. D. Bjorken for useful discussions.

*Work supported in part by National Science Foundation Grants No. PHY 75-18444 and No. PHY 75-20427.

†A. P. Sloan Foundation fellow. On leave from Physics Department, Harvard University, Cambridge, Mass. 02138.

[1]G. 't Hooft, Phys. Rev. Lett. 37, 8 (1976), and Phys. Rev. D 14, 3432 (1976).

[2]This fact has been previously discussed in other contexts by L. Schulman, Phys. Rev. 176, 1558 (1968).

[3]R. Jackiw and C. Rebbi, Phys. Rev. Lett. 37, 172 (1976); C. G. Callan, R. F. Dashen, and D. J. Gross, Phys. Lett. 63B, 334 (1976).

[4]This result can be naively guessed by making a widely separated pseudoparticle approximation and imagining one fermion zero mode centered about each individual pseudoparticle. A more rigorous discussion will be presented by L. S. Brown, R. Carlitz, and C. K. Lee in a forthcoming paper (L. S. Brown, private communication). See also J. Kiskis, Phys. Rev. D 15, 2329 (1977); B. Grossman, to be published.

[5]This is strictly speaking only true after properly integrating over all possible gauge choices for the pseudoparticle configurations, a step referred to by 't Hooft as gauge averaging.

[6]R. D. Peccei and Helen R. Quinn, to be published.

[7]S. Weinberg, Phys. Rev. Lett. 31, 494 (1973).

NATURALLY WEAK *CP* VIOLATION

Ann NELSON
Lyman Laboratory of Physics, Harvard University, Cambridge, MA 02138, USA

Received 14 December 1983

An SU(5) GUT model is proposed in which the observed *CP* violation is due to spontaneous symmetry breakdown. The effective field theory below the GUT scale is simply the standard model. Matching conditions at the GUT scale give a non-zero contribution to $\bar{\theta}$ at the one-loop level. This contribution is proportional to ratios of superheavy fermion masses over the GUT scale, which can be naturally small, in the sense of 't Hooft.

1. Introduction. How can we naturally explain the absence of strong *CP* violation? In the standard model, the strong interactions contain a physical *CP* violating parameter $\bar{\theta} = \theta_{\text{QCD}} + \arg\det M_q$ where M_q is the quark mass matrix and θ_{QCD} is the coefficient of $g^2/32\pi^2 F\tilde{F}$. Experimental measurement of the electric dipole moment of the neutron determines $\bar{\theta}$ to be less than 10^{-9} [1]. 't Hooft has convincingly argued [2] that a physical parameter may only be very small if replacing it by zero increases the symmetry of the theory. Then, any renormalization of the parameter will be proportional to the parameter itself. Otherwise, the theory is said to be unnatural. In the standard model, however, *CP* is explicitly broken in the weak interactions by dimension-four operators. $\bar{\theta}$ probably receives an infinite renormalization even if it is zero at the tree level [3]. Thus it seems $\bar{\theta}$ cannot be naturally small.

One possible explanation for the observed absence of a neutron electric dipole moment is that the up quark is massless. A massless quark in the theory means that $\bar{\theta}$ is undefined and so can have no physical effect. However, current algebra estimates of the up quark mass make this explanation unlikely [4].

Another possible solution is due to Peccei and Quinn [5]. If the lagrangian possesses a spontaneously broken global U(1) symmetry with a color anomaly, $\bar{\theta}$ is dynamically determined to be zero. Such a theory predicts the existence of a pseudo-Goldstone boson, called an axion, whose couplings are inversely proportional to the Peccei–Quinn symmetry breaking scale. Experimental searches for the axion and astrophysical requirements force this scale to be larger than 10^9 GeV [6]. Recent cosmological arguments [7] place an upper limit on this scale of 10^{12} GeV. An unnatural fine tuning seems to be required to give the axion just the right amount of invisibility.

A third and promising approach is that *CP* is violated softly or spontaneously. $\bar{\theta}$ then receives only finite, calculable contributions. It is difficult but not impossible to find models where these contributions are sufficiently small. Several rather complicated examples exist in the literature [8]. Some of these contain a low energy left–right symmetry, and most rely on additional charged scalars in the low energy theory to produce the observed *CP* violation in the weak interactions. In this paper we present an example of a model with spontaneously broken *CP* where the low energy effective theory is the standard model with no extra undesirable symmetries or scalar fields.

Our model is essentially the same as one proposed by the author in a recent paper [9] on flavor symmetry and proton decay. We extend the Georgi–Glashow SU(5) model [10] to include an SO(3) flavor symmetry and some additional heavy fermions. We impose a U(1) global chiral symmetry and *CP* invariance. At the GUT scale SU(5) spontaneously breaks to SU(3) ⊗ SU(2) ⊗ U(1) and all the global symmetries get broken. All fermions and scalars except for the three light families and the Weinberg–Salam Higgs doublet get masses.

0.370-2693/84/$ 03.00 © Elsevier Science Publishers B.V.
(North-Holland Physics Publishing Division)

This model contains the usual fine tuning to keep the Higgs doublet light. At the tree level, the heavy fermion masses are all real. The light quark–Higgs Yukawa coupling matrices are complex but have real determinants. $\bar{\theta}$ is induced in the effective theory by one-loop and higher order matching conditions at the GUT scale. However, these corrections turn out to be proportional to ratios of the heavy fermion masses over the GUT scale. These masses can naturally be small because when they are zero some chiral symmetries are restored. In practice the heavy fermions need to have masses $\lesssim 10^{-(3-4)}$ times the GUT scale for $\bar{\theta}$ to be within experimental limits.

2. The model. The symmetry group of our model is $SU(5)_{Gauge} \otimes SO(3) \otimes U(1)_{Global} \otimes CP$. Under this group the fermion transform as

$$[10,3,0] \oplus [5,3,0] \oplus [10,1,1] \oplus [\overline{10},1,-1] \oplus [\bar{5},1,1] \oplus [5,1,-1]\,.$$

The usual Higgs doublet is in a $(5,1,0)$ rep, along with a superheavy $SU(3)_C$ triplet. The symmetry breaking at the GUT scale is done by scalars in reps $(r_i,3,1)$, where the r_i are SU(5) reps containing $SU(3) \otimes SU(2) \otimes U(1)$ singlets, such as **1**, **24**, and **75**.

The scalar–fermion Yukawa couplings and fermion masses are:

$$\begin{aligned}&\lambda_u[10,3,0][10,3,0](5,1,0)\\&+\lambda_d[10,3,0][\bar{5},3,0](5,1,0)^*\\&+h_i[10,3,0][\overline{10},1,-1](r_i,3,1)\\&+f_i[\bar{5},3,0][5,1,-1](r_i,3,1)\\&+m_1[10,1,1][\overline{10},1,-1]+m_2[\bar{5},1,1][5,1,-1]\\&+\text{h.c.}\end{aligned} \tag{1}$$

Note that we have used square brackets to designate fermions and round ones for scalars. Since we assume *CP* to be a symmetry of the theory, all masses and coupling constants are real. The VEVs of the complex fields $(r_i,3,1)$ will break SU(5), SO(3), U(1), and *CP* invariance, and cause mixing between SO(3) triplet and singlet fermions. This mixing will give the structure of the light quark and lepton mass matrices, which are examined in detail in ref. [9]. Each $SU(3) \otimes SU(2) \otimes U(1)$ multiplet in the $[\overline{10},1,-1]$ and $[5,1,-1]$ will pair up with a linear combination of fields in the $[10,3,0]$, $[10,1,1]$, $[\bar{5},3,0]$, and $[\bar{5},1,1]$ reps to get a mass.

We can make a chiral rotation of the quark fields to give the heavy quarks real diagonal masses. We can choose this rotation to have determinant 1 so as not to contribute to $F\tilde{F}$. In the appendix we show that the Yukawa coupling matrix of the quarks to the light Higgs is complex *but has a real determinant.* $\bar{\theta}$ receives no contribution at tree level.

3. Effective field theory calculation of $\bar{\theta}$. At energies sufficiently below the GUT scale, the effective theory is the standard $SU(3) \otimes SU(2) \otimes U(1)$ gauge theory, containing only renormalizable interactions of light fields. Above M_W, $\bar{\theta}$ is defined to be $\theta_{QCD} + \arg(\det \lambda^u \det \lambda^d)$ [‡1], which is invariant under chiral transformations of the quarks. Below M_W the quarks are massive, and $\bar{\theta}$ is defined in the usual way. Ellis and Gaillard have shown [3] that finite corrections to $\bar{\theta}$ in the standard model are very small, of order 10^{-16}. Contributions to the $\bar{\theta}$ β-function do not occur until at least the seven-loop order and so any running of $\bar{\theta}$ is incredibly slow and may be neglected. Thus the $\bar{\theta}$ which we calculate from matching conditions at the GUT scale is, to a very good approximation, the $\bar{\theta}$ we measure from the neutron electric dipole moment.

We have shown in the previous section that there is no tree level contribution to the matching conditions for $\bar{\theta}$. In this section we will estimate the higher order corrections. In general there are two sources of *CP* violation in the full theory which contribute to the matching conditions for $\bar{\theta}$. One source of *CP* violation is complex gauge and scalar couplings to the mass eigenstates. The other source is complex contributions to the heavy SO(3) triplet scalar mass matrix from the VEVs of the scalar fields. When this mass matrix is made real by redefinitions of the scalar fields, more complex phases are introduced into the couplings of the SO(3) triplet scalars to the fermions.

In calculating the corrections to θ_{QCD} we will use the method of Georgi et al. [11]. At the one-loop level there is no contribution as all our heavy quarks have real tree level masses.

There will, however, be one-loop contributions to

‡1 Here and below λ^u and λ^d are the Yukawa coupling *matrices* defined in eq. (3) of the appendix.

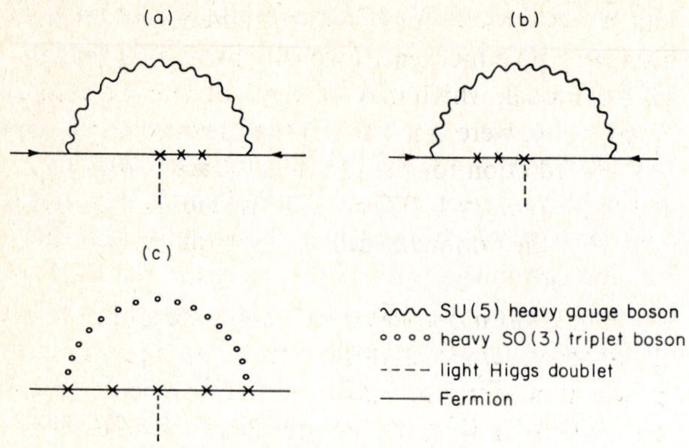

Fig. 1. One-loop contribution to λ_{ij}^{u} and λ_{ij}^{d}. The crosses on the fermion lines represent the fact that these graphs only contribute to arg det λ if there are at least two mass insertions.

det λ^u and det λ^d from the graphs in fig. 1. These contributions could be infinite if there were no fermion mass insertions and yet we know all contributions to $\bar{\theta}$ are finite. However, it is only the chiral rotations performed to diagonalize the fermion masses which introduce *CP* violating phases into the couplings of the heavy gauge bosons and the light Higgs doublet. Therefore if there are no mass insertions in figs. 1a and 1b, these graphs conserve *CP*, and simply give a real renormalization of λ^u and λ^d. The graph in fig. 1c vanishes with no mass insertions, because the SO(3) triplet scalars only couple the light particles to the SO(3) singlet fermions, which have no tree level couplings to the light Higgs.

In the limit that the heavy fermions are lighter than the GUT scale the sum of all contributions to $\bar{\theta}$ from fig. 1 type graphs with arbitrary numbers of mass insertions gives:

$$\delta\bar{\theta} = \sum_{i=1,2,3,4} A_i \frac{g^2}{8\pi^2}\frac{M_i}{\mu^2}\ln\frac{M_i^2}{\mu^2}$$

$$+ B\frac{f^2}{32\pi^2}\frac{M_1M_2}{\mu^2}\ln\frac{M_1M_2}{\mu^2} + C\frac{fh}{32\pi^2}\frac{M_3M_4}{\mu^2}\ln\frac{M_3M_4}{\mu^2}.$$

A_i, B and C are factors of O(1) coming from phases in the couplings of the gauge and Higgs bosons to mass eigenstates. M_1, M_2, M_3 and M_4 are the masses of the heavy [3,2,1/6], [$\bar{3}$,1,−2/3], [1,2,−1/2] and [1,1,1] fermions. μ is the GUT scale.

For these contributions to be less than 10^{-9} requires that the heavy fermion masses be less than the GUT scale by a factor of 10^{-3}–10^{-4}. This is not unnatural! From (1) we can see that these fermions are light if m_1 and m_2 are $\ll M_{\mathrm{GUT}}$ and $h_i, f_i \ll g$. In the limit h_i and $f_i \to 0$ we gain a new anomaly free chiral symmetry under which the [10,1,1] has charge 1 and the [$\overline{10}$,1,−1] has charge −1. When m_1 and $m_2 \to 0$ we gain two more. Under one chiral symmetry the [10,1,1] has charge 1 and the [$\bar{5}$,1,1] has charge −3. Under the other, the [$\overline{10}$,1,−1] and [5,1,−1] also have charges of 1 and −3. Thus at the one-loop level $\bar{\theta}$ can naturally be small.

The same arguments can be extended to the two-loop level and beyond. We will argue that all contributions to $\bar{\theta}$ are suppressed by fermion masses or Yukawa couplings. For example, let us look at the two-loop

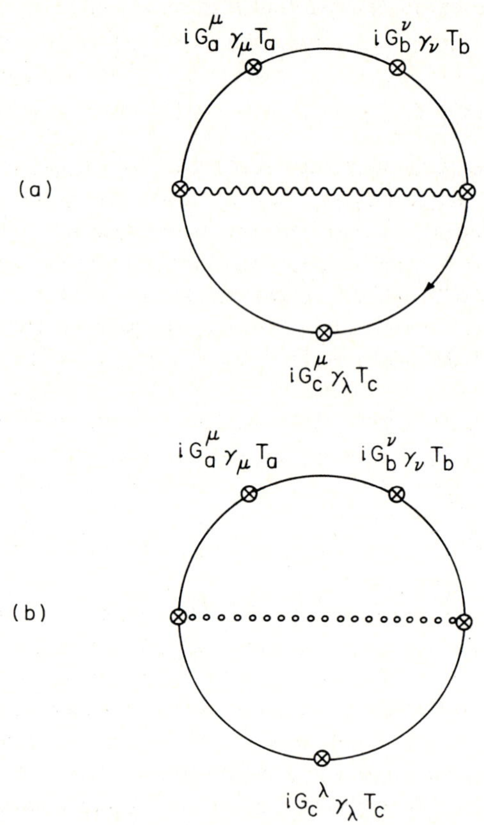

Fig. 2. Two-loop contribution to $F\tilde{F}$. The circles with crosses in them represent space dependence of the gluon potentials and of the *CP* violating phases in the couplings. The details of this calculational method are given in ref. [11]. This space dependence must be included because otherwise the graph vanishes.

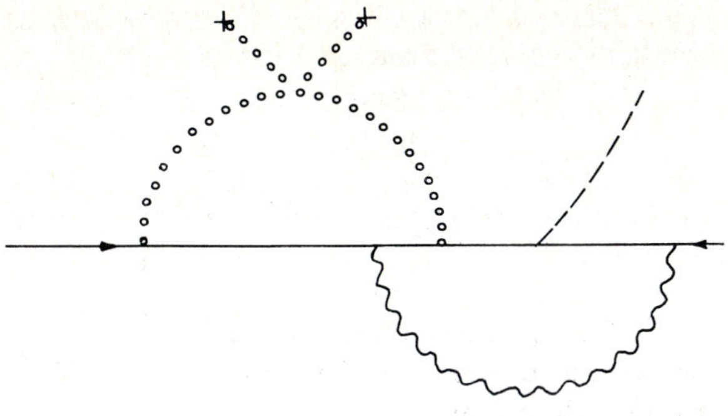

Fig. 3. This is an example of a complex two-loop contribution to λ_{ij} which is suppressed by small Yukawa couplings instead of fermion masses, fondly known as the dead duck graph.

contribution to $F\tilde{F}$ in fig. 2. The graph in fig. 2a with no mass insertions does not contribute to $\bar{\theta}$ for the same reason the graphs in figs. 1a and 1b do not contribute – the gauge boson couplings do not violate *CP* unless the effects of the fermion masses are included. Fig. 2b at first seems a little more dangerous since there may be complex contributions to the heavy scalar–fermion couplings from diagonalizing the complex scalar mass matrix. With no mass insertions, however, any complex coupling on one side of the graph gets multiplied by its complex conjugate on the other side. Thus both of these graphs give contributions to $\bar{\theta}$ which are suppressed by powers of fermion masses over boson masses.

The graph in fig. 3 gives a complex renormalization of λ_{ij} even if no fermion masses are inserted because of the complex scalar masses. It is, however, suppressed by a couple of powers of the SO(3) triplet scalar Yukawa couplings, which are small if the fermion masses are small.

Clearly any other contributions to the $\bar{\theta}$ matching conditions from the complex SO(3) triplet scalar mass matrix will also be proportional to their small Yukawa couplings. Similarly, any contributions which are due to the complex couplings of the heavy gauge boson fields to the mass eigenstates will be suppressed by powers of fermion masses over gauge boson masses. In this model, $\bar{\theta}$ can naturally be as small as desired.

4. Conclusion. The model presented above and in ref. [9] has some remarkable characteristics. It was originally proposed to explain the experimental absence of the proton decay mode $p \to \pi e$ and it turned out to also be able to explain the absence of a neutron electric dipole moment. Two experimental signatures of this model, which may or may not rule it out in the near future, were discussed in the previous paper. One is the prediction for the top quark mass of $m_t \simeq \theta_c(m_b/m_d)\, m_c \simeq 150$ GeV, and another is the prediction that the dominant decay modes of the proton are $p \to Ke$ and $p \to \pi\mu$.

With slight modifications, such as the addition of more heavy fermions, this model loses its predictive power about fermion masses and proton decay modes, but can continue to explain the smallness of $\bar{\theta}$. I believe it would still be simpler than other models of spontaneously broken *CP*.

I thank Howard Georgi, Andrew Cohen, and David Kaplan for useful conversations and David Kaplan for naming the dead duck graph.

This work was supported in part by the National Science Foundation under Grant No. PHY-82-15249.

Appendix. The heavy linear combination of fields in the [10,3,0], [10,1,1], [$\bar{5}$,3,0] and [$\bar{5}$,1,1] reps may be represented by a four-vector with the first three components corresponding to an SO(3) three-vector and the fourth to an SO(3) singlet. For example, we will label the linear combination of the [3,2, 1/6]'s in the [10,3,0] and [10,1,1] which combine with the [$\bar{3}$,2,−1/6] in the [$\overline{10}$,1,−1] by a four-vector q_μ^{SH}

$$q_\mu^{SH} = \sum_i (f_i v_i^{1*}/M_1, f_i v_i^{2*}/M_1, f_i v_i^{3*}/M_1, m_1/M_1), \quad (1)$$

where

$$M_1 = \Big(\sum_{a=1,2,3} \Big|\sum_i f_i v_i^a\Big|^2 + m_1^2\Big)^{1/2}$$

= the tree level mass of the heavy [3,2,1/6] quark.

The v_i^a are complex VEVs of the SO(3) triplet scalar fields contributing to the heavy [3,2,1/6] quark mass. Since m_1 is real, q_μ^{SH} may be written without loss of generality as:

$$(s_3 s_2 s_1, s_3 s_2 c_1, s_3 c_2, c_3)\,,$$

where $c_i = c_i^*$ and $|s_i|^2 + |c_i|^2 = 1$. If m_1 were complex, we could not require both c_3 and the heavy quark mass to be real.

We can find three directions orthogonal to q_μ^{SH}, which correspond to light quarks

$$q_\mu^i = \begin{pmatrix} c_3 s_2 s_1, & c_3 s_2 c_1, & c_3 c_2, & -s_3^* \\ s_1 c_2, & c_1 c_2, & -s_2^*, & 0 \\ c_1, & -s_1^*, & 0, & 0 \end{pmatrix}. \qquad (2)$$

Note that this chiral transformation has determinant 1 and so does not contribute to $F\tilde{F}$. The coupling constant matrices of the light quarks to the Higgs doublet are:

$$\lambda_{ij}^u = \lambda_u \sum_{\mu=1,2,3} q_\mu^i \bar{u}_\mu^j, \quad \lambda_{ij}^d = \lambda_d \sum_{\mu=1,2,3} q_\mu^i \bar{d}_\mu^j. \qquad (3)$$

By inspection of (2) and (3) we can see that λ_{ij}^u and λ_{ij}^d are the products of three-by-three matrices having real determinants, and so they also have real determinants. If m_1 were not real and so c_3 were complex, λ_{ij}^u and λ_{ij}^d would no longer have real determinants and $\bar{\theta}$ would get a tree level contribution.

References

[1] B.J. Crewther et al., Phys. Lett. 88B (1979) 123; V. Baluni, Phys. Rev. D19 (1979) 2227.
[2] G. 't Hooft, Lecture Cargèse Summer Institute (1979).
[3] J. Ellis and M.K. Gaillard, Nucl. Phys. B150 (1979) 141.
[4] S. Weinberg, unpublished.
[5] R. Peccei and H. Quinn, Phys. Rev. Lett. 38 (1977) 1440; Phys. Rev. D16 (1977) 1791.
[6] S. Weinberg, Phys. Rev. Lett. 40 (1978) 223;
F. Wilczek, Phys. Rev. Lett. 46 (1978) 279;
M. Dine, W. Fischler and M. Srednicki, Phys. Lett. 104B (1981) 199;
J. Donnely et al., Phys. Rev. D18 (1978) 1607;
D. Dicus et al., Phys. Rev. D18 (1978) 1827.
[7] J. Preskill, M.B. Wise and F. Wilczek, Phys. Lett. 120B (1983) 127.
[8] H. Georgi, Hadronic J. 1 (1978) 155;
M. Bég and H. Tsao, Phys. Rev. Lett. 41 (1978) 278;
R. Mohapatra and G. Senjanovic, Phys. Lett. 79B (1978) 283;
G. Segre and H. Weldon, Phys. Rev. Lett. 42 (1979) 1191;
V. Goffin, G. Segre and H. Weldon, Phys. Rev. D21 (1980) 1410;
S. Barr, Phys. Rev. D23 (1981) 2434;
S. Barr and D. Seckel, University of Washington preprint (1983).
[9] A. Nelson, Harvard University preprint HUTP-83/A075.
[10] H. Georgi and S.L. Glashow, Phys. Rev. Lett. 32 (1974) 438;
H. Georgi, H. Quinn and S. Weinberg, Phys. Rev. Lett. 33 (1974) 451.
[11] H. Georgi, T. Tomaras and A. Pais, Phys. Rev. D23 (1981) 469.